T0326396

Periphyton

Periphyton

Functions and Application in Environmental Remediation

Yonghong Wu
Institute of Soil Science
Chinese Academy of Sciences
Nanjing, China

ELSEVIER
AMSTERDAM • BOSTON • HEIDELBERG • LONDON • NEW YORK • OXFORD
PARIS • SAN DIEGO • SAN FRANCISCO • SINGAPORE • SYDNEY • TOKYO

Elsevier
Radarweg 29, PO Box 211, 1000 AE Amsterdam, Netherlands
The Boulevard, Langford Lane, Kidlington, Oxford OX5 1GB, United Kingdom
50 Hampshire Street, 5th Floor, Cambridge, MA 02139, United States

Library of Congress Cataloging-in-Publication Data
A catalog record for this book is available from the Library of Congress

British Library Cataloguing-in-Publication Data
A catalogue record for this book is available from the British Library

ISBN: 978-0-12-801077-8

Publisher: Candice G. Janco
Acquisition Editor: Laura S. Kelleher
Editorial Project Manager: Emily Thomson
Production Project Manager: Mohanapriyan Rajendran
Designer: Greg Harris

Typeset by TNQ Books and Journals

Contents

5. Periphytic Biofilm and Its Functions in Aquatic Nutrient Cycling

6. Periphyton Functions in Adjusting P Sinks in Sediments

7. The Evaluation of Phosphorus Removal Processes and Mechanisms From Surface Water by Periphyton

List of Figures

List of Tables

Author's Biography

Professor Yonghong Wu works at the Institute of Soil Science, Chinese Academy of Sciences. He obtained his bachelor's degree from Northwest A & F University, China. He finished his master's study at the Institute of Hydrobiology, Chinese Academy of Science, and obtained his PhD from the University of Chinese Academy of Sciences. He is currently the head of the Soil–Water Interface Ecology Research Group in the Institute of Soil Science, Chinese Academy of Sciences, and the head of the Zigui Ecological Experiment Station of the Three Gorge Project, Chinese Academy of Sciences.

Dr Wu's research goal is to understand the role of periphyton in soil and water and to manipulate periphyton to control nonpoint source pollution and remediate dysfunctional ecosystems. In the past 10 years, a series of technologies for periphyton culture, in situ collection and characterizing have been built by Dr Wu's research group. The group has since innovatively extended the use of periphyton to the purification of nonpoint source wastewater and the management of nutrients in paddy fields. The six patented technologies based on periphyton have been sold to commercial companies and are widely applied. Dr Wu has more than 70 publications and nine patents and has been acknowledged by the Chinese Academy of Sciences, the Ministry of Agriculture, and Yunnan Province, China, awards for excellence in soil–water interface ecology studies.

Preface

Biotechnology based on native periphyton uses the integrated functions of microorganisms in the periphytic assembly to improve environmental quality. These improvements include interrupting the fluxes of pollutants in natural waters, purifying wastewater and natural water ways, and producing valuable resources for human society. The biotechnology based on native periphyton is essential to society and is truly unique as a technical discipline.

Biotechnology based on native periphyton has been recently developed and now complements several typical microbial treatment technologies such as activated sludge, biofilm, and anaerobic digestion, which have been used since the 20th century. The new periphyton technologies, such as periphyton bioreactors, were developed to address contemporary problems, such as purification of nonpoint source pollutant wastewater. The biotechnology based on native periphyton is also used to acquire nutrients from wastewater and reuse these nutrients in agricultural lands. The biotechnology based on native periphyton follows well established ecological principles of microbial communities and ecological engineering. The application of these principles, however, usually requires some degree of empiricism because native periphyton is inherently complex, varying in its composition and structure both spatially and temporally. Moreover, the complexity of the microbial communities makes it difficult to create biotechnology based on native periphyton that can reliably and effectively remove contaminants from waters and/or reuse nutrients.

In *Periphyton: Functions and Application in Environmental Remediation,* I connect these different biotechnology facets based on native periphyton. The basic concepts and study methods are developed in the first two chapters, which comprise the principles section of the book. I consistently refer to these principles as when describing the functions in Chapters 3–9. Each application has its own unique features and these special application cases are summarized in Chapters 10–16.

This book is an overview combined with my last research results describing the functions of periphyton and its valuable application in environmental remediation. Understanding periphyton, that is,the biochemical interactions amongst periphyton component organisms and their environmental functions, is a significant interdisciplinary subject that currently does not have a suitable introductory work. Other introductory periphyton works include *Periphyton of Freshwater Ecosystems* (R G. Wetzel, 1983; Springer) and *Periphyton: Ecology,*

Exploitation and Management (M. Ekram Azim, 2005; CABI), but these books focus on the ecology of periphyton rather than the function and application of periphyton in environmental remediation. In the face of rapidly degrading environmental quality, especially degenerated aquatic ecosystems, many biological measures for environmental remediation, such as the technologies based on periphyton, are attractive to environmental managers and the general public.

The material in *Periphyton: Functions and Application in Environmental Remediation* can be used in several courses. For students not already having a solid background in microbial treatments, Chapters 1 and 2 can serve as a text for an introductory course in environmental microbiology or can be used as a resource for students who need to refresh their knowledge in preparation for a more process-oriented course, research, or practice. The functions and application of periphyton summarized in Chapters 3–16, directly link principles to practice providing a reference for upper-level undergraduate students, graduate students studying environmental science, environmental engineering, ecology, ecological engineering, or hydrobiology. Green material researchers will see how their subject involves interpreting biological phenomena and bioscientists will see how their subject has a hydrobiology base. Ecologists and environmental engineers should see the benefits of integrating disciplines for environmental remediation.

Periphyton: Functions and Application in Environmental Remediation features many sample problems. During the introduction of specific cases, these problems illustrate the step-by-step procedures for understanding how native periphyton systems work and/or designing a treatment process based on native periphyton.

Some of the content in this book has been previously published by my research group members in journals including *Environmental Science and Pollution Research* (Wan et al., 2016a,b; Lu et al., 2014), *PLoS One* (Lu et al., 2014), *Chemosphere* (Lu et al., 2016 is like a reference citation. Please provide complete details.), *Journal of Hazardous Materials* (Yang et al., 2016; Wu et al., 2011), and *Bioresource Technology* (Shangguan et al., 2015; Wu et al., 2011, 2012, 2014; Yan et al., 2011). This published information has been systematically collated and organized in this book to better illustrate the link between different aspects of periphyton research and application in environmental remediation.

If readers find this book half as interesting to read as I found it to prepare, then my efforts will have been amply rewarded.

Yonghong Wu
Nanjing, China
March 21, 2016

REFERENCES

Lu, H., Yang, L., Zhang, S., Wu, Y., 2014. The behaviour of organic phosphorus under non-point source wastewater in the presence of phototrophic periphyton. PloS one. http://dx.doi.org/10.1371/journal.pone.0085910.

Lu, H., Wan, J., Li, J., Shao, H., Wu, Y., 2016. Periphytic biofilm: A buffer for phosphorus precipitation and release between sediments and water. Chemosphere 144, 2058–2064.

Shangguan, S., Liu, J., Zhu, Y., Tong, Z., Wu, Y., 2015. Start-up of a spiral periphyton bioreactor (SPR) for removal of COD and the characteristics of the associated microbial community. Bioresource Technology 193, 456–462.

Wan, J., Liu, X., Kerr, P.G., Wu, C., Wu, Y., 2016. Comparison of the properties of periphyton attached to modified agro-waste carriers. Environmental Science and Pollution Research. http://dx.doi.org/10.1007/s11356-015-5541-0.

Wan, J., Liu, X., Wu, C., Wu, Y., 2016. Nutrient capture recycling by periphyton attached to modified agrowaste carriers. Environmental Science and Pollution Research. http://dx.doi.org/10.1007/s11356-015-5988-z.

Wu, Y., Hu, Z., Yang, L., Graham, B., Kerr, P.G., 2011. The removal of nutrients from non-point source wastewater by a hybrid bioreactor. Bioresource Technology 102, 2419–2426.

Wu, Y., Li, T., Yang, L., 2012. Mechanisms of removing pollutants from aqueous solutions by microorganisms and their aggregates: A review. Bioresource Technology 107, 10–18.

Wu, Y., Xia, L., Liu, N., Gou, S., Nath, B., 2014. Cleaning and regeneration of periphyton biofilm in surface water treatment system. Water Science and Technology 69, 235–243.

Yang, J., Tang, C., Wang, F., Wu, Y., 2016. Co-contamination of Cu and Cd in paddy fields: using periphyton to entrap heavy metals. Journal of Hazardous Materials. http://dx.doi.org/10.1016/j.jhazmat.2015.10.051.

Yan, R., Yang, F., Wu, Y., Hu, Z., Nath, B., Yang, L., Fang, Y., 2011. Cadmium and mercury removal from non-point source wastewater by a hybrid bioreactor. Bioresource Technology 102 (21), 9927–9932.

Acknowledgments

This work was supported by the State Key Development Program for Basic Research of China (2015CB158200), the National Natural Science Foundation of China (41422111), and the National Science Foundation of Jiangsu Province, China (BK20150066).

Dr Wu greatly appreciates the many wonderful students and colleagues from the Research Group of Soil–Water Interface Ecological Process, Institute of Soil Science, Chinese Academy of Sciences, who have taught him new ideas, inspired him to look further and deeper, and corrected his frequent errors: Linzhang Yang, Sadaf Shabbir, Juanjuan Wan, Zhengyi Hu, Philip Kerr, Hiaying Lu, Jiali Yang, Shuzhi Liu, Haidong Shangguang, Junzhuo Liu, Yan Zhu, and Lizhong Xia. Thank you for everything.

Acknowledgments

[illegible]

[illegible]

Table of Abbreviations

Abbreviations	Full Names
ΔG°	Gibbs free energy (kJ mol^{-1})
a_1	Initial adsorption rate (mg g^{-1} h^{-1})
b_1	Extent of surface coverage and activation energy for chemisorption (mg g^{-1})
C	Constant of boundary layer effect
C_0	Initial P (or Cu) concentration (mg P L^{-1})
C_e	P (or Cu) concentration at equilibrium (mg L^{-1})
C_t	P (or Cu) concentration at time t (mg L^{-1})
E	Adsorption energy (kJ mol^{-1})
k_1	Pseudo-first-order kinetic model rate constant (h^{-1})
k_2	Pseudo-second-order kinetic model rate constant (mg $g^{-1}h^{-1}$)
K_F	Freundlich adsorption constant (L mg^{-1})
K_{id}	Rate constant of intraparticle diffusion (mg g^{-1} $h^{-0.5}$)
K_L	Langmuir adsorption constant (L mg^{-1})
m	Mass of adsorbent (g)
P_i	Inorganic phosphorus (mg P L^{-1}); the proportion of the relative absorbance value of well i to the total plate's wells
P_o	Organic phosphorus (mg P L^{-1})
P_{total}	Total phosphorus content (mg P L^{-1})
q_e	Amount of P (or Cu) adsorbed onto the periphyton at equilibrium (mg g^{-1})
$q_{e,cal}$	Amount of P (or Cu) adsorbed onto the periphyton calculated by model at equilibrium (mg g^{-1})
q_m	Maximum adsorption capacity for the periphyton (mg g^{-1})
q_t	Amount of P (or Cu) adsorbed onto the periphyton at time t (mg g^{-1})
R	Ideal gas constant (8.314 J mol^{-1} K^{-1})
R^2	Linear regression coefficient
R_L	Adsorption intensity
T	Temperature (K)
t	Time (h)
$T^{0.5}$	Square root of the time ($h^{0.5}$)
V	Volume of solutions (L)
α	Initial adsorption rate (mg g^{-1} h^{-1})
β	Desorption constant (g mg^{-1})
ε	Polanyi potential
AAM	Artificial aquatic mat
A^2/O	Anaerobic, anoxic and aerobic
AFDW	Ash free dry mass
AGNPs	Agricultural nonpoint source pollution

AI	Autotrophic index
AOPs	Advanced-oxidation processes
AP	Alkaline phosphates
APS	Acidified peanut shell
ARH	Acidified rice husk
AWCD	Average well color development
BDI	Biological diatom index
BFM	Biofertilizer microorganism
BOD	Biochemical oxygen demand
BMPs	Best management practice
B_t	Boyd model constant
C	Ceramsite
Cs	Dice index
CA	Concentration addition
CAT	Catalase activity
Chl a	Chlorophyll a
CLPPs	Community-level physiological profiles
COD	Chemical oxygen demand
CSLM	Confocal scanning laser microscopy
D	Simpson index
DAI	Diatom assemblage index for organic pollution
DGGE	Denaturing gradient gel electrophoresis
DIP	Dissolved inorganic phosphorus
DPS	Decomposed peanut shell
DO	Dissolved oxygen
DOC	Dissolved organic carbon
DOM	Dissolved organic matter
DOP	Dissolved organic phosphorus
EDCs	Endocrine disrupting compounds
EEC	European Economic Community
EPA	Environmental Protection Agency
EPI	Eutrophication pollution index
EPS	Extracellular polymeric substances
FA	Fatty acid
FCs	Fiber carriers
FGR	*Fragilaria* group richness
FPOM	Fine particulate organic matter
FTIR	Fourier transform infrared
GA	Gibberellic acid
GCC	Global climatic change
GDI	Generic diatom index
H	Shannon index
HPLC	High-performance liquid chromatography
HRT	Hydraulic retention time
HTS	High-throughput sequencing
HUFA	Highly unsaturated fatty acid
IAA	Indole acetic acid
IAP	Immediately available phosphorus
IBD	Indice Biologique Diatomées
ICP-MS	Inductively coupled plasma mass spectrometry
ILM	Leclercq and Maquet index
IPS	Indice de Polluosensibilité Spécifique

ISC	Industrial soft carriers
MF	Microbial aggregate
MO	Methyl orange
P	Phosphorus
PP	Particular phosphorus
PAP	Potential available phosphorus
PCA	Principal component analysis
PCBs	Polychlorinated biphenyls
PCM	Phase contrast microscope
PCR	Polymerase chain reaction
PLFA	Phospholipid fatty acid
PICT	Pollution-induced community tolerance
PS	Peanut shell
PSB	Phosphate-solubilizing bacteria
PTI	Pollution tolerance index
PUFA	Polyunsaturated fatty acid
RDI	River diatom index
RH	Rice husk
ROT	Saprobic index of Rott
SAFA	Saturated fatty acid
SEM	Scanning electric microscope
SLA	Czech Republic, the Sládeček index
SPI	Specific polluosensitivity index
SPR	Spiral periphyton bioreactor
SRP	Soluble reactive phosphate
TBT	Ri-*n*-butyl tin
TDP	Total dissolved phosphorus
TDI	Trophic diatom index
TEM	Transmission electron microscope
TEP	Transparent exopolymer particle
TP	Total phosphorus
TN	Total nitrogen
TNDT	Total number of diatom taxa
U	McIntosh
WFD	Water Framework Directive
MUFA	Monounsaturated fatty acid
NMR	Nuclear magnetic resonance
NPS	Nonopoint source
NR	Nitrate reductuse
NRDA	Natural Resource Damage Assessments
XREAN	X-ray absorption near-edge structure spectroscopy

Chapter 1

Periphyton and Its Study Methods

1.1 INTRODUCTION

Rapid urbanization and population growth have enhanced the number and complexity of lethal and sublethal environmental changes (Tusseau-Vuillemin et al., 2007). Human activities, intentional or accidental, have profoundly caused deterioration of fluvial ecosystems. The modification and degradation of aquatic environments affect water quality, habitat of organisms, and different biochemical pathways (Strayer et al., 2003; Townsend et al., 2003). Urban and agricultural runoffs are leading causes of increasing nutrient levels (particularly phosphorus) due to fertilizer, animal waste, and land erosion (Allan and Castillo, 2007).

The contamination of open waters by anthropogenic activities may lead to eutrophication or nutrient enrichment resulting in dense mats of periphytic algae and plants, and the subsequent biomanipulation of open waters. Excessive algal growth deteriorates the color and odor of water and diminishes the biodiversity of open waters (Wu et al., 2010). The strong metabolic interdependence of periphytic microorganisms permits nutrient recycling and the utilization of organic matter by bacteria and protists in respiratory activities. The intrinsic values of periphytic phototrophic benthic biofilms are recognized for many reasons:

- Acts as a source of carbon in flood plain systems
- Are the major primary producers and source of food for fish and invertebrates in fluvial ecosystems and form the basis of aquatic food chains
- Plays an important role in nutrient recycling and capture and trophic transfer thus controlling nutrient cycles
- Acts as an important indicator of ecological contamination and decreases in water quality through biomass and compositional changes
- Purifies water by absorption and adsorption of nutrients (especially nitrogen and phosphorus) and metals and has been successfully applied in studies of the pollution-induced community tolerance (PICT) effect

Periphyton. http://dx.doi.org/10.1016/B978-0-12-801077-8.00001-6

TABLE 1.1 Instream Advantages and Disadvantages of Periphyton Accrual Mass

Instream Value	Issue
Esthetic	Degradation of scenery, odor production
Biodiversity	Habitat alteration damages sensitive taxa and reduces benthic biodiversity
Contact recreation	Odor production and swimming and wading impairments
Industrial use	Odor and taste problems
Irrigation	Clogging problem
Monitoring structure	Contamination of sensor surface
Potable supply	Clogging problem, odor and taste problems
Native fish conservation	Impairment of spawning and living habit
Stock and domestic animal health	Cyanobacterial blooms producing toxicity
Waste assimilation	Disturbs stream flow, reduced ability to absorb ammonia, loss of ability to degrade organic matter without depletion of DO (dissolved oxygen)
Water quality	Increased suspended detritus, ammonia toxicity and pH, interstitial anoxia in the stream bed
Whitebait fishing	Clogging nets

Reproduced from Biggs, B.J., Hickey, C.W., 1994. Periphyton responses to a hydraulic gradient in a regulated river in New Zealand. Freshwater Biology 32 (1), 49–59.

Habitually, periphyton are ubiquitous like phytoplankton, found in almost every aquatic ecosystem ranging from small ponds to large oceans and from oligotrophic to eutrophic zones with the exception of muddy and sandy bottomed streams (Azim and Asaeda, 2005). They are important ecosystem components due to competition with macrophytes for carbon and with phytoplankton for light and nutrients (Dijk, 1993; Jones et al., 2002; Sand-Jensen, 1977). Although periphytic algal communities are typically beneficial, overproduction of periphyton can have detrimental effects on aquatic ecosystems (Table 1.1).

1.2 DEFINITION OF PERIPHYTON

Over the past few decades, many definitions have been given for "periphyton," and there is now a need for a standard definition to reflect more recent developments in the field.

According to (Sládečková, 1962), many names or terminologies like *Nereiden, Aufwuchs, Bewuchs, Lasion, Be-lag*, and *Besatz* were adapted by hydrobiologists over the years to refer to periphyton or periphyton-like organisms. The term "periphyton" (*peri* = round; *phyton* = plant) was introduced in 1928 for the first time.

Reflecting the pattern seen with the many different terminologies, a series of definitions and descriptions have been used for periphyton. Young (1945) first defined periphyton as the organisms on submerged objects with a slimy coating. Hunt (1953) refined the definition to algae and tiny animals with a slimy coating ranging from attached sedentary organisms to free living organisms upon submerged objects. Lakatos (1989) defined the term as sessile algal communities of water clinging to submerged objects. Later he proposed the term benthos for organisms attached to submerged objects and used the term biofilm synonymously with periphyton (Wetzel, 2001). MacIntyre et al. (1996) used the term microphytobenthos for periphyton (Stevenson, 1996). described periphyton as a microbial community of algae and bacteria associated with substrates. But this definition did not include all components like algae, bacteria, protozoa and other fluvial animal components as defined by the German term *Aufwuchs* (meaning to grow upon). The term *Aufwuchs* was displaced by the term biofilm after 2000 (Kalff, 2002). In aquatic environments, algae are found either in suspended form as phytoplanktons or sedentary form as periphyton or benthic algae (Hutchinson, 1975; Kalff, 2002). On the basis of these definitions pertiphyton can be identified as an important part of food chains in aquatic ecosystems.

Periphyton (also known as biofilm) is a microecosystem composed of a complex mucopolysaccharide matrix with embedded autotrophic and heterotrophic microorganisms and has the natural ability to respond and recuperate from stress (Sabater et al., 2007). The periphytic community is comprised of green algae, diatoms, bacteria, fungi, protozoans, zooplankton and smaller invertebrates (Azim and Asaeda, 2005); red algae, chrysophyceans, and tribophyceans are also found sporadically (Graham and Wilcox, 2000). This biological entity also contains detritus and calcium carbonate along with organisms. It usually develops on submerged entities like rocks, plants and sediments and can freely move among these substrates.

The color of periphytic biofilms is due to the dominant algal types found in them, ranging from attached or green biofilm to large floating mats of greenish or brownish color. Proximate factors influencing periphyton taxonomic diversity and composition are (1) submersion time (Azim and Asaeda, 2005), (2) water current (Stelzer and Lamberti, 2001), (3) substrate (Azim et al., 2002; Murdock and Dodds, 2007), (4) chemical composition of water (Chessman et al., 1999), (5) grazing (Peterson et al., 2001), (6) nutrient availability (Biggs and Close, 1989), (7) light intensity and quality (Goldsborough et al., 2005) and (8) temperature (DeNicola, 1996). Other factors affecting periphyton biomass accrual are topography, geology, land use, and vegetation type (Biggs, 1996).

Due to its remarkable heterogeneity, periphyton is found in streams year round (Muller, 1999). In winter periphyton shows extensive growth, sometimes exceeding that of summer, if light availability and temperature are relatively high. In aquatic ecosystems, studies of periphyton facilitate the understanding and investigation of various natural processes, such as productivity and interactions among food webs, as well as anthropogenic acidification and contamination (Chaessman, 1985; Cuker, 1983; Cushing and Wolf, 1984; Lowe and Hogg, 1986). Global warming and excessive use of chemicals and nutrients due to dumping of wastes from agricultural activities, WWTPs (wastewater treatment plants), deforestation and soil instability lead to the excessive growth of periphyton in open waters (Bojsen and Jacobsen, 2003; Cascallar et al., 2003; Chessman et al., 1999; Giorgi and Malacalza, 2002; Winter and Duthie, 1998).

In aquatic ecosystems, nutrient cycling is of great importance and has always been the focus of ecological research. Since the regulatory requirement of the European Water Framework Directive (WFD) to achieve ‘good’ ecological status in a range of water bodies by 2015, understanding the delivery process and multiple sources of nutrients entering waters and how they circulate in aquatic ecosystems is urgently needed (Neal and Jarvie, 2005). Before the mid-20th century, most research focused on the role of biotic components such as macrophytes, plankton (zooplankton and phytoplankton) and invertebrates (benthos, nekton and neuston) in aquatic nutrient cycling. At that time, periphytic biofilm or periphyton was not given any attention for its role in aquatic ecosystems. In 1963, Wetzel remarked on the importance of periphyton in aquatic ecosystems as primary producers in his revolutionary review paper (Wetzel, 1963). Periphyton communities are solar-powered biogeochemical reactors, with assemblages of photoautotrophic algae, heterotrophic and chemoautotrophic bacteria, fungi, protozoans, metazoans, and viruses (Larned, 2010). They are commonly attached to submerged surfaces in most aquatic ecosystems and play a significant role in natural aquatic ecosystems through their influence on primary production, food chains, organic matter, and nutrient recycling (Battin et al., 2003; Cantonati and Lowe, 2014; Saikia, 2011). Unfortunately, periphytic biofilm is still ignored as a major contributor of most nutrients to aquatic ecological cycles (Saikia et al., 2013). Thus, in-depth research into periphytic biofilm in aquatic ecosystems is important not only in terms of understanding the significance of periphytic biofilm in aquatic nutrient migration and transformation, but also in improving our comprehension of nutrient biogeochemical process in aquatic ecosystems. With this background, answering certain basic questions regarding the functions and roles of periphytic biofilm in nutrient cycling in aquatic ecosystems are critical. Consequently, this chapter is an effort to (1) characterize periphyton biofilm; (2) state the factors controlling the growth and death of periphyton; (3) evaluate the important functions of periphytic biofilm on aquatic nutrient (C, N and P) cycling; and (4) provide some valuable directions for future research on periphytic biofilm.

Periphytic biofilm has a series of synonyms such as biofilm, periphyton, periphyton assemblages, periphyton mat, surface coatings, microbial aggregate, microbial biofilm, phototrophic biofilms, and microalgae biofilm (Azim, 2009; Azim et al., 2005; Battin et al., 2003; de Beer and Stoodley, 2006; DeNicola and Kelly, 2014; Dong et al., 2002; Ellwood et al., 2012; Roeselers et al., 2008; Saikia et al., 2013; Saikia, 2011; Wu, 2013; Wu et al., 2014). Generally, the term 'periphyton' and 'biofilm' are used most frequently. Historically, Young defined periphyton as the assemblage of organisms growing upon the free surfaces of submerged objects in water, and covering them with a slimy coat. He further described periphyton as that slippery brown or green layer usually found adhering to the surfaces of water plants, wood, stones or other objects immersed in water which may gradually develop from a few gelatinous plants to culminate in a woolly, felted coating that may be slippery or crusty containing marl or sand (Young, 1945). Wetzel (1983) expanded the definition of periphyton as the microfloral community living attached to any substrate under water like submerged plants or plant parts, rocks, and sediment. Some authors (particularly algal biologists), were inclined to define periphyton as attached algae or blue-green algae which disregarded the many other organisms that live in periphyton assemblages (Azim, 2009). In 1996, Stevenson described the periphyton as algae and bacteria growing in association with substrata, including all the organisms attached to, or moving upon, a submerged substrate, but which do not penetrate into it (Stevenson et al., 1996). Among these definitions, periphyton was used mostly in relation to the ecology of aquatic ecosystems, especially lakes and streams, with the terms 'euperiphyton', 'pseudoperiphyton' or 'metaphyton' sometimes used (Azim, 2009). Groups of organisms in periphyton include fungi, bacteria, protozoa, and algae. The most conspicuous group is the algae and this group is usually the focus of studies (especially stream ecology). According to the literature, most previous papers focus on periphyton ecology that can be classified into seven broad topics: (1) effects of physical disturbances; (2) effects of exposure to stressors; (3) limiting abiotic factors; (4) competitive interactions; (5) effects of herbivores; (6) periphytic algae as environmental indicators; and (7) the roles of periphyton in nutrient cycling in foodwebs and between abiotic pools (Larned, 2010).

The term 'biofilm' was also often used to describe microbial assemblages or aggregates growing in association with substrata (Wu et al., 2012), and mostly appears in medical science and environmental science studies such as biological wastewater treatment systems. In 1978, Costerton described biofilm as an aggregation of bacteria, algae, fungi and protozoa enclosed in a matrix consisting of a mixture of polymeric compounds, primarily polysaccharides, generally referred to as extracellular polymeric substance (EPS) (Costerton et al., 1978). When bacteria and microalgae are associated with surfaces or substrate, they secrete a matrix of mucilaginous EPS to form a microbial biofilm that is a prerequisite and essential step for the existence and survival of all microbial aggregates (Boulêtreau et al., 2011; Flemming and Wingender, 2001; Sutherland, 2001; van Hullebusch et al., 2003). In comparison to periphyton, the true definition of biofilm differs from a compositional and functional point of view. This definition 'may' exclude eukaryotic primary producers, and thereby mostly includes decomposers and pioneer colonizing groups of early successional stages, can develop on a number of different surfaces, such as natural aquatic and soil environments, living tissues (e.g., gut lumen), medical devices or industrial or potable water piping systems (Donlan, 2002; Flemming and Wingender, 2001).

Given the abovementioned background, periphyton or periphytic biofilm is defined in this book as an integral and independent micro-ecosystem in aquatic ecosystems, harboring biotic components like algae, fungi, bacteria, protozoans, metazoans and abiotic components like substrata, EPS and detritus. The photoautotrophs (e.g., cyanobacteria and diatoms) and heterotrophs (e.g., bacteria, fungi and protozoa) co-inhabit a self-created common micro-ecosystem and undergo mutualistic, predatory and competitive interactions. Most importantly, periphytic biofilm plays an important role in water bodies, not only by being primary producers and serving as an energy source for higher trophic levels, but also by affecting nutrient turnover and the transfer of nutrients between the water–sediment interface.

1.3 EFFECTS ON WATER ECOLOGY

Periphyton is an essential part of healthy ecosystems and can retain nutrients and chemicals by various processes. Phototrophic periphyton is the focal nutrient source for invertebrates and benthic consumers (Hansson, 1992; Munoz et al., 2001; Pusch et al., 1998). It controls the drift of energy and material by fabricating biomass, thus subsidizing biogeochemical cycles (Roman and Sabater, 1999; Woodruff et al., 1999). It also plays a pivotal role in the bioproductivity of aquatic ecosystems, especially in the coastal zone with sufficient light availability (Nelson et al., 1999; Wolfstein et al., 2000). Periphyton can remove toxic substances, nutrients and metals from open water (Hill et al., 2000; Vymazal, 1988) and can be useful in ecotoxicological studies. According to Adey et al. (1993), retention of nutrients in periphyton is facilitated by four processes. Firstly, periphyton creates a flux between water and sediments. In the second step, advective transport of nutrients occurs by slow water exchange, followed by interception of the nutrients through diffusion into sediments and macrophytes. Lastly, entrapment of particulate matter occurs.

Periphyton rapidly and predictably responds to nutrient changes at observable spatial scales, and subsequently can be used as a pre-indicator of drastic changes in the ecosystem (Gaiser et al., 2005). Understanding the food web structure can provide information about the factors affecting community structure and ecology in waterways, which can be further used as an indicator of disturbance and deteriorating conditions in open waters.

Dense periphytic blooms affect invertebrates, vacillate water pH and DO (dissolved oxygen) levels and can adversely affect the ecology of water. Due to differences between different periphytic algal communities, these blooms can detrimentally affect the ecological and esthetic aspects of water by changing the color, odor and physical appearance (Biggs and Hickey, 1994). Cyanobacterial blooms can decrease water quality by releasing toxins making it unsafe for human and animal consumption.

1.4 COLONIZATION OF PERIPHYTON

Wahl (1989) proposed a stepwise phenomenon for colonization by periphyton (Fig. 1.1).

1.4.1 Phase 1 (Surface Conditioning)

The conditioning phase starting from the adsorption of nutrients on the submerged surfaces with a state of dynamic equilibrium reached within hours after a number of biochemical reactions. The condition process is similar in all types of environments.

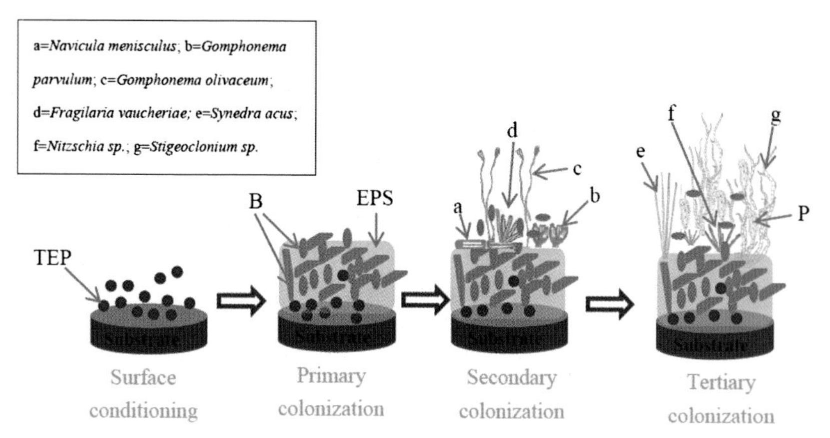

FIGURE 1.1 The sequential colonization and algal succession (a–g) of a periphytic community. *B*, Bacteria; *EPS*, extracellular polymeric substances; *P*, protozoan; *TEP*, transparent exopolymer particle. *Reproduced from Saikia, S., Nandi, S., Majumder, S., 2013. A review on the role of nutrients in development and organization of periphyton. Journal of Research in Biology 3 (1), 780–788 and Azim, M.E., 2009. Photosynthetic periphyton and surfaces. In: Likens, G.E. (Ed.), Encyclopedia of Inland Waters. Academic Press, Oxford, pp. 184–191.*

1.4.2 Phase 2 (Primary Colonization)

The bacterial absorption process is a combination of adsorption and adhesion. Adsorption is purely a physical process involving Brownian motion, electrostatic interaction, gravity, and Van-der-Waal forces leading to the production of polysaccharide fibrils followed by macromolecules. The initial periphyton composed of dead and live bacteria then excretes slime forming primary periphyton.

1.4.3 Phase 3 (Secondary Colonization)

Bacterial colonization is followed by protozoa, diatoms, algae and cyanobacteria. Colonization by diatoms leads to several biochemical reactions on the substrate surface. Protozoans are either sessile in nature or predators of other microorganisms.

1.4.4 Phase 4 (Tertiary Colonization)

Colonization by multicellular organisms and is the last and longest phase leading to a three dimensional periphyton structure.

According to Saikia et al. (2013), there is also a fifth phase, the tertiary (Sládečková) phase. In this phase the bacterioplankton form a cover on secondary periphyton and bacteria reside on algal surfaces.

Generally, the development of a periphytic community on a clean substrate/surface is initiated by the deposition of a coating of dissolved organic substances (mainly amino acids and mucopolysaccharides) by electrostatic forces. Within hours, a coating of bacteria begins to form by hydrophobic reactions. The dissolved and nonliving particulate organic matters serve as support for attachment and as a substrate for the metabolism of bacteria. Then, bacteria actively attach to the substrates using mucilaginous strands. Bacterial colonization is a prerequisite for the subsequent attachment of other organisms. In a primarily physical process, the mucilage produced by bacteria offers potential binding sites for a variety of colloidal, organic and inorganic elements. Bacteria also produce extracellular enzymes that make a significant contribution to the processing of dissolved organic substances, especially for degrading larger fractions of organic molecules into assimilable lower weight molecules as well as inorganic substances. When sufficient light is available, algae will subsequently attach and form the major structure of the developing periphyton. After a few days of bacterial colonization, small pennate diatoms (often *Cocconeis, Navicula*) can adhere to the organic matrix secreted by the bacteria. These are followed by erected or stalked and long stalked species and then by diatoms with rosettes and mucilage pads. During the climax stages of development, green or red algae with upright filaments or long strands can grow forming a layered community within weeks (Azim, 2009). The successional process in periphytic algal communities depends on a complex suite of interactions between physical habitat characteristics, allogenic factors like light and temperature, autogenic changes in the community and species composition (Borduqui and Ferragut, 2012; dos Santos and Ferragut, 2013; Ferragut and de Campos Bicudo, 2010; França et al., 2011).

1.5 COMPOSITION OF PERIPHYTON

1.5.1 Taxonomic Composition

The community composition of periphyton is dependent upon multiple factors such as substrate type and nutrient availability in the external environment. In running water, the periphytic communities are usually found in upper stream sections and are comprised of organisms from three kingdoms, (1) photosynthetic protists including diatoms, chrysophyta (yellow brown algae) and Euglenophyta (euglenoids); (2) chlorophyta and rhodophyta; and (3) cyanobacteria along with other autotrophic bacteria (Allan, 1995). Sometimes waterways are categorized on the basis of the composition and structure of the periphytic community (DeNicola et al., 2004). The composition of the periphyton also varies in different zones of different streams. Allan (1995) mentions the three zones originally described by Margalef in the 1980s, (1) an upper region with fast flowing water, (2) middle, and (3) downstream regions. In rivers and streams, periphytic biofilms are the major primary producers and act as biological models due to their ability to adapt to variable ecological conditions and environments (EPA, 1999) resulting from the generally sessile nature of periphyton (Guckert et al., 1992; Hill et al., 2000).

Periphyton are cosmopolitan in nature and found in almost all aquatic environments where light and temperature conditions are suitable. Many physicochemical and biological factors are involved in the composition, distribution and biomass of periphyton (Chindah, 2004). Composition and biomass vary with water quality due to the sedentary nature of periphyton, with any drastic or ecological changes leading to community transformations (Kiss et al., 2002; McCormick and Stevenson, 1998). Highly diverse periphyton communities are found in healthy waters whereas excessive nutrients and higher temperature results in stagnant conditions leading to excessive algal communities and reduced periphyton biodiversity.

The community composition of the biofilm in a specific zone and under specific weather conditions is an intrinsic property of periphyton. The periphytic community composition also varies seasonally, even during constant temperature conditions, due to light variations (Hynes, 1970). Diatoms form the dominant community in spring and early summer while cyanobacteria are more copious during summer (Allan, 1995). Periphyton is either loosely attached to a submerged object or forms a strong vertical attachment with the object depending on the community composition (Hudon et al., 1987). It may contain "pseudoperiphyton" (motile forms) and "euperiphyton" (sessile forms) (Hynes, 1970).

Periphyton retains more contaminants than the surrounding environment (Behra et al., 2002; Meylan et al., 2003; Wright and Mason, 1999) and shows variable sensitivity to different contaminants (Barranguet et al., 2000).

1.5.2 Nutrient Composition

Periphytic biofilm mainly consists of polysaccharides, proteins, nucleic acids, lipids and humic substances (Saikia, 2011). Several previous studies reported approximate protein, lipid, and carbohydrate compositions of 2–63%, 1–22%, and 10–57%, respectively (Becker, 2007; Ledger and Hildrew, 1998; Montgomery and Gerking, 1980). Polysaccharides and proteins are usually the major components of EPS. EPS is a complex high-molecular-weight mixture of polymers secreted from microorganisms in biofilm, which can be subdivided into bound EPS (sheaths, capsular polymers, condensed gels, loosely bound polymers, and attached organic materials) and soluble EPS (soluble macromolecules, colloids, and slimes) (Laspidou and Rittmann, 2002; Nielsen et al., 1996). EPS can affect various physicochemical characteristics of biofilm by influencing mass transfer, surface charge, flocculation ability, settle ability, dewatering ability, adhesion ability and biofilm formation (Sheng et al., 2010). The content and composition of the EPS in microbial aggregates are heterogeneous, which is attributed to many factors, such as the type of microorganisms, age of the biofilms, surrounding environmental conditions, extraction method, and the analytical tools used (Mayer et al., 1999).

The EPS of biofilm accounts for 50–90% of the total organic matter (Donlan, 2002; Nielsen et al., 1997), therefore, a high amount of carbohydrate. In aquatic environments, bacterial EPS, which is a precondition for periphyton colonization on natural substrates, was proposed as an important supplier of organic C for many organisms that feed on periphytic aggregates (Decho and Moriarty, 1990; Hoskins et al., 2003). This is because EPS is polyanionic in nature and can permit nutrient entrapment through ion exchange processes (Costerton et al., 1978; Freeman and Lock, 1995), which allows the storage of organic C in periphyton. In terms of protein, many periphyton communities have nitrogenase that is able to fix atmospheric N_2 (Costerton et al., 1978; Inglett et al., 2009). Baldwin et al. showed that periphyton may act as a sink for nitrogen (mainly nitrate) in riverine systems (Baldwin et al., 2006). Polunin estimated that the average protein content of periphyton collected from coral reefs was 15% (Polunin, 1988). Dempster et al. (1993) reported 28–55% protein and 5–18% lipid in some algal species of periphytic biofilm. Azim et al. (2001) estimated 27.19% crude protein from periphyton grown on bamboo substrate, 14.63% protein on Hizol (*Barringtonia* sp.) branches, 18.74% on Kanchi (bamboo side shoot), and 12.69% protein on jute sticks. Keshavanath et al. (2004) recorded protein levels of 19.27–35.56% in periphyton.

Being an integral assemblage, all microorganisms present in the periphyton regime represent a complementary food source, providing essential nutrients like polyunsaturated fatty acids (PUFA), sterols, amino acids, vitamins and pigment that help the development of post successional organisms (Thompson et al., 2002). A study on algal bacterial interactions revealed that in the case of submerged plant surfaces, bacterial abundance is significantly higher in areas of diatom colonization (Donnelly and Herbert, 1998). The bacteria involved in the community metabolism of periphyton can trap not only dissolved organic materials and debris drifting in the water body but also the metabolic products released by bacteria and algal species (Makk et al., 2003). Such algal–bacteria interactions turn the periphytic organic matrix into a source of polysaccharides, proteins, nucleic acid and other polymers (Davey and O'toole, 2000). As periphytic biofilm consists of heterogeneous prokaryotic and eukaryotic epiphytic microbial communities, the interactions within the periphyton microorganism may be more intraspecific than interspecific. Such interactions could provide variable food sources to the periphytic community as a whole. Due to its nutritional characteristics, periphyton alone can support fish production of 5000 kg ha^{-1} $year^{-1}$ (Azim et al., 2001). The consumption of periphyton by the pink-shrimp *Farfantepenaeus paulensis* enhanced shrimp survival and growth rates by satisfying 49% of the carbon and 70% of the nitrogen demands of the shrimp (Abreu et al., 2007; Ballester et al., 2007).

1.6 THE TYPES AND STRUCTURE OF PERIPHYTON

In aquatic ecosystems, especially in shallow ecosystems, periphytic biofilm can accumulate on dead or living surface/substrates ranging from clay particles, fine sand, pebbles, and rocks to short-lived filamentous algae, macrophytes, and animal bodies. Thus, periphytic biofilm is widely distributed in water bodies like ponds, lakes, marsh/wetlands, stream/rivers, and estuaries (Azim, 2009). Characteristics of habitats and assemblages vary strongly with water depth. The periphyton, especially in shallow lakes, is often subjected to high-energy conditions, and species must cope with high radiation and water-level fluctuations. In contrast, the deep zone is a much more stable environment and may host a distinct subset of lentic periphyton. Several studies have identified depth distribution zones (shallow, mid-depth, and deep) while analyzing the spatial structure of diatom/algal assemblages along littoral depth gradients (Hawes and Smith, 1994; Kingston et al., 1983; Stevenson and Stoermer, 1981). Disturbance factors are most important in the shallow zone, whereas the mid-depth and deep zones are stable, and the deep zone is affected by extreme light reduction. Light and wave action affect biofilm development and productivity in opposite ways, such that optimal conditions occur below 1 m where disturbance is low but light levels are still relatively high (Vadeboncoeur et al., 2014).

Periphyton can grow on a variety of submerged objects and can influence benthic organisms and water quality. Periphyton biomass and occurrence on various objects help categorize it into various types.

1.6.1 On the Basis of Substrate Type

Depending on the substrate, periphytic biofilm can be divided into natural substrate (such as plants, rock, sediment, etc.) periphyton and artificial substrate (such as glass, industrial soft carriers, polyethylene sheets, etc.) periphyton (Cattaneo and Amireault, 1992; Danilov and Ekelund, 2001; Tuchman and Stevenson, 1980). Some studies have shown that periphyton on natural substrates have an advantage in primary production over periphyton on artificial substrates. This implies that using artificial substrates would lead to an underestimation of periphyton productivity in flowing waters (Nielsen et al., 1984).

Allan and Castillo (2007) categorized periphyton growing on natural substrates into five types on the basis of the submerged objects in the aquatic environment. These are epiphyton, epilithon, epipelon, epipsammon and epixylon (Figs. 1.2 and 1.3).

- Epiphyton are attached to plants, plant parts, macrophytes or large microalgae.
- Epilithon are attached to hard substrates e.g., rock
- Epipelon live on the surface of organic (muddy) sediments
- Epipsammon grow on sand grains
- Epixylon are attached to wood.

Biggs and Hickey (1994) described two more types:

- Epizoic reside on the surface of animal shells and larvae
- Metaphyton originate on sediment and are freely floating algal mates not attached to any substrate.

(Poulíčková et al., 2008) presented several other types including those within animals (Endozoon), within plants (Endophyton) and within rocks (Endolithon) (see Fig. 1.3).

FIGURE 1.2 Frequently encounted periphytic biofilm types in natural water bodies.

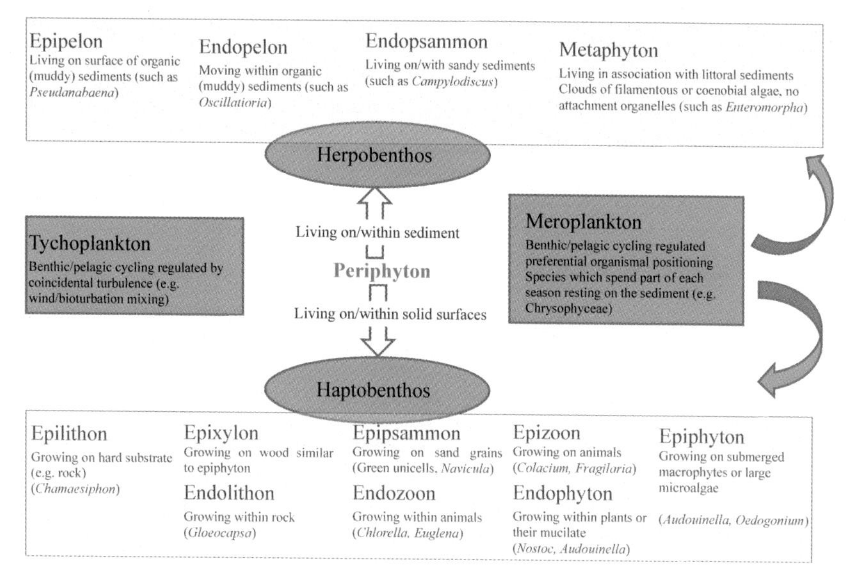

FIGURE 1.3 Classification of algal communities on the basis of substrate (Poulíčková et al., 2008).

Two taxa, epipelon and epipsammon, are quick colonizers even on artificial media due to their motile nature allowing them to be easily dispersed by the water current. Epiphyton and epilithon taxa are more firmly attached to macrophytes and attachment is enhanced by mucilaginous secretions (Allan and Castillo, 2007). Differences in stream energy levels also affect periphyton growth and distribution. Epilithon reside in high energy zones whereas epipelon exist in low energy zones where sediments are not easily disturbed. Epiphyton usually lives on plant parts but does not parasitize plants. Excessive epiphyton growth however, can affect plants (Martin, 2013). Epipelon is one of the most extensively studied types due to its ease in colonizing the surface of artificial substrates used for sampling and its consumption of carbon, nitrogen and phosphorus as nutrients (Hunt, 1953).

Meroplankton, tychoplankton, or metaphyton are the three categories of the organisms co-existing between benthos and water column depending upon their life histories and other factors (Poulíčková et al., 2008).

Periphytic biofilm can also be categorized as micro- or macro-periphyton depending on its size. Macro-periphyton is mostly filamentous algae, representing a primary source of plant food material for aquatic animals, including insects, other invertebrates, and fish. In general, these morphologically large, attached algae cannot be completely consumed by aquatic animals and therefore can develop very dense populations. Macro-periphyton can also serve as the attachment surface for micro-periphyton (Azim, 2009).

1.6.2 On the Basis of Method of Contact With the Substrate

Periphyton attach to substrates either as adpressed or as contrast forms. Contrasts are the forms in which periphyton attach to the substrate erectly using basal parts (e.g., stalked or filamentious forms in Fig. 1.4). In adpressed forms the cell wall, colony or sometimes the whole filamentous system becomes attached to the substrate (e.g., crustose or prostrate forms in Fig. 1.4). A variety of forms exist in the benthic region displaying structural diversity of periphyton (Allan, 1995) (Fig. 1.4).

1.6.3 On the Basis of Photosynthetic Ability

There are two periphyton subdivisions on the basis of photosynthetic capability.

- Microorganisms with visible organelles, e.g., green, yellow-green and red algae, and diatoms.
- Microorganisms without visible organelles, e.g., cyanobacteria.

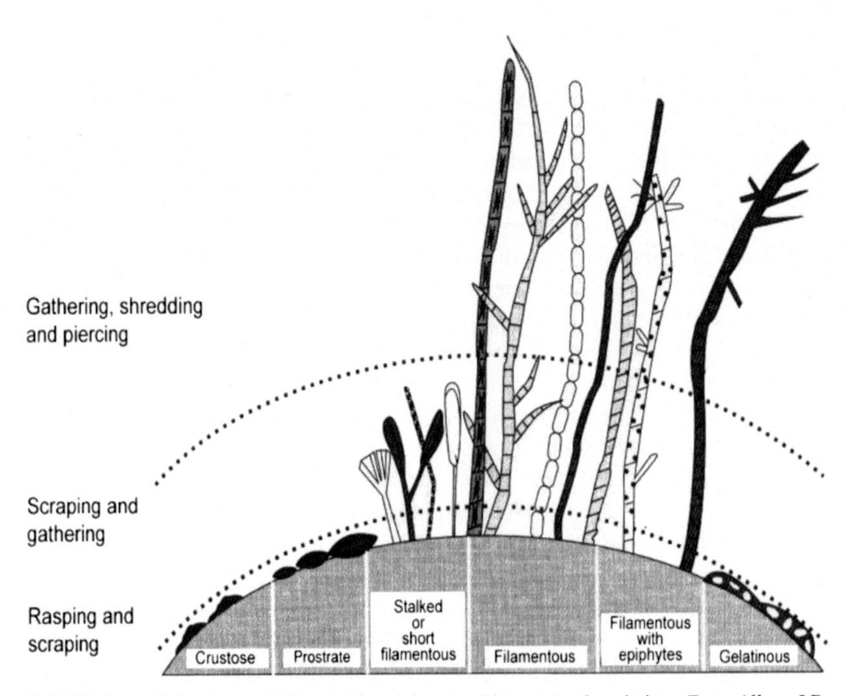

FIGURE 1.4 Various growth forms of periphyton with structural variation. *From Allan, J.D., 1995. Autotrophs. In: Stream Ecology. Springer, Netherlands, pp. 83–108. After Steinman, A.D., Lamberti G.A., 1996. Biomass and pigments of benthic algae, In: Hauer F.R. Lamberti G. A.(Eds.), Methods in Stream Ecology, Academic Press, New York, pp. 295–314.*

1.6.4 Periphyton Structure

Periphytic biofilm structure is the spatial arrangement of bacteria, cell clusters, EPS and particulates, which significantly determine the activity of the biofilm by affecting transport resistance. Various conceptual and mathematical models have been proposed to describe the structure and function of periphytic biofilms (Monds & O'Toole, 2009; Rittmann and Manem, 1992). The conceptual model divides the biofilm system into specific compartments: the substratum, the biofilm, the bulk liquid and a possible head space. The biofilm compartment is further subdivided into a base film and a surface film (de Beer and Stoodley, 2006). Microscopic observations with confocal scanning laser microscopy (CSLM) indicates that periphyton are not flat and the distribution of microorganisms is not uniform. Instead, multispecies biofilms are heterogeneous with complex structures containing voids, channels, cavities, pores, and filaments with cells arranged in clusters or layers (de Beer et al., 1994; Lu et al., 2014). The presence of voids has considerable consequences for mass transfer inside biofilms (advection) and exchange of substrates and products with the water phase (effective exchange surface) (de Beer and Stoodley, 2006), as discussed in more detail in Chapter 4.

The morphological structure of periphyton, differs among groups due to the broad diversity of habitats and surfaces that periphytic algae are adapted to. The size of periphytic algae ranges from that of a single cell of about one thousandth of a millimeter in diameter to a large alga (*Cladophora*) up to many tens of centimeters in length. The shape varies from simple non motile single cells to motile, multicellular, filamentous structures. Non motile forms of unicellular, colonial, and filamentous algae attach to substrates by specially adapted cells and mucilaginous secretions. Some taxa (e.g., *Stigeoclonium*) have morphologically distinct basal and filamentous cells. The basal cells form broad horizontal expanses of cells across the substratum surface and filaments develop vertically from the basal cells. Mucilaginous secretions can be amorphous for unicellular cyanobacteria and green algae or organized into special pads, stalks, or tubes for diatoms (Azim, 2009).

In conclusion, the structural features of biofilms can be viewed in terms of a hierarchical arrangement, the basic components of the biofilm being the cells and the EPS. The structural arrangement of biofilms is not only spatially but also temporally complex. These can combine to form secondary structures such as discrete cell clusters (which may take on various forms and dimensions) and a base film. Finally, the arrangement of base film, cell clusters, and the void areas between the clusters provides the overall biofilm architecture. Many other factors can influence the structural characteristics of periphyton such as hydrodynamic conditions (flow velocity and shear stress) and electrochemical conditions (Lewandowski and Walser, 1991; Stoodley and Lappin-Scott, 1997; Stoodley et al., 1999; Van Loosdrecht et al., 1995).

1.7 THE DISTRIBUTION OF PERIPHYTON AND ITS ROLE IN ENVIRONMENTS

Periphyton are cosmopolitan in nature and thrive in polluted and unpolluted freshwater streams, river and even in sea. If the water is not contaminated, then periphyton exist naturally in the mid water zones, where water current, erosion and deposition do not create any daunting conditions for growth. The submerged objects not only serve as a media for growth of periphyton but also serve in 'trophic Upgrading' and transfer of nutrients. Usually the nutrients are dumped from upstream are used by periphyton attached to rocks or in an area with light availability. Many overwhelming factors affect the growth in upstream areas of stream such as erosion and allochthonous input, dearth of nutrients whereas opposite situation is found in downstream. Phytoplankton is the dominant community in nutrient rich zones (Welcomme and Food, 1985). Littoral zone and coral reefs are densely covered with periphytic biofilm. Coral reefs with high organic matter contents and nitrogen fixation properties have an over-arching growth of periphyton with a combination of algae and cyanophytes leading to higher fish productivity in oligotrophic water (Longhurst and Pauly, 1987).

According to previous study (Gaiser et al., 2005), periphyton serve as an indicator of bio-manipulation by responding quickly to environmental changes and regulate water quality. In return, the environmental changes could change the physiological characteristics and distribution of periphyton. In Everglades, periphyton interact with the physicochemical environment and effect the features like physiology, structure and species composition of the ecosystem (Gaiser et al., 2006; McCormick and O'Dell, 1996; Pan et al., 2000; Swift and Nicholas, 1987).

Physiological properties of periphyton helps to measure the effect of pollutant to the community level by functional tolerance measurement (Blanck et al., 1988) relatively tolerant communities replace the sensitive ones. Pollution induced community tolerance has already been observed in periphyton by different researchers (Blanck and Dahl, 1996; Knauer et al., 1999). It may be the result of some allogeneic or autogenic changes in the surrounding waters that change the adaptation capability of a species phenotypically as well as genotypically (Blanck, 2002) and might be caused by selection pressure of the toxicant on the community (Blanck, 2002). Although uptake of toxic chemicals by periphyton is affected by many environmental factors but the selection process would have some important role to play in the retention process.

Anthropogenic activities sometimes have confounding effects on the distribution and abundance of periphyton by disturbing flow regime and nutrient concentrations (Biggs, 2000; Dodds et al., 1998). Although the physiological properties of periphyton make it essential component for normal functioning of the ecosystem but under certain circumstances it can lead to deterioration of ecosystem by disturbing esthetic and natural biodiversity of the fluvial ecosystems. Distribution and growth rates of periphyton are principally affected by nitrogen and phosphorus, light and temperature which are influenced by anthropogenic activities such as deforestation and vegetation management (Boothroyd et al., 2004; Davies-Colley and Quinn, 1998). The rate of abundance of periphyton can influence many resources and can sometimes alter water physiology as well; accrual rate of biomass and stability period can help to determine maximum abundance of biomass.

Periphyton has played a major role in the primary productivity of the water body which is the ability of a community to produce itself and produce organic matter. The growth of the periphyton on different objects is usually quantitatively described in terms of standing crop which is number of organism per unit area regardless of time. Another terminology biomass is also applied to describe productivity, if standing crops are expressed as AFDW (ash free dry weight), Chl a (chlorophyll a), biogenic elements and rate of oxygen consumption (Larned et al., 2004). explained a model of nutrient uptake involving membrane kinetics or DBL (diffusive boundary Layers). The nutrient uptake by periphyton follows three criteria; (1) the Thickness of DBL control the uptake regime when canopy is submerged covering substratum; (2) when canopy height is greater than DBL, but comparable with the substratum DBL thickness, uptake is controlled jointly by the substratum DBL and by individual DBLs surrounding the periphyton elements that protrude above the substratum DBL; and (3) when the substratum DBL is very thin and most of the canopy protrudes above it, uptake is controlled by the DBLs surrounding periphyton elements.

1.8 GROWTH AND PRODUCTIVITY

The growth of periphytic biofilm usually involves two phases. Initially, periphyton biomass increases exponentially through colonization and growth and reaches a peak which is termed as the 'accrual phase'. There is then a shift to the dominance of loss processes through death, emigration, sloughing, and grazing, which is termed the 'loss phase' (Biggs, 1996). To meet the nutritional requirements of growth, periphyton assemblages constantly assimilate ambient macronutrients by trapping particulate material from the water column (Adey et al., 1993). This macronutrient assimilation is mainly controlled by three principal processes: (1) diffusion from the enviroment into the viscous sub layer of the periphytic boundary layer; (2) slower transport, dominated by molecular diffusion, through the inner portion of the viscous sub layer (the diffusive boundary layer, or DBL) to periphyton cell surfaces; and (3) membrane transport from cell surfaces into cells (Saikia, 2011). A major proportion of the periphyton in waters is composed of algae. These algae capture the energy of sunlight via their chlorophyll molecules, absorb carbon dioxide and other nutrients such as phosphorus and nitrogen from the surrounding water, and then synthesise organic carbon in the form of new or enlarged cells. Algae commonly secrete a portion of this carbon, and a host of other organisms such as communities of bacteria, fungi and protozoa then live off this material. Large algal filaments are often substrates for smaller algal filaments or unicellular algae. The periphyton biomass then increases and reaches a peak biomass that often requires a few days to several months, depending on the availability of light and nutrients and on grazing intensity. When immigration (incoming movement) equals emigration (outgoing movement), an equilibrium periphyton biomass is reached. This equilibrium is influenced by environmental factors regulating growth such as nutrients, light, and temperature and those regulating biomass loss, including disturbances resulting from substratum instability, current velocity, suspended solids, grazing by invertebrates and fish, disease, parasitism, and age. As the community increases in size, algae close to the substrate may senesce as a result of light and or nutrient limitation and become detached from the substrate. Abrupt water motion in the form of spates can dislodge the outer parts of living periphyton layers from their base. This process is called sloughing (autogenic). Dense growths of periphyton (such as filamentous green algae) may become several centimeters in thickness, particularly on gravel and rocks in eutrophic waters. Such dense growths may effectively outshade the lower layers, which then deteriorate and may become susceptible to sloughing.

Periphytic communities in aquatic systems, especially in shallow ecosystems, make significant contributions to primary productivity and oxygen levels due to photosynthesis (Pratiwi et al., 2012). In many water bodies, the contribution of the periphyton community to productivity is greater than that of phytoplankton. Some earlier studies reported that periphyton contributed 42–97% of total annual productivity, especially in shallow lakes (2–3 m). Such productivity in shallow water environments provides an abundant, rapidly renewed, easily assimilated food resource that can be more important than macrophytes (Azim, 2009). However, the contribution of periphyton to annual primary productivity differs among various aquatic systems. For example, periphyton productivity on coral reef systems has been found to range between 1 and 3 g C m^{-2} day^{-1}, while productivity was 1.0–1.7 g C m^{-2} day^{-1} in pond systems and 7.9 g C m^{-2} day^{-1} in acadia systems (Azim, 2009).

Very few investigations have examined photosynthetic rates of algal periphyton under natural conditions (Saikia et al., 2013). Traditionally, measurement of oxygen variations has been used to provide accurate estimates of periphyton photosynthesis and respiration rates. Such measurements require an accurate estimate of the reaeration coefficient between the water and the atmosphere. Recently, microelectrodes, Pulse Amplitude-Modulated (PAM) and High Performance Liquid Chromatography (HPLC) techniques have been used to measure photosynthesis of algal groups, but the method needs refinement for periphyton studies.

1.9 THE STUDY METHODS OF PERIPHYTON

There are different methods and procedures for studying periphyton depending on periphyton type and the research objectives. There are however, a few common steps:

- Collection of the sample and storage for further analysis in laboratory
- Analysis of water to determine the optimal growth conditions for periphyton
- Determination of periphyton biomass and productivity
- Taxonomic analysis

1.9.1 Sample Collection

Periphyton are grown on both natural and artificial substrates but to date, glass slides are the most extensively used artificial substrates for experimental purposes (Albay and Akcaalan, 2003; Pizarro et al., 2002). There are two types of methods typically used for the collection of samples from natural and artificial substrates.

1.9.1.1 Natural Substrates

Natural substrates include all submerged objects in the aquatic ecosystem. The collection of different types of periphyton from natural waters necessitates specific methodologies so periphytic biofilms can be obtained without damage. Four types of quantitative sampling methods are usually employed for different types of natural substrates; cobble and gravel sample, scraping sample with brush, underwater bed rock method, and sand/silt sampling.

Small et al. (2008) used periphyton scraped from rocks in streams to study uranium exposure and after sieving, the scraped contents were allowed to grow on glass slides under laboratory conditions. Kiss et al. (2003) collected epiphyton from different plant species and calculated the surface area of the scraped area of plant. Scinto and Reddy (2003) collected epipelon, epiphyton, and metaphyton from the field and augmented the periphyton in the laboratory to get denser growth for experiments.

1.9.1.2 Artificial Substrates

Periphyton grow on plants, stones and submerged objects in nutrient rich lakes but some researchers have also employed artificial substrates. The ideal sampling technique should involve a substrate suitable for every microorganism, be of suitable size and dimensions for growth dynamics, be convenient for placement in water and retrievable for scraping off the sample. Many kinds of artificial substrates are allowed to stand in the aquatic system and after a certain time interval, the sampling material is taken out and analyzed in the laboratory. Many kinds of artificial substrates are used such as glass slides, bamboo and plexiglass.

Recently Murdock et al. (2013) used 7 × 15 cm unglazed clay tiles for a period of four weeks to study the effect of nutrients and biodegradation ability. Four types of substrate (wood, fiber glass, mosquito screen and garden netting) were used by Richard et al. (2009) with the highest diversity found on mosquito screen. Sugarcane bagasse (Keshavanath et al., 2001), bamboo poles, side shoots, hizel (Azim et al., 2003a), glass slides (Azim et al., 2003b; Schmitt-Jansen and Altenburger, 2005), strips of tape (Liboriussen and Jeppesen, 2006), PVC pipes, plastic sheets, fibrous scrubbers, and ceramic tiles (Khatoon et al., 2007) have been used by different researchers at a temperature range of 20–35°C with the highest biomass and productivity obtained using bamboo shoots as a growth media (Liboriussen et al., 2005; Vermaat, 2005). Some of these researchers placed the artificial media in the natural environment while others used artificial streams in the laboratory.

Glass slides are the most frequently used growth substrates but are unsuitable for large scale experiments. For large scale studies polyethylene sheets (Jöbgen et al., 2004; Liboriussen and Jeppesen, 2006) and more recently Artificial Aquatic Mats (AAM) and Industrial Soft Carriers (ISC) Wu et al. (2010) were found to be very effective due to their light weight and flexibility.

1.9.2 Analysis of Water

Physical and biochemical analyses of natural waters is done for two reasons: to understand the growth conditions for periphyton in natural environments and to provide the same growth conditions for periphyton during enrichment processes in the laboratory. These analyses involved measurement of TN (total nitrogen), inorganic nitrogen, TP (total phosphorus), inorganic phosphorus (Pizarro et al., 2002), suspended substances, dissolved oxygen, conductivity, alkalinity, depth, light and temperature (Azim et al., 2003a; Gold et al., 2003; Liboriussen and Jeppesen, 2006; Vézina and Vincent, 1997; Wu et al., 2010).

The methods used for the analysis of water chemistry and physical conditions differ between researchers, even sometimes for the same contaminant at times, depending on the availability of facilities and equipment.

1.9.3 Determination of Biomass and Productivity

The mass of the communities in the periphyton biofilms is usually determined by two methods.

- Chlorophyll a (chl a)
- Ash Free Dry Mass (AFDW)

Chlorophyll a is used to determine the amount of autotrophic organisms in the biofilm i.e. the primary productivity, whereas AFDW is used to determine the total organic content including heterotrophic and autotrophic biota, dead periphyton and allochthonous substances. Three methods are typically used to determine the chl a content.

- Spectrophotometry
- Fluorometry and
- High performance liquid chromatography (HPLC).

For AFDW, all the water in the sample must be removed by drying. The sample must then be weighed before and after burning at high temperature furnace (usually at 500–550°C) for 1–2 h (Eaton, 1995). The difference in the two weights is the AFDW.

AFDW (g per sample) = [(weight of crucible + filter + sample after drying) – (weight of crucible + filter + sample after ashing)] × [sample volume/volume of filtered sub-sample]

Another parameter used to determine biomass is the autotrophic index, AI (Webb-Robertson et al., 2011). AI is basically the ratio of chl a to AFDW and gives the ratio of heterotrophic to autotrophic communities in the sample. An AI value between 50 and 100 indicates that the water is not polluted. If the value of AI is <100 the dominant community is viable algae. Values between 100 and 200 indicate autotrophs whereas values >200 indicate heterotrophs dominate the community (Murdock et al., 2013). Values >400 show that organic detritus and heterotrophs form the dominant proportion of the community.

% organic matter = (AFDW × 100)/(dry mass)

Autotrophic index (Webb-Robertson et al.) = AFDM (in mg/m^2)/chlorophyll a (mg/m^2)

Lakatos (1989) devised a similar classification of periphyton based on AFDW and chlorophyll a concentrations (Table 1.2). Based on the calculation of AFDW content (and chlorophyll a content), four types of periphyton are divided and their corresponding AFDW content percentage are also presented.

1.9.4 Taxonomic Analysis

Taxonomic analysis is used to identify the composition and diversity of the community at the species and genus level and changes in community composition as a result of exposure to different nutrients and chemicals. There are two types of/steps in taxonomic analysis, (1) observation under a microscope and (2) advanced microbiological methods.

1.9.4.1 Microscopic Examination

The basic characterization of the periphyton community is usually performed in the laboratory using a microscope after the sample is collected. If sampling is done outside the laboratory, then the sample is usually preserved in buffered formalin until taxonomic analysis. Before mounting the sample on a

TABLE 1.2 Periphyton Classified on the Basis of AFDW and Chl a

Type	AFDW Content	Percentage
I	Inorganic periphyton	More than 75
II	Inorganic-organic periphyton	50–75
III	Organic-inorganic periphyton	25–50
IV	Organic periphyton	Less than 25
Type	**Chlorophyll a**	**Percentage**
I	Autotrophic periphyton	More than 0.60
II	Auto-heterotrophic periphyton	0.25–0.60
III	Hetero-autotrophic periphyton	0.10–0.25
IV	Heterotrophic periphyton	Less than 0.10

From Lakatos, G., 1989. Composition of reed periphyton (biotecton) in the Hungarian part of lake Fertö. BFB-Bericht 71, 125–134.

microscopic slide, the sample is washed either with H_2SO_4, distilled water and KNO_3 to remove organic matter or for acid digestion with hydrochloric acid (37% HCl) and hydrogen peroxide (30% H_2O_2).

The relative abundance of different taxa is usually done using an inverted light microscope whereas a compound microscope is required for quantification of cells. For the analysis of diatoms and soft bodied algae, phase-contrast microscope is utilized after mounting on permanent slides using Naphrax and Palmer-Maloney counting chambers, respectively. Identification of the diatoms to species level is performed by using different keys (Dellamano-Oliveira et al., 2008; Ferragut et al., 2005; Fontana and Bicudo, 2009; Lange-Bertalot and Krammer, 2001; Round, 1974; Silva and Cecy, 2004).

The information obtained from microscopic and biomass analyses can be used to explain many metrics like species richness, generic richness, community diversity, community evenness, percent community similarity, simple diagnostic metrics and pollution tolerance indices.

Two other types of microscopy frequently used for more than a decade are Scanning Electron Scanning Microscopy (SEM) and Transmitting Electron Microscopy (TEM) to obtain fine and three dimensional structures.

1.9.4.2 Molecular Analysis

Molecular analysis helps to identify the biofilm composition to genus level and to determine shifts in the community due to environmental changes. For more than a decade, different analysis techniques have been introduced for DNA fingerprinting including Denaturing Gradient Gel Electrophoresis (DGGE), automated ribosomal spacer analysis or Terminal-Restriction Fragment Length Polymorphism (T-RFLP) (Amann et al., 1995; Dorigo et al., 2002; SzabÓ et al., 2008).

PLFA (phospholipid fatty acid) assists in differentiating the metabolic and phylogenetic groups in a community of different microbial species on the basis of membrane components (Guckert et al., 1985; Hedrick et al., 2000). PLFA can characterize living species due to rapid degradation of phospholipids after cell death (Small et al., 2008) so can be used as a biomarker to detect the functional groups in an unidentified community and monitors any perturbations in community structure (Boston and Hill, 1991; Webb-Robertson et al., 2011; White and Ringelberg, 1997). In PLFA lipids are extracted from the periphyton biofilm and then fatty acids into methyl esters (FAMEs) are identified and quantified by using GC−MS (Small et al., 2008).

DGGE describes the community composition at a more superlative level. By using DDGE techniques all the organisms in the community can be genetically described at the genus and species level (Small et al., 2008). DGGE is the combination of amplification of DNA after extraction using primers (Diez et al., 2001) and then loading the PCR product onto gel in the DGGE system.

Some other important tests are typically used to determine productivity and tolerance of periphytic communities.

1.9.4.3 PICT (Pollution-Induced Community Tolerance)

Periphyton shows remarkable tolerance to contaminants like metals and pesticides as determined by short term toxicity tests. The concept of PICT was first proposed by Blanck et al. (1988). PICT can be determined by determining and comparing the EC50 values of the reference and exposed biofilms (Blanck and Dahl, 1996).

1.9.4.4 Gross Primary Productivity

Gross primary productivity of periphyton can be calculated from the photosynthesis and respiration rates using the oxygen evolution method (Wetzel and Likens, 2000).

1.10 CONCLUSION

Periphyton are ubiquitous in aquatic systems and play a large role in primary production and nutrient cycling. Periphyton community structure and compostion are influenced by a number of biotic (e.g., species interactions, herbivory, etc.) and abiotic (e.g., light, temperature, substrate type, etc.) factors. As a result of the diversity of organsims involved in periphyton communities, periphyton is able to rapidly and predictably respond to changes in environmental conditions. This makes periphyton an ideal bio-indicator as well as enhances its ability to remove a range of pollutants from water.

REFERENCES

Abreu, P.C., Ballester, E.L.C., Odebrecht, C., 2007. Importance of biofilm as food source for *shrimp* evaluated by stable isotopes (^{13}C and ^{15}N). Journal of Experimental Marine Biology and Ecology 347 (1), 88–96.

Adey, W., Luckett, C., Jensen, K., 1993. Phosphorus removal from natural waters using controlled algal production. Restoration Ecology 1 (1), 29–39.

Albay, M., Akcaalan, R., 2003. Comparative study of periphyton colonisation on common reed (*Phragmites australis*) and artificial substrate in a shallow lake, Manyas, Turkey. Hydrobiologia 506–509 (1–3), 531–540.

Allan, J.D., Castillo, M.M., 2007. Stream Ecology: Structure and Function of Running Waters. Springer.

Allan, J.D., 1995. Autotrophs. In: Stream Ecology. Springer, Netherlands, pp. 83–108.

Amann, R.I., Ludwig, W., Schleifer, K.H., 1995. Phylogenetic identification and in situ detection of individual microbial cells without cultivation. Microbiological Reviews 59 (1), 143–169.

Azim, M.E., Asaeda, T., 2005. Periphyton structure, diversity and colonization. In: Azim, M.E., Beveridge, M.C.M., van Dam, A.A. (Eds.), Periphyton: Ecology, Exploitation and Management. CABI Publishing, pp. 15–49.

Azim, M.E., Wahab, M.A., Van Dam, A.A., 2001. The potential of periphyton-based culture of two Indian major carps, rohu *Labeo rohita* (Hamilton) and gonia *Labeo gonius* (Linnaeus). Aquaculture Research 32 (3), 209–216.

Azim, M.E., Verdegem, M.C.J., Khatoon, H., Wahab, M.A., van Dam, A.A., Beveridge, M.C.M., 2002. A comparison of fertilization, feeding and three periphyton substrates for increasing fish production in freshwater pond aquaculture in Bangladesh. Aquaculture 212 (1–4), 227–243.

Azim, M.E., Milstein, A., Wahab, M.A., Verdegam, M.C.J., 2003a. Periphyton–water quality relationships in fertilized fishponds with artificial substrates. Aquaculture 228 (1–4), 169–187.

Azim, M.E., Verdegem, M.C.J., Singh, M., Van Dam, A.A., Beveridge, M.C.M., 2003b. The effects of periphyton substrate and fish stocking density on water quality, phytoplankton, periphyton and fish growth. Aquaculture Research 34 (9), 685–695.

Azim, M.E., Verdegem, M.C., van Dam, A.A., Beveridge, M.C., 2005. Periphyton: Ecology, Exploitation and Management. CABI.

Azim, M.E., 2009. Photosynthetic periphyton and surfaces. In: Likens, G.E. (Ed.), Encyclopedia of Inland Waters. Academic Press, Oxford, pp. 184–191.

Baldwin, D.S., Mitchell, A., Rees, G., Watson, G., Williams, J., 2006. Nitrogen processing by biofilms along a lowland river continuum. River Research and Applications 22 (3), 319–326.

Ballester, E.L.C., Wasielesky, J.W., Cavalli, R.O., 2007. Nursery of the *pink shrimp* in cages with artificial substrates: biofilm composition and *shrimp* performance. Aquaculture 269 (1), 355–362.

Barranguet, C., Charantoni, E., Plans, M., Admiraal, W., 2000. Short-term response of monospecific and natural algal biofilms to copper exposure. European Journal of Phycology 35 (4), 397–406.

Battin, T.J., Kaplan, L.A., Denis Newbold, J., Hansen, C.M.E., 2003. Contributions of microbial biofilms to ecosystem processes in stream mesocosms. Nature 426 (27), 439–442.

Becker, E.W., 2007. Micro-algae as a source of protein. Biotechnology Advances 25 (2), 207–210.

Behra, R., Landwehrjohann, R., Vogel, K., Wagner, B., Sigg, L., 2002. Copper and zinc content of periphyton from two rivers as a function of dissolved metal concentration. Aquatic Sciences 64 (3), 300–306.

Biggs, B.J.F., Close, M.E., 1989. Periphyton biomass dynamics in gravel bed rivers: the relative effects of flows and nutrients. Freshwater Biology 22 (2), 209–231.

Biggs, B.J., Hickey, C.W., 1994. Periphyton responses to a hydraulic gradient in a regulated river in New Zealand. Freshwater Biology 32 (1), 49–59.

Biggs, B., 1996. Patterns in benthic algae of streams. In: Stevenson, R.J., Bothwell, M.L., Lowe, R.L. (Eds.), Algal Ecology: Freshwater Benthic Ecosystems. Academic Press, San Diego, California.

Biggs, B.J., 2000. Eutrophication of streams and rivers: dissolved nutrient-chlorophyll relationships for benthic algae. Journal of the North American Benthological Society 19 (1), 17–31.

Blanck, H., Dahl, B., 1996. Pollution-induced community tolerance (PICT) in marine periphyton in a gradient of tri-n-butyltin (TBT) contamination. Aquatic Toxicology 35 (1), 59–77.

Blanck, H., Molander, S., Wängberg, S., 1988. Pollution-induced community tolerance-a new ecotoxicological tool. In: Pratt, J.R., Cairns, J. (Eds.), Functional Testing of Aquatic Biota for Estimating Hazards of Chemicals. American Society for Testing and Materials, pp. 219–230, 1989.

Blanck, H., 2002. A critical review of procedures and approaches used for assessing pollution-induced community tolerance (PICT) in biotic communities. Human and Ecological Risk Assessment: An International Journal 8 (5), 1003–1034.

Bojsen, B.H., Jacobsen, D., 2003. Effects of deforestation on macroinvertebrate diversity and assemblage structure in Ecuadorian Amazon streams. Archiv für Hydrobiologie 158 (3), 317–342.

Boothroyd, I.K.G., Quinn, J.M., Langer, E.R., Costley, K.J., Steward, G., 2004. Riparian buffers mitigate effects of pine plantation logging on New Zealand streams: 1. Riparian vegetation structure, stream geomorphology and periphyton. Forest Ecology and Management 194 (1–3), 199–213.

Borduqui, M., Ferragut, C., 2012. Factors determining periphytic algae succession in a tropical hypereutrophic reservoir. Hydrobiologia 683 (1), 109–122.

Boston, H., Hill, W., 1991. Photosynthesis-light relations of stream periphyton communities. Limnology and Oceanography 36 (4), 644–656.

Boulêtreau, S., Charcosset, J.-Y., Gamby, J., Lyautey, E., Mastrorillo, S., Azémar, F., Moulin, F., Tribollet, B., Garabetian, F., 2011. Rotating disk electrodes to assess river biofilm thickness and elasticity. Water Research 45 (3), 1347–1357.

Cantonati, M., Lowe, R.L., 2014. Lake benthic algae: toward an understanding of their ecology. Freshwater Science 33 (2), 475–486.

Cascallar, L., Mastranduono, P., Mosto, P., Rheinfeld, M., Santiago, J., Tsoukalis, C., Wallace, S., 2003. 19 Periphytic algae as bioindicators of nitrogen inputs in lakes. Journal of Phycology 39, 7–8.

Cattaneo, A., Amireault, M.C., 1992. How artificial are artificial substrata for periphyton? Journal of the North American Benthological Society 1 (2), 244–256.

Chaessman, B., 1985. Estimates of ecosystem metabolism in the La Trobe river, Victoria. Marine and Freshwater Research 36 (6), 873–880.

Chessman, B., Growns, I., Currey, J., Plunkett-Cole, N., 1999. Predicting diatom communities at the genus level for the rapid biological assessment of rivers. Freshwater Biology 41 (2), 317–331.

Chindah, A.C., 2004. Response of periphyton community to salinity gradient in tropical estuary, Niger Delta. Polish Journal of Ecology 52 (1), 83–89.

Costerton, J.W., Geesey, G.G., Cheng, K.J., 1978. How bacteria stick. Scientific American 238 (1), 86–95.

Cuker, B.E., 1983. Competition and coexistence among the grazing snail Lymnaea, Chironomidae, and mircrocrustacea in an Arctic epilithic lacustrine community. Ecology 64 (1), 10–15.

Cushing, C., Wolf, E., 1984. Primary production in Rattlesnake Springs, a cold desert spring-stream. Hydrobiologia 114 (3), 229–236.

Danilov, R.A., Ekelund, N., 2001. Comparison of usefulness of three types of artificial substrata (glass, wood and plastic) when studying settlement patterns of periphyton in lakes of different trophic status. Journal of Microbiological Methods 45 (3), 167–170.

Davey, M.E., O'toole, G.A., 2000. Microbial biofilms: from ecology to molecular genetics. Microbiology and Molecular Biology Reviews 64 (4), 847–867.

Davies-Colley, R.J., Quinn, J.M., 1998. Stream lighting in five regions of North Island, New Zealand: control by channel size and riparian vegetation. New Zealand Journal of Marine and Freshwater Research 32 (4), 591–605.

de Beer, D., Stoodley, P., 2006. Microbial biofilms. Prokaryotes 1, 904–937.

de Beer, D., Stoodley, P., Lewandowski, Z., 1994. Liquid flow in heterogeneous biofilms. Biotechnology and Bioengineering 44 (5), 636–641.

Decho, A.W., Moriarty, D.J.W., 1990. Bacterial exopolymer utilization by a harpacticoid copepod: a methodology and results. Limnology and Oceanography 35 (5), 1039–1104.

Dellamano-Oliveira, M.J., Sant'Anna, C.L., Taniguchi, G.M., Senna, P.A.C., 2008. Os gêneros Staurastrum, Staurodesmus e Xanthidium (Desmidiaceae, Zygnemaphyceae) da Lagoa do Caçó, Estado do Maranhão, Nordeste do Brasil. Hoehnea 35, 333–350.

Dempster, P.W., Beveridge, M.C.M., Baird, D.J., 1993. Herbivory in the tilapia *Oreochromis niloticus*: a comparison of feeding rates on phytoplankton and periphyton. Journal of Fish Biology 43 (3), 385–392.

DeNicola, D.M., Kelly, M., 2014. Role of periphyton in ecological assessment of lakes. Freshwater Science 33 (2), 619–638.

DeNicola, D.M., Eyto, E.d., Wemaere, A., Irvine, K., 2004. Using epilithic algal communities to assess trophic status in Irish Lakes. Journal of Phycology 40 (3), 481–495.

DeNicola, D.M., 1996. Periphyton responses to temperature at different ecological levels. In: Stevenson, R.J., Bothwell, M.L., Lowe, R.L. (Eds.), Algal Ecology: Freshwater Benthic Ecosystems. Academic Press, San Diego, pp. 149–181.

Diez, B., Pedros-Alio, C., Massana, R., 2001. Study of genetic diversity of eukaryotic picoplankton in different oceanic regions by small-subunit rRNA gene cloning and sequencing. Applied and Environmental Microbiology 67 (7), 2932–2941.

Dijk, G., 1993. Dynamics and attenuation characteristics of periphyton upon artificial substratum under various light conditions and some additional observations on periphyton upon *Potamogeton pectinatus* L. Hydrobiologia 252 (2), 143–161.

Dodds, W.K., Jones, J.R., Welch, E.B., 1998. Suggested classification of stream trophic state: distributions of temperate stream types by chlorophyll, total nitrogen, and phosphorus. Water Research 32 (5), 1455–1462.

Dong, D., Hua, X., Li, Y., Li, Z., 2002. Lead adsorption to metal oxides and organic material of freshwater surface coatings determined using a novel selective extraction method. Environmental Pollution 119, 317–321.

Donlan, R.M., 2002. Biofilms: microbial life on surfaces. Emerging Infectious Diseases Journal 8, 881–890.

Donnelly, A.P., Herbert, R.A., 1998. Bacterial interactions in the rhizosphere of seagrass communities in shallow coastal lagoons. Journal of Applied Microbiology 85 (S1), 151S–160S.

Dorigo, U., Berard, A., Humbert, J.F., 2002. Comparison of eukaryotic phytobenthic community composition in a polluted river by partial 18S rRNA gene cloning and sequencing. Microbial Ecology 44 (4), 372–380.

dos Santos, T.R., Ferragut, C., 2013. The successional phases of a periphytic algal community in a shallow tropical reservoir during the dry and rainy seasons. Limnetica 32 (2), 337–352.

Eaton, A.D., Clesceri, L.S., Greenberg, A.E. (Eds.), 1995. Standard Methods for the Examination of Water and Wastewater, 19th ed. American Public Health Association. America Water Works Association, and Water Enviroment Federation.

Ellwood, N.T.W., Di Pippo, F., Albertano, P., 2012. Phosphatase activities of cultured phototrophic biofilms. Water Research 46 (2), 378–386.

EPA, U., 1999. Biological Criteria. National Program Guidance for Surface Waters (Washington, DC).

Ferragut, C., de Campos Bicudo, D., 2010. Periphytic algal community adaptive strategies in N and P enriched experiments in a tropical oligotrophic reservoir. Hydrobiologia 646 (1), 295–309.

Ferragut, C., Lopes, M.R.M., Bicudo, D.C., Bicudo, C.E.M., Vercellino, I.S., 2005. Ficoflórula perifítica e planctônica (exceto Bacillariophyceae) de um reservatório oligotrófico raso (Lago do IAG, São Paulo). Hoehnea 32 (2), 137–184.

Flemming, H., Wingender, J., 2001. Relevance of microbial extracellular polymeric substances (EPSs)-part I: structural and ecological aspects. Water Science & Technology 43 (6), 1–8.

Fontana, L., Bicudo, D.C., 2009. Diatomáceas (Bacillariophyceae) de sedimentos superficiais dos reservatórios em cascata do Rio Paranapanema (SP/PR, Brasil): Coscinodiscophyceae e Fragilariophyceae. Hoehnea 36 (3).

França, R., Lopes, M.R.M., Ferragut, C., 2011. Structural and successional variability of periphytic algal community in a Amazonian lake during the dry and rainy season (Rio Branco, Acre). Acta Amazonica 41 (2), 257–266.

Freeman, C., Lock, M.A., 1995. The biofilm polysaccharide matrix: a buffer against changing organic substrate supply? Limnology and Oceanography 40 (2), 273–278.

Gaiser, E.E., Trexler, J.C., Richards, J.H., Childers, D.L., Lee, D., Edwards, A.L., Scinto, L.J., Jayachandran, K., Noe, G.B., Jones, R.D., 2005. Cascading ecological effects of low-level phosphorus enrichment in the Florida Everglades. Journal of Environmental Quality 34 (2), 717–723.

Gaiser, E.E., Richards, J.H., Trexler, J.C., Jones, R.D., Childers, D.L., 2006. Periphyton responses to eutrophication in the Florida Everglades: cross-system patterns of structural and compositional change. Limnology and Oceanography 51 (1).

Giorgi, A., Malacalza, L., 2002. Effect of an industrial discharge on water quality and periphyton structure in a pampeam stream. Environmental Monitoring and Assessment 75 (2), 107–119.

Gold, C., Feurtet-Mazel, A., Coste, M., Boudou, A., 2003. Impacts of Cd and Zn on the development of periphytic diatom communities in artificial streams located along a river pollution gradient. Archives of Environmental Contamination and Toxicology 44 (2), 189–197.

Goldsborough, L., McDougal, R., North, A., Azim, M., Verdegem, M., van Dam, A., Beveridge, M., 2005. Periphyton in freshwater lakes and wetlands. Periphyton: Ecology, Exploitation, and Management 71–89.

Graham, L.E., Wilcox, L.W., 2000. Algae—Prentice Hall. Upper Saddle River, New Jersey.

Guckert, J.B., Antworth, C.P., Nichols, P.D., White, D.C., 1985. Phospholipid, ester-linked fatty acid profiles as reproducible assays for changes in prokaryotic community structure of estuarine sediments. FEMS Microbiology Letters 31 (3), 147–158.

Guckert, J.B., Nold, S.C., Boston, H.L., White, D.C., 1992. Periphyton response in an industrial receiving stream: lipid-based physiological stress analysis and pattern recognition of microbial community structure. Canadian Journal of Fisheries and Aquatic Sciences 49 (12), 2579–2587.

Hansson, L.-A., 1992. The role of food chain composition and nutrient availability in shaping algal biomass development. Ecology 73 (1), 241–247.

Hawes, I., Smith, R., 1994. Seasonal dynamics of epilithic periphyton in oligotrophic Lake Taupo, New Zealand. New Zealand Journal of Marine and Freshwater Research 28 (1), 1–12.

Hedrick, D.B., Peacock, A., Stephen, J.R., Macnaughton, S.J., Brüggemann, J., White, D.C., 2000. Measuring soil microbial community diversity using polar lipid fatty acid and denaturing gradient gel electrophoresis data. Journal of Microbiological Methods 41 (3), 235–248.

Hill, B.H., Willingham, W.T., Parrish, L.P., McFarland, B.H., 2000. Periphyton community responses to elevated metal concentrations in a Rocky Mountain stream. Hydrobiologia 428 (1), 161–169.

Hoskins, D.L., Stancyk, S.E., Decho, A.W., 2003. Utilization of algal and bacterial extracellular polymeric secretions (EPS) by the deposit-feeding brittlestar *Amphipholis gracillima (Echinodermata).* Marine Ecology Progress Series 247, 93–101.

Hudon, C., Duthie, H.C., Paul, B., 1987. Physiological modifications related to density increase in periphyton assemblages. Journal of Phycology 23 (3), 393–399.

Hunt, B.P., 1953. Food relationships between Florida spotted gar and other organisms in the Tamiami Canal, Dade County, Florida. Transactions of the American Fisheries Society 82 (1), 13–33.

Hutchinson, G.E., 1975. A Treatise on Limnology: Limnological Botany. Wiley.

Hynes, H.B.N., 1970. The Ecology of Running Waters. University of Toronto Press.

Inglett, P.W., D'Angelo, E.M., Reddy, K.R., McCormick, P.V., Hagerthey, S.E., 2009. Periphyton nitrogenase activity as an indicator of wetland eutrophication: spatial patterns and response to phosphorus dosing in a northern Everglades ecosystem. Wetlands Ecology and Management 17 (2), 131–144.

Jöbgen, A., Palm, A., Melkonian, M., 2004. Phosphorus removal from eutrophic lakes using periphyton on submerged artificial substrata. Hydrobiologia 528 (1–3), 123–142.

Jones, J.I., Young, J.O., Eaton, J.W., Moss, B., 2002. The influence of nutrient loading, dissolved inorganic carbon and higher trophic levels on the interaction between submerged plants and periphyton. Journal of Ecology 90 (1), 12–24.

Kalff, J., 2002. Limnology: Inland Water Ecosystems. Prentice Hall.

Keshavanath, P., Gangadhar, B., Ramesh, T.J., Van Rooij, J.M., Beveridge, M.C.M., Baird, D.J., Verdegem, M.C.J., Van Dam, A.A., 2001. Use of artificial substrates to enhance production of freshwater herbivorous fish in pond culture. Aquaculture Research 32 (3), 189–197.

Keshavanath, P., Gangadhar, B., Ramesh, T.J., 2004. Effects of bamboo substrate and supplemental feeding on growth and production of hybrid red *tilapia fingerlings*. Aquaculture 235 (1), 303–314.

Khatoon, H., Yusoff, F., Banerjee, S., Shariff, M., Bujang, J.S., 2007. Formation of periphyton biofilm and subsequent biofouling on different substrates in nutrient enriched brackishwater shrimp ponds. Aquaculture 273 (4), 470–477.

Kingston, J.C., Lowe, R.L., Stoermer, E.F., Ladewski, T.B., 1983. Spatial and temporal distribution of benthic diatoms in northern Lake Michigan. Ecology 1566–1580.

Kiss, M.K., Lakatos, G., Keresztúri, P., Borics, G., Szilágyi, E.K., 2002. Investigation of macrophyte-periphyton complex in Tisza reservoir. In: Gallé, L., Körmöczi, L. (Eds.), Ecology of River Valleys, vol. 5. Department of Ecology, University of Szeged, p. 211, 2000.

Kiss, M., Lakatos, G., Borics, G., Gidó, Z., Deák, C., 2003. Littoral macrophyte–periphyton complexes in two Hungarian shallow waters. Hydrobiologia 506–509 (1–3), 541–548.

Knauer, K., Behra, R., Hemond, H., 1999. Toxicity of inorganic and methylated arsenic to algal communities from lakes along an arsenic contamination gradient. Aquatic Toxicology 46 (3–4), 221–230.

Lakatos, G., 1989. Composition of reed periphyton (biotecton) in the Hungarian part of lake Fertö. BFB-Bericht 71, 125–134.

Lange-Bertalot, H., Krammer, K., 2001. Diatoms of Europe: Navicula Sensu Stricto, 10 Genera Separated from Navicula Sensu Lato, Frustulia. A.R.G. Ganter.

Larned, S.T., Nikora, V.I., Biggs, B.J., 2004. Mass-transfer-limited nitrogen and phosphorus uptake by stream periphyton: a conceptual model and experimental evidence. Limnology and Oceanography 49 (6), 1992–2000.

Larned, S.T., 2010. A prospectus for periphyton: recent and future ecological research. Journal of the North American Benthological Society 29 (1), 182–206.

Laspidou, C.S., Rittmann, B.E., 2002. A unified theory for extracellular polymeric substances, soluble microbial products, and active and inert biomass. Water Research 36 (11), 2711–2720.

Ledger, M.E., Hildrew, A.G., 1998. Temporal and spatial variation in the epilithic biofilm of an acid stream. Freshwater Biology 40 (4), 655–670.

Lewandowski, Z., Walser, G., 1991. Influence of hydrodynamics on biofilm accumulation. Environmental Engineering 1991, 619–624. ASCE.

Liboriussen, L., Jeppesen, E., 2006. Structure, biomass, production and depth distribution of periphyton on artificial substratum in shallow lakes with contrasting nutrient concentrations. Freshwater Biology 51 (1), 95–109.

Liboriussen, L., Jeppesen, E., Bramm, M., Lassen, M., 2005. Periphyton-macroinvertebrate interactions in light and fish manipulated enclosures in a clear and a turbid shallow lake. Aquatic Ecology 39 (1), 23–39.

Longhurst, A.R., Pauly, D., 1987. Ecology of Tropical Oceans. Academic Press.

Lowe, D.J., Hogg, A.G., 1986. Tephrostratigraphy arid chronology of the Kaipo Lagoon, an 11,500 year-old montane peat bog in Urewera National Park, New Zealand. Journal of the Royal Society of New Zealand 16 (1), 25–41.

Lu, H., Yang, L., Zhang, S., Wu, Y., 2014. The behavior of organic phosphorus under non-point source wastewater in the presence of phototrophic periphyton. PLoS One 9 (1), e85910.

MacIntyre, H.L., Geider, R.J., Miller, D.C., 1996. Microphytobenthos: the ecological role of the "secret garden" of unvegetated, shallow-water marine habitats. I. Distribution, abundance and primary production. Estuaries 19 (2), 186–201.

Makk, J., Beszteri, B., Ács, É., 2003. Investigations on diatom-associated bacterial communities colonizing an artificial substratum in the River Danube. Archiv für Hydrobiologie Supplementband. Large Rivers 14 (3–4), 249–265.

Martin, J.L., 2013. Hydro-environmental Analysis: Freshwater Environments. Taylor & Francis Publishing.

Mayer, C., Moritz, R., Kirschner, C., 1999. The role of intermolecular interactions: studies on model systems for bacterial biofilms. International Journal of Biological Macromolecules 26 (1), 3–16.

McCormick, P.V., O'Dell, M.B., 1996. Quantifying periphyton responses to phosphorus in the Florida Everglades: a synoptic-experimental approach. Journal of the North American Benthological Society 450–468.

McCormick, P.V., Stevenson, R.J., 1998. Periphyton as a tool for ecological assessment and management in the Florida Everglades. Journal of Phycology 34 (5), 726–733.

Meylan, S., Behra, R., Sigg, L., 2003. Accumulation of copper and zinc in periphyton in response to dynamic variations of metal speciation in freshwater. Environmental Science & Technology 37 (22), 5204–5212.

Monds, R.D., O'Toole, G.A., 2009. The developmental model of microbial biofilms: ten years of a paradigm up for review. Trends in Microbiology 17 (2), 73–87.

Montgomery, W.L., Gerking, S.D., 1980. Marine macroalgae as foods for fishes: an evaluation of potential food quality. Environmental Biology of Fishes 5 (2), 143–153.

Muller, U., 1999. The vertical zonation of adpressed diatoms and other epiphytic algae on *Phragmites australis*. European Journal of Phycology 34 (5), 487–496.

Munoz, I., Real, M., Guasch, H., Navarro, E., Sabater, S., 2001. Effects of atrazine on periphyton under grazing pressure. Aquatic Toxicology 55 (3–4), 239–249.

Murdock, J.N., Dodds, W.K., 2007. Linging benthic algal biomass to stream substratum topography. Journal of Phycology 43 (3), 449–460.

Murdock, J.N., Shields Jr., F.D., Lizotte Jr., R.E., 2013. Periphyton responses to nutrient and atrazine mixtures introduced through agricultural runoff. Ecotoxicology 22 (2), 215–230.

Neal, C., Jarvie, H.P., 2005. Agriculture, community, river eutrophication and the water framework directive. Hydrological Processes 19 (9), 1895–1901.

Nelson, J.R., Eckman, J.E., Robertson, C.Y., Marinelli, R.L., Jahnke, R.A., 1999. Benthic microalgal biomass and irradiance at the sea floor on the continental shelf of the South Atlantic Bight: spatial and temporal variability and storm effects. Continental Shelf Research 19 (4), 477–505.

Nielsen, T.S., Funk, W.H., Gibbons, H.L., Duffner, R.M., 1984. A comparison of periphyton growth on artificial and natural substrates in the upper Spokane River. Northwest Science 58 (4), 243–248.

Nielsen, P.H., Frølund, B., Keiding, K., 1996. Changes in the composition of extracellular polymeric substances in activated sludge during anaerobic storage. Applied Microbiology and Biotechnology 44 (6), 823–830.

Nielsen, P.H., Jahn, A., Palmgren, R., 1997. Conceptual model for production and composition of exopolymers in biofilms. Water Science and Technology 36 (1), 11–19.

Pan, Y., Stevenson, R.J., Vaithiyanathan, P., Slate, J., Richardson, C.J., 2000. Changes in algal assemblages along observed and experimental phosphorus gradients in a subtropical wetland, USA. Freshwater Biology 44 (2), 339–353.

Peterson, C.G., Horton, M.A., Marshall, M.C., Valett, H.M., Dahm, C.N., 2001. Spatial and temporal variation in the influence of grazing macroinvertebrates on epilithic algae in a montane stream. Archiv für Hydrobiologie 153 (1), 29–54.

Pizarro, H., Vinocur, A., Tell, G., 2002. Periphyton on artificial substrata from three lakes of different trophic status at Hope Bay (Antarctica). Polar Biology 25 (3), 169–179.

Polunin, N.V.C., 1988. Efficient uptake of algal production by a single resident herbivorous fish on the reef. Journal of Experimental Marine Biology and Ecology 123 (1), 61–76.

Poulíčková, A., Hašler, P., Lysáková, M., Spears, B., 2008. The ecology of freshwater epipelic algae: an update. Phycologia 47 (5), 437–450.

Pratiwi, N.T.M., Hariyadi, S., Tajudin, R., 2012. Photosynthesis of periphyton and diffusion process as source of oxygen in rich-riffle upstream waters. Microbiology Indonesia 5 (4), 182.

Pusch, M., Fiebig, D., Brettar, I., Eisenmann, H., Ellis, B.K., Kaplan, L.A., Lock, M.A., Naegeli, M.W., Traunspurger, W., 1998. The role of micro-organisms in the ecological connectivity of running waters. Freshwater Biology 40 (3), 453–495.

Richard, M., Trottier, C., Verdegem, M.C.J., Hussenot, J.M.E., 2009. Submersion time, depth, substrate type and sampling method as variation sources of marine periphyton. Aquaculture 295 (3–4), 209–217.

Rittmann, B.E., Manem, J.A., 1992. Development and experimental evaluation of a steady-state, multispecies biofilm model. Biotechnology and Bioengineering 39 (9), 914–922.

Roeselers, G., Van Loosdrecht, M., Muyzer, G., 2008. Phototrophic biofilms and their potential applications. Journal of Applied Phycology 20 (3), 227–235.

Roman, A.M., Sabater, S., 1999. Effect of primary producers on the heterotrophic metabolism of a stream biofilm. Freshwater Biology 41 (4), 729–736.

Round, F.E., 1974. The Biology of the Algae. Edward Arnold.

Sabater, S., Guasch, H., Ricart, M., Romani, A., Vidal, G., Klunder, C., Schmitt-Jansen, M., 2007. Monitoring the effect of chemicals on biological communities. The biofilm as an interface. Analytical and Bioanalytical Chemistry 387 (4), 1425–1434.

Saikia, S., Nandi, S., Majumder, S., 2013. A review on the role of nutrients in development and organization of periphyton. Journal of Research in Biology 3 (1), 780–788.

Saikia, S.K., 2011. Review on periphyton as mediator of nutrient transfer in aquatic ecosystems. Ecologia Balkanica 3 (2), 65–78.

Sand-Jensen, K., 1977. Effect of epiphytes on eelgrass photosynthesis. Aquatic Botany 3 (0), 55–63.

Schmitt-Jansen, M., Altenburger, R., 2005. Toxic effects of isoproturon on periphyton communities – a microcosm study. Estuarine, Coastal and Shelf Science 62 (3), 539–545.

Scinto, L.J., Reddy, K.R., 2003. Biotic and abiotic uptake of phosphorus by periphyton in a subtropical freshwater wetland. Aquatic Botany 77 (3), 203–222.

Sheng, G.-P., Yu, H.-Q., Li, X.-Y., 2010. Extracellular polymeric substances (EPS) of microbial aggregates in biological wastewater treatment systems: a review. Biotechnology Advances 28 (6), 882–894.

Silva, S.R.V.F., Cecy, I.I.T., 2004. Desmídias (*Zygnemaphyceae*) da área de abrangência da Usina Hidrelétrica de Salto Caxias, Paraná, Brasil, I: Gênero Cosmarium. Iheringia 59 (1), 13–26.

Sládečková, A., 1962. Limnological investigation methods for the periphyton ("Aufwuchs") community. The Botanical Review 28 (2), 286–350.

Small, J.A., Bunn, A., McKinstry, C., Peacock, A., Miracle, A.L., 2008. Investigating freshwater periphyton community response to uranium with phospholipid fatty acid and denaturing gradient gel electrophoresis analyses. Journal of Environmental Radioactivity 99 (4), 730–738.

Steinman, A.D., Lamberti, G.A., 1996. Biomass and pigments of benthic algae. In: Hauer, F.R., Lamberti, G.A. (Eds.), Methods in Stream Ecology. Academic Press, New York, pp. 295–314.

Stelzer, R.S., Lamberti, G.A., 2001. Effects of N:P ratio and total nutrient concentration on stream periphyton community structure, biomass, and elemental composition. Limnology and Oceanography 46 (2), 356–367.

Stevenson, R.J., Stoermer, E.F., 1981. Quantitative differences between benthic algal communities along a depth gradient in Lake Michigan. Journal of Phycology 17 (1), 29–36.

Stevenson, R.J., Bothwell, M.L., Lowe, R.L., Thorp, J.H., 1996. Algal Ecology: Freshwater Benthic Ecosystem. Academic Press.

Stevenson, R.J., 1996. An introduction to algal ecology in freshwater benthic habitats. In: Bothwell, M.L., Stevenson, R.J., Lowe, R.L., Thorp, J.H. (Eds.), Algal Ecology: Freshwater Benthic Ecosystem. Academic Press, New York, NY, USA, pp. 1–27.

Stoodley, P., Lappin-Scott, H., 1997. Influence of electric fields and pH on biofilm structure as related to the bioelectric effect. Antimicrobial Agents and Chemotherapy 41 (9), 1876–1879.

Stoodley, P., Lewandowski, Z., Boyle, J.D., Lappin-Scott, H.M., 1999. Structural deformation of bacterial biofilms caused by short-term fluctuations in fluid shear: an in situ investigation of biofilm rheology. Biotechnology and Bioengineering 65 (1), 83–92.

Strayer, D.L., Beighley, R.E., Thompson, L.C., Brooks, S., Nilsson, C., Pinay, G., Naiman, R.J., 2003. Effects of land cover on stream ecosystems: roles of empirical models and scaling issues. Ecosystems 6 (5), 407–423.

Sutherland, I.W., 2001. Biofilm exopolysaccharides: a strong and sticky framework. Microbiology 147 (1), 3–9.

Swift, D.R., Nicholas, R.B., 1987. Periphyton and Water Quality Relationships in the Everglades Water Conservation Areas, 1978–1982. Environmental Sciences Division, Resource Planning Department, South Florida Water Management District.

SzabÓ, K.É., Makk, J., Kiss, K.T., Eiler, A., ÁCs, É., TÓTh, B., Kiss, Á.K., Bertilsson, S., 2008. Sequential colonization by river periphyton analysed by microscopy and molecular fingerprinting. Freshwater Biology 53 (7), 1359–1371.

Thompson, F.L., Abreu, P.C., Wasielesky, W., 2002. Importance of biofilm for water quality and nourishment in intensive shrimp culture. Aquaculture 203 (3), 263–278.

Townsend, C.R., Dolédec, S., Norris, R., Peacock, K., Arbuckle, C., 2003. The influence of scale and geography on relationships between stream community composition and landscape variables: description and prediction. Freshwater Biology 48 (5), 768–785.

Tuchman, M.L., Stevenson, R.J., 1980. Comparison of clay tile, sterilized rock, and natural substrate diatom communities in a small stream in southeastern Michigan, USA. Hydrobiologia 75 (1), 73–79.

Tusseau-Vuillemin, M.H., Gourlay, C., Lorgeoux, C., Mouchel, J.M., Buzier, R., Gilbin, R., Seidel, J.L., Elbaz-Poulichet, F., 2007. Dissolved and bioavailable contaminants in the Seine river basin. Science of the Total Environment 375 (1–3), 244–256.

Vadeboncoeur, Y., Devlin, S.P., McIntyre, P.B., Vander Zanden, M.J., 2014. Is there light after depth? Distribution of periphyton chlorophyll and productivity in lake littoral zones. Freshwater Science 33, 524–536.

van Hullebusch, E.D., Zandvoort, M.H., Lens, P.N., 2003. Metal immobilisation by biofilms: mechanisms and analytical tools. Reviews in Environmental Science and Biotechnology 2 (1), 9–33.

Van Loosdrecht, M., Eikelboom, D., Gjaltema, A., Mulder, A., Tijhuis, L., Heijnen, J., 1995. Biofilm structures. Water Science and Technology 32 (8), 35–43.

Vermaat, J., 2005. Periphyton Dynamics and Influencing Factors. Periphyton Ecology, Exploitation and Management. CABI Publishing, Cambridge, pp. 35–49.

Vézina, S., Vincent, W.F., 1997. Arctic cyanobacteria and limnological properties of their environment: Bylot Island, Northwest Territories, Canada (73°N, 80°W). Polar Biology 17 (6), 523–534.

Vymazal, J., 1988. The use of periphyton communities for nutrient removal from polluted streams. Hydrobiologia 166 (3), 225–237.

Wahl, M., 1989. Marine epibiosis. I. Fouling and antifouling: some basic aspects. Marine Ecology Progress Series 58, 175–189.

Webb-Robertson, B.J., Bunn, A.L., Bailey, V.L., 2011. Phospholipid fatty acid biomarkers in a freshwater periphyton community exposed to uranium: discovery by non-linear statistical learning. Journal of Environmental Radioactivity 102 (1), 64–71.

Welcomme, R.L., Food and Agriculture Organization of the United Nations, 1985. River Fisheries. Food and Agriculture Organization of the United Nations.

Wetzel, R.G., Likens, G.E., 2000. Limnological Analyses. Springer.

Wetzel, R.G., 1963. Primary productivity of periphyton. Nature 197 (4871), 1026–1027.
Wetzel, R.G., (Ed.), 1983. Dynamics of periphytic communities, In: Periphyton of Freshwater Ecosystems, Springer, pp. 14–17.
Wetzel, R.G., 2001. Limnology: Lake and River Ecosystems. Academic Press.
White, D.C., Ringelberg, D.B., 1997. Utility of the signature lipid biomarker analysis in determining the in situ viable biomass, community structure and nutritional/physiologic status of deep subsurface microbiota. In: Penny, D.L.H., Amy, S. (Eds.), Microbiology of the Terrestrial Deep Subsurface, vol. 4. CRC Press, pp. 119–136.
Winter, J.G., Duthie, H.C., 1998. Effects of urbanization on water quality, periphyton and invertebrate commmunities in a southern Ontario stream. Canadian Water Resources Journal 23 (3), 245–257.
Wolfstein, K., Colijn, F., Doerffer, R., 2000. Seasonal dynamics of microphytobenthos biomass and photosynthetic characteristics in the northern German Wadden Sea, obtained by the photosynthetic light dispensation system. Estuarine, Coastal and Shelf Science 51 (5), 651–662.
Woodruff, S.L., House, W.A., Callow, M.E., Leadbeater, B.S.C., 1999. The effects of biofilms on chemical processes in surficial sediments. Freshwater Biology 41 (1), 73–89.
Wright, P., Mason, C.F., 1999. Spatial and seasonal variation in heavy metals in the sediments and biota of two adjacent estuaries, the Orwell and the Stour, in eastern England. Science of the Total Environment 226 (2–3), 139–156.
Wu, Y., Zhang, S., Zhao, H., Yang, L., 2010. Environmentally benign periphyton bioreactors for controlling cyanobacterial growth. Bioresource Technology 101 (24), 9681–9687.
Wu, Y., Li, T., Yang, L., 2012. Mechanisms of removing pollutants from aqueous solutions by microorganisms and their aggregates: a review. Bioresource Technology 107 (0), 10–18.
Wu, Y., Xia, L., Yu, Z., Shabbir, S., Kerr, P.G., 2014. In situ bioremediation of surface waters by periphytons. Bioresource Technology 151 (0), 367–372.
Wu, Y., 2013. The studies of periphyton: from waters to soils. Hydrology Current Research 4, e107.
Young, O.W., 1945. A limnological investigation of periphyton in Douglas Lake, Michigan. Transactions of the American Microscopical Society 1–20.

Chapter 2

The Living Environment of Periphyton

2.1 INTRODUCTION

One of the leading causes of detrimental anthropogenic environmental modification is urbanization, which affects aquatic ecosystems by altering (1) the morphology of waterways, (2) soil permeability resulting in hydrographic changes, and (3) nutrient concentrations in waterways (Walsh et al., 2005). Hydrographic changes accelerate eutrophication and subsequently affect the structure of aquatic communities through changes in physical, chemical, and biological parameters like current flow, nutrient availability, and herbivory.

One of the devastating effects of eutrophication is the appearance of algal blooms in lakes and open waterbodies. Cyanobacterial blooms are toxic not only for aquatic environments but also for humans (Chen et al., 2009; Yin et al., 2005) and result in the destruction of foodwebs (Huisman et al., 2006). Although periphytic communities mostly appear as a result of eutrophication, similar communities are naturally found in undisturbed tropical freshwater marshes (McCormick and Stevenson, 1998) or mountain bogs and mires (Negro et al., 2003).

Periphyton are a cornerstone for aquatic ecosystems due to their role as primary producers and energy sources for invertebrates as well as providing a habitat for many organisms (Stevenson et al., 1996). They form the foundation of foodwebs in aquatic environments and are sensitive to physical, chemical, and biological variations in aquatic ecosystems including temperature, nutrient levels, current regimens, and grazing (Horner and Welch, 1981; Pringle and Bowers, 1984; Squires et al., 1979; Steinaman and McIntire, 1986). This sensitivity enables them to be used as a tool for assessing environmental conditions in aquatic environments.

The abundance and structure of periphytic biofilms vary spatially and temporally through natural processes and with human manipulation. The factors controlling accrual can be divided into factors and processes causing mass gain and factors and processes causing mass loss. Algal growth resulting in the accrual of periphytic biomass generally depends on nutrient supply, light ,and temperature, while mass loss primarily depends

Periphyton. http://dx.doi.org/10.1016/B978-0-12-801077-8.00002-8

on water flow rates. All these factors vary between aquatic ecosystems (Biggs, 1996).

2.2 FACTORS AFFECTING GROWTH OF PERIPHYTIC COMMUNITIES

The growth of periphytic biofilms depends on many biotic and abiotic factors (Fig. 2.1). Abiotic factors can be further categorized into hydrological, physical, and chemical factors.

2.2.1 Abiotic Factors

2.2.1.1 Substrates

Periphytic biofilms are an ecological group of organisms found at the interface of water and hard surfaces and usually develop on a range of natural substrates available in natural environments. These substrates include rocks (epilithon), plants (epiphyton), and sediments such as sand (epipsammon) and silt (epipelon). Algal mucilage provides the stability required for biofilm growth on a substrate (Madsen et al., 1993; Smith and Underwood, 2000), while unstable substrates preclude the development of thicker biofilm mats on the substrate surface (Eriksson, 2001). Substrates play a very diverse role in the characteristics and mode of periphytic community development such as colonization, succession, productivity dynamics, and pollution assessment (Tuchman and Stevenson, 1980).

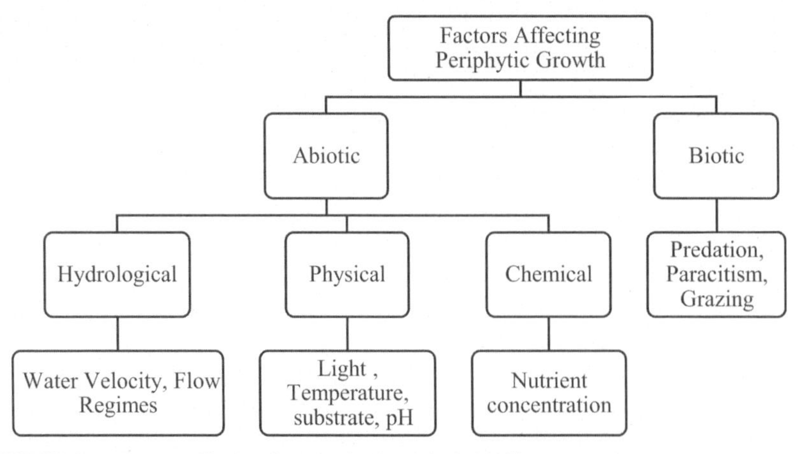

FIGURE 2.1 Factors affecting the growth of periphytic biofilms.

The composition and abundance of periphytic biofilms can be affected by the physical and chemical nature of the substrates. The substrate provides a physical surface for growth, while its chemical and physical properties can affect periphytic biofilm at microenvironmental levels (Burkholder, 1996; Bergey, 2005; Murdock and Dodds, 2007). Physical surfaces submerged in water always provide a more nutrient-rich microhabitat than water itself (Burkholder, 1996), with periphytic biofilm grown on living aquatic substrates, such as macrophytes, affected in terms of biomass and nutrient status (Bergey, 2005; Guariento et al., 2007, 2009; Murdock and Dodds, 2007). The chemical composition of nonliving substrates does not have a great effect on periphyton community structure and rate of biomass accumulation (Bergey, 2008), whereas the physical characteristics of these nonliving substrates, including microtopography and orientation, have a great effect on periphytic community and its structure (Bergey, 2005; Murdock and Dodds, 2007). While substrate type can affect growth, it is thought to have little influence on the taxonomic characteristics of periphyton (Rodrigues and Bicudo, 2001).

The topography and roughness of the substrate surface help to minimize the influence of water currents on periphytic community development in the overlying water (Dodds and Biggs, 2002). A rough substrate harbors a greater density of organisms compared with smooth substrates (Richard et al., 2009). Over time, periphytic algal communities can change the topography of the substrate by forming a gelatinous mat of up to 1-cm thickness that changes the shape and structure of substrate (Biggs, 1996) and reduces the effect of topography on periphyton development. The physical properties of substrates should be considered during sampling, especially when periphytic communities from different areas and surfaces are compared (Murdock and Dodds, 2007).

The crevices present on the substrate surface not only influence the development of the periphytic community but also help to prevent grazing (Dudley and D'Antonio, 1991; Figueiredo et al., 1996) and damage due to desiccation (Gosselin and Chia, 1995). The roughness and smoothness of a substrate are useful indicators for determining the value of a substrate as a refuge. According to Bergey (1999), the abrasiveness of a substrate plays a very important role in protecting the biofilm from grazing and other damaging factors. The protective value of a substrate depends on the size of the crevices and the relative size of the organisms in the crevices.

A variety of natural substrates have been used for research into the growth of periphytic biofilms, including bamboo and tree branches, jute sticks, fire wood, hyacinth, bamboo mats, and bundles of sugarcane bagasse (Azim et al., 2003; Keshavanath et al., 2001; Ramesh et al., 1999). Artificial substrates used include PVC pipes, plastic sheets, and materials such as Aquamats (Bratvold and Browdy, 2001; Keshavanath et al., 2001). Some studies demonstrate structural differences in periphyton community structure between periphyton grown on natural versus artificial substrates (Bergey, 2005; Murdock and

Dodds, 2007) while some indicate little (Rodrigues and Bicudo, 2001) or no differences (Lane et al., 2003). On the whole, the role of natural and artificial substrates in the growth and accrual rate of periphytic biofilms is controversial.

2.2.1.2 pH

Although pH is one of the important abiotic factors influencing periphytic biofilm growth and structure, there are few studies of the impacts of pH on periphyton in natural waters. The diversity and distribution of periphytic algal communities are affected by the concentration of hydrogenions (Planas, 1996). The effect of pH is more profound in lakes and waterways where anthropogenic activities are more intense and in areas where the surface waters receive natural nutrient inputs such as from volcanic emissions and bog water drainage inflow (Turner et al., 1991, 1987). Thawing permafrost due to climatic change results in the expansion of water bodies due to increased water runoff from uplands (Osterkamp et al., 2000) and weathering and alkalization of the surrounding lakes (Schindler, 1998). Decreases in pH can decrease overall species richness in fluvial ecosystems (Turner et al., 1991) and subsequently increases the abundance of certain filamentous algae from the Zygnemataceae family (Müller, 1980; Turner et al., 1995a,b). Periphytic algae are one of the frequently described aquatic microbe groups that grow at lower pH (Van Dam and Mertens, 1995).

An increase in pH (generally due to the restoration of acidified lakes and surface waters by liming) results in alkalization of algal communities (Hörnström, 2002; Hultberg and Andersson, 1982). Bloom-forming filamentous green algae, particularly *Mougeotia* (Chlorophyta, Zygnemataceae), become less dominant as the pH increases from acidic to neutral (Fairchild and Sherman, 1992; Hultberg and Andersson, 1982; Jackson et al., 1990). The increase in periphyton biomass accrual at higher pH is due to an increase in nutrients. After liming to increase the pH in an acidic lake, the amount of nutrients, particularly phosphorus, increased resulting in an increase in algal biomass (Olem, 1990). In summary, pH is an important factor in the development of periphytic communities but is also affected by the nutrient levels of waterbodies before and after alkalization.

2.2.1.3 Light

It is a common observation in ecology that the rate of biomass production and densities of autotrophs usually increase with light. Light is a key factor controlling stream ecology through its influence on temperature and growth of instream organisms and plants (Rutherford et al., 1997; Vannote et al., 1980). Nutrients and light in small streams are altered by natural and artificial disturbances in the environment, which, in turn, may influence the grazer–periphyton relationship in water.

Light also has a significant effect on chlorophyll a (Chl a) accrual in aquatic environments. Larned and Santos (2000) observed that Chl a density on tiles placed in lightly shaded pools was three times higher than that on tiles in highly shaded areas of the pool. More light results in higher conversion rates of inorganic material into organic living biomass. Variation in the availability and intensity of light may have a profound effect on the structure, physiology, and growth of periphytic biofilms (Petersen et al., 2003). Streams that are located in the vicinity of small forests have a higher periphyton biomass (measured in terms of Chl a) compared with urban or agricultural streams (Smart et al., 1985) because the urban or agricultural streams have no or less canopy cover than forest streams. According to Cashman et al. (2013), greater light availability in open canopies results in greater periphyton biomass but decreases the nitrogen, phosphorus, polyunsaturated fatty acids (PUFAs), and highly unsaturated fatty acids (HUFAs) in periphyton. Greater light also decreases the proportion of 18C–22C PUFAs in many diatoms, which may be due to photoacclimation (Leu et al., 2010; Thompson et al., 1993) and oxidative damage (Guschina and Harwood, 2009). Seasonal changes in light have been found to have no effect on algal growth rates (Bothwell, 1988). The carbon proportion of cells is directly related to light whereas phosphorus is inversely related to light. An increase in light intensity can reduce N:C and P:C ratios as well as Chl a:carbon ratios (Hessen et al., 2002).

At the beginning of periphyton development, a young biofilm is composed of bacteria and capsular extracellular polymeric substances (EPS). If it is not disturbed for up to 3 weeks, it shows a higher number of taxa and biomass, and a substantial relationship develops between organic carbon, Chl a, and EPS production. Biofilm grown in light-limited conditions lack this coupling and is dominated by diatoms, which are more efficient than green algae in low irradiance (Barranguet et al., 2005). The photosynthetic machinery of algae has the capacity to adapt to different light conditions. During photoadaptation in low light conditions, the demand for N increases due to increases in the protein content of the photosynthetic machinery (Raven, 1984). Due to N limitation during this process, the rate of Chl a production is also inhibited (Prézelin and Matlick, 1983). If we assume that food quality depends on the amount of light and nutrients, this Chl a inhibition highlights an important problem regarding the quality of food and its variability in different seasons and at different depths within a lake (Hessen et al., 2002).

High light and UV radiation may have detrimental effects on periphyton growth. Data suggest that greater light levels may result in decreased PUFA concentrations in phytoplankton through oxidative damage (Guschina and Harwood, 2009). Some algae can tolerate high light or UV radiation; one of the passive defense strategies of some cyanoprokaryote genera is production and storage of the UVA (UVB) absorbing pigment, scytonemin, in the sheath. Scytonemin provides UV protection in periods of high irradiation associated with reduced cellular metabolism (Garcia-Pichel and Castenholz, 1991). Light could also influence the biochemical composition of periphytic algal. For example, an investigation into the relationship between light and the nutritional quality of stream periphyton showed that greater light increased periphyton biomass (Chl a, ash-free dry mass [AFDM]), periphyton carbon concentrations, and monounsaturated fatty acids (MUFAs) but decreased saturated fatty acids (SAFAs). Greater light availability also increased levels of short-chain (<20°C) PUFAs but decreased quantities of several long-chain (20–22°C), highly unsaturated fatty acids (HUFAs) (Cashman et al., 2013). In addition, the influence of riparian tree canopies on light transmission to stream periphyton has been a prominent research topic since the 1950s (Ambrose et al., 2004; Bowes et al., 2012; DeNicola et al., 1992; Hill and Dimick, 2002; Lange et al., 2011; Larned and Santos, 2000; McConnell and Sigler, 1959). Most of these studies categorically compared periphyton under canopy gaps (high light) or dense canopies (low light). Generally, the changes in light transmission or intensity significantly affected the structural and functional properties of periphytic biofilms, which can be reflected at the whole stream ecosystem level.

2.2.1.4 Flow

The effect of change in flow on the growth patterns of algae, invertebrates, and vegetation is well documented (Dufford et al., 1987; Petts et al., 1993; Ward and Stanford, 1979). Modifications in flow can affect the fluvial ecology in four ways: (1) rivers with reduced flow and low current velocities have an overall impoverished stream habitat, (2) streams with seasonal flow and uniform flow velocities have enhanced riparian and aquatic vegetation, (3) rivers with high flow and high current velocities change the stream bed to a coarse substratum, and (4) short-term fluctuations due to physical interventions, like the production of hydroelectricity, result in more drastic changes due to modifications in flow, depth, and stability in banks and beds. The effect of flow on periphyton communities has not been as well studied as in other communities, but it is clear that flow changes affect diatoms (Reiter, 1986; Reiter and Carlson, 1986; Stevenson, 1984). Most of the studies on the effects of flow velocities have been carried out in the Northern Hemisphere (Growns and Growns, 2001).

Streams have transient zones, that is, the zones with zero or near-zero flow, that provide refuge to many kinds of organisms that are not adapted to higher velocities. The stream sections with higher velocities usually have higher nutrient concentrations and very different taxonomic compositions of organisms compared with transient zones. In the high-flow zones, nutrients are always available to be taken up by organisms and converted into organic matter. In the transient zones, the organisms are dependent on organic matter produced by the organisms from high-flow zones. The stream segment displayed in Fig. 2.2 has two sections: a transient storage zone near the bottom and a free-flowing zone closer to the surface (DeAngelis et al., 1995).

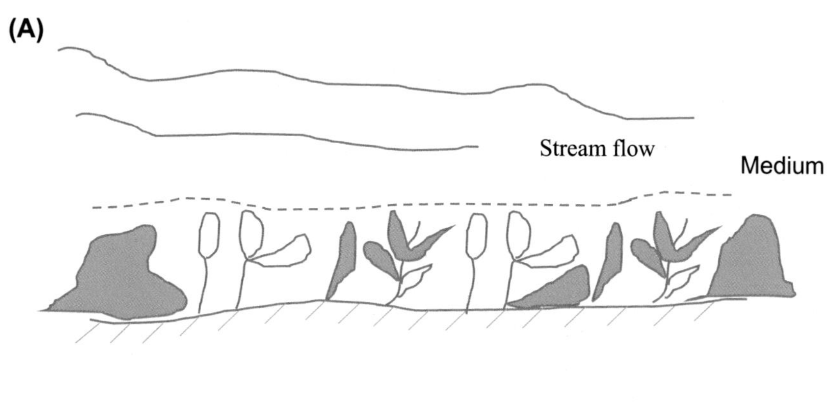

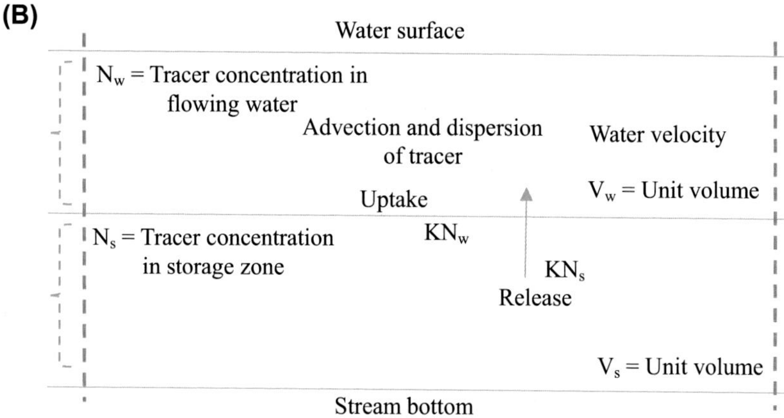

FIGURE 2.2 (A) Longitudinal sketch of a stream segment, and (B) schematic of nutrient transfer between stream segments. *From DeAngelis, D.L., Loreau, M., Neergaard, D., Mulholland, P.J., Marzolf, E.R., 1995. Modelling nutrient-periphyton dynamics in streams: the importance of transient storage zones. Ecological Modelling 80 (2–3), 149–160.*

For long filamentous green algae, it is well documented that increasing the flow velocity results in increased nutrient uptake and photosynthesis, respiration, and metabolism rates due to the reduction of the thickness of the diffusion boundary of the biofilms (Jorgensen and Des Marais, 1990; Kühl et al., 1996; McIntire, 1966). The taxonomic composition, structure, and rate of nutrient uptake by a biofilm are restricted by hydrodynamic pressure at velocities greater than 10–15 cm^{-1} (Bergey et al., 1995; Biggs et al., 1998; Hondzo and Wang, 2002). Only small variations in nutrient supply to biofilm are observed under laminar flow conditions, and the biofilm is not under any shear stress (McCormick and Stevenson, 1991; Peterson and Stevenson, 1990). Some cyanobacterial and mucilaginous thick diatom mats have good diffusion capabilities and are resistant to shear forces at higher velocities. Higher flow rates are good for these biofilm types and increase the rate of biomass production in these communities (Biggs et al., 1998). Whitford and Schumacher (1961) elucidated a direct relationship between velocity and respiration rates in fresh water ecosystems. With an increase in velocity from 8 to 15 cm s^{-1}, respiration rate increased 40.5–57.1%. Briefly, flow velocities can have a shearing or assimilating effect on periphytic communities depending on the types of organisms found in the biofilm and the thickness of the diffusion boundary layer.

2.2.1.5 Temperature

Temperature can affect reproduction rate, structure, growth ability, succession, and metabolism of periphytic communities. If the production of CO_2 is not repressed in this century, the doubling of atmospheric CO_2 concentrations predicted by climate models will result in temperature increases of 1.1–6.4°C (Solomon, 2007), followed by an increase of 0.3–0.9°C in open waters for every 1°C rise in temperature (Koycheva and Karney, 2009; Langan et al., 2001), with an additional 6°C rise if the riparian canopy is lost (Stefan and Sinokrot, 1993). Streams with a riparian canopy have a lower mean temperature and lower diurnal variations than open streams (DeNicola, 1996). Local ground water discharges to streams can increase water temperature in winter and decrease water temperature in summer (Petersen et al., 2003). Although algal growth occurs at a wide range of temperatures (Hawkes, 1969), freshwater benthic algal communities can experience thermal stress after significant temperature changes due to mixing of thermal water discharges into streams (Krenkel et al., 1969; Langford, 1990).

Temperature variation also has a profound effect on SAFAs with SAFA production higher at higher temperatures (Ravet et al., 2010). Increases in water temperature can change the rate of metabolism of aquatic communities, which may, in turn, influence community structure, species distribution, interspecific relationships, biodiversity (Castella et al., 2001; Mouritsen et al., 2005; Mouthon and Daufresne, 2006), and processes like carbon mineralization, primary production, and denitrification (AcuÑA et al., 2008; Demars et al., 2011).

If the temperature decreases, the uptake of N by algal communities will also decrease (Reay et al., 1999), thus affecting the biogeochemical N cycle. Temperature fluctuations can also have species-specific effects on the pigment concentration of periphytic communities (Larras et al., 2013). Pigment contents can either increase or decrease depending on the temperature change and the species (Chalifour and Juneau, 2011).

Temperature increase has a major influence on the main processes of periphytic communities such as respiration and photosynthesis (Baulch et al., 2005), enzymatic activity (Bonet et al., 2013; Bouletreau et al., 2012), and biofilm formation (Rao, 2010). Increases in temperature have led to increases in respiration rates and primary productivity in streams through photosynthesis

(Bott et al., 1985; Demars et al., 2011) due to increases in enzymatic activity (Brown et al., 2004). Short-term increases in temperature increase the respiration rate in periphytic biofilms, and if the increases continue in the long term, higher accrual biomass and higher rates of production of Chl a will occur. This can lead to higher respiration and primary productivity rates in both oligotrophic and eutrophic streams (Rosa et al., 2013). Variability in the behavior, structure, function, and density of biofilms due to temperature fluctuations depends on the enzymatic activities of the periphytic biofilms. Higher temperatures typically produce more persistent changes compared with lower temperatures due to decrease in enzymatic activities at lower temperatures.

2.2.1.6 Nutrients

The relationship between nutrients and periphyton is a popular topic in periphyton ecology (Bowes et al., 2012; Cashman et al., 2013; Greenwood and Rosemond, 2005; Schiller et al., 2007; Steinman et al., 2011). Nutrient availability, especially N and P, plays a significant role in periphyton biomass and productivity. Generally, enhanced nutrient availability leads to shifts in taxonomic composition and increases in periphyton density and thickness (Fermino et al., 2011; Ferragut and de Campos Bicudo, 2012; Ferragut and de Campos Bicudo, 2010). Since periphytic communities are often dominated by diatoms, the availability of S_i should also be considered. S_i concentrations often increase with depth in lakes, usually because of the dissolution of diatom frustules. Large diatom species with well-silicified valves (e.g., large species of *Epithemia*, *Eunotia*, *Diploneis*, *Aulacoseira*) have been observed in the deep zone of carbonate Lake Tovel (Hawes and Smith, 1994; St. Jónsson, 1992).

An important way to study the influence of nutrients on periphyton is by experimental nutrient addition. Nutrient addition experiments have been used in periphyton studies for over 60 years starting with Huntsman in 1948 (Huntsman, 1948) and generally focus on two issues: (1) whether periphyton growth is nutrient limited (and by which nutrients) and (2) what are the effects of nutrient limitation on periphyton community composition and succession (Larned, 2010). Nutrient availability could also affect the biodiversity of periphyton. One of the clearest examples of these effects comes from a comparison of NO_3-enriched and unenriched substrata in Sycamore Creek, Arizona (Peterson and Grimm, 1992). Results showed that unenriched substrata were dominated by diatoms with N_2-fixing endosymbiontic cyanobacteria during early successional stages and by N_2-fixing cyanobacteria during later stages. NO_3-enriched substrata were initially colonized by non–N_2-fixing diatoms and filamentous chlorophytes and, later, by cyanobacteria. Seral stages were more apparent, and diversity higher, on NO_3-enriched substrata than on unenriched substrata. The relationships between periphyton diversity and the availability of limiting nutrients are, however, still controversial, with various enrichment experiments resulting in increased, decreased, and/or unchanged diversity (Larned, 2010).

Nutrient loading into open waters is intensifying due to urbanization, industrialization, and agricultural activities and results in higher primary productivity. These authorized nutrient additions have already been reported in Alaska, where nutrient concentration, in terms of sulfate and phosphate, was increased 6-fold in seawater (Hinzman et al., 2006). Escalation in periphytic biomass as a result of nutrient loading is detrimental to stream ecology. It can lower dissolved oxygen concentrations (Horne and Goldman, 1994), reduce aesthetic properties and recreational use, and increase the cost of water extraction (Biggs, 1996). Excessive nutrient supply also influences the integrity of ecological communities—periphytic biofilms with excessive nutrient supply are dominated by cyanobacteria, are less consumed by grazers (Ghadouani et al., 2003; Roelke et al., 1997, 2004), and produce chemical toxins harmful to aquatic foodwebs (Hay and Kubanek, 2002; Lehtiniemi et al., 2002).

Increases in nutrient contents for periphyton is usually estimated in terms of P and N. The nutrient contents of streams usually increase due to geological, meteorological, and, increasingly, anthropogenic factors (Schindler, 1998; Sterner et al., 1997). According to Dodds et al. (2002), nutrient levels and algal biomass are strongly correlated highlighting the need to study N and P as limiting nutrients for periphyton in lotic ecosystems. N and P also limit the growth and productivity of primary producers (Carey et al., 2007; Downing et al., 1999; Stephens et al., 2012). N and P can affect different types of streams; P is the limiting nutrient factor in Boreal, tropical, and temperate streams (Newbold et al., 1983; Pringle and Bowers, 1984; Pringle and Triska, 1991). N is important in temperate streams in landscapes of volcanic or tectonic origin (Grimm and Fisher, 1986; Ludwig et al., 2008; Peterson and Grimm, 1992; Tank and Dodds, 2003) and some tropical streams (Downing et al., 1999; Flecker et al., 2002).

Making predictions about stream conditions on the basis of nutrient levels is not an easy task because it also depends on the morphology of periphytic biofilms, seasonality, sediments, light, and flow regimens (Biggs and Close, 1989). Optimum concentrations of N and P can vary between the members of different species constituting a biofilm, with some members in the biofilm being N limited while others are P limited (Borchardt, 1996a). According to Bothwell (1989), the threshold limit of total P (TP) that can shift the periphytic community to a distressed condition through N limitation is 10 $\mu g\ L^{-1}$, and the TP concentration required to relieve the periphyton from nutritional stress is 1.0 $\mu g\ L^{-1}$. The balanced N:P ratio for the growth of algae is 7:1 (Francoeur, 2001) with the periphytic community structure more sensitive to N:P than periphyton biomass (Stelzer and Lamberti, 2001). Higher stream nutrient levels also result in increased phytoplankton biomass, which has a detrimental effect on periphytic communities due to lower water transparencies (Hansson, 1992; Vadeboncoeur et al., 2001). According to Biggs and Smith (2002), excessive nutrients are the most common factors influencing river ecosystems and can totally change the community structure through "bottom-up" effects.

In most studies, the effects of nutrients and light have been studied simultaneously. This might be due to the fact that periphyton production rate and ultimately nutrient consumption rates are dependent on light. The primary production of streams depends on photosynthetically active radiation (Krenkel et al., 1969) and nutrient supply, and the qualitative output in terms of elemental ratios can be determined by a combination of light and nutrients (Sterner et al., 1997; Urabe and Sterner, 1996). In riparian canopy areas, light is the limiting factor and affects periphytic communities more than nutrients (Hill, 1988). Areas with low or no low canopies are nutrient limited (Dodds et al., 1996; Lowe, 1986).

2.2.1.7 Sedimentation

To date, laboratory and field studies of the effects of sedimentation on periphyton are insufficient to describe the individual and combined effects. Although there has been little work in the last few decades about sedimentation, the importance of the process and its effect on periphyton should not be overlooked. Sediments usually accumulate in dead zones with very low current velocities (Tipping et al., 1993) and can be easily trapped by gravel beds (Carling, 1984; Petts, 1988) and periphyton (Graham, 1990). Sediments can make open waters more turbid, thus decreasing light penetration, ultimately reducing the primary productivity of the streams (Davies-Colley et al., 1992; Van Nieuwenhuyse and LaPerriere, 1986). Sedimentation cannot only modify the substrate in the streams (Graham, 1990); it can also modify the morphology of the river channel (Doeg and Koehn, 1994; Nuttall, 1972), diminish the aquatic flora (Brookes, 1986; Edwards, 1969), clog the spaces between the substrates, and reduce habitat availability for aquatic organisms (Richards and Bacon, 1994; Schälchli, 1992). Suspended sediments and their deposition can affect aquatic environment producers in four ways: (1) by reducing the penetration of light into water, resulting in decreases in primary productivity and photosynthesis, (2) by reducing the organic content of streams, (3) by abrasive damage to stream macrophytes, and (4) by preventing the attachment of algal cells to substrates and, in extreme cases, eliminating periphyton and macrophytes (Wood and Armitage, 1997). Although sedimentation does not have a striking influence on periphytic biofilms like other physical factors, high sedimentation levels can destroy the whole community structure within a waterbody by affecting other factors.

2.2.1.8 Hydrological Regimens

Hydrological regimens have a profound effect on the development of biofilm in streams depending on the disturbance frequency. They affect periphyton development and community structure by altering moisture and nutrient availability and types of available substrates (Gaiser, 2009). Various studies on the effect of flow regimens on algal biomass and periphytic species composition are available (Battin et al., 2003a,b; Francoeur and Biggs, 2006). The number of events over time affecting the responsiveness of an ecosystem is called the disturbance frequency (Pickett and White, 1985). Rapidly growing filamentous green algae or cyanobacteria cause difficulties in periphytic growth in eutrophic waters (Davis et al., 1990), whereas floods cause changes in periphyton biomass and communities by completely removing periphyton from substrates, thus restarting the succession process. A streams flow regimen therefore affects the accumulation and composition of periphytic biofilms (Biggs and Close, 1989). Changes in hydrological regimens pose a serious threat to the biological integrity of biota in floodplain rivers either directly or indirectly by affecting various aspects of habitat (Agostinho et al., 2004, p. 160). These changes lead to structural and functional community alterations of the flood plain rivers, especially in terms of periphytic algal communities (Davidson et al., 2012; Gottlieb et al., 2006; Luttenton and Baisden, 2006). Flood pulses affect many factors that in turn can influence periphytic communities such as propagules (Rodrigues and Bicudo, 2001), primary productivity in terms of biomass (Leandrini et al., 2008; Leandrini and Rodrigues, 2008), and taxonomic composition (Algarte et al., 2009; Rodrigues and Bicudo, 2001).The impacts of desiccation on periphytic biofilm depend on the biofilm thickness, taxonomic composition, and amount of mucilage produced (Hawes et al., 1992; Ledger et al., 2008; McKnight et al., 2007; Stanley et al., 2004). Variations in flow regimens can change the structure and dynamics of periphytic algal communities and limit the development of the community to the initial stages of development (Zanon et al., 2013).

Change in hydrological regimens also alters nutrient availability in open waters. Organic matter and nutrients are attached to the substrate (boulders, cobbles, and sediments) in the streams, and sudden changes in hydrological regimens or floods can result in detachment of nutrients and suspension of the material in the water column (Larned and Santos, 2000). Therefore, changes in hydrological regimes affect biofilm thickness and augment phosphate and nitrate enrichment (Borchardt, 1994, 1996a). Phosphate-enriched substrates always have a higher amount of Chl a attached to the surface compared with nitrate-enriched substrates, but these results are highly modified when flow regimens are changed (Larned and Santos, 2000). More frequent changes in flow regimens can result in the extinction of many intolerant species and system instability (Collins et al., 2001). According to the "intermediate disturbance hypothesis," diversity should be higher at moderate fluctuation levels (Connell, 1978). In Verdon River, southeastern France, the hydrological regimen was the key factor affecting stream ecology and algal communities, in response to level and variability in flow. The level of disturbance varied throughout the year due to anthropogenic disturbances from irrigation and drinking water activities. The algal communities were typically less abundant along the canal where the hydrological disturbances were more frequent and abrupt (Bertrand et al., 2001).

2.2.1.9 Organic Matter

Organic matter is important in the development and primary production rate of periphyton in open waters, and the amount of organic matter is directly related to the growth rate of periphytic biofilms. Dissolved organic matter (DOM) in streams can serve as a source of nutrients enhancing algal growth (Tuchman, 1996) and the growth of heterotrophic bacteria (Bernhardt and Likens, 2002). There are generally two sources of DOM in streams: (1) from the catchment and terrestrial sources (Kaushik and Hynes, 1971) and (2) within the stream channel (Minshall, 1978). The source and relative importance of different types of organic matter depend on the land use in the stream catchment (Quinn et al., 1992). If the canopy in the riparian zone is removed, the amount of leaf litter decreases (Delong and Brusven, 1994), and the amount of light reaching the water increases, thus increasing the rate of primary productivity in the stream (Feminella et al., 1989). Different types of leaves have different effects due to seasonal variability in degradation rates (Webster, 1986). Increases in DOM will reduce the amount of solar radiation penetrating freshwater (Xenopoulos and Schindler, 2001), thus affecting the impact of solar radiations on primary producers (Kelly et al., 2001; Vinebrooke and Leavitt, 1998). Increases in DOM can also affect the C:N:P biomass ratio in periphytic biofilm (Cross et al., 2005). In acutely high DOM conditions, the photosynthesis rate is higher due to higher nutrient uptake rates, thus lowering C:N and C:P ratios of periphyton (Frost et al., 2002).

The contribution and effect of various sources of organic matter to foodwebs can be determined by analyzing the rate of metabolism in terms of primary productivity and community respiration in streams. There are two types of methods used to determine community metabolism: enclosed chambers and "open-system method." In enclosed channels, the amount of DO is measured by using small chambers enclosing a portion of streams community. In the "open-system method," natural changes in DO are measured (Young and Huryn, 1999), providing a more reliable method for determining the rate of metabolism in streams (Bott et al., 1978; Marzolf et al., 1994).

Organic matter not only directly affects aquatic communities; it also indirectly exacerbates its effects through interactions with other factors. DOM concentration can increase heterotrophic algal communities (Tuchman, 1996). DOM might be increased due to some dominant algal communities that can increase the DOM concentration due to changes in N:P ratios (Klug, 2002; Romani et al., 2004). Generally, higher DOM can increase C:P and N:P ratios and algal biomass in periphytic biofilms, which may cause changes in the dynamics of benthic food webs and stream biochemistry such as higher rates of secondary production (Feminella and Hawkins, 1995), changes in export of particulate organic carbon (Battin et al., 2003b), and providing a sink for N and/or P (Bernhardt and Likens, 2002). The C:N:P ratios of organic compounds form the basis of plankton (Sterner et al., 1997) and benthos (Bowman et al., 2005; Frost and Elser, 2002) foodwebs. The living algal cells in periphytic communities also contribute to periphytic organic matter. The organic matter of periphyton is able to influence the effects of light and nutrient supplies on C:N:P ratios (Frost et al., 2002). For instance, the ratios of epilimnetic light to nutrient depends on the C:P ratios of organic matter in lakes as these ratios reflect balanced net photosynthetic C fixation and P-uptake (Sterner et al., 1997). The extent of the effect of algal C:P ratios under different biomass production rates on C:P ratios in bulk organic matter remains unclear (Hessen et al., 2002).

At higher DOM concentrations, the rate of uptake of carbon by algal and bacterial cells also increases, leading to the production of higher amounts of EPS (Frost et al., 2007), a well-known component of periphytic biofilm (Hoagland et al., 1993). In nutrient-limited conditions, the EPS decomposition rate is inhibited (Cebrian and Duarte, 1995), and its accumulation rate is enhanced.

2.2.1.10 Disturbances

Disturbances are defined as unpredictable events in periphyton ecology, including desiccation, anoxia, freezing, rapid changes in osmotic potential, acute contaminant exposure, substrate movement, and rapid increases in hydraulic forces, heat, and light. Periphyton responses to disturbances are typically assessed in two phases: (1) responses to the onset of disturbance and (2) responses to the cessation of disturbance. Responses of periphyton to the onset of disturbance are related to susceptibility or resistance (e.g., periphyton biomass loss and changes in metabolism and taxonomic composition). Highly susceptible periphyton growth forms include chain-forming diatoms, uniseriate filaments, and loosely attached cyanobacterial mats. Highly resistant forms include prostrate diatoms and chlorophytebasal cells and rhizoids (Benenati et al., 2000; Biggs et al., 1998; Grimm and Fisher, 1989; Passy, 2007). The susceptibility of periphyton to hydraulic disturbance generally increases as intervals between disturbances increase. When disturbances are infrequent, thick mats develop with weak attachment to substrata because of basal cell senescence; these mats can be removed by small increases in shear stress (Peterson, 1996). Periphyton responses to the onset of desiccation vary with mat or biofilm thickness, taxonomic composition, and production of extracellular mucilage (Ledger et al., 2008; McKnight et al., 2007; Stanley et al., 2004). In general, desiccation resistance is high in mucilaginous cyanobacteria and diatoms and low in chlorophytes, rhodophytes, and nonmucilaginous diatoms. Periphyton responses to the cessation of disturbance are related to recovery and resilience.

Periphyton communities recover from hydrodynamic disturbances mainly through recolonization or by regrowth from persistent cells. Many studies have partially evaluated the relative importance of these pathways by comparing periphyton biomass accrual on newly exposed substrata (colonization only) with that on substrata containing persistent tissue (colonization plus regrowth) (Dodds et al., 1996; Downes and Street, 2005). Recovery by regrowth appears to be more important at sites with gradual water loss caused by seepage or evaporation.

Field and laboratory stream observations demonstrated that disturbances like flood were always accompanied by elevated nutrient concentrations in waters. In these cases, periphyton experiences both the negative effects of sediment movement and the positive effects of nutrient enrichment. There may be two general consequences in these cases: (1) nutrient limitation increases periphyton susceptibility to disturbances and (2) nutrient enrichment hastens recovery following floods (Biggs and Smith, 2002; Riseng et al., 2004).

2.2.2 Biotic Factors

Grazing and competition are the most common biotic factors influencing periphyton. Heavy grazing of periphyton by animals (insect larvae, crayfish, and certain fishes) can result in reductions in periphyton biomass, an altered rate of primary productivity, and/or a change in taxonomic composition and community structure (Christofoletti et al., 2011; Hill et al., 1992; Hill and Knight, 1988; Hillebrand, 2008; Walton et al., 1995). Moderate grazing however, can help regeneration of clonal vegetation and stimulates growth rates by (1) improved light availability within the periphyton communities and (2) enhanced nutrient availability from the water as well as from the feeding activities of the herbivores (Azim, 2009).

Close packing among organisms and steep resource gradients within periphyton communities create suitable conditions for interspecific and intraspecific competition. McCormick and Stevenson (1991) reported negative relationships between growth rates of some benthic diatom taxa and biovolumes of other taxa. However, direct evidence for competitive inhibition or exclusion in periphyton is rare (Larned, 2010). Allelopathy is a form of interference in which chemicals produced by some organisms inhibit colonization and growth of other organisms and is a widespread competitive strategy. Interspecific allelopathy has been documented in benthic cyanobacteria, charophytes, and periphyton (Jüttner and Wu, 2000; Smith and Doan, 1999; Wu et al., 2010) Nutrient uptake rates can exceed rates of nutrient input or remineralization and cause nutrient depletion and competition in closed systems. These conditions favor taxa with low half-saturation constants, high nutrient storage, and efficient conversion of nutrients to cellular material (Borchardt, 1996b). Longitudinal nutrient depletion and the corresponding changes in periphyton composition have been observed in natural and artificial streams with high residence time, high periphyton biomass, and low nutrient input (Mulholland et al., 1995; Mulholland and Webster, 2010; Vis et al., 2008).

2.2.2.1 Parasitism

Parasitism is a nonmutual relationship between two organisms in which one benefits at the expense of the other. There are two types of parasites affecting living organisms: ectoparasites (living on the surface of host) and endoparasites (living in the body of host). Periphyton parasites can alter the physiology and behavior of the host, resulting in changes in density, survival rates, life span, and growth rates (Thomas et al., 2011). The effect of parasitism on aquatic communities, especially on periphytic communities, is poorly described and studied, which might be due to their small size and difficulty in determining the energy transfer between host and parasite (Combes, 2001; Dobson and Hudson, 1986). Thus, the role of one of the most influential communities in aquatic foodwebs is sometimes ignored (Lafferty et al., 2006; Marcogliese and Cone, 1997).

The periphytic community is either affected by organisms infected with parasites or they are themselves affected by a parasite. To determine the rate of parasitism in periphytic communities and zooplanktons, Wolska (2013) studied two lakes in Drawa National Park, Poland, for 3 years and observed that 0.8% of *Chydorus sphaericus* (Cladocera, Chydoridae) was infected by *Saprolegnia* sp. (chromistan fungal analogues, kingdom Chromista) and 1.2% of *Brachionus calyciflorus* (Rotifera) was infected by protozoan *Microsporidium* sp. (Fig. 2.3), while 5% of a population of nematodes was infected by *Pythium* sp. They did not explain the effect of these parasites on the periphytic communities.

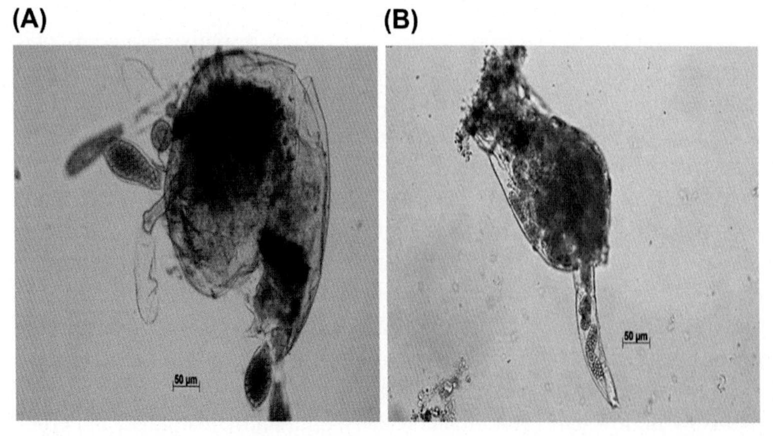

FIGURE 2.3 (A) *Chydorus sphaericus* (Cladocera) infected with *Saprolegnia* sp. (B) *Brachionus calyciflorus* (Rotifera) infected with *Microsporidium* sp. (Wolska, 2013).

Periphytic communities are also affected by the parasitism of grazers and other organisms feeding on them. Snails are the dominant grazers of periphytic communities and can affect algal biomass, community composition, and primary production (Bernot and Turner, 2001; Lamberti et al., 1989). The snail *Physa acuta* infected with trematodes *Posthodiplostomum minimum* had higher rates of grazing on periphyton (almost 20% higher) compared with the uninfected snails. *Cladophora glomerata* (algae) was more dominant after grazing by infected snails, whereas diatoms and blue-green algae were dominant in the biofilms grazed by snails with lower infection levels (Bernot and Lamberti, 2008).

2.2.2.2 Predation

Predation has not been extensively studied in relation to periphytic biofilms, but according to a few studies, it has a negative effect on periphytic communities (Blanchet et al., 2008). *M. tuberculatus* is a dominant mollusk in Brazilian aquatic ecosystems (Santos & Eskinazi-Sant'Anna, 2010). Periphytic communities subjected to predation by *M. tuberculatus* had lower densities of diatoms, which resulted in the disappearance of some species from the periphytic biofilm. This change in the periphytic community is due to predation pressure by the mollusks (Rosenberg and Resh, 1992). In predator–prey relationships, the predator not only changes the mean biomass but also affects the spatial heterogeneity of biomass distribution (Hillebrand, 2008). Predators have strong and negative effects on the average biomass, growth, and survival of the prey assemblage. Variation in the effects produced by the consumer can help us understand ecological patterns in natural communities (Benedetti-Cecchi, 2000).

2.2.3 Grazing

The removal of periphytic communities from the substrate as a source of food by invertebrates and vertebrates is known as grazing. In the late 1970s, predation and grazing were the main factors used to determine the biomass and productivity of different trophic levels, and for any given trophic level these values were determined by grazing and predation at higher levels (i.e., top-down effects) (Steinman, 1996). The distribution and abundance of riverine algal communities are estimated from higher to lower trophic levels depending on time, place, and environmental conditions (Allan and Castillo, 2007; Rosemond et al., 1993). Consumer-induced heterogeneity is especially important in aquatic ecosystems, where grazers can change community physiognomy, rate of nutrient uptake, and competition (Hillebrand and Kahlert, 2001; Steinman, 1996). According to Benedetti-Cecchi et al. (2005), both average grazing pressure and variability in grazing pressure have a strong influence on spatial distribution of algae.

If the periphytic biomass is patchily distributed, it can affect the availability of nutrients (Riber and Wetzel, 1987), competition for light (Hillebrand, 2005), and recognition and selection as food by grazers (Kawata and Agawa, 1999). The effect of grazers on the spatial heterogeneity of periphyton is not consistent. In some studies, it was increased (Alvarez and Peckarsky, 2005; Sommer, 2000), and in other studies, a decrease in spatial heterogeneity was observed (Gelwick and Matthews, 1997; Sommer, 2000). The increase in heterogeneity can be due to the higher mobility of grazers (Alvarez and Peckarsky, 2005). According to Wellnitz and Leroy Poff (2006), several factors can influence postgrazing periphytic development: (1) current velocity, which can regulate the effect of stress by simultaneously regulating nutrient uptake rates and shear forces (Biggs et al., 1998), and (2) grazing duration, which can affect the rate of removal of periphytic mats and characteristics of biofilm regrowth (McCormick, 1994).

Grazing has a more profound effect on periphyton than nutrient concentration. Sometimes, however, the amount of nutrients is also increased due to excretion by grazers, thus increasing the amount of algal P (Hillebrand and Kahlert, 2001). Selective grazing and excretion by grazers can change the nutrient ratios in terms of C, N, and P (Evans-White and Lamberti, 2005, 2006; Hillebrand et al., 2008). As the grazer community increases, it give rise to intraspecific or interspecific density-dependent interactions due to competition of grazers for algal resources (Hart, 1987; Lamberti et al., 1987), and this competition for food can limit the density of grazers by reducing periphyton growth rates and community shifts in temperate streams (Lamberti and Resh, 1983). The location, morphology (Geddes and Trexler, 2003; Horner and Welch, 1981), and nutrient contents can produce diverse responses of algae (positive or negative) to grazing. Grazing is also helpful in removing the old algal cells from surface waters, thus increasing the amount of light penetrating to underlying cells (Hillebrand and Kahlert, 2001; Rosemond et al., 1993) and changing nutritional demands (Dodds et al., 1996; Dodds, 1989). There are many types of grazers in aquatic environments feeding on periphyton such as crustaceans, insects, snails, herbivorous fish, and frogs.

2.2.3.1 Crustaceans and Insects

The highest proportion of periphytic biomass is grazed by crustaceans (amphipods and isopods) and trichopteran larvae (insects) (Hill et al., 1992; Lamberti et al., 1987). The effects of grazers increase with higher biomass, higher temperature, and lower resource availability. When three algal genera (cyanobacteria, diatoms, and chlorophytes) were used, a source of food for chironomids, its larvae used diatoms as the main food source with food sources varying seasonally (Tarkowska-Kukuryk, 2013). Chironomids can also feed on microalgae (Goldfinch and Carman, 2000) and stemmed diatoms (Tall et al.,

2006). Hill (1988) studied the effect of grazing by caddisfly larvae (*Neophylax*) and mayfly nymphs (*Amulets*) in northern California. *Neophylax* had a negative effect on the algal species *Epithelia*, which was found in its gut, whereas *Epithelia* reduced *Nitzschia* spp., *Synedra* spp., *Amphipleura pellucida*, and *Melosira varians* abundances but had no significant effect on overall biomass or volume.

2.2.3.2 Snails

Pomacea canaliculata are omnivorous snails that feed on periphyton, macrophytes, detritus, and animal matter (Dillon, 2000). According to Villanueva et al. (2004), in a combined experiment for mayfly *Meridialaris chiloeensis* and the snail *Chilina dombeiana*, snails had a stronger overall effect on Chl a and AFDM, whereas in spring and summer, both species had the same effect. *M. chiloeensis* changed the taxonomic composition of periphytic biofilm by depressing rosette-forming algae and favoring the formation of prostrate forms, whereas grazing by snails did not change the community to a large extent. When periphyton was exposed to low, medium, and high rates of grazing by freshwater snails *Physella* sp., the periphytic community showed different responses. At higher grazing pressure, filamentous taxa were suppressed, while prostrate forms were enhanced at moderate to high grazing levels, and low-profile algae were also enhanced under moderate grazing (Swamikannu and Hoagland, 1989).

2.2.3.3 Frogs

The tadpoles of *Rana tigrina* are algal feeders with green algae (Dickman, 1968), epiphytic diatoms (Kupferberg, 1997), detritus (including decomposed higher plants), and various algae (Jenssen, 1967) being a source of food for these grazers. Grazing by tadpoles can lead to reductions in algal biomass (Brönmark et al., 1991). Algae also benefit from nutrient increases due to excretions by grazing tadpoles (Osborne and McLachlan, 1985). Tadpole growth is also dependent on the availability of the algae (Leips and Travis, 1994). These effects of food availability are also found in tadpoles of other species such as *Rana sylvatica* (wood frog) (Murray, 1990; Wilbur, 1977) and *Scaphiopus couchii* (spade-foot toad) (Newman, 1994). The growth rates of tadpoles are indirectly affected by light and nutrient concentrations because these two factors influence the rate of periphyton biomass production. Increases in light can augment the primary production of periphytic biofilms and provide more food to the grazer while nutrients have smaller effect (Kim and Richardson, 1999).

2.2.3.4 Herbivores

Grazing by herbivores can also affect the periphytic community biomass and diversity. Grazing rates are inversely proportional to biomass and number of taxa in the community (Gelwick and Matthews, 1997; Power et al., 1985). In late summer and fall, the abundance of algal communities and herbivorous central stoneroller (*Campostoma anomalum*) are inversely related to each other (Power et al., 1985), whereas according to Matthews et al. (1987), the algal community is mainly composed of tightly attached blue-green algae (i.e., *Calothrix*), after grazing by *Campostoma*.

Herbivorous fish have a strong effect on the nutrient availability for algal biomass and production in different lakes (Flecker et al., 2002; McIntyre et al., 2006).

2.2.4 Periphyton Accrual in Streams

Periphytic communities in natural and artificial waters always follow a generalized pattern of algal biomass accrual (Fig. 2.4). According to Stevenson et al. (1996), there are usually four stages in accrual of periphytic biomass:

(a) Colonization: the early stage and usually involves the immigration of periphyton
(b) Exponential phase: the time when the biomass reaches at its peak
(c) Autogenic sloughing: during this stage, degradation of the periphytic community starts
(d) Loss phase: when autogenic sloughing becomes the dominant process

Immigration or colonization is the process of settlement of algal cells and depends on the nature and texture of substrate and water velocity. In the second phase, diatoms are the dominant community after 7 days (Godillot et al., 2001). After 1–2 weeks, classic growth occurs, and after 3 weeks, current velocity determines the composition of biofilm. After 4 weeks, the mat is composed of filamentous algae independent of other factors (Graba et al., 2012).

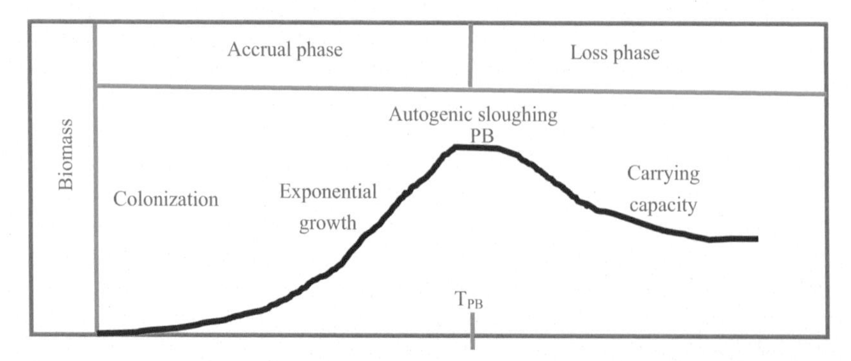

FIGURE 2.4 An idealized benthic algal accrual curve with different phases shown. PB (peak of biomass) is the maximum accrual cycle biomass, and TPB is the time to PB from commencement of colonization (Stevenson et al., 1996).

2.3 CONCLUSION

Periphytic community is a complex combination of different types of organisms, organic matter, and detritus. In aquatic ecosystems, these communities can be manipulated by a variety of factors including physical, chemical, and biological factors. These factors sometimes intensify the structural changes in communities that can result in complete destruction of periphytic biofilms. Therefore, finding the balance of these factors is the cornerstone in the development of periphytic biofilms.

REFERENCES

AcuÑA, V., Wolf, A., Uehlinger, U., Tockner, K., 2008. Temperature dependence of stream benthic respiration in an Alpine river network under global warming. Freshwater Biology 53 (10), 2076–2088.

Agostinho, A., Gomes, L., Veríssimo, S., Okada, E.K., 2004. Flood regime, dam regulation and fish in the Upper Paraná River: effects on assemblage attributes, reproduction and recruitment. Reviews in Fish Biology and Fisheries 14 (1), 11–19.

Algarte, V.M., Siqueira, N.S., Murakami, E.A., Rodrigues, L., 2009. Effects of hydrological regime and connectivity on the interannual variation in taxonomic similarity of periphytic algae. Brazilian Journal of Biology 69 (2 Suppl.), 609–616.

Allan, J.D., Castillo, M.M., 2007. Stream Ecology: Structure and Function of Running Waters. Springer.

Alvarez, M., Peckarsky, B.L., 2005. How do grazers affect periphyton heterogeneity in streams? Oecologia 142 (4), 576–587.

Ambrose, H.E., Wilzbach, M.A., Cummins, K.W., 2004. Periphyton response to increased light and salmon carcass introduction in northern California streams. Journal of the North American Benthological Society 23 (4), 701–712.

Azim, M.E., 2009. Photosynthetic periphyton and surfaces. In: Likens, G.E. (Ed.), Encyclopedia of Inland Waters. Academic Press, Oxford, pp. 184–191.

Azim, M.E., Milstein, A., Wahab, M.A., Verdegam, M.C.J., 2003. Periphyton–water quality relationships in fertilized fishponds with artificial substrates. Aquaculture 228 (1–4), 169–187.

Barranguet, C., Veuger, B., Van Beusekom, S.A.M., Marvan, P., Sinke, J.J., Admiraal, W., 2005. Divergent composition of algal-bacterial biofilms developing under various external factors. European Journal of Phycology 40 (1), 1–8.

Battin, T.J., Kaplan, L.A., Denis Newbold, J., Hansen, C.M.E., 2003a. Contributions of microbial biofilms to ecosystem processes in stream mesocosms. Nature 426 (6965), 439–442.

Battin, T.J., Kaplan, L.A., Newbold, J.D., Cheng, X., Hansen, C., 2003b. Effects of current velocity on the nascent architecture of stream microbial biofilms. Applied and Environmental Microbiology 69 (9), 5443–5452.

Baulch, H.M., Schindler, D.W., Turner, M.A., Findlay, D.L., Paterson, M.J., Vinebrooke, R.D., 2005. Effects of warming on benthic communities in a boreal lake: implications of climate change. Limnology and Oceanography 50 (5), 1377–1392.

Benedetti-Cecchi, L., 2000. Variance in ecological consumer-resource interactions. Nature 407 (6802), 370–374.

Benedetti-Cecchi, L., Vaselli, S., Maggi, E., Bertocci, I., 2005. Interactive effects of spatial variance and mean intensity of grazing on algal cover in rock pools. Ecology 86 (8), 2212–2222.

Benenati, E.P., Shannon, J.P., Blinn, D.W., Wilson, K.P., Hueftle, S.J., 2000. Reservoir–river linkages: lake powell and the Colorado River, Arizona. Journal of the North American Benthological Society 19 (4), 742–755.

Bergey, E., 1999. Crevices as refugia for stream diatoms: effect of crevice size on abraded substrates. Limnology and Oceanography 44 (6), 1522–1529.

Bergey, E., 2008. Does rock chemistry affect periphyton accrual in streams? Hydrobiologia 614 (1), 141–150.

Bergey, E.A., 2005. How protective are refuges? Quantifying algal protection in rock crevices. Freshwater Biology 50 (7), 1163–1177.

Bergey, E.A., Boettiger, C.A., Resh, V.H., 1995. Effects of water velocity on the architecture and epiphytes of *Cladophora glomerata* (Chlorophyta). Journal of Phycology 31 (2), 264–271.

Bernhardt, E.S., Likens, G.E., 2002. Dissolved organic carbon enrichment alters nitrogen dynamics in a forest stream. Ecology 83, 1689–1700.

Bernot, R., Turner, A., 2001. Predator identity and trait-mediated indirect effects in a littoral food web. Oecologia 129 (1), 139–146.

Bernot, R.J., Lamberti, G.A., 2008. Indirect effects of a parasite on a benthic community: an experiment with trematodes, snails and periphyton. Freshwater Biology 53 (2), 322–329.

Bertrand, C., Siauve, V., Fayolle, S., Cazaubon, A., 2001. Effects of hydrological regime on the drift algae in a regulated Mediterranean river (River Verdon, Southeastern France). Regulated Rivers: Research & Management 17 (4–5), 407–416.

Biggs, B.J., Goring, D.G., Nikora, V.I., 1998. Subsidy and stress responses of stream periphyton to gradients in water velocity as a function of community growth form. Journal of Phycology 34 (4), 598–607.

Biggs, B.J., Smith, R.A., 2002. Taxonomic richness of stream benthic algae: effects of flood disturbance and nutrients. Limnology and Oceanography 47 (4), 1175–1186.

Biggs, B.J.F., 1996. Patterns in benthic algae of streams. In: Stevenson, R.J., Bothwell, M.L., Lowe, R.L. (Eds.), Algal Ecology. Academic Press, San Diego, pp. 31–56.

Biggs, B.J.F., Close, M.E., 1989. Periphyton biomass dynamics in gravel bed rivers: the relative effects of flows and nutrients. Freshwater Biology 22 (2), 209–231.

Blanchet, S., Loot, G., Dodson, J.J., 2008. Competition, predation and flow rate as mediators of direct and indirect effects in a stream food chain. Oecologia 157 (1), 93–104.

Bonet, B., Corcoll, N., Acuna, V., Sigg, L., Behra, R., Guasch, H., 2013. Seasonal changes in antioxidant enzyme activities of freshwater biofilms in a metal polluted Mediterranean stream. Science of the Total Environment 444, 60–72.

Borchardt, M.A., 1994. Effects of flowing water on nitrogen- and phosphorus-limited photosynthesis and optimum N:P ratios by *Spirogyra fluviatilis* (Charophyceae). Journal of Phycology 30 (3), 418–430.

Borchardt, M.A., 1996a. Nutrients. In: Stevenson, R.J., Bothwell, M.L., Lowe, R.L. (Eds.), Algal Ecology. Academic Press, San Diego, pp. 183–227.

Borchardt, M.A., 1996b. Nutrients. In: Algal Ecology: Freshwater Benthic Ecosystems. Academic Press, San Diego.

Bothwell, M.L., 1988. Growth rate responses of lotic periphytic diatoms to experimental phosphorus enrichment: the influence of temperature and light. Canadian Journal of Fisheries and Aquatic Sciences 45 (2), 261–270.

Bothwell, M.L., 1989. Phosphorus–limited growth dynamics of lotic periphytic diatom communities: areal biomass and cellular growth rate responses. Canadian Journal of Fisheries and Aquatic Sciences 46 (8), 1293–1301.

Bott, T.L., Brock, J.T., Cushing, C.E., Gregory, S.V., King, D., Petersen, R.C., 1978. A comparison of methods for measuring primary productivity and community respiration in streams. Hydrobiologia 60 (1), 3–12.

Bott, T.L., Brock, J.T., Dunn, C.S., Naiman, R.J., Ovink, R.W., Petersen, R.C., 1985. Benthic community metabolism in four temperate stream systems: an inter-biome comparison and evaluation of the river continuum concept. Hydrobiologia 123 (1), 3–45.

Bouletreau, S., Salvo, E., Lyautey, E., Mastrorillo, S., Garabetian, F., 2012. Temperature dependence of denitrification in phototrophic river biofilms. Science of the Total Environment 416, 323–328.

Bowes, M., Ings, N., McCall, S., Warwick, A., Barrett, C., Wickham, H., Harman, S., Armstrong, L., Scarlett, P., Roberts, C., 2012. Nutrient and light limitation of periphyton in the River Thames: implications for catchment management. Science of the Total Environment 434, 201–212.

Bowman, M.F., Chambers, P.A., Schindler, D.W., 2005. Changes in stoichiometric constraints on epilithon and benthic macroinvertebrates in response to slight nutrient enrichment of mountain rivers. Freshwater Biology 50 (11), 1836–1852.

Brönmark, C., Simon, D.R., Erlandsson, A., 1991. Interactions between freshwater snails and tadpoles: competition and facilitation. Oecologia 87 (1), 8–18.

Bratvold, D., Browdy, C.L., 2001. Effects of sand sediment and vertical surfaces (AquaMats™) on production, water quality, and microbial ecology in an intensive *Litopenaeus vannamei* culture system. Aquaculture 195 (1–2), 81–94.

Brookes, A., 1986. Response of aquatic vegetation to sedimentation downstream from river channelisation works in England and Wales. Biological Conservation 38 (4), 351–367.

Brown, J.H., Gillooly, J.F., Allen, A.P., Savage, V.M., West, G.B., 2004. Toward a metabolic theory of ecology. Ecology 85 (7), 1771–1789.

Burkholder, J.M., 1996. Interactions of benthic algae with their substrata. In: Lowe, R.L., Stevenson, R.J., Bothwell, M.L. (Eds.), Algal Ecology. Academic Press, San Diego, pp. 253–297.

Carey, R.O., Vellidis, G., Lowrance, R., Pringle, C.M., 2007. Do nutrients limit algal periphyton in small blackwater coastal plain streams? JAWRA: Journal of the American Water Resources Association 43 (5), 1183–1193.

Carling, P.A., 1984. Deposition of fine and coarse sand in an open-work gravel bed. Canadian Journal of Fisheries and Aquatic Sciences 41 (2), 263–270.

Cashman, M.J., Wehr, J.D., Truhn, K., 2013. Elevated light and nutrients alter the nutritional quality of stream periphyton. Freshwater Biology 58 (7), 1447–1457.

Castella, E., Adalsteinsson, H., Brittain, J.E., Gislason, G.M., Lehmann, A., Lencioni, V., Lods-Crozet, B., Maiolini, B., Milner, A.M., Olafsson, J.S., Saltveit, S.J., Snook, D.L., 2001. Macrobenthic invertebrate richness and composition along a latitudinal gradient of European glacier-fed streams. Freshwater Biology 46 (12), 1811–1831.

Cebrian, J., Duarte, C.M., 1995. Plant growth-rate dependence of detrital carbon storage in ecosystems. Science 268 (5217), 1606–1608.

Chalifour, A., Juneau, P., 2011. Temperature-dependent sensitivity of growth and photosynthesis of *Scenedesmus obliquus*, *Navicula pelliculosa* and two strains of *Microcystis aeruginosa* to the herbicide atrazine. Aquatic Toxicology 103 (1–2), 9–17.

Chen, J., Xie, P., Li, L., Xu, J., 2009. First identification of the hepatotoxic microcystins in the serum of a chronically exposed human population together with indication of hepatocellular damage. Toxicological Sciences 108 (1), 81–89.

Christofoletti, R.A., Almeida, T.V., Ciotti, A.M., 2011. Environmental and grazing influence on spatial variability of intertidal biofilm on subtropical rocky shores. Marine Ecology-Progress Series 424, 15.

Collins, B., Wein, G., Philippi, T., 2001. Effects of disturbance intensity and frequency on early old-field succession. Journal of Vegetation Science 12 (5), 721–728.

Combes, C., 2001. Parasitism: The Ecology and Evolution of Intimate Interactions. University of Chicago Press.

Connell, J.H., 1978. Diversity in tropical rain forests and coral reefs. Science 199 (4335), 1302–1310.

Cross, W.F., Benstead, J.P., Frost, P.C., Thomas, S.A., 2005. Ecological stoichiometry in freshwater benthic systems: recent progress and perspectives. Freshwater Biology 50 (11), 1895–1912.

Davidson, T.A., Mackay, A.W., Wolski, P., Mazebedi, R., Murray-Hudson, M., Todd, M., 2012. Seasonal and spatial hydrological variability drives aquatic biodiversity in a flood-pulsed, subtropical wetland. Freshwater Biology 57 (6), 1253–1265.

Davies-Colley, R., Hickey, C., Quinn, J., Ryan, P., 1992. Effects of clay discharges on streams. Hydrobiologia 248 (3), 215–234.

Davis, L.S., Hoffmann, J.P., Cook, P.W., 1990. Seasonal succession of algal periphyton from a wastewater treatment facility. Journal of Phycology 26 (4), 611–617.

DeAngelis, D.L., Loreau, M., Neergaard, D., Mulholland, P.J., Marzolf, E.R., 1995. Modelling nutrient-periphyton dynamics in streams: the importance of transient storage zones. Ecological Modelling 80 (2–3), 149–160.

Delong, M., Brusven, M., 1994. Allochthonous input of organic matter from different riparian habitats of an agriculturally impacted stream. Environmental Management 18 (1), 59–71.

Demars, B.O.L., Russell Manson, J., ÓLafsson, J.S., GÍSlason, G.M., GudmundsdÓTtir, R., Woodward, G.U.Y., Reiss, J., Pichler, D.E., Rasmussen, J.J., Friberg, N., 2011. Temperature and the metabolic balance of streams. Freshwater Biology 56 (6), 1106–1121.

DeNicola, D.M., 1996. Periphyton responses to temperature at different ecological levels. In: Lowe, R., Stevenson, R.J., Bothwell, M.L. (Eds.), Algal Ecology. Academic Press, San Diego, pp. 149–181.

DeNicola, M., Hoagland, K.D., Roemer, S.C., 1992. Influences of canopy cover on spectral irradiance and periphyton assemblages in a prairie stream. Journal of the North American Benthological Society 391–404.

Dickman, M., 1968. The effect of grazing by tadpoles on the structure of a periphyton community. Ecology 49 (6), 1188–1190.

Dillon, R.T., 2000. The Ecology of Freshwater Molluscs. Cambridge University Press.

Dobson, A.P., Hudson, P.J., 1986. Parasites, disease and the structure of ecological communities. Trends in Ecology & Evolution 1 (1), 11–15.

Dodds, W., Hutson, R., Eichem, A., Evans, M., Gudder, D., Fritz, K., Gray, L., 1996. The relationship of floods, drying, flow and light to primary production and producer biomass in a prairie stream. Hydrobiologia 333 (3), 151–159.

Dodds, W.K., 1989. Microscale vertical profiles of N_2 fixation, photosynthesis, O_2, chlorophyll a, and light in a cyanobacterial assemblage. Applied and Environmental Microbiology 55 (4), 882–886.

Dodds, W.K., Biggs, B.J.F., 2002. Water velocity attenuation by stream periphyton and macrophytes in relation to growth form and architecture. Journal of the North American Benthological Society 21 (1), 2–15.

Dodds, W.K., Smith, V.H., Lohman, K., 2002. Nitrogen and phosphorus relationships to benthic algal biomass in temperate streams. Canadian Journal of Fisheries and Aquatic Sciences 59 (5), 865–874.

Doeg, T.J., Koehn, J.D., 1994. Effects of draining and desilting a small weir on downstream fish and macroinvertebrates. Regulated Rivers: Research & Management 9 (4), 263–277.

Downes, B.J., Street, J.L., 2005. Regrowth or dispersal? Recovery of a freshwater red alga following disturbance at the patch scale. Austral Ecology 30 (5), 526–536.

Downing, J.A., McClain, M., Twilley, R., Melack, J.M., Elser, J., Rabalais, N.N., Lewis Jr., W.M., Turner, R.E., Corredor, J., Soto, D., Yanez-Arancibia, A., Kopaska, J.A., Howarth, R.W., 1999. The impact of accelerating land-use change on the N-cycle of tropical aquatic ecosystems: current conditions and projected changes. Biogeochemistry 46 (1–3), 109–148.

Dudley, T.L., D'Antonio, C.M., 1991. The effects of substrate texture, grazing, and disturbance on macroalgal establishment in streams. Ecology 72 (1), 297–309.

Dufford, R.G., Zimmermann, H.J., Cline, L.D., Ward, J.V., 1987. Responses of epilithic algae to regulation of rocky mountain streams. In: Craig, J., Kemper, J.B. (Eds.), Regulated Streams. Springer, USA, pp. 383–390.

Edwards, D., 1969. Some effects of siltation upon aquatic macrophyte vegetation in rivers. Hydrobiologia 34 (1), 29–38.

Eriksson, P., 2001. Interaction effects of flow velocity and oxygen metabolism on nitrification and denitrification in biofilms on submersed macrophytes. Biogeochemistry 55 (1), 29–44.

Evans-White, M.A., Lamberti, G.A., 2005. Grazer species effects on epilithon nutrient composition. Freshwater Biology 50 (11), 1853–1863.

Evans-White, M.A., Lamberti, G.A., 2006. Stoichiometry of consumer-driven nutrient recycling across nutrient regimes in streams. Ecological Letters 9 (11), 1186–1197.

Fairchild, G.W., Sherman, J.W., 1992. Linkage between epilithic algal growth and water column nutrients in softwater Lakes. Canadian Journal of Fisheries and Aquatic Sciences 49 (8), 1641–1649.

Feminella, J.W., Hawkins, C.P., 1995. Interactions between stream herbivores and periphyton: a quantitative analysis of past experiments. Journal of the North American Benthological Society 14 (4), 465–509.

Feminella, J.W., Power, M.E., Resh, V.H., 1989. Periphyton responses to invertebrate grazing and riparian canopy in three northern California coastal streams. Freshwater Biology 22 (3), 445–457.

Fermino, F.S., Bicudo, D.d.C., Bicudo, C., 2011. Seasonal influence of nitrogen and phosphorus enrichment on the floristic composition of the algal periphytic community in a shallow tropical, mesotrophic reservoir (São Paulo, Brazil). Oecologia Australis 15 (3), 476–493.

Ferragut, C., de Campos Bicudo, D., 2012. Effect of N and P enrichment on periphytic algal community succession in a tropical oligotrophic reservoir. Limnology 13 (1), 131–141.

Ferragut, C., de Campos Bicudo, D., 2010. Periphytic algal community adaptive strategies in N and P enriched experiments in a tropical oligotrophic reservoir. Hydrobiologia 646 (1), 295–309.

Figueiredo, M.A.d.O., Kain, J.M., Norton, T.A., 1996. Biotic interactions in the colonization of crustose coralline algae by epiphytes. Journal of Experimental Marine Biology and Ecology 199 (2), 303–318.

Flecker, A.S., Taylor, B.W., Bernhardt, E.S., Hood, J.M., Cornwell, W.K., Cassatt, S.R., Vanni, M.J., Altman, N.S., 2002. Interactions between herbivous fishes and limiting nutrients in a tropical stream ecosystem. Ecology 83 (7), 1831–1844.

Francoeur, S., Biggs, B.F., 2006. Short-term effects of elevated velocity and sediment abrasion on benthic algal communities. Hydrobiologia 561, 59–69.

Francoeur, S.N., 2001. Meta-analysis of lotic nutrient amendment experiments: detecting and quantifying subtle responses. Meta 20 (3), 358–368.

Frost, P.C., Cherrier, C.T., Larson, J.H., Bridgham, S., Lamberti, G.A., 2007. Effects of dissolved organic matter and ultraviolet radiation on the accrual, stoichiometry and algal taxonomy of stream periphyton. Freshwater Biology 52 (2), 319–330.

Frost, P.C., Elser, J.J., 2002. Effects of light and nutrients on the net accumulation and elemental composition of epilithon in boreal lakes. Freshwater Biology 47 (2), 173–183.

Frost, P.C., Stelzer, R.S., Lamberti, G.A., Elser, J.J., 2002. Ecological stoichiometry of trophic interactions in the benthos: understanding the role of C:N:P ratios in lentic and lotic habitats. Journal of the North American Benthological Society 21, 515–528.

Gaiser, E., 2009. Periphyton as an indicator of restoration in the Florida Everglades. Ecological Indicators 9 (6 Suppl.), S37–S45.

Garcia-Pichel, F., Castenholz, R.W., 1991. Characterization and biological implications of scytonemin, a cyanobacterial sheath pigment. Journal of Phycology 27 (3), 395–409.

Geddes, P., Trexler, J.C., 2003. Uncoupling of omnivore-mediated positive and negative effects on periphyton mats. Oecologia 136 (4), 585–595.

Gelwick, F.P., Matthews, W.J., 1997. Effects of algivorous minnows (*Campostoma*) on spatial and temporal heterogeneity of stream periphyton. Oecologia 112 (3), 386–392.

Ghadouani, A., Pinel-Alloul, B., Prepas, E.E., 2003. Effects of experimentally induced cyanobacterial blooms on crustacean zooplankton communities. Freshwater Biology 48 (2), 363–381.

Godillot, R., Caussade, B., Ameziane, T., Capblancq, J., 2001. Interplay between turbulence and periphyton in rough open-channel flow. Journal of Hydraulic Research 39 (3), 227–239.

Goldfinch, A., Carman, K., 2000. Chironomid grazing on benthic microalgae in a Louisiana salt marsh. Estuaries 23 (4), 536–547.

Gosselin, L.A., Chia, F.S., 1995. Distribution and dispersal of early juvenile snails: effectiveness of intertidal microhabitats as refuges and food sources. Marine Ecology Progress Series 128 (1–3), 213–223.

Gottlieb, A., Richards, J., Gaiser, E., 2006. Comparative study of periphyton community structure in long and short-hydroperiod Everglades marshes. Hydrobiologia 569 (1), 195–207.

Graba, M., Kettab, A., Sauvage, S., Sanchez-Pérez, J.M., 2012. On modeling chronic detachment of periphyton in artificial rough, open channel flow. Desalination and Water Treatment 41 (1–3), 79–87.

Graham, A.A., 1990. Siltation of stone-surface periphyton in rivers by clay-sized particles from low concentrations in suspension. Hydrobiologia 199 (2), 107–115.

Greenwood, J.L., Rosemond, A.D., 2005. Periphyton response to long-term nutrient enrichment in a shaded headwater stream. Canadian Journal of Fisheries and Aquatic Sciences 62 (9), 2033–2045.

Grimm, N.B., Fisher, S.G., 1986. Nitrogen limitation in a Sonoran Desert stream. Journal of the North American Benthological Society 5 (1), 2–15.

Grimm, N.B., Fisher, S.G., 1989. Stability of periphyton and macroinvertebrates to disturbance by flash floods in a desert stream. Journal of the North American Benthological Society 293–307.

Growns, I.O., Growns, J.E., 2001. Ecological effects of flow regulation on macroinvertebrate and periphytic diatom assemblages in the Hawkesbury–Nepean River, Australia. Regulated Rivers: Research & Management 17 (3), 275–293.

Guariento, R., Caliman, A., Esteves, F., Enrich-Prast, A., Bozelli, R., Farjalla, V., 2007. Substrate-mediated direct and indirect effects on periphytic biomass and nutrient content in a tropical coastal lagoon, Rio de Janeiro, Brazil. Acta Limnologica Brasiliensia 19, 331–340.

Guariento, R.D., Caliman, A., Esteves, F.A., Bozelli, R.L., Enrich-Prast, A., Farjalla, V.F., 2009. Substrate influence and temporal changes on periphytic biomass accrual and metabolism in a tropical humic lagoon. Limnologica - Ecology and Management of Inland Waters 39 (3), 209–218.

Guschina, I., Harwood, J., 2009. Algal lipids and effect of the environment on their biochemistry. In: Kainz, M., Brett, M.T., Arts, M.T. (Eds.), Lipids in Aquatic Ecosystems. Springer, New York, pp. 1–24.

Hörnström, E., 2002. Phytoplankton in 63 limed lakes in comparison with the distribution in 500 untreated lakes with varying pH. Hydrobiologia 470 (1–3), 115–126.

Hansson, L.A., 1992. Factors regulating periphytic algal biomass. Limnology and Oceanography 37 (2), 322–328.

Hart, D.D., 1987. Experimental studies of exploitative competition in a grazing stream insect. Oecologia 73 (1), 41–47.

Hawes, I., Howard-Williams, C., Vincent, W., 1992. Desiccation and recovery of antarctic cyanobacterial mats. Polar Biology 12 (6–7), 587–594.

Hawes, I., Smith, R., 1994. Seasonal dynamics of epilithic periphyton in oligotrophic Lake Taupo, New Zealand. New Zealand Journal of Marine and Freshwater Research 28 (1), 1–12.

Hawkes, H.A., 1969. Ecological changes of applied significance induced by the discharge of heated waters. In: Parker, F.L., Krenkel, P.A. (Eds.), Engineering Aspects of Thermal Pollution. Vanderbilt University Press, pp. 15–57.

Hay, M.E., Kubanek, J., 2002. Community and ecosystem level consequences of chemical cues in the plankton. Journal of Chemical Ecology 28 (10), 2001–2016.

Hessen, D.O., Færøvig, P.J., Andersen, T., 2002. Light, nutrients, and P:C ratios in algae: grazer performance related to food quality and quantity. Ecology 83 (7), 1886–1898.

Hill, G.E., 1988. Age, plumage brightness, territory quality, and reproductive success in the black-headed grosbeak. Condor 379–388.

Hill, W.R., Boston, H.L., Steinman, A.D., 1992. Grazers and nutrients simultaneously limit lotic primary productivity. Canadian Journal of Fisheries and Aquatic Sciences 49 (3), 504–512.

Hill, W.R., Dimick, S.M., 2002. Effects of riparian leaf dynamics on periphyton photosynthesis and light utilisation efficiency. Freshwater Biology 47 (7), 1245–1256.

Hill, W.R., Knight, A.W., 1988. Concurrent grazing effects of two stream insects on periphyton. Limnology and Oceanography 33 (1), 15–26.

Hillebrand, H., 2008. Grazing regulates the spatial variability of periphyton biomass. Ecology 89 (1), 165–173.

Hillebrand, H., 2005. Light regime and consumer control of autotrophic biomass. Journal of Ecology 93 (4), 758–769.

Hillebrand, H., Frost, P., Liess, A., 2008. Ecological stoichiometry of indirect grazer effects on periphyton nutrient content. Oecologia 155 (3), 619–630.

Hillebrand, H., Kahlert, M., 2001. Effect of grazing and nutrient supply on periphyton biomass and nutrient stoichiometry in habitats of different productivity. Limnology and Oceanography 46 (8), 1881–1898.

Hinzman, L.D., Viereck, L.A., Adams, P.C., Romanovsky, V.E., Yoshikawa, K., 2006. Climate and Permafrost Dynamics of the Alaskan Boreal Forest. Oxford University Press, New York.

Hoagland, K.D., Rosowski, J.R., Gretz, M.R., Roemer, S.C., 1993. Diatom extracellular polymeric substances: function, fine structure, chemistry, and physiology. Journal of Phycology 29 (5), 537–566.

Hondzo, M., Wang, H., 2002. Effects of turbulence on growth and metabolism of periphyton in a laboratory flume. Water Resources Research 38 (12), 1277.

Horne, A.J., Goldman, C.R., 1994. Limnology. McGraw-Hill.

Horner, R.R., Welch, E.B., 1981. Stream periphyton development in relation to current velocity and nutrients. Canadian Journal of Fisheries and Aquatic Sciences 38 (4), 449–457.

Huisman, J., Matthijs, H.C.P., Visser, P.M., 2006. Harmful Cyanobacteria. Springer.

Hultberg, H., Andersson, I., 1982. Liming of acidified lakes: induced long-term changes. Water, Air, and Soil Pollution 18 (1–3), 311–331.

Huntsman, A., 1948. Fertility and fertilization of streams. Journal of the Fisheries Board of Canada 7 (5), 248–253.

Jüttner, F., Wu, J.-T., 2000. Evidence of allelochemical activity in subtropical cyanobacterial biofilms of Taiwan. Archiv für Hydrobiologie 147 (4), 505–517.

Jackson, M.B., Vandermeer, E.M., Lester, N., Booth, J.A., Molot, L., Gray, I.M., 1990. Effects of neutralization and early reacidification on filamentous algae and macrophytes in Bowland Lake. Canadian Journal of Fisheries and Aquatic Sciences 47 (2), 432–439.

Jenssen, T.A., 1967. Food habits of the green frog, *Rana clamitans*, before and during metamorphosis. Copeia 1967 (1), 214–218.

Jorgensen, B.B., Des Marais, D.J., 1990. The diffusive boundary layer of sediments: oxygen microgradients over a microbial mat. Limnology and Oceanography 35 (6), 1343–1355.

Kühl, M., Glud, R.N., Ploug, H., Ramsing, N.B., 1996. Microenvironmental control of photosynthesis and photosynthesis-coupled respiration in an epilithic cyanobacterial biofilm. Journal of Phycology 32 (5), 799–812.

Kaushik, N.K., Hynes, H.B.N., 1971. The fate of the dead leaves that fall into streams. Archiv für Hydrobiologie 68, 465–515.

Kawata, Agawa, 1999. Perceptual scales of spatial heterogeneity of periphyton for freshwater snails. Ecology Letters 2 (4), 210–214.

Kelly, D.J., Clare, J.J., Bothwell, M.L., 2001. Attenuation of solar ultraviolet radiation by dissolved organic matter alter benthic colonization patterns in streams. Journal of the North American Benthological Society 20, 96–108.

Keshavanath, P., Gangadhar, B., Ramesh, T.J., Van Rooij, J.M., Beveridge, M.C.M., Baird, D.J., Verdegem, M.C.J., Van Dam, A.A., 2001. Use of artificial substrates to enhance production of freshwater herbivorous fish in pond culture. Aquaculture Research 32 (3), 189–197.

Kim, M.A., Richardson, J.S., 1999. Effects of light and nutrients on grazer–periphyton interactions. In: Biology and Management of Species and Habitats at Risk, 15–19 February, 1999, Kamloops, BC, pp. 497–502.

Klug, J.L., 2002. Positive and negative effects of allochthonous dissolved organic matter and inorganic nutrients on phytoplankton growth. Canadian Journal of Fisheries and Aquatic Sciences 59 (1), 85–95.

Koycheva, J., Karney, B., 2009. Stream water temperature and climate change-an ecological perspective. In: International Symposium on Water Management and Hydraulic Engineering, Ohrid Macedonia.

Krenkel, P.A., Parker, F.L., United States. Federal Water Pollution Control Administration, Vanderbilt University, 1969. Biological Aspects of Thermal Pollution: Proceedings. Vanderbilt University Press.

Kupferberg, S., 1997. Facilitation of periphyton production by tadpole grazing: functional differences between species. Freshwater Biology 37 (2), 427–439.

Lafferty, K.D., Dobson, A.P., Kuris, A.M., 2006. Parasites dominate food web links. Proceedings of the National Academy of Sciences 103 (30), 11211–11216.

Lamberti, G.A., Ashkenas, L.R., Gregory, S.V., Steinman, A.D., 1987. Effects of three herbivores on periphyton communities in laboratory streams. Journal of the North American Benthological Society 6 (2), 92–104.

Lamberti, G.A., Gregory, S.V., Ashkenas, L.R., Steinman, A.D., McIntire, C.D., 1989. Productive capacity of periphyton as a determinant of plant-herbivore interactions in streams. Ecology 70 (6), 1840–1856.

Lamberti, G.A., Resh, V.H., 1983. Stream periphyton and insect herbivores: an experimental study of grazing by a caddisfly population. Ecology 64 (5), 1124–1135.

Lane, C., Taffs, K., Corfield, J., 2003. A comparison of diatom community structure on natural and artificial substrata. Hydrobiologia 493 (1–3), 65–79.

Langan, S.J., Johnston, L., Donaghy, M.J., Youngson, A.F., Hay, D.W., Soulsby, C., 2001. Variation in river water temperatures in an upland stream over a 30-year period. Science of the Total Environment 265 (1–3), 195–207.

Lange, K., Liess, A., Piggott, J.J., Townsend, C.R., Matthaei, C.D., 2011. Light, nutrients and grazing interact to determine stream diatom community composition and functional group structure. Freshwater Biology 56 (2), 264–278.

Langford, T., 1990. Ecological Effects of Thermal Discharges. Springer.

Larned, S., Santos, S., 2000. Light- and nutrient-limited periphyton in low order streams of Oahu, Hawaii. Hydrobiologia 432 (1–3), 101–111.

Larned, S.T., 2010. A prospectus for periphyton: recent and future ecological research. Journal of the North American Benthological Society 29 (1), 182–206.

Larras, F., Lambert, A.S., Pesce, S., Rimet, F., Bouchez, A., Montuelle, B., 2013. The effect of temperature and a herbicide mixture on freshwater periphytic algae. Ecotoxicology and Environmental Safety 98, 162–170.

Leandrini, J., Fonseca, I., Rodrigues, L., 2008. Characterization of habitats based on algal periphyton biomass in the upper Parana River floodplain, Brazil. Brazilian Journal of Biology 68 (3), 503–509.

Leandrini, J., Rodrigues, L., 2008. Temporal variation of periphyton biomass in semilotic environments of the upper Paraná River floodplain. Acta Limnologica Brasiliensia 20 (1), 21–28.

Ledger, M.E., Harris, R.M., Armitage, P.D., Milner, A.M., 2008. Disturbance frequency influences patch dynamics in stream benthic algal communities. Oecologia 155 (4), 809–819.

Lehtiniemi, M., Engström-Öst, J., Karjalainen, M., Kozlowsky-Suzuki, B., Viitasalo, M., 2002. Fate of cyanobacterial toxins in the pelagic food web: transfer to copepods or to faecal pellets? Marine Ecology Progress Series 241, 13–21.

Leips, J., Travis, J., 1994. Metamorphic responses to changing food levels in two species of hylid frogs. Ecology 75 (5), 1345–1356.

Leu, E., Wiktor, J., Søreide, J.E., Berge, J., Falk-Petersen, S., 2010. Increased irradiance reduces food quality of sea ice algae. Marine Ecology Progress Series 411, 49–60.

Lowe, R.L., 1986. Periphyton response to nutrient manipulation in streams draining clearcut and forested watersheds. Journal of the North American Benthological Society 5 (3), 221–229.

Ludwig, A., Matlock, M., Haggard, B.E., Matlock, M., Cummings, E., 2008. Identification and evaluation of nutrient limitation on periphyton growth in headwater streams in the Pawnee Nation, Oklahoma. Ecological Engineering 32 (2), 178–186.

Luttenton, M.R., Baisden, C., 2006. The relationships among disturbance, substratum size and periphyton community structure. Hydrobiologia 561 (1), 111–117.

Müller, P., 1980. Effects of artificial acidification on the growth of periphyton. Canadian Journal of Fisheries and Aquatic Sciences 37 (3), 355–363.

Madsen, K.N., Nilsson, P., Sundbäck, K., 1993. The influence of benthic microalgae on the stability of a subtidal sediment. Journal of Experimental Marine Biology and Ecology 170 (2), 159–177.

Marcogliese, D.J., Cone, D.K., 1997. Food webs: a plea for parasites. Trends in Ecology and Evolution 12 (8), 320–325.

Marzolf, E.R., Mulholland, P.J., Steinman, A.D., 1994. Improvements to the diurnal upstream–downstream dissolved oxygen change technique for determining whole-stream metabolism in small streams. Canadian Journal of Fisheries and Aquatic Sciences 51 (7), 1591–1599.

Matthews, W.J., Stewart, A.J., Power, M.E., 1987. Grazing fishes as components of North American stream ecosystems—effects of *Campostoma anomalum*. In: Matthews, W.J., Heins, D.C. (Eds.), Community and Evolutionary Ecology of North American Stream Fishes, first ed. University of Oklahoma Press, pp. 128–135.

McConnell, W.J., Sigler, W.F., 1959. Chlorophyll and productivity in a mountain river. Limnology and Oceanography 4 (3), 335–351.

McCormick, P., 1994. Evaluating the multiple mechanisms underlying herbivore-algal interactions in streams. Hydrobiologia 291 (1), 47–59.

McCormick, P.V., Stevenson, R.J., 1991. Mechanisms of benthic algal succession in lotic environments. Ecology 72 (5), 1835–1848.

McCormick, P.V., Stevenson, R.J., 1998. Periphyton as a tool for ecological assessment and management in the Florida Everglades. Journal of Phycology 34 (5), 726–733.

McIntire, C.D., 1966. Some effects of current velocity on periphyton communities in laboratory streams. Hydrobiologia 27 (3–4), 559–570.

McIntyre, P.B., Michel, E., Olsgard, M., 2006. Top-down and bottom-up controls on periphyton biomass and productivity in Lake Tanganyika. Limnology and Oceanography 51 (3), 1514–1523.

McKnight, D., Tate, C., Andrews, E., Niyogi, D., Cozzetto, K., Welch, K., Lyons, W., Capone, D.G., 2007. Reactivation of a cryptobiotic stream ecosystem in the McMurdo Dry Valleys, Antarctica: a long-term geomorphological experiment. Geomorphology 89 (1), 186–204.

Minshall, G.W., 1978. Autotrophy in stream ecosystems. BioScience 28 (12), 767–771.

Mouritsen, K.N., Tompkins, D.M., Poulin, R., 2005. Climate warming may cause a parasite-induced collapse in coastal amphipod populations. Oecologia 146 (3), 476–483.

Mouthon, J., Daufresne, M., 2006. Effects of the 2003 heatwave and climatic warming on mollusc communities of the Saône: a large lowland river and of its two main tributaries (France). Global Change Biology 12 (3), 441–449.

Mulholland, P.J., Marzolf, E.R., Hendricks, S.P., Wilkerson, R.V., Baybayan, A.K., 1995. Longitudinal patterns of nutrient cycling and periphyton characteristics in streams: a test of upstream-downstream linkage. Journal of the North American Benthological Society 357–370.

Mulholland, P.J., Webster, J.R., 2010. Nutrient dynamics in streams and the role of J-NABS. Journal of the North American Benthological Society 29 (1), 100–117.

Murdock, J.N., Dodds, W.K., 2007. Linging benthic algal biomass to stream substratum topography. Journal of Phycology 43 (3), 449–460.

Murray, D.L., 1990. The effects of food and density on growth and metamorphosis in larval wood frogs (*Rana sylvatica*) from central Labrador. Canadian Journal of Zoology 68 (6), 1221–1226.

Negro, A.I., De Hoyos, C., Aldasoro, J.J., 2003. Diatom and desmid relationships with the environment in mountain lakes and mires of NW Spain. Hydrobiologia 505 (1–3), 1–13.

Newbold, J.D., Elwood, J.W., O'Neill, R.V., Sheldon, A.L., 1983. Phosphorus dynamics in a woodland stream ecosystem: a study of nutrient spiralling. Ecology 64 (5), 1249–1265.

Newman, R.A., 1994. Effects of changing density and food level on metamorphosis of a desert amphibian, *Scaphiopus Couchii*. Ecology 75 (4), 1085–1096.

Nuttall, P.M., 1972. The effects of sand deposition upon the macroinvertebrate fauna of the River Camel, Cornwall. Freshwater Biology 2 (3), 181–186.

Olem, H., 1990. Liming Acidic Surface Waters. Taylor & Francis.

Osborne, P.L., McLachlan, A.J., 1985. The effect of tadpoles on algal growth in temporary, rain-filled rock pools. Freshwater Biology 15 (1), 77–87.

Osterkamp, T.E., Viereck, L., Shur, Y., Jorgenson, M.T., Racine, C., Doyle, A., Boone, R.D., 2000. Observations of thermokarst and its impact on boreal forests in Alaska, USA. Arctic,-Antarctic,-and-Alpine-Research 32 (3), 303–315.

Passy, S.I., 2007. Diatom ecological guilds display distinct and predictable behavior along nutrient and disturbance gradients in running waters. Aquatic Botany 86 (2), 171–178.

Petersen, J.C., Femmer, S.R., Geological, S., National Water-Quality Assessment Program, 2003. Periphyton Communities in Streams of the Ozarks Plateaus and Their Relations to Selected Environmental Factors. U.S. Dept. of the Interior, U.S. Geological Survey; Branch of Information Services, Little Rock, Arkansas: Denver, Colo.

Peterson, C.G., 1996. Response of Benthic Algal Communities to Natural Physical Disturbance. Algal Ecology: Freshwater Benthic Ecosystems. Academic Press, San Diego, California, pp. 375–401.

Peterson, C.G., Grimm, N.B., 1992. Temporal variation in enrichment effects during periphyton succession in a nitrogen-limited desert stream ecosystem. Journal of the North American Benthological Society 11 (1), 20–36.

Peterson, C.G., Stevenson, R.J., 1990. Post-spate development of epilithic algal communities in different current environments. Canadian Journal of Botany 68 (10), 2092–2102.

Petts, G., Armitage, P., Castella, E., 1993. Physical habitat changes and macroinvertebrate response to river regulation: the river Rede, UK. Regulated Rivers: Research & Management 8 (1–2), 167–178.

Petts, G.E., 1988. Accumulation of fine sediment within substrate gravels along two regulated rivers, UK. Regulated Rivers: Research & Management 2 (2), 141–153.

Pickett, S.T.A., White, P.S., 1985. Chapter 21-Patch dynamics: a synthesis. In: Pickett, S.T., White, P.S. (Eds.), The Ecology of Natural Disturbance and Patch Dynamics. Academic Press, San Diego, pp. 371–384.

Planas, D., 1996. Acidification effects. In: Stevenson, R.J., Bothwell, M.L., Lowe, R.L., Thorp, J.H. (Eds.), Algal Ecology: Freshwater Benthic Ecosystem. Elsevier Science.

Power, M.E., Matthews, W.J., Stewart, A.J., 1985. Grazing minnows, piscivorous bass, and stream algae: dynamics of a strong interaction. Ecology 66 (5), 1448–1456.

Prézelin, B.B., Matlick, H.A., 1983. Nutrient-dependent low-light adaptation in the dinoflagellate *Gonyaulax polyedra*. Marine Biology 74 (2), 141–150.

Pringle, C.M., Bowers, J.A., 1984. An in situ substratum fertilization technique: diatom colonization on nutrient-enriched, sand substrata. Canadian Journal of Fisheries and Aquatic Sciences 41, 1247–1251.

Pringle, C.M., Triska, F.J., 1991. Effects of geothermal groundwater on nutrient dynamics of a lowland Costa Rican stream. Ecology 72 (3), 951–965.

Quinn, J.M., Williamson, R.B., Smith, R.K., Vickers, M.L., 1992. Effects of riparian grazing and channelisation on streams in Southland, New Zealand. 2. Benthic invertebrates. New Zealand Journal of Marine and Freshwater Research 26 (2), 259–273.

Ramesh, M.R., Shankar, K.M., Mohan, C.V., Varghese, T.J., 1999. Comparison of three plant substrates for enhancing carp growth through bacterial biofilm. Aquacultural Engineering 19 (2), 119–131.

Rao, T.S., 2010. Comparative effect of temperature on biofilm formation in natural and modified marine environment. Aquatic Ecology 44 (2), 463–478.

Raven, J.A., 1984. A cost-benefit analysis of photo absorption by photosynthetic unicells. New Phytologist 98 (4), 593–625.

Ravet, J.L., Brett, M.T., Arhonditsis, G.B., 2010. The effects of seston lipids on zooplankton fatty acid composition in Lake Washington, Washington, USA. Ecology 91 (1), 180–190.

Reay, D.S., Nedwell, D.B., Priddle, J., Ellis-Evans, J.C., 1999. Temperature dependence of inorganic nitrogen uptake: reduced affinity for nitrate at suboptimal temperatures in both algae and bacteria. Applied and Environmental Microbiology 65 (6), 2577–2584.

Reiter, M.A., 1986. Interactions between the hydrodynamics of flowing water and the development of a benthic algal community. Journal of Freshwater Ecology 3 (4), 511–517.

Reiter, M.A., Carlson, R.E., 1986. Current velocity in streams and the composition of benthic algal mats. Canadian Journal of Fisheries and Aquatic Sciences 43 (6), 1156–1162.

Riber, H.H., Wetzel, R.G., 1987. Boundary-layer and internal diffusion effects on phosphorus fluxes in lake periphyton. Limnology and Oceanography 32 (6), 1181–1194.

Richard, M., Trottier, C., Verdegem, M.C.J., Hussenot, J.M.E., 2009. Submersion time, depth, substrate type and sampling method as variation sources of marine periphyton. Aquaculture 295 (3–4), 209–217.

Richards, C., Bacon, K.L., 1994. Influence of fine sediment on macroinvertebrate colonization of surface and hyporheic stream substrates. The Great Basin Naturalist 54 (2), 106–113.

Riseng, C., Wiley, M., Stevenson, R., 2004. Hydrologic disturbance and nutrient effects on benthic community structure in midwestern US streams: a covariance structure analysis. Journal of the North American Benthological Society 23 (2), 309–326.

Rodrigues, L., Bicudo, D.D.C., 2001. Similarity among periphyton algal communities in a lentic-lotic gradient of the upper Paraná river floodplain, Brazil. Brazilian Journal of Botany 24, 235–248.

Roelke, D., Buyukates, Y., Williams, M., Jean, J., 2004. Interannual variability in the seasonal plankton succession of a shallow, warm-water lake. Hydrobiologia 513 (1–3), 205–218.

Roelke, D.I., Cifuentes, L.A., Eldridge, P.M., 1997. Nutrient and phytoplankton dynamics in a sewage-impacted Gulf coast estuary: a field test of the PEG-model and equilibrium resource competition theory. Estuaries 20 (4), 725–742.

Romani, A.M., Giorgi, A., Acuna, V., Sabater, S., 2004. The influence of substratum type and nutrient supply on biofilm organic matter utilization in streams. Limnology and Oceanography 49 (5), 1713–1721.

Rosa, J., Ferreira, V., Canhoto, C., Graça, M.A.S., 2013. Combined effects of water temperature and nutrients concentration on periphyton respiration – implications of global change. International Review of Hydrobiology 98 (1), 14–23.

Rosemond, A.D., Mulholland, P.J., Elwood, J.W., 1993. Top-down and bottom-up control of stream periphyton: effects of nutrients and herbivores. Ecology 74 (4), 1264–1280.

Rosenberg, D.M., Resh, V.H., 1992. Freshwater Biomonitoring and Benthic Macroinvertebrates. Springer.

Rutherford, J.C., Blackett, S., Blackett, C., Saito, L., Davies-Colley, R.J., 1997. Predicting the effects of shade on water temperature in small streams. New Zealand Journal of Marine and Freshwater Research 31 (5), 707–721.

Santos, C.M., Eskinazi-Sant'Anna, E., 2010. The introduced snail *Melanoides tuberculatus* (Muller, 1774) (Mollusca: Thiaridae) in aquatic ecosystems of the Brazilian Semiarid Northeast (Piranhas-Assu River basin, State of Rio Grande do Norte). Brazilian Journal of Biology 70 (1), 1–7.

Schälchli, U., 1992. The clogging of coarse gravel river beds by fine sediment. Hydrobiologia 235–236 (1), 189–197.

Schiller, D.V., Martí, E., Riera, J.L., Sabater, F., 2007. Effects of nutrients and light on periphyton biomass and nitrogen uptake in Mediterranean streams with contrasting land uses. Freshwater Biology 52 (5), 891–906.

Schindler, D.W., 1998. Whole-ecosystem experiments: replication versus realism: the need for ecosystem-scale experiments. Ecosystems 1 (4), 323–334.

Smart, M.M., Jones, J.R., Sebaugh, J.L., 1985. Stream-watershed relations in the Missouri Ozark Plateau province. Journal of Environmental Quality 14 (1), 77–82.

Smith, D.J., Underwood, G.J.C., 2000. The production of extracellular carbohydrates by estuarine benthic diatoms: the effects of growth phase and light and dark treatment. Journal of Phycology 36 (2), 321–333.

Smith, G.D., Doan, N.T., 1999. Cyanobacterial metabolites with bioactivity against photosynthesis in cyanobacteria, algae and higher plants. Journal of Applied Phycology 11 (4), 337–344.

Solomon, S., 2007. Climate Change 2007. The Physical Science Basis: Working Group I Contribution to the Fourth Assessment Report of the IPCC. Cambridge University Press.

Sommer, U., 2000. Benthic microalgal diversity enhanced by spatial heterogeneity of grazing. Oecologia 122 (2), 284–287.

Squires, L., Rushforth, S., Brotherson, J., 1979. Algal response to a thermal effluent: study of a power station on the provo river, Utah, USA. Hydrobiologia 63 (1), 17–32.

St. Jónsson, G., 1992. Photosynthesis and production of epilithic algal communities in Thingvallavatn. Oikos 222–240.

Stanley, E.H., Fisher, S.G., Jones Jr., J.B., 2004. Effects of water loss on primary production: a landscape-scale model. Aquatic Sciences 66 (1), 130–138.

Stefan, H.G., Sinokrot, B.A., 1993. Projected global climate change impact on water temperatures in five north central U.S. streams. Climatic Change 24 (4), 353–381.

Steinaman, A.D., McIntire, C.D., 1986. Effects of current velocity and light energy on the structure of periphyton assemblage in laboratory streams. Journal of Phycology 22 (3), 352–361.

Steinman, A.D., 1996. 12-Effects of grazers on freshwater benthic algae. In: Stevenson, R.J., Bothwell, M.L., Lowe, R.L. (Eds.), Algal Ecology. Academic Press, San Diego, pp. 341–373.

Steinman, A.D., Ogdahl, M.E., Wessell, K., Biddanda, B., Kendall, S., Nold, S., 2011. Periphyton response to simulated nonpoint source pollution: local over regional control. Aquatic Ecology 45 (4), 439–454.

Stelzer, R.S., Lamberti, G.A., 2001. Effects of N:P ratio and total nutrient concentration on stream periphyton community structure, biomass, and elemental composition. Limnology and Oceanography 46 (2), 356–367.

Stephens, S., Brasher, A.D., Smith, C., 2012. Response of an algal assemblage to nutrient enrichment and shading in a Hawaiian stream. Hydrobiologia 683 (1), 135–150.

Sterner, R.W., Elser, J.J., Fee, E.J., Guildford, S.J., Chrzanowski, T.H., 1997. The light: nutrient ratio in lakes: the balance of energy and materials affects ecosystem structure and process. The American Naturalist 150 (6), 663–684.

Stevenson, R.J., 1984. Epilithic and epipelic diatoms in the Sandusky River, with emphasis on species diversity and water pollution. Hydrobiologia 114 (3), 161–175.

Stevenson, R.J., Bothwell, M.L., Lowe, R.L., Thorp, J.H., 1996. Algal Ecology: Freshwater Benthic Ecosystem. Elsevier Science.

Swamikannu, X., Hoagland, K.D., 1989. Effects of snail grazing on the diversity and structure of a periphyton community in a eutrophic pond. Canadian Journal of Fisheries and Aquatic Sciences 46 (10), 1698–1704.

Tall, L., Cattaneo, A., Cloutier, L., Dray, S., Legendre, P., 2006. Resource partitioning in a grazer guild feeding on a multilayer diatom mat. Journal of the North American Benthological Society 25 (4), 800–810.

Tank, J.L., Dodds, W.K., 2003. Nutrient limitation of epilithic and epixylic biofilms in ten North American streams. Freshwater Biology 48 (6), 1031–1049.

Tarkowska-Kukuryk, M., 2013. Periphytic algae as food source for grazing chironomids in a shallow phytoplankton-dominated lake. Limnologica - Ecology and Management of Inland Waters 43 (4), 254–264.

Thomas, S.H., Housley, J.M., Reynolds, A.N., Penczykowski, R.M., Kenline, K.H., Hardegree, N., Schmidt, S., Duffy, M.A., 2011. The ecology and phylogeny of oomycete infections in *Asplanchna rotifers*. Freshwater Biology 56 (2), 384–394.

Thompson, P.A., Guo, M., Harrison, P.J., 1993. The influence of irradiance on the biochemical composition of three phytoplankton species and their nutritional value for larvae of the Pacific Oyster (*Crassostrea gigas*). Marine Biology 117 (2), 259–268.

Tipping, E., Woof, C., Clarke, K., 1993. Deposition and resuspension of fine particles in a riverine 'dead zone'. Hydrological Processes 7 (3), 263–277.

Tuchman, M., Stevenson, R.J., 1980. Comparison of clay tile, sterilized rock, and natural substrate diatom communities in a small stream in Southeastern Michigan, USA. Hydrobiologia 75 (1), 73–79.

Tuchman, N.C., 1996. The role of heterotrophy in algae. In: Stevenson, R.J., Bothwell, M.L., Lowe, R.L. (Eds.), Algal Ecology. Academic Press, San Diego, pp. 299–319.

Turner, M.A., Todd Howell, E., Summerby, M., Hesslein, R.H., Findlay, D.L., Jackson, M.B., 1991. Changes in epilithon and epiphyton associated with experimental acidification of a lake to pH 5. Limnology and Oceanography 36 (7), 1390–1405.

Turner, M.A., Jackson, M.B., Findlay, D.L., Graham, R.W., DeBruyn, E.R., Vandermeer, E.M., 1987. Early responses of periphyton to experimental lake acidification. Canadian Journal of Fisheries and Aquatic Sciences 44 (S1), s135–s149.

Turner, M.A., Schindler, D.W., Findlay, D.L., Jackson, M.B., Robinson, G.G., 1995a. Disruption of littoral algal associations by Experimental Lake acidification. Canadian Journal of Fisheries and Aquatic Sciences 52 (10), 2238–2250.

Turner, M.A., Townsend, B.E., Robinson, G.G.C., Hann, B.J., Amaral, J.A., 1995b. Ecological effects of blooms of filamentous green algae in the littoral zone of an acid lake. Canadian Journal of Fisheries and Aquatic Sciences 52 (10), 2264–2275.

Urabe, J., Sterner, R.W., 1996. Regulation of herbivore growth by the balance of light and nutrients. Proceedings of the National Academy of Sciences of the United States of America 93 (16), 8465–8469.

Vadeboncoeur, Y., Lodge, D.M., Carpenter, S.R., 2001. Whole-lake fertilization effects on distribution of primary production between benthic and pelagic habitats. Ecology 82 (4), 1065–1077.

Van Dam, H., Mertens, A., 1995. Long-term changes of diatoms and chemistry in headwater streams polluted by atmospheric deposition of sulphur and nitrogen compounds. Freshwater Biology 34 (3), 579–600.

Van Nieuwenhuyse, E.E., LaPerriere, J.D., 1986. Effects of placer gold mining on primary production in subarctic streams of Alaska. JAWRA: Journal of the American Water Resources Association 22 (1), 91–99.

Vannote, R.L., Minshall, G.W., Cummins, K.W., Sedell, J.R., Cushing, C.E., 1980. The river continuum concept. Canadian Journal of Fisheries and Aquatic Sciences 37 (1), 130–137.

Villanueva, V.D., Albariño, R., Modenutti, B., 2004. Grazing impact of two aquatic invertebrates on periphyton from an Andean-Patagonian stream. Archiv für Hydrobiologie 159 (4), 455–471.

Vinebrooke, R.D., Leavitt, P.R., 1998. Direct and interactive effects of allochthonous dissolved organic matter, inorganic nutrients, and ultraviolet radiation on an alpine littoral food web. Limnology and Oceanography 43 (6), 1065–1081.

Vis, C., Cattaneo, A., Hudon, C., 2008. Shift from chlorophytes to cyanobacteria in benthic macroalgae along a gradient of nitrate depletion. Journal of Phycology 44 (1), 38–44.

Walsh, C.J., Roy, A.H., Feminella, J.W., Cottingham, P.D., Groffman, P.M., Morgan, R.P., 2005. The urban stream syndrome: current knowledge and the search for a cure. Journal of the North American Benthological Society 24 (3), 706–723.

Walton, S.P., Welch, E.B., Horner, R.R., 1995. Stream periphyton response to grazing and changes in phosphorus concentration. Hydrobiologia 302 (1), 31–46.

Ward, J.V., Stanford, J.A., 1979. The Ecology of Regulated Streams. Plenum Press.

Webster, J.R., 1986. Vascular plant breakdown in freshwater ecosystems. Annual Review of Ecology Systems 17, 567–594.

Wellnitz, T., Leroy Poff, N., 2006. Herbivory, current velocity and algal regrowth: how does periphyton grow when the grazers have gone? Freshwater Biology 51 (11), 2114–2123.

Whitford, L.A., Schumacher, G.J., 1961. Effect of current on mineral uptake and respiration by a fresh-water Alga. Limnology and Oceanography 6, 423–425.

Wilbur, H.M., 1977. Interactions of food level and population density in Rana Sylvatica. Ecology 58 (1), 206–209.

Wolska, M., 2013. Parasites of zooplankton and periphyton assemblages in the littoral zone of lakes in Drawa National Park, Poland. Acta Mycologia 48 (1), 51–59.

Wood, P.J., Armitage, P.D., 1997. Biological effects of fine sediment in the lotic environment. Environmental Management 21 (2), 203–217.

Wu, Y., Zhang, S., Zhao, H., Yang, L., 2010. Environmentally benign periphyton bioreactors for controlling cyanobacterial growth. Bioresource Technology 101 (24), 9681–9687.

Xenopoulos, M., Schindler, D., 2001. Physical factors determining ultraviolet radiation flux into ecosystems. In: Cockell, C.S., Blaustein, A.R. (Eds.), Ecosystems, Evolution, and Ultraviolet Radiation. Springer, New York, pp. 36–62.

Yin, L., Huang, J., Li, D., Liu, Y., 2005. Microcystin-RR uptake and its effects on the growth of submerged macrophyte *Vallisneria natans* (lour.) hara. Environmental Toxicology 20 (3), 308–313.

Young, R.G., Huryn, A.D., 1999. Effects of land use on stream metabolism and organic matter turnover. Ecological Applications 9 (4), 1359–1376.

Zanon, J.E., Simoes, N.R., Rodrigues, L., 2013. Effects of recurrent disturbances on the periphyton community downstream of a dammed watercourse. Brazilian Journal of Biology 73 (2), 253–258.

Chapter 3

Indicators for Monitoring Aquatic Ecosystem

3.1 INTRODUCTION

Aquatic environments have been disturbed by human activities for millennia, increasingly resulting in deterioration of food chains and, ultimately, the destruction of aquatic food webs. During the past few decades, research has focused on the development of robust and feasible methods to accurately reflect changes in natural aquatic systems and to estimate the rates of inundation of toxic chemicals into these environments. A number of physical, chemical, and biological methods have already been established for these purposes, but biological assessment methods are generally the most proficient.

As periphytic biofilms are widely distributed in aquatic ecosystems and form the basis of all aquatic food webs, they are increasingly being used to assess and monitor the health of these systems.

3.2 ASSESSMENT OF WATER QUALITY

Historically, water quality was assessed by comparing the chemistry of the water downstream of point sources with that of the point source itself. This approach, however, ignores many instream processes such as the response of biological entities to chemical changes, the conversion of these chemicals into other compounds, variations in chemical concentration in different parts of streams, and physical alterations. Ideally, water quality should be determined using physical, chemical, and biological variables, but these traditional processes are arduous and complicated by the fact that wastewater may consist of various types of organic and inorganic chemicals such as pesticides, pharmaceuticals, nutrients, and metallic compounds. Biological methods are generally more reliable, save time, and are cost-effective.

Water quality in fluvial ecosystems is influenced by anthropogenic activities, and determining the extent of the damage is necessary for remediating streams. Water quality and quantity deteriorate with increases in pollution

Periphyton. http://dx.doi.org/10.1016/B978-0-12-801077-8.00003-X

(Douterelo et al., 2004). The use of biological methods to estimate water quality was introduced a century ago (Kireta et al., 2012) but was not practically adopted until 40 years later. The United States Environmental Protection Agency (US EPA) recommended the establishment of criteria to biologically estimate water quality and to identify pollutants (EPA, 1999, 2000). Biological assessment of aquatic environments is based on monitoring the concentration of different chemicals and their link to biological entities, that is, the effect of the chemicals at suborganism and organism levels under standardized conditions (Rotter et al., 2011). Periphyton composed of bacteria, algae, protozoa, and detritus grown on different substrates in open water bodies are used for biomonitoring of water quality. Periphyton are preferred as a biological tool for several reasons (Biggs, 1985; Feminella, 2000; Jan Stevenson and Peterson, 1991) such as:

1. Sessile habitat, making it impossible for periphyton to avoid contaminants,
2. Miscellaneous/biodiverse nature, meaning it will respond to a wide range of contaminants,
3. Rapid rate of recolonization after being disturbed by external forces such as changes in flow or water quality.
4. Easy to handle and analyze during monitoring.
5. Ubiquitous in aquatic environments.

Periphyton have been used as indicators for decades, with an aboriginal saprobien system reported from 1908 (Hill et al., 2000b). The taxonomic structure of algal communities in periphytic biofilms is influenced by many environmental variables including substrate type, temperature variability, light vacillations, current regimens, nutrient levels, grazing pressure (Cattaneo, 1983; Fairchild and Sherman, 1992; Horner and Welch, 1981; Hudon and Bourget, 1983; Steinaman and McIntire, 1986), and the chemical characteristics of water.

The historic use of biological integrities to determine water quality is described in Feminella (2000). Water quality has been monitored by various kinds of biological entities, such as fish (An et al., 2002; Karr, 1981), plants (Rothrock et al., 2008), diatoms (Kireta et al., 2012; Seele et al., 2000), macrophytes (Moore et al., 2012), phytoplankton (Fano et al., 2003; Xu et al., 2001), birds (Sorace et al., 2002), macrobenthos (Gabriels et al., 2010), periphyton (Griffith et al., 2005), and other taxa (Wefering et al., 2000). Periphyton are usually preferred to other taxonomic groups due to their short generation times (24 h to several weeks) (Rott, 1991), compared with weeks to years in macroinvertebrates (Wallace and Anderson, 1996) and more than 1 year for fish (Schlosser, 1990), birds, and plants. Diatoms, cyanobacteria, and/or biofilm as a whole have been used by different researchers to evaluate water quality using different biological indices. Some of the research on these different periphyton components is described next.

3.2.1 Diatoms

Diatoms are one of the most extensively used biological indicators for wastewater assessment due to their wide distribution and variability relating to water quality. Diatoms are commonly used as bioindicators due to their (1) importance in food webs, (2) role in oxygenation and biogeochemical linkages; (3) effectiveness as bioindicators for numerous physical, chemical, and biological contaminants, (4) ease of use and identification compared with other algal groups, and (5) inexpensive methodologies. Diatoms have been used globally to determine river acidification (ter Braak and van Dame, 1989), organic pollution (Descy and Coste, 1991), eutrophication (Kelly and Whitton, 1995), and metal pollution gradients (Genter and Lehman, 2000; Rushforth et al., 1981; Sabater, 2000) for more than a decade.

Diatoms have been used as indicators in various European countries (Kelly and Whitton, 1998; Whitton et al., 1991), North America (Lowe and Pan, 1996; Stevenson et al., 1999), and Australia (Chessman et al., 1999). Diatom indicators fall into three basic groups: (1) those based on the indicator species concept (Hill et al., 2000a), (2) those based on the concept that pristine environments foster greater biodiversity than contaminated environments (Hill et al., 2000a), and (3) multimetric indicators based on several indices (Tang et al., 2006). Depending on the approach, different metrics are used in different studies. Species richness, species dominance, acidobiontic diatoms, eutraphentic diatoms, mobile diatoms, chlorophyll, biomass, and phosphatases were used to monitor various streams in the eastern United States (Hill et al., 2003), while the Kentucky Department for Environmental Protection used the total number of diatom taxa (TNDT), Shannon diversity (H′), pollution tolerance index (PTI), Siltation Index (%NNS), Fragilaria group richness (FGR), and Cymbella group richness (Division, 2002).

Higher rates of perturbation in fluvial ecosystems cause many changes; one of the foremost changes is eutrophication. Eutrophication can cause obscured water conditions, making it difficult to estimate water quality. Some dominant diatom taxa can be used in these perturbed conditions (Hofmann, 1996), such as *Achnanthes minutissimai* and/or *Amphora pediculus*. These diatoms are naturally found in eutrophic conditions and can be used to discern organic pollutants in aquatic environments (Steinberg and Schiefele, 1988) and improvements in water quality (Kwandrans et al., 1998). Periphytic diatoms are also valuable in assessing changes in fluvial ecosystems as a result of point source inorganic nutrient pollution from sewage discharge. This type of pollution can result in community shifts toward more tolerant diatom taxa and less variability in diatom diversity, especially in winter (Dela-Cruz et al., 2006). Diatoms are very sensitive to changes in pH, salinity ,and other chemical aspects of their environment, so these physical and chemical variables must be considered when selecting diatom taxa for monitoring water quality.

3.2.2 Cyanobacteria

Cyanobacteria, often referred to as blue green algae, are slimy films found in both clean and contaminated water. Aboal (1989) identified cyanobacteria as one of the dominant organisms in fluvial ecosystems, and they were later identified as dominant organisms in aquatic systems perturbed by eutrophication (Pizzolon et al., 1999) as well as in urban wastewater with high coliform levels (Vis et al., 1998). This variation is due to the diversity of organisms within cyanobacteria whose abundances vary in different environments. Some of these species dominate clean water environments, whereas the dominance of other species vary according to type of pollutant and the pollution index. For example, *Oscillatoria putrida*, *O. tenuis*, and *O. lauterboni* are used as pollution indicators, while *Rivularia* is the dominant genus in clean water environments (Douterelo et al., 2004).

Nutrient perturbations also have a profound effect on cyanobacterial biofilm. According to Douterelo et al. (2004), a decrease in species richness and diversity was observed in areas with higher nutrient concentrations and higher pollution levels, usually downstream of the pollution source, with Oscillatoriales the dominant organisms. In upstream areas, with relatively lower nutrient concentrations, Nostocales were dominant. Soluble reactive phosphate can disrupt the concentration of phycobiliprotein, indicating a decline in cyanobacterial biomass. Using phototrophs is a feasible mechanism for elucidating stream contamination. In mining areas, low abundance and diversity of epilithic diatoms indicated stressed conditions and high metal pollution (Nunes et al., 2003). At high metal concentrations, some morphological abnormalities have been observed in diatom valves (Nunes et al., 2003). These abnormalities will be discussed in more detail later in this chapter.

3.2.3 Periphytic Ciliates

Little work has been done on the use of periphytic ciliates for water quality assessment due to the fact that this method is more time consuming and restrictive than methods involving diatoms and cyanobacteria. Periphytic ciliates are relatively immobile and have been used to evaluate the ecological impacts of aquatic contamination in a number of studies (Gong and Song, 2005; Mieczan, 2010; Norf et al., 2009; Xu et al., 2009).

3.2.4 Biofilm

Periphytic biofilms are sensitive to perturbations in fluvial ecosystems with the resultant community shifts dependent on the contaminant. These shifts have been observed in laboratory and large scale studies of a variety of natural and artificial contaminants. When biofilm dominated by diatoms was exposed to polychlorinated biphenyls, it shifted to a community co-dominated by different kinds of cyanobacteria (blue-green algae) and a single diatom taxon (Kostel et al., 1999). According to this study, structural changes are better indicators than biomass (chlorophyll a and biovolume), cell number , or the presence/absence of particular species for monitoring contamination and its impacts. Most of the studies involved in monitoring ecosystem quality are based on the determination of pollution-induced community tolerance (PICT), which will be discussed in more detail later in this chapter.

3.2.5 Biological Indices

Contamination of the fluvial ecosystem leads to variability in certain periphytic biofilm characteristics and it is variations that allow periphyton to be used as a valuable water quality monitoring tool. These variations include biomass (Biggs, 1989), diversity indices (Stevenson, 1984; Weitzel and Bates, 1981), biotic indices (Descy and Coste, 1991; Kelly et al., 1995), algal community size (Cattaneo et al., 1995), and taxonomic composition (Archibald, 1972). Biological indices are increasingly preferred due to their ability to reflect real-world conditions and form the foundation of river management (Meng et al., 2009).

An index can be defined as a numerical value that helps identify changes in biological characteristics in relation to disturbances and pollution (Barbour et al., 1999). A variety of biological indices are used to determine the biological quality of aquatic environments such as multimetric indices, univariate indices, standard zoological and botanical indicators, and predictive models. Biological indicators can be used to predict multiple characteristics of the aquatic environment at a large scale in natural environments, identify biological restoration goals, and monitor implementation of remediation measures and acquiescence to environmental regulations, for establishing standard for Natural Resource Damage Assessments and for framing predictive (Simon, 2000).

Multimetric biotic indices and multivariate analyses are increasingly used to monitor the response of specific stressors on diatoms (Gerritsen, 1995; Norris, 1995). Multimetric indices are used to elucidate the biological factor most affected by specific contaminant and this factor can then be used for the assessment of damage caused by certain contaminants. Multivariate indices are used with environmental data rather than for impact assessment. Multivariate analysis techniques are more objective when the history of the contamination site and information about the biota is scarce (Gerritsen, 1995), while multimetric indices are based on the history of monitoring approaches and the response of biological factors to human impact (Karr and Chu, 1997). Some researchers have argued that multimetric indices reduce the data into a single unit with no clear evidence about statistical behavior and distribution (Norris, 1995; Suter, 1993). The possibility of inappropriate results is more common in multivariate indices. According to Karr and Chu (1997), however, multimetric indices are a significant tool for monitoring ecological systems and can be tailored with other indices to increase their usefulness. The approach used for a particular bioassessment should be the one that demonstrates a realistic relationship between environmental stressors and metrics with the metrics fostering a conclusive composite index (Hill et al., 2003). The river diatom index (RDI) is a multimetric index, used for nine different attributes of diatoms (percentage of valves belonging to species sensitive to disturbance, percentage of valves belonging to species very tolerant of disturbance, eutrophic species richness, percentage of valves belonging to nitrogen heterotrophic species, percentage of valves from motile genera and percentage of deformed valves). This index has been successfully used for measuring the effect of human disturbances at site and catchment levels (Fore and Grafe, 2002).

The practicality of diatom indices for monitoring and management of streams is related to the index attributes and indices should respond to both single and multiple variables (Hill et al., 2003). Different diatom-based models have been implemented to estimate the effect of different stressors such as pH (Birks et al., 1990), total phosphorus (Bennion et al., 1996), water clarity (Dixit and Smol, 1994), water temperature (Pienitz et al., 1995), and chlorophyll a (Jones and Juggins, 1995). A number of diatom indices have been developed in European countries to assess the biological quality of running waters. The most common indices are, in France, the Specific Polluosensitivity Index (SPI0, the Generic Diatom Index (GDI), and the standardized Biological Diatom Index (BD); in Belgium, the Leclercq and Maquet Index (ILM); in the United Kingdom, the Trophic Diatom Index (TDI); in Germany, the SHE index; in the Czech Republic, the Sládeček index (SLA); in Italy, the Eutrophication Pollution Index using Diatoms (EPI); in Austria, the Saprobic Index of Rott (ROT); and at the European level, European Economic Community Index (EEC) along with Diatom Assemblage Index for organic pollution (DAI) (Alvarez and Peckarsky, 2014; Barranguet et al., 2005). ROT and SLA are used for the detection of organic pollutants, whereas TDI, EPI, and EEC are used for trophic pollution (Debenest et al., 2009). According to Reavie et al. (2010) and Lavoie et al. (2011), the results of metrics are more accurate and appropriate when absolute (cell density or biovolume) rather than relative (percentage) abundance data are used. According to Feio et al. (2009), EEC is more efficient in monitoring changes caused by land use and nutrients, BDI is useful for useful for nutrients and organic contamination, and SPI is useful for nutrients and organic contamination. EEC and BDI are more efficient than SPI and GDI. Porter et al. (2008) implemented a number of algal indices to estimate the quality of water in 976 streams and rivers in the US Geological Survey's National Water-Quality Assessment Program during 1993–2001 as mentioned in Table 3.1.

Almeida et al. (2014) attempted to overcome the problems found when intercalibrating data from different European countries by using diatoms for

TABLE 3.1 Algal Metrics With One or More Significant Correlations With Nutrient and Suspended-Sediment Concentrations

Algal Metric	Definition	Attribute	Class
Cellden	Cell density (total)	Standing crop	Cell density (cells/cm^2)
Biovol	Biovolume (total)	Standing crop	Biovolume (cm^3/m^2)
Taxarich	Taxa richness	Diversity	Taxa richness
SP_OL	Saprobity, oligo	Saprobity	Oligosaprobous diatoms
SP_BM	Saprobity, beta-meso	Saprobity	β-Mesosaprobous diatoms
SP_AM	Saprobity, alpha-meso	Saprobity	α-Mesosaprobous diatoms
SP_AP	Saprobity, alpha-meso	Saprobity	α-Mesoolysaprobous diatoms)
SP_PS	Saprobity, polysaprobous	Saprobity	Polysaprobous diatoms
ON_AL	Organic nitrogen, autotrophic, low	Nitrogen-uptake metabolism	Low nitrogen autotrophic diatoms
ON_AH	Organic nitrogen, autotrophic, high	Nitrogen-uptake metabolism	High nitrogen autotrophic diatoms
ON_NH	Organic nitrogen, nitrogen heterotrophic	Nitrogen-uptake metabolism	Nitrogen heterotrophic diatoms
NF_YS	Nitrogen fixer	Nitrogen-fixing algae	Nitrogen fixer
TR_O	Trophic, oligotrophic	Trophic state	Oligotrophic diatoms
TR_M	Trophic, mesotrophic	Trophic state	Mesotrophic diatoms
TR_E	Trophic, eutrophic	Trophic state	Eutrophic diatoms
ES_SF	Eutrophic soft algae	Trophic state	Eutrophic soft algae
Eutrophic	Eutrophic algae	Trophic state	Eutrophic algae
PC_MT	Pollution class, most tolerant	Pollution class	Most tolerant diatoms
PC_LT	Pollution class, less tolerant	Pollution class	Less tolerant diatoms
PC_SN	Pollution class, sensitive	Pollution class	Sensitive diatoms
PC_VT	Pollution tolerance, very tolerant	Pollution tolerance	Very tolerant diatoms
PC_TA	Pollution tolerance, tolerant (A)	Pollution tolerance	Tolerant diatoms

Continued

TABLE 3.1 Algal Metrics With One or More Significant Correlations With Nutrient and Suspended-Sediment Concentrations—cont'd

Algal Metric	Definition	Attribute	Class
PC_TB	Pollution tolerance, tolerant (B)	Pollution tolerance	Tolerant diatoms
PC_LA	Pollution tolerance, less tolerant (A)	Pollution tolerance	Less tolerant diatoms
PC_LB	Pollution tolerance, less tolerant (B)	Pollution tolerance	Less tolerant diatoms
SL_FR	Salinity, fresh	Salinity	Fresh water diatoms
SL_FB	Salinity, fresh-brackish	Salinity	Fresh-brackish water diatoms
SL_HB	Salinity, brackish	Salinity	Brackish water diatoms
BS_SE	Bentho/sestonic, sestonic	Habitat	Sestonic algae
MT_YS	Motility	Motility	Motile algae
OT_AH	Oxygen tolerance, always high	Oxygen requirements	Continuously high (diatoms)
OT_FH	Oxygen tolerance, fairly high	Oxygen requirements	Fairly high (diatoms)
OT_MD	Oxygen tolerance, moderate high	Oxygen requirements	Moderate (diatoms)
OT_LW	Oxygen tolerance, low	Oxygen requirements	Low (diatoms)
OT_VL	Oxygen tolerance, very low	Oxygen requirements	Very low (diatoms)

From Porter, S.D., Mueller, D.K., Spahr, N.E., Munn, M.D., Dubrovsky, N.M., 2008. Efficacy of algal metrics for assessing nutrient and organic enrichment in flowing waters. Freshwater Biology 53 (5), 1036–1054.

assessments of Mediterranean rivers. Four diatom indices (Indice de Polluosensibilité Spécifique (IPS), Indice Biologique Diatomées (IBD), Intercalibration Common Metric Italy (ICMi), and Slovenian Ecological Status assessment system) were intercalibrated using data from six European Mediterranean countries (Cyprus, France, Italy, Portugal, Slovenia, and Spain). They described high/good and good/moderate quality on the basis of the Intercalibration Common Metric (ICM). In good water quality conditions *Achnanthidium minutissimum* sensu lato was the dominant species, whereas in moderate conditions, *Planothidium frequentissimum, Gomphonema parvulum,* and *Nitzschia palea* were dominant. This work provides the benchmark for attaining a common boundary for the assessment of water quality among different European countries.

There are many indices used to estimate the water quality, but other approaches can also be used to analyze certain characteristics and toxins that are not detectable by taxonomic approaches such as ash-free dry mass and chlorophyll a. For example, some toxic pollutants cause lethal effects that cannot be readily detected by taxonomic analysis and might be better detected with lower biomass and chlorophyll a contents (Hill et al., 2000b). Chlorophyll a and the Autotrophic Index (AI) have been consistently used for more than 20 years to investigate water quality in Eagle River. AI, however, has certain drawbacks such as results being overexaggerated by the accumulation of allochthonous organic debris and heterotrophs being the dominant forms during the initial stages of community succession (Biggs, 1989).

3.3 EFFECTS OF TOXINS

In the past few decades, many pollutants and commercial chemicals have been demonstrated to affect aquatic communities. These contaminants include pesticides, herbicides, and personal care products that become part of fluvial ecosystems either due to run off or human disturbance. Toxicity is a part of environmental risk assessments that helps to determine the effects of toxicants on ecological communities (Cowan et al., 1995). Single species concepts to determine toxicity have been criticized due to the lack of realism in the predictions of consequences after exposure to toxicants (Belanger, 1994). Most recent research has focused on the effects of toxicants on multispecies aquatic communities. As microbial communities form the foundation of all food webs and biogeochemical cycles and can alternate and perturb other communities, it is imperative to determine the effects of toxins on this group of organisms (Guckert, 1996).

Chemical pollution is one of the leading problems in aquatic environments (Schwarzenbach et al., 2006) and can affect all trophic levels in aquatic ecosystems (Geiszinger et al., 2009). Microorganisms are ideal for community-level studies due to their size and short generation time (Petchey et al., 2002). Thick periphytic biofilm mats of microorganisms on different substrates in the phototrophic zone are composed of bacteria, fungi, algae, protozoa, and metazoa (Cooke, 1956). These biofilms are microecosystems, with clearly defined trophic structures, biotic diversity, and material cycles (Hoagland et al., 1982). One of most drastic effects of toxicants observed in periphytic biofilms is a shift in species dominance and abundance (Hansen, 1987; Powers et al., 1982). According to Kostel et al. (1999), when a periphytic community was exposed to polychlorinated biphenyls, it shifted from a community dominated by diatoms to one co-dominated by diatoms and cyanobacteria. The decrease in the proportion of diatoms from 81% to 68% was due to inhibition of the diatoms ability to divide and multiply in number.

PICT is a concept used to determine the extent of toxicity to microbial communities as a result of exposure to contaminants (Blanck and Dahl, 1996; Wängberg, 1995). Community structure and PICT are more reliable in defining the effect of toxicity than photosynthesis and net production of periphyton. A decrease in the sensitivity of marine periphytic communities was observed in conjunction with a decrease in concentration of TBT (due to its ban) on the Swedish west coast demonstrating that PICT can be specifically used to detect even minor community changes (Blanck and Dahl, 1996). Tropical estuarine biofilms composed of diatoms, filamentous brown algae, and cyanobacteria developed a community level tolerance to diuron, a herbicide known to inhibit photosystem II. Four weeks' exposure to diuron resulted in community shifts toward diatom dominance. Chronic exposure (6.5 μg/L) led to PICT, and these community shifts could not be recovered in uncontaminated water (Magnusson et al., 2012).

Community-level tests are better than single species tests as they increase the probability of detecting sensitive species (Blanck et al., 2009). Irgarol 1051 is a degradation-resistant algicide that is highly noxious to marine algae. Due to its high stability, it can cause significant damage to nontarget marine algal communities (Dahl and Blanck, 1996). In periphytic communities, the no-effect concentration of Irgarol is 44–215 ng/L. After long-term exposure, periphytic communities become tolerant to Iragol and survive even at 10 nmol/L concentrations. This increase in tolerance in some species might be due to two reasons: selection pressure or mutagenic effects of Iragol (Blanck et al., 2009).

Metabolic changes, such as photosynthesis and respiration rates, are more appropriate for determining short-term changes in periphytic communities than structural changes that need prolonged exposure at low concentrations. For arsenate contamination, photosynthetic changes are more evident than respiratory changes as arsenate targets photosynthetic machinery and phosphorylation (Melandri et al., 1970; Slooten and Nuyten, 1983). When periphyton sensitivity (based on photosynthetic inhibition) was determined in the presence of the pesticide prometryn, more tolerant species were found at polluted sites. The EC_{50} value of polluted sites was five times lower than control sites (Rotter et al., 2011).

Flowing waters are exposed to multiple contaminants and, currently, many researchers are focusing on the consequences of multiple toxins on biological communities in fluvial ecosystems. Concentration addition (CA) (Arrhenius et al., 2004) and independent action (IA) (Bliss, 1939) are two important concepts that help predict combined toxicity from the toxic effects of individual compounds (Greco et al., 1992). In a mixture of substances, those affecting similar mechanisms can be substituted by each other without changing the inclusive effects (Arrhenius et al., 2004). CA has been used for individual species such as the marine bacterium *Vibrio fischeri* (Backhaus et al., 2000a,b; Faust et al., 2000), the unicellular green freshwater alga *Scenedesmus vacuolatus* (Backhaus et al., 2004; Faust et al., 2000), water flea *Daphnia magna* (Hermens et al., 1984a), and guppies (*Poecilia reticulata*) (Hermens and Leeuwangh, 1982; Hermens et al., 1984b; Könemann, 1980, 1981). IA has been used for multiple species, such as the toxicity of dissimilar-acting chemicals on *V. fischeri* and *Scenedesmus vacuolatus* (Backhaus et al., 2000a; Faust et al., 2000, 2003) and for chemically heterogeneous mixtures (Hermens et al., 1985; Walter et al., 2002). According to Arrhenius et al. (2004), CA was more useful than IA for predicting the effects of 12 congeneric phenylurea toxic compounds (Buturon, Chlorbromuron, Chlortoluron, Diuron, Fenuron, Fluometuron, Isoproturon, Linuron, Metobromuron, Metoxuron, Monolinuron, Monuron), but questions have been raised about its applicability for mixture toxicity. These questions require further investigation on the applicability of CA for noncongeneric chemicals and the long-term effects of toxic compound mixtures.

Active biomonitoring of a system is based on translocation of organisms and their biochemical, physiological, and/or organismal responses (de Kock and Kramer, 1994). Biomonitoring has already been conducted on a variety of algal communities to examine recovery and stress responses (Dorigo et al., 2010a,b; Gold et al., 2002; Hirst et al., 2004; Morin et al., 2010; Rimet et al., 2005). In situ studies have provided useful information regarding fluctuations in exposure and stressor interactions, which are not possible using laboratory-scale studies of a single species (Rotter et al., 2011). Therefore, in situ studies of indigenous communities are more realistic approaches for biomonitoring (Crane et al., 2007).

The development, structure, taxonomic diversity, and functioning of periphytic biofilms are affected by both physical and chemical factors in a stream (Biggs, 2000; Sabater et al., 1998; Villeneuve et al., 2010). Chemical modification takes place via either nutrient (Ludwig et al., 2008; Luttenton and Baisden, 2006) or toxic pollutant inputs (Guasch et al., 2003; Pesce et al., 2008). Pesticides (insecticides, herbicides, fungicides, etc.) are the most commonly found organic compounds in flowing waters (Azevedo et al., 2000; Saenz and Di Marzio, 2009), and their harmful effects on taxonomic composition and species richness, as a result of selection pressure, have been discussed by many authors (Dahl and Blanck, 1996; DeLorenzo et al., 1999; Peres et al., 1996).

The type of response of periphytic biofilm to a certain contaminant also depends on the community composition at the time of contamination (Guasch et al., 1997; Pesce et al., 2006; Tlili et al., 2008). According to the "biodiversity-stability hypothesis," ecosystem resistance and resilience depend on the trophic interactions among community members (McCann, 2000). A similar hypothesis was described by Yachi and Loreau (1999), known as the "biological insurance hypothesis." According to this concept, ecosystems with higher biodiversity are more resistant to disruptions and impairments and, according to their second hypothesis, more diverse biofilms are more resistant to pesticide contamination. This hypothesis was verified by Pesce et al. (2006), who postulated that the response of autotrophic periphytic communities to diuron depends on the initial composition of the biofilm rather than on number of species present. According to (Villeneuve et al., 2011), an increase in species richness has no effect on the tolerance level of periphyton to pesticide mixes at low concentrations.

Resistance of a population against a certain contaminant is the result of selection pressure and decreased resistance to a second tolerant might result from the damage caused by the first contaminant (Luoma, 1977). Arsenate exposure produced decreases in biomass accrual and species composition resulting in the development of a more tolerant community through the replacement of sensitive species (Blanck and Wängberg, 1988a). In fresh waters, species are more sensitive to short- or long-term arsenate exposure than marine communities (i.e., freshwater communities are affected by arsenate levels 10 times lower than those affecting marine species) (Wängberg et al., 1991).

3.3.1 PICT

The basic concept behind PICT is the replacement of sensitive species by more tolerant species, otherwise known as toxicant-induced succession, or TIS (Blanck, 2002). Time of exposure to the toxicant is one of the major factors affecting selection pressure on the different species (Blanck and Wängberg, 1988b). PICT determination using periphyton has proven to be a robust process for differentiating between contaminated and unpolluted sites and helps to improve the stressor–effect diagnosis. PICT has the potential to improve the accuracy and relevance of weight-of-evidence–based frameworks as described by European Water Framework Directive (Rotter et al., 2011).

PICT can be affected by (1) target community, (2) intensity of toxicant exposure, (3) the species successional stage, and (4) the physicochemical characteristics of the study site (Tlili et al., 2011a). PICT is also affected by factors inherent in the natural succession of biofilm. Different species and communities (e.g., heterotrophs vs. autotrophs) in a biofilm respond differently to the same contaminants. Tlili et al. (2011a) reported that autotrophic and heterotrophic communities had different responses to Zn contamination and that the responses also differed between different components within the same communities. Mature communities are more tolerant to metal contamination and metal has the same effects on short- or long-term pollution tolerance.

The mechanisms involved in pollution tolerance are mainly of two types: abundance of tolerant and reduction of sensitive species (Pesce et al., 2010a; Schmitt-Jansen and Altenburger, 2005) and tolerance acquisition of species to pollution (Ivorra et al., 2002). According to the PICT concept, biological communities undergo certain changes due to selection pressure when exposed to contaminants at critical concentrations for specific time periods, which result in the elimination of sensitive species (Tlili et al., 2010). There is a paucity of information on the toxic effects of contaminants at higher biological levels; therefore, it is difficult to extrapolate results from single species tests (Seitz and Ratte, 1991). Community-based studies are more appropriate for understanding the ecologically relevant information from aquatic environments using PICT.

The PICT concept has been widely accepted and used in the past few decades. The remainder of this section will discuss the effects of herbicides and metals on periphyton including the relationship with PICT.

3.3.2 Herbicides and Pesticides

Biofilm either absorb toxic chemicals or degrade them into intermediate products. Atrazine and diclofop methyl are absorbed and then degraded by periphytic biofilm acting as a sink (Lawrence et al., 2001; Margoum et al., 2007). These studies reveal the effect of chronic contamination of biofilm in a small stream by pesticide diffusion. According to Molander and Blanck (1992), the effect of diuron on biofilm is a three-stage process. Initially, no effects were observed; then, an effect was observed in biomass and structure followed by the elimination of sensitive species and decline in biomass and diversity, leading to easy detection of PICT.

When isoproturon and *S*-metolachlor, two major herbicides used in French agriculture, were tested on periphytic biofilm at low (5 μg/L) and high (30 μg/L) concentrations with different recovery periods, complete decreases in chlorophyll a concentration and live cell density were observed. The recovery time was shorter for biofilms exposed to *S*-metolachlor than to isoproturon. *Melosira varians*, *Nitzschia dissipata,* and *Cocconeis placentula* were resistant species, while *Eolimna minima* and *Navicula reichardtiana* were sensitive. Isoproturon favored facultative heterotroph species by inhibiting photosynthesis, while *S*-metolachlor disturbed cell multiplication more than photosynthesis (Debenest et al., 2009).

The response of periphytic diatoms to different pesticides is variable, dependent on compounds (Eullaffroy and Vernet, 2003; Nyström et al., 1999) and their combinations (Hatakeyama et al., 1994), concentrations (Abdel Hamid et al., 1996), the species and strains involved (Lockert et al., 2006; Nyström et al., 1999), and experimental conditions (Tlili et al., 2008).

Physicochemical and environmental conditions of streams also affect the response of diatom communities to contaminants such as pesticides (Morin et al., 2009). Toxicity is also influenced by nutrient inputs, especially phosphorus (Ivorra et al., 2002; Kamaya et al., 2004). Periphytic communities show a relationship between light and atrazine toxicity (i.e., increases in sensitivity under higher light conditions) (Guasch et al., 2003, 1998, 1997; Guasch and

Sabater, 1998). An increase in pulse frequency of light resulted in a decrease in periphyton dry weight and chlorophyll a (Bere and Tundisi, 2012). Global climate warming can affect the toxicity of compounds by altering the periphyton community composition before exposure to the toxicant, resulting in changes to organism resistance and recovery from toxicant exposure (Moe et al., 2013).

Villeneuve et al. (2011) investigated the effect of different hydraulic regimens on structure, diversity, and functioning of periphytic biofilms after exposure to a mixture of two pesticides: diuron and azoxystrobin. Two types of hydraulic regimes were taken into account: turbulent with high variation vs. laminar with low variation. Pesticides affected the growth dynamics and primary production of the reference communities, but changes in flow regimens had a more profound effect than pesticides; a denser bacterial community was associated with laminar flow. When the combined effect of these two factors was observed, more diversified communities were found in turbulent conditions, but this greater variability in periphytic biofilm did not enhance the tolerance of the community to pesticides.

Limiting the availability of phosphorous increased the tolerance of biofilm to Cu toxicity (Guasch et al., 2004) but had no effect on community tolerance to atrazine (Guasch et al., 2007) or diuron (Pesce et al., 2010c). Gustavson et al. (2003) determined the effect of exposure to four herbicides—metribuzin, hexazinone, isoproturon, and pendimethalin—on periphytic biofilm and its recovery duration. Biofilms were affected at lower concentrations of isoproturon, metribuzin, and hexazinone than concentrations noted for single-species growth tests. The EC_{50} of isoproturon and metribuzin declined 1- to 2-fold after 24 h, while hexazinone stimulated photosynthetic activity after the first hour and inhibited it after 24 h.

The extent of disturbance on periphyton dominance and diversity depends on the time of exposure to contaminants. Acute exposure for shorter periods can have a more significant impact on the community than exposure to low levels of contaminants for a longer duration. Lower concentrations (0.056–0.56 mmol/L) of the herbicide atrazine fostered a shift in algal community, with more sensitive species disappearing and tolerant species dominating the community structure. Exposure to higher concentrations disturbed all species in the community, resulting in a more sensitive algal community (Nyström et al., 2000). Similar results were postulated in a study when periphyton within limnocorals were exposed to atrazine. Within 3 weeks of exposure, a 36–57% decline in organic matter, chlorophyll a, and algal biomass was observed followed by a decrease in carbon assimilation. Cyanophyta disappeared completely, while some species of Chlorophyta decreased and others increased. Biomass level was maintained at higher levels than the control due to members of Bacillariophyceae. A 36–67% decrease in biomass, oxygen demand, and NO_3–NO_2–N due to photosynthetic inhibition and amplified demand of nutrients was observed after the second exposure to atrazine. This study also demonstrated the inability of the algal biomass to recover to original conditions up to 294 days after the second exposure (Herman et al., 1986).

When diuron was used as a toxicant against periphyton at a concentration of 1 μg/L, chlorophyll a concentration and carbon intake increased (Tlili et al., 2008). Diuron also affected the photosynthetic process by disturbing electron flow in PSII during light reactions, but phototrophs are able to adjust intracellular light-harvesting pigments (e.g., chlorophyll a) and convert light energy to chemical energy (Tlili et al., 2011b). Physiological characteristics of photoautotrophic communities (microalgae, cyanobacteria) make them more vulnerable to herbicides that mainly inhibit photosynthetic processes (Pesce et al., 2010b).

Exposure to the herbicide metachlor resulted in community alterations due to greater development of tolerant species such as *Planothidium frequentissimum*, *P. lanceolatum*, *Amphora montana*, *Surirella brebissonii,* and *Nitzschia gracilis*. Some of the species showed prominent deformities, such as frustules of species *Surirella angusta* (Fig. 3.1). The teratogenic effects of

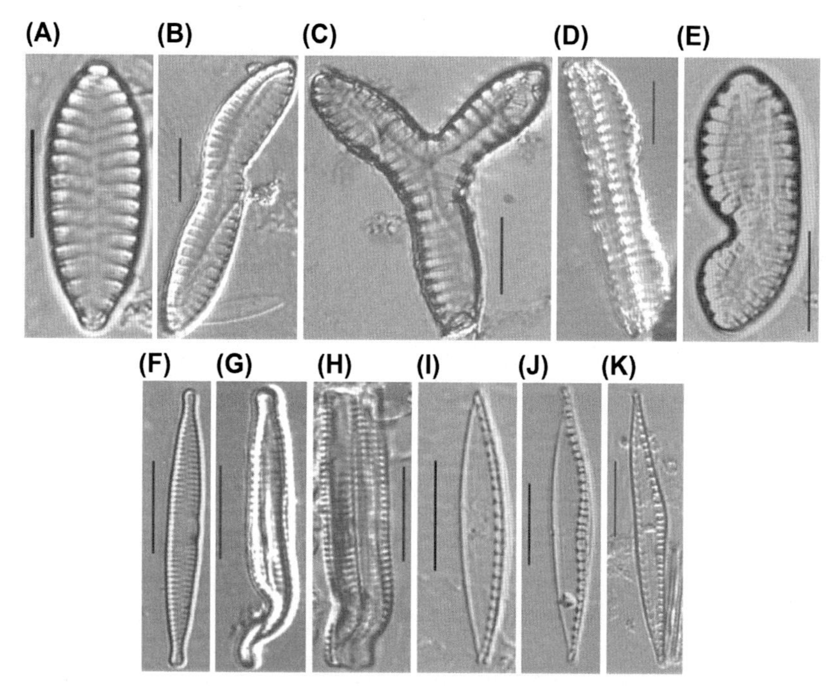

FIGURE 3.1 Abnormal diatom frustules observed in the contaminated experimental units for the species *Surirella angusta* (A–E), *Fragilaria capucina* (F–H), and *Nitzschia sociabilis* (I–K). (A), (F) and (I): normal form of each species; (D) and (H): pair of deformed frustules following vegetative reproduction. *Taken from Roubeix, V., Mazzella, N., Méchin, B., Coste, M., Delmas, F., 2011. Impact of the herbicide metolachlor on river periphytic diatoms: experimental comparison of descriptors at different biological organization levels. Annales de Limnologie – International Journal of Limnology (47), 239–249.*

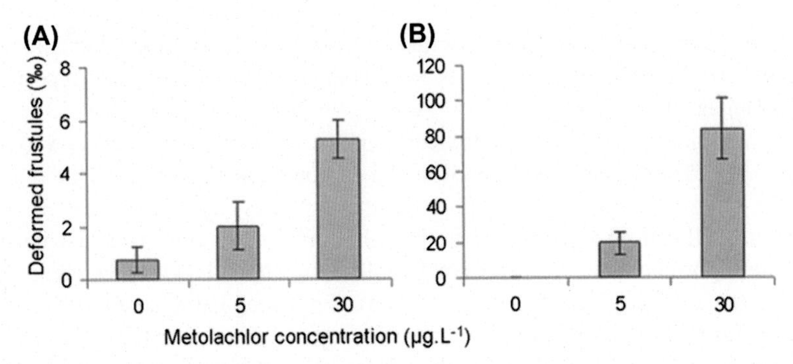

FIGURE 3.2 Relative abundance of deformed frustules in the whole diatom community (A) and within *Surirella angusta* populations (B). *From Roubeix, V., Mazzella, N., Méchin, B., Coste, M., Delmas, F., 2011. Impact of the herbicide metolachlor on river periphytic diatoms: experimental comparison of descriptors at different biological organization levels. Annales de Limnologie – International Journal of Limnology (47), 239–249.*

metachlor on *S. angusta* were confirmed by monospecific acute toxicity tests (Roubeix et al., 2011) with the concentration of metachlor being directly proportional to rate of deformities (Fig. 3.2). The frequency of abnormal forms has already been used as an indicator for heavy metal contamination (Cattaneo et al., 2004), but the use of deformed forms for monitoring contamination is difficult due to the small percentage of deformities (Morin et al., 2009) and greater effort.

3.3.3 Metals

Metal contamination in aquatic environments is an increasingly common and significant problem. The sources of contamination are natural processes and anthropogenic disturbances such as mining, smelting, and agricultural fertilization (Audry et al., 2004; Ruangsomboon and Wongrat, 2006). Periphytic communities are well known for their metal accumulation properties (Clements, 1991; Ramelow et al., 1992). Major processes involved in metal accumulation by algae are (1) binding to EPS, (2) cell surface adsorption (Patrick et al., 1983; Solomon, 2007), and (3) intracellular uptake (Holding et al., 2003). Most studies relating to toxicity of metals for biofilms have considered diatom communities (Gold et al., 2003a,b; Guasch et al., 2003; Soldo and Behra, 2000). In some cases, accumulation was detected in natural periphytic biofilms even when the metal was not detectable in open waters. One such case was reported by Duong et al. (2008), who detected the presence of cadmium (Cd) in biofilm with the Cd accumulation being seasonally dependent. *Eolimna minima*, *Nitzschia palea*, *Encyonema minutum*, *Surirella angusta*, and *Gomphonema parvulum* were Cd-tolerant species, while

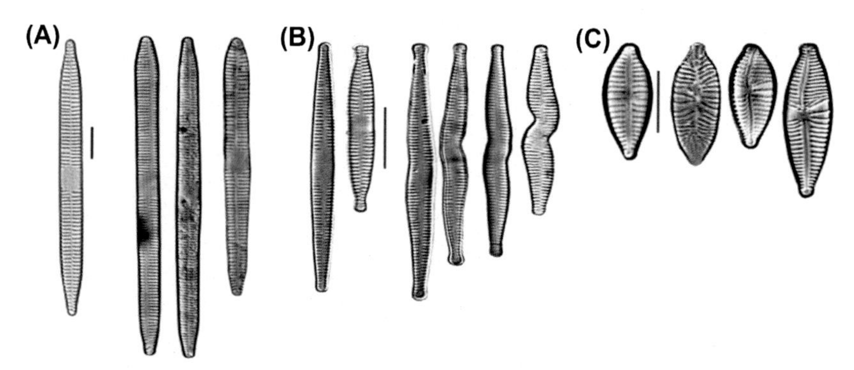

FIGURE 3.3 Abnormal forms of *Ulnaria ulna* (A), *Fragilaria capucina* (B), and *Gomphonema parvulum* (C) found in Joanis Station (first member of each group is a normal form). *From Duong, T.T., Morin, S., Herlory, O., Feurtet-Mazel, A., Coste, M., Boudou, A., 2008. Seasonal effects of cadmium accumulation in periphytic diatom communities of freshwater biofilms. Aquatic Toxicology 90 (1), 19–28.*

C. meneghiniana, *N. gregaria*, *Navicula lanceolata*, *Melosira varians,* and *Nitzschia dissipata* were Cd-sensitive species. Some species had undergone teratological changes at contaminated sites such as *Ulnaria ulna*, *Fraglaria capucina,* and *Comphonema parvulum* (Fig. 3.3).

Copper is a component of algicides and plant herbicides and is one of the most toxic metals in aquatic ecosystems due to its high stability and the fact that its fate is determined by numerous factors. The extent of toxicity of copper to aquatic environments depends on the physicochemical conditions of the stream (Town and Filella, 2000). Cyanobacteria and diatoms are well known for regulating both internal and external copper through the synthesis of organic compounds externally by phytochelatins (Bossuyt and Janssen, 2004; McKnight and Morel, 1979) and internally by P-type ATPase, proteins like copper chaperones (Bossuyt and Janssen, 2004), phytochelatins, and polyphosphates (Knauer et al., 1997). The effects of copper on biofilm are also dependent on physical factors such as current flow; it is more marked at intermediate velocity than at high or slow velocities (Sabater et al., 2002).

Eutrophication and metal contamination are two major problems in many developed and developing parts of the world (Wang and Dei, 2006). The main source of phosphorus input is agricultural activities (Ekholm et al., 2000). There are multiple hypotheses regarding the relationship between metal and nutrient concentrations; some support the concept that increasing nutrients like P can decrease metal toxicity (Harding and Whitton, 1977; Say and Whitton, 1977), while others report nutrient limitation by metals results in abridged algal growth (Paulsson et al., 2002). According to Serra et al. (2010), the addition of P can lead to a 1.6-fold reduction in Cu toxicity and P–Cu interactions decrease Cu availability. pH is another important factor that can influence the toxicity of Cu for periphyton (Guasch et al., 2002) and algae (Starodub et al., 1987) as the amount of available Cu increases at low pH (Guasch et al., 2002).

The harmful effects of Cd on periphytic community development and structure are well described (Bere and Tundisi, 2011, 2012; Duong et al., 2008, 2010; Gold et al., 2003b; Ivorra et al., 2000; Morin et al., 2008a,b). Cd is more toxic to periphytic biofilms than Pb and Pb is more toxic than Cr III (Genter, 1996). When periphytic communities are exposed to all three metals together, Pb and Cr inhibit the toxic effects of Cd. Cd metal is retained within the cells of the periphytic community even after the exposure ceases. As the main place for Cd accumulation and storage are cell walls of periphytic communities, Cd can also indirectly affect periphyton predators (Bere et al., 2012).

Algae are more sensitive to Zn in field studies than in laboratory studies (Say and Whitton, 1977). One of the reasons for the decrease in algal communities after exposure to Zn (1–9 mg/L) is the increase in fungal communities (Williams and Mount, 1965). Zn can change the periphytic community even at concentration levels of 0.05 mg/L, resulting in shifts from diatoms to blue-green algae (Genter et al., 1987).

When periphytic communities from a relatively uncontaminated river (Göta Älv, Sweden) were exposed to zinc (Zn), photosynthesis and thymidine incorporation were inhibited. PICT was determined in terms of EC_{50}, and marked changes were observed at higher concentrations resulting in an increase in tolerance levels due to selection pressure at higher Zn concentrations. Another factor observed in this study was the co-tolerance of the communities to two different metals (Paulsson et al., 2000). The communities in the river were also exposed to high levels of Cd so biofilms developed multitolerance properties, a result observed by other researchers (Gustavson and Wängberg, 1995; Hashemi et al., 1994; Pennanen et al., 1996). Phosphorus concentration also affects this process; at higher phosphorus concentrations, periphyton face less Zn toxicity (Paulsson et al., 2000).

Zn accumulation increased after increasing exposure time, leading to physiological disruptions followed by increases in carotenoids and decreases in biomass (Bonet et al., 2012). Increases in carotenoids might be due to oxidative stress by chemicals as observed in many algae and plants (Ledford and Niyogi, 2005). After a week of exposure, cyanobacteria and green algae dominated diatoms, and a higher catalase activity (CAT) was observed. This increase in CAT might be due to direct exposure to the metal or due to secondary succession resulting from the increase in cyanobacterial and algal communities and inhibition of the diatom community (Bonet et al., 2012).

Bioavailability and toxicity of heavy metals depend on the physiological structure and chemical properties of periphyton (Lombardi et al., 2002). Specification of metal for periphyton can be useful in predicting its toxicity effects for algae (Lombardi et al., 2002, 2007). It has long been established that metal toxicity can inhibit photosynthetic processes by affecting the enzymes involved in the process (Rai et al., 1981). Long-term exposure to Cu can enhance chlorophyll a production but has a negative effect on photosynthetic activity. Long-term diuron exposure results in inhibition of both chlorophyll a concentrations and photosynthetic efficiency. Due to the different modes of action of these two contaminants, bacterial cell density increased after exposure to Cu and showed no effect after exposure to diuron in the absence of P. Under P enrichment conditions, the opposite was observed with an increase in bacterial density after exposure to diuron (Tlili et al., 2010).

Arsenate can induce tolerance in periphytic communities when exposed to concentrations above 0.1–0.3 μmol/L. The tolerance was amplified more than 17,000-fold under higher arsenate stress conditions. This may be due to replacement of arsenate-sensitive species by arsenate tolerant species. This community shift was also discerned in terms of fluctuations in species structure and biomass accretion, one of the basic requirements of PICT (Blanck and Wängberg, 1988a).

3.4 OCCURRENCE OF NUISANCE ORGANISMS

It is well known that eutrophication due to human intervention such as municipal wastewater and agricultural runoff may cause an increase in nutrient levels, usually nitrogen and phosphorus. These nutrient increases are one of the main causes of increases in primary production followed by changes in species composition and nuisance growth of benthic algae. Human interventions have resulted in an increase in the frequency, magnitude, and duration of harmful phytoplankton blooms (Hudnell, 2008). These blooms have adverse effects on humans and animals through anoxia and toxins and affect aquatic biodiversity (Briand et al., 2003; Carmichael et al., 2001; Pflugmacher, 2004). Cyanobacteria (blue-green algae) are one of the most morphologically, physiologically, and metabolically diverse group among prokaryotes (Mankiewicz et al., 2003). Increases in nutrient concentrations can result in "nuisance" cyanobacterial blooms and periphytic scums, which deteriorate water quality through odor and taste effects and impede aesthetic use (Welch, 2002) if more than 40% of the substrate in lakes are covered by macroalgal cells (Barbour et al., 1999). The percentage of algal community in a water body has been used to estimate water quality based on the degree of nuisance (Barbour et al., 1999; Biggs, 1996). In the Baltic Sea, size of algal mats can range from a few square meters to a hectare with thickness exceeding 0.5 m (Norkko and Bonsdorff, 1996). The earthy and musty odor that spoil drinking water is produced by many cyanobacterila genera such as *Anabaena*, *Aphanizomenon*, *Lyngbya*, *Microcystis*, *Oscillatoria*, *Phormidium*, *Schizothrix*, and *Symploca* (Persson, 1983). Hyenstrand et al. (1998) postulated nine mechanisms involved in the production of nuisance blooms: resource ratio (N:P) competition. differential light requirements. CO_2 competition. buoyancy. high temperature tolerance. avoidance by herbivores. superior cellular nutrient storage. ammonium–N exploitation. and trace element competition. A variety of algal cells co-exist with each other in lakes and this diversity diminishes at the end of summer with the dominant species being cyanobacteria such as *Microcystis*, *Planktothrix*, *Anabaena,* and *Nodularia* (Dittmann and Wiegand, 2006).

Contact or ingestion of water containing cyanobacterial toxins can result in skin irritations, allergic responses, blistering of mucosa, hay fever symptoms, diarrhea, acute gastroenteritis, and liver and kidney damage (Bell and Codd, 1996; Pilotto et al., 1997; Ressom et al., 1994). On the basis of toxicity, cyanobacterial toxins are categorized into hepatotoxins (microcystin, and nodularin), neurotoxins (anatoxin-a, anatoxin-as, homanatoxin), cytotoxins (cylindrospermopsin), dermatotoxins (lyngbyatoxin), and irritant toxins (lipopolysaccharides) (Carmichael, 2001). The most extensive toxins produced by blue-green algae are hepatotoxins, microcystin, and nodularin.

One of the best-known and -studied nuisance algal species is *Cladophora glomerata*. Nuisance levels for this species increase with phosphorus addition (Chételat et al., 1999; Ponader et al., 2007; Stevenson et al., 2006). This species is reported to grow at a rate of 0.6 day^{-1} (i.e., an increase of 60% per day when phosphorus is not limited) (Auer and Canale, 1982) and is adaptable to changes in irradiance (Bautista and Necchi-Junior, 2008; Ensminger et al., 2000a,b). The high abundance of this species is also due to its multifaceted branching, repugnance to herbivores (Patrick et al., 1983), constant production of asexual zoospores (Graham et al., 2009), conformation to hydrodynamic pressure (Bergey et al., 1995; Dodds and Gudder, 1992), great tensile strength (9–35 MPa) (Johnson et al., 1996), holdfast attachment, and cell wall rich in cellulose, rendering it a crystalline structure (Mihranyan, 2011).

Another well-known and -studied nuisance algae found in many oligotrophic water bodies is *Didymosphenia geminata* (Lyngbye). *D. geminate* is a diatom, single-celled algae that forms epilithic algal mats, and it is reported in many regions of North America and Europe and New Zealand (Spaulding and Elwell, 2007). It is the only freshwater diatom that can produce stalk material to a level to cause nuisance (Fig. 3.4; Blanco and Ector, 2009). Under favorable conditions, *D. geminate* form thick mats over the water surface that can cover an area more than 1 km and have been described as appearing like fiberglass insulation, tissue paper, "rock snot," brown shag carpet, or sheep skins covering the streambed (Spaulding and Elwell, 2007). The dynamic state of *D. geminata* is also influenced by abiotic factors (Cullis et al., 2012). This species is negatively related to nitrate concentrations, macroinvertebrate richness, shear stress, and multiple tropical storms and floods (Richardson et al., 2014). In Iceland, *D. geminata* has been reported to adversely affect spawning of salmon and might be one of the reasons in the decline in salmon populations (Spaulding and Elwell,

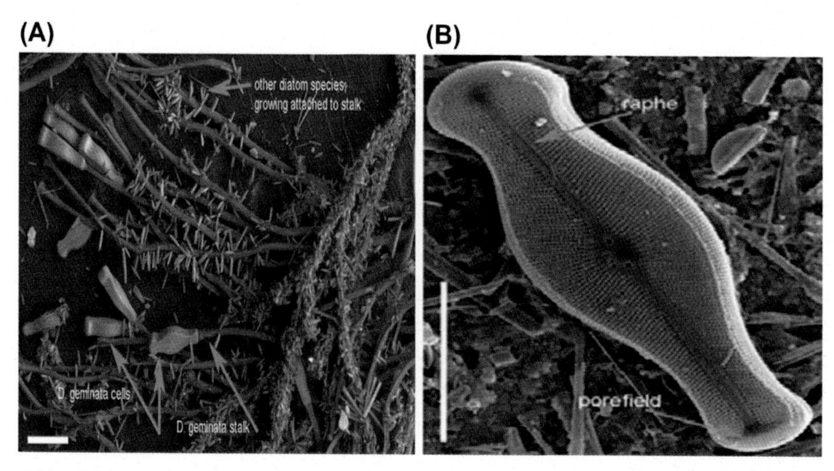

FIGURE 3.4 (A) Scanning electron micrograph of *D. geminata* cells and their mucopolysaccharide stalks. (B) Scanning electron micrograph of the silica cell wall of *D. geminata*. The raphe is composed of the two slits that run along the apical axis of the cell. The cell secretes mucopolysaccarides through the raphe in order to move on surfaces. At the base of the cell is the porefield, through which the stalk is secreted. *From Spaulding, S.A., Elwell, L., 2007. Increase in Nuisance Blooms and Geographic Expansion of the Freshwater Diatom* Didymosphenia geminata.

2007). In Opuha Dam, the frequency and magnitude of floods were altered due to an increase in mean annual cover and proliferation of *D. geminata* (Lessard et al., 2013).

According to McComb and Davis (1993), the total biomass of four perennial algal species (*Cladophora*, *Chaetomorpha*, *Ulva,* and *Enteromorpha*) increased in areas near agricultural fields and sites where nutrients entered the estuary in Peel Inlet. Their growth was affected by temperature, light intensity, and salinity. Species dominance changed time in Peel Inlet, and during the most intense conditions, the nuisance was due to a ball-forming species of *Cladophora*, *C. montagneana* Kutz (McComb and Davis, 1993).

Planktothrix is a filamentous cyanobacteria genus that is repoted in Grand Lake St. Marys (GLSM) (Agency, 2010). Later, concentration of microcystine in this place was found to be greater than 500 mg/g dry weight algal biomass (Belanger, 1994; Schmidt et al., 2013; Steffen et al., 2012). Steffen et al. (2014) proved that nuisance in the lake was the result of toxin production by *Microcystis* and hypereutrophic accumulation of *Planktothrix,* and further evaluation of this problem can suggest the clear cause and remedies against cyanobacterial blooms. Peretyatko et al. (2010) studied 42 Brussels ponds for 5 years and found that cyanobacteria have a threshold relationship with environmental variables. So cyanobacterial risk at a place can be quantified, it was a significant approach toward facilitation of monitoring planning, remediation efforts, and setting restoration priorities.

3.5 CONCLUSION

Damage to fluvial ecosystems due to anthropogenic influences deteriorates water quality and quantity available for aquatic biota as well as for humans and animals. Economically feasible and socially acceptable methods of monitoring and assessing water quality are needed to return disturbed aquatic systems to healthy states. In this regard, periphyton are an excellent bio-indicator for monitoring and predicting water quality. They are an ideal indicator of contamination by multiple contaminants including organic and inorganic toxins. If nutrient levels in contaminated aquatic systems are not remediated however, these periphytic communities can become nuisance organisms in aquatic environments and produce toxins that can affect aquatic and terrestrial organisms living in/on these ecosystems.

REFERENCES

Abdel Hamid, M.I., Källqvist, T., Hessen, D.O., Berge, D., 1996. The use of field enclosure experiments to study the effect of pesticides on lake phytoplankton. Lakes & Reservoirs: Research & Management 2 (3–4), 199–209.

Aboal, M., 1989. Epilithic algal communities from River Segura Basin, Southeastern Spain. Archiv fur Hydrobiologie 116, 113–124.

Agency, U.S.E.P., 2010. National Lakes Assessment: A Collaborative Survey of the Nation's Lakes.

Almeida, S.F., Elias, C., Ferreira, J., Tornes, E., Puccinelli, C., Delmas, F., Dorflinger, G., Urbanic, G., Marcheggiani, S., Rosebery, J., Mancini, L., Sabater, S., 2014. Water quality assessment of rivers using diatom metrics across Mediterranean Europe: a methods intercalibration exercise. Science of the Total Environment 476-477, 768–776.

Alvarez, M., Peckarsky, B.L., 2014. Cascading effects of predatory fish on the composition of benthic algae in high-altitude streams. Oikos 123 (1), 120–128.

An, K.-G., Park, S.S., Shin, J.-Y., 2002. An evaluation of a river health using the index of biological integrity along with relations to chemical and habitat conditions. Environment International 28 (5), 411–420.

Archibald, R.E.M., 1972. Diversity in some South African diatom associations and its relation to water quality. Water Research 6 (10), 1229–1238.

Arrhenius, A., Gronvall, F., Scholze, M., Backhaus, T., Blanck, H., 2004. Predictability of the mixture toxicity of 12 similarly acting congeneric inhibitors of photosystem II in marine periphyton and epipsammon communities. Aquatic Toxicology 68 (4), 351–367.

Audry, S., Schäfer, J., Blanc, G., Bossy, C., Lavaux, G., 2004. Anthropogenic components of heavy metal (Cd, Zn, Cu, Pb) budgets in the Lot-Garonne fluvial system (France). Applied Geochemistry 19 (5), 769–786.

Auer, M.T., Canale, R.P., 1982. Ecological studies and mathematical modeling of *Cladophora* in Lake Huron: 3. The dependence of growth rates on internal phosphorus pool size. Journal of Great Lakes Research 8 (1), 93–99.

Azevedo, A.S., Kanwar, R.S., Pereira, L.S., 2000. Atrazine transport in irrigated heavy- and coarse-textured soils, Part II: simulations with the root zone water quality model. Journal of Agricultural Engineering Research 76 (4), 341–354.

Backhaus, T., Altenburger, R., Boedeker, W., Faust, M., Scholze, M., Grimme, L.H., 2000a. Predictability of the toxicity of a multiple mixture of dissimilarly acting chemicals to Vibrio fischeri. Environmental Toxicology and Chemistry 19 (9), 2348–2356.

Backhaus, T., Scholze, M., Grimme, L.H., 2000b. The single substance and mixture toxicity of quinolones to the bioluminescent bacterium *Vibrio fischeri*. Aquatic Toxicology 49 (1–2), 49–61.

Backhaus, T., Faust, M., Scholze, M., Gramatica, P., Vighi, M., Grimme, L.H., 2004. Joint algal toxicity of phenylurea herbicides is equally predictable by concentration addition and independent action. Environmental Toxicology and Chemistry 23 (2), 258–264.

Barbour, M.T., Gerritsen, J., Snyder, B., Stribling, J., 1999. Rapid Bioassessment Protocols for Use in Streams and Wadeable Rivers. USEPA, Washington.

Barranguet, C., Veuger, B., Van Beusekom, S.A.M., Marvan, P., Sinke, J.J., Admiraal, W., 2005. Divergent composition of algal-bacterial biofilms developing under various external factors. European Journal of Phycology 40 (1), 1–8.

Bautista, A.I., Necchi-Junior, O., 2008. Photoacclimation in a tropical population of *Cladophora glomerata* (L.) Kutzing 1843 (Chlorophyta) from southeastern Brazil. Brazilian Journal of Biology 68 (1), 129–136.

Belanger, S.E., 1994. Review of experimental microcosm,mesocosm, and field tests used to evaluate the potential hazard of surfactants to aquatic life and the relation to single species data. In: Hill, I.R., Heimbach, F., Leeuwangh, P., Matthiessen, P. (Eds.), Freshwater Field Tests for Hazard Assessment of Chemicals. Taylor & Francis, pp. 287–314.

Bell, S.G., Codd, G.A., 1996. Detection analysis and risk assessment of cyanobacterial toxins. In: Hester, R.E., Harrison, R.M. (Eds.), Agricultural Chemicals and the Environment. The Royal Society of Chemistry, pp. 109–122.

Bennion, H., Juggins, S., Anderson, N.J., 1996. Predicting epilimnetic phosphorus concentrations using an improved diatom-based transfer function and its application to lake eutrophication management. Environmental Science & Technology 30 (6), 2004–2007.

Bere, T., Chia, M.A., Tundisi, J.G., 2012. Effects of Cr III and Pb on the bioaccumulation and toxicity of Cd in tropical periphyton communities: implications of pulsed metal exposures. Environmental Pollution 163, 184–191.

Bere, T., Tundisi, J., 2012. Cadmium and lead toxicity on tropical freshwater periphyton communities under laboratory-based mesocosm experiments. Hydrobiologia 680 (1), 187–197.

Bere, T., Tundisi, J.G., 2011. Toxicity and sorption kinetics of dissolved cadmium and chromium III on tropical freshwater phytoperiphyton in laboratory mesocosm experiments. Science of the Total Environment 409 (22), 4772–4780.

Bergey, E.A., Boettiger, C.A., Resh, V.H., 1995. Effects of water velocity of the architecture and epiphytes of *Cladophora glomerata* (Chlorophyta). Journal of Phycology 31 (2), 264–271.

Biggs, B.J., 2000. New Zealand periphyton Guideline: Detecting, Monitoring and Managing Enrichment of Streams, New Zealand.

Biggs, B.J.F., 1989. Biomonitoring of organic pollution using periphyton, South Branch, Canterbury, New Zealand. New Zealand Journal of Marine and Freshwater Research 23 (2), 263–274.

Biggs, B.J.F., 1996. Patterns in benthic algae of streams. In: Stevenson, R.J., Bothwell, M.L., Lowe, R.L. (Eds.), Algal Ecology. Academic Press, San Diego, pp. 31–56.

Biggs, B.J.F., 1985. The use of periphyton in the monitoring of water quality. In: Pridmore, R.D., Cooper, A.B., Water, N.Z.N., Authority, S.C., Water, N.Z., Directorate, S. (Eds.), Biological Monitoring in Freshwaters: Proceedings of a Seminar, Hamilton, November 21–23, 1984, vol. 82. National Water and Soil Conservation Authority, New Zealand.

Birks, H.J.B., Line, J.M., Juggins, S., Stevenson, A.C., Braak, C.J.F.T., 1990. Diatoms and pH reconstruction. Philosophical Transactions of the Royal Society of London. Series B, Biological Sciences 327 (1240), 263–278.

Blanck, H., 2002. A critical review of procedures and approaches used for assessing pollution-induced community tolerance (PICT) in biotic communities. Human and Ecological Risk Assessment 8 (5), 1003–1034.

Blanck, H., Dahl, B., 1996. Pollution-induced community tolerance (PICT) in marine periphyton in a gradient of tri-*n*-butyltin (TBT) contamination. Aquatic Toxicology 35 (1), 59–77.

Blanck, H., Eriksson, K.M., Gronvall, F., Dahl, B., Guijarro, K.M., Birgersson, G., Kylin, H., 2009. A retrospective analysis of contamination and periphyton PICT patterns for the antifoulant irgarol 1051, around a small marina on the Swedish west coast. Marine Pollution Bulletin 58 (2), 230–237.

Blanck, H., Wängberg, S.-Å., 1988a. Induced community tolerance in marine periphyton established under arsenate stress. Canadian Journal of Fisheries and Aquatic Sciences 45 (10), 1816–1819.

Blanck, H., Wängberg, S.-Å., 1988b. Validity of an ecotoxicological test system: short-term and long-term effects of arsenate on marine periphyton communities in laboratory systems. Canadian Journal of Fisheries and Aquatic Sciences 45 (10), 1807–1815.

Blanco, S., Ector, L., 2009. Distribution, ecology and nuisance effects of the freshwater invasive diatom *Didymosphenia geminata* (Lyngbye) M. Schmidt: a literature review. Nova Hedwigia 88 (3–4), 347–422.

Bliss, C., 1939. The toxicity of poisons applied jointly. Annals of Applied Biology 26 (3), 585–615.

Bonet, B., Corcoll, N., Guasch, H., 2012. Antioxidant enzyme activities as biomarkers of Zn pollution in fluvial biofilms. Ecotoxicology and Environmental Safety 80, 172–178.

Bossuyt, B.T., Janssen, C.R., 2004. Long-term acclimation of *Pseudokirchneriella subcapitata* (Korshikov) Hindak to different copper concentrations: changes in tolerance and physiology. Aquatic Toxicology 68 (1), 61–74.

Briand, J.F., Jacquet, S., Bernard, C., Humbert, J.F., 2003. Health hazards for terrestrial vertebrates from toxic cyanobacteria in surface water ecosystems. Veterinary Research 34 (4), 361–377.

ter Braak, C.F., van Dame, H., 1989. Inferring pH from diatoms: a comparison of old and new calibration methods. Hydrobiologia 178 (3), 209–223.

Carmichael, W.W., 2001. Health effects of toxin-producing cyanobacteria: "the cyanoHABs". Human and Ecological Risk Assessment: An International Journal 7 (5), 1393–1407.

Carmichael, W.W., Azevedo, S.M., An, J.S., Molica, R.J., Jochimsen, E.M., Lau, S., Rinehart, K.L., Shaw, G.R., Eaglesham, G.K., 2001. Human fatalities from cyanobacteria: chemical and biological evidence for cyanotoxins. Environmental Health Perspectives 109 (7), 663–668.

Cattaneo, A., 1983. Grazing on epiphytes. Limnology and Oceanography 28 (1), 124–132.

Cattaneo, A., Couillard, Y., Wunsam, S., Courcelles, M., 2004. Diatom taxonomic and morphological changes as indicators of metal pollution and recovery in Lac Dufault (Québec, Canada). Journal of Paleolimnology 32 (2), 163–175.

Cattaneo, A., Methot, G., Pinel-Alloul, B., Niyonsenga, T., Lapierre, L., 1995. Epiphyte size and taxonomy as biological indicators of ecological and toxicological factors in Lake Saint-Francois (Quebec). Environmental Pollution 87 (3), 357–372.

Chételat, J., Pick, F.R., Morin, A., Hamilton, P.B., 1999. Periphyton biomass and community composition in rivers of different nutrient status. Canadian Journal of Fisheries and Aquatic Sciences 56 (4), 560–569.

Chessman, B., Growns, I., Currey, J., Plunkett-Cole, N., 1999. Predicting diatom communities at the genus level for the rapid biological assessment of rivers. Freshwater Biology 41 (2), 317–331.

Clements, W.H., 1991. Community responses of stream organisms to heavy metals: a review of observational and experimental approaches. In: Metal Ecotoxicology: Concepts and Applications. Lewis Publishers, Chelsea, Michigan, pp. 363–391.

Cooke, W.B., 1956. Colonization of artificial bare areas by microorganisms. The Botanical Review 22 (9), 613–638.

Cowan, C.E., Versteeg, D.J., Larson, R.J., Kloepper-Sams, P.J., 1995. Integrated approach for environmental assessment of new and existing substances. Regulated Toxicology and Pharmacology 21 (1), 3–31.

Crane, M., Burton, G.A., Culp, J.M., Greenberg, M.S., Munkittrick, K.R., Ribeiro, R., Salazar, M.H., St Jean, S.D., 2007. Review of aquatic in situ approaches for stressor and effect diagnosis. Integrated Environmental Assessment and Management 3 (2), 234–245.

Cullis, J.D.S., Gillis, C.A., Bothwell, M.L., Kilroy, C., Packman, A., Hassan, M., 2012. A conceptual model for the blooming behavior and persistence of the benthic mat – forming diatom *Didymosphenia geminata* in oligotrophic streams. Journal of Geophysical Research: Biogeosciences 117 (G2).

Dahl, B., Blanck, H., 1996. Toxic effects of the antifouling agent irgarol 1051 on periphyton communities in coastal water microcosms. Marine Pollution Bulletin 32 (4), 342–350.

de Kock, W.C., Kramer, K.J.M., 1994. Active Biomonitoring (ABM) by translocation of bivalve molluscs. In: Kramer, K.J.M. (Ed.), Biomonitoring of Coastal Waters and Estuaries. CRC Press, pp. 51–84.

Debenest, T., Pinelli, E., Coste, M., Silvestre, J., Mazzella, N., Madigou, C., Delmas, F., 2009. Sensitivity of freshwater periphytic diatoms to agricultural herbicides. Aquatic Toxicology 93 (1), 11–17.

Dela-Cruz, J., Pritchard, T.I.M., Gordon, G., Ajani, P., 2006. The use of periphytic diatoms as a means of assessing impacts of point source inorganic nutrient pollution in south-eastern Australia. Freshwater Biology 51 (5), 951–972.

DeLorenzo, M.E., Scott, G.I., Ross, P.E., 1999. Effects of the agricultural pesticides atrazine, deethylatrazine, endosulfan, and chlorpyrifos on an estuarine microbial food web. Environmental Toxicology and Chemistry 18 (12), 2824–2835.

Descy, J.P., Coste, M., 1991. A test of methods for assessing water quality based on diatoms. Verhandlungen des Internationalen Verein Limnologie 24, 2112–2116.

Dittmann, E., Wiegand, C., 2006. Cyanobacterial toxins–occurrence, biosynthesis and impact on human affairs. Molecular Nutrition & Food Research 50 (1), 7–17.

Division, K.D.F.E.P., 2002. Methods for Assessing Biological Integrity of Surface Waters. Kentucky Department for Environmental Protection Division of Water Ecological Support Section.

Dixit, S.S., Smol, J.P., 1994. Diatoms as indicators in the Environmental Monitoring and Assessment Program-Surface Waters (EMAP-SW). Environmental Monitoring and Assessment 31 (3), 275–307.

Dodds, W.K., Gudder, D.A., 1992. The ecology of Cladophora. Journal of Phycology 28 (4), 415–427.

Dorigo, U., Bérard, A., Bouchez, A., Rimet, F., Montuelle, B., 2010a. Transplantation of microbenthic algal assemblages to assess structural and functional recovery after diuron exposure. Archives of Environmental Contamination and Toxicology 59 (4), 555–563.

Dorigo, U., Berard, A., Rimet, F., Bouchez, A., Montuelle, B., 2010b. *In situ* assessment of periphyton recovery in a river contaminated by pesticides. Aquatic Toxicology 98 (4), 396–406.

Douterelo, I., Perona, E., Mateo, P., 2004. Use of cyanobacteria to assess water quality in running waters. Environmental Pollution 127 (3), 377–384.

Duong, T.T., Morin, S., Coste, M., Herlory, O., Feurtet-Mazel, A., Boudou, A., 2010. Experimental toxicity and bioaccumulation of cadmium in freshwater periphytic diatoms in relation with biofilm maturity. Science of the Total Environment 408 (3), 552–562.

Duong, T.T., Morin, S., Herlory, O., Feurtet-Mazel, A., Coste, M., Boudou, A., 2008. Seasonal effects of cadmium accumulation in periphytic diatom communities of freshwater biofilms. Aquatic Toxicology 90 (1), 19–28.

Ekholm, P., Kallio, K., Salo, S., Pietiläinen, O.P., Rekolainen, S., Laine, Y., Joukola, M., 2000. Relationship between catchment characteristics and nutrient concentrations in an agricultural river system. Water Research 34 (15), 3709–3716.

Ensminger, I., Hagen, C., Braune, W., 2000a. Strategies providing success in a variable habitat: I. Relationships of environmental factors and dominance of *Cladophora glomerata*. Plant, Cell & Environment 23 (10), 1119–1128.

Ensminger, I., Hagen, C., Braune, W., 2000b. Strategies providing success in a variable habitat: II. Ecophysiology of photosynthesis of *Cladophora glomerata*. Plant, Cell & Environment 23 (10), 1129–1136.

EPA, U., 1999. Biological Criteria. National Program Guidance for Surface Waters, Washington, DC.

EPA, U., 2000. National Water Quality Inventory: 1998 Report to Congress.

Eullaffroy, P., Vernet, G., 2003. The F684/F735 chlorophyll fluorescence ratio: a potential tool for rapid detection and determination of herbicide phytotoxicity in algae. Water Research 37 (9), 1983–1990.

Fairchild, G.W., Sherman, J.W., 1992. Linkage between epilithic algal growth and water column nutrients in softwater Lakes. Canadian Journal of Fisheries and Aquatic Sciences 49 (8), 1641–1649.

Fano, E.A., Mistri, M., Rossi, R., 2003. The ecofunctional quality index (EQI): a new tool for assessing lagoonal ecosystem impairment. Estuarine, Coastal and Shelf Science 56 (3–4), 709–716.

Faust, M., Altenburger, R., Backhaus, T., Bödeker, W., Scholze, M., Grimme, L., 2000. Predictive assessment of the aquatic toxicity of multiple chemical mixtures. Journal of Environmental Quality 29 (4), 1063–1068.

Faust, M., Altenburger, R., Backhaus, T., Blanck, H., Boedeker, W., Gramatica, P., Hamer, V., Scholze, M., Vighi, M., Grimme, L.H., 2003. Joint algal toxicity of 16 dissimilarly acting chemicals is predictable by the concept of independent action. Aquatic Toxicology 63 (1), 43–63.

Feio, M.J., Almeida, S.F., Craveiro, S.C., Calado, A.J., 2009. A comparison between biotic indices and predictive models in stream water quality assessment based on benthic diatom communities. Ecological Indicators 9 (3), 497–507.

Feminella, J.W., 2000. Correspondence between stream macroinvertebrate assemblages and 4 ecoregions of the southeastern USA. Journal of the North American Benthological Society 19 (3), 442–461.

Fore, L.S., Grafe, C., 2002. Using diatoms to assess the biological condition of large rivers in Idaho (U.S.A.). Freshwater Biology 47 (10), 2015–2037.

Gabriels, W., Lock, K., De Pauw, N., Goethals, P.L.M., 2010. Multimetric Macroinvertebrate Index Flanders (MMIF) for biological assessment of rivers and lakes in Flanders (Belgium). Limnologica – Ecology and Management of Inland Waters 40 (3), 199–207.

Geiszinger, A., Bonnineau, C., Faggiano, L., Guasch, H., Lopez-Doval, J., Proia, L., Ricart, M., Ricciardi, F., Romani, A., Rotter, S., 2009. The relevance of the community approach linking chemical and biological analyses in pollution assessment. TrAC Trends in Analytical Chemistry 28 (5), 619–626.

Genter, R.B., 1996. Ecotoxicology of inorganic chemical stress to algae. In: Stevenson, R.J., Bothwell, M.L., Lowe, R.L. (Eds.), Algal Ecology. Academic Press, San Diego, pp. 403–468.

Genter, R.B., Cherry, D.S., Smith, E.P., Cairns Jr., J., 1987. Algal-periphyton population and community changes from zinc stress in stream mesocosms. Hydrobiologia 153 (3), 261–275.

Genter, R.B., Lehman, R.M., 2000. Metal toxicity inferred from algal population density, heterotrophic substrate use, and fatty acid profile in a small stream. Environmental Toxicology and Chemistry 19 (4), 869–878.

Gerritsen, J., 1995. Additive biological indices for resource management. Journal of the North American Benthological Society 451–457.

Gold, C., Feurtet-Mazel, A., Coste, M., Boudou, A., 2002. Field transfer of periphytic diatom communities to assess short-term structural effects of metals (Cd, Zn) in rivers. Water Research 36 (14), 3654–3664.

Gold, C., Feurtet-Mazel, A., Coste, M., Boudou, A., 2003a. Impacts of Cd and Zn on the development of periphytic diatom communities in artificial streams located along a river pollution gradient. Archives of Environmental Contamination and Toxicology 44 (2), 0189–0197.

Gold, C., Feurtet Mazel, A., Coste, M., Boudou, A., 2003b. Effects of cadmium stress on periphytic diatom communities in indoor artificial streams. Freshwater Biology 48 (2), 316–328.

Gong, J., Song, W., 2005. Periphytic ciliate colonization: annual cycle and responses to environmental conditions. Aquatic Microbial Ecology 39, 159–170.

Graham, L.E., Graham, J.M., Wilcox, L.W., 2009. Algae. Benjamin Cummings.

Greco, W.R., Unkelbach, H.-D., Pöch, G., Suhnel, J., Kundi, M., Boedeker, W., 1992. Consensus on concepts and terminology for combined-action assessment: the Saariselkä agreement. Archive in Complex Environmental Studies 4 (3), 65–69.

Griffith, M.B., Hill, B.H., McCormick, F.H., Kaufmann, P.R., Herlihy, A.T., Selle, A.R., 2005. Comparative application of indices of biotic integrity based on periphyton, macroinvertebrates, and fish to southern Rocky Mountain streams. Ecology Indicator 5 (2), 20.

Guasch, H., Admiraal, W., Sabater, S., 2003. Contrasting effects of organic and inorganic toxicants on freshwater periphyton. Aquatic Toxicology 64 (2), 165–175.

Guasch, H., Ivorra, N., Lehmann, V., Paulsson, M., Real, M., Sabater, S., 1998. Community composition and sensitivity of periphyton to atrazine in flowing waters: the role of environmental factors. Journal of Applied Phycology 10 (2), 203–213.

Guasch, H., Lehmann, V., Van Beusekom, B., Sabater, S., Admiraal, W., 2007. Influence of phosphate on the response of periphyton to atrazine exposure. Archives of Environmental Contamination and Toxicology 52 (1), 32–37.

Guasch, H., Muñoz, I., Rosés, N., Sabater, S., 1997. Changes in atrazine toxicity throughout succession of stream periphyton communities. Journal of Applied Phycology 9 (2), 137–146.

Guasch, H., Navarro, E., Serra, A., Sabater, S., 2004. Phosphate limitation influences the sensitivity to copper in periphytic algae. Freshwater Biology 49 (4), 463–473.

Guasch, H., Paulsson, M., Sabater, S., 2002. Effect of copper on algal communities from oligotrophic calcareous streams. Journal of Phycology 38 (2), 241–248.

Guasch, H., Sabater, S., 1998. Light history influences the sensitivity to atrazine in periphytic algae. Journal of Phycology 34 (2), 233–241.

Guckert, J.B., 1996. Toxicity assessment by community analysis. Journal of Microbiological Methods 25 (2), 101–112.

Gustavson, K., Mohlenberg, F., Schluter, L., 2003. Effects of exposure duration of herbicides on natural stream periphyton communities and recovery. Archives of Environmental Contamination and Toxicology 45 (1), 48–58.

Gustavson, K., Wängberg, S.-Å., 1995. Tolerance induction and succession in microalgae communities exposed to copper and atrazine. Aquatic Toxicology 32 (4), 283–302.

Hansen, L., 1987. Environmental toxicology of polychlorinated biphenyls. In: Polychlorinated Biphenyls (PCBs): Mammalian and Environmental Toxicology. Springer, pp. 15–48.

Harding, J.P.C., Whitton, B.A., 1977. Environmental factors reducing the toxicity of zinc to Stigeoclonium tenue. British Phycological Journal 12 (1), 17–21.

Hashemi, F., Leppard, G.G., Kushnert, D.J., 1994. Copper resistance in *Anabaena variabilis*: effects of phosphate nutrition and polyphosphate bodies. Microbial Ecology 27 (2), 159–176.

Hatakeyama, S., Fukushima, S., Kasai, F., Shiraishi, H., 1994. Assessment of herbicide effects on algal production in the Kokai River (Japan) using a model stream and Selenastrum bioassay. Ecotoxicology 3 (2), 143–156.

Herman, D., Kaushik, N.K., Solomon, K.R., 1986. Impact of atrazine on periphyton in freshwater enclosures and some ecological consequences. Canadian Journal of Fisheries and Aquatic Sciences 43 (10), 1917–1925.

Hermens, J., Busser, F., Leeuwangh, P., Musch, A., 1985. Quantitative structure-activity relationships and mixture toxicity of organic chemicals in *Photobacterium phosphoreum*: the microtox test. Ecotoxicology and Environmental Safety 9 (1), 17–25.

Hermens, J., Canton, H., Janssen, P., De Jong, R., 1984a. Quantitative structure-activity relationships and toxicity studies of mixtures of chemicals with anaesthetic potency: acute lethal and sublethal toxicity to *Daphnia magna*. Aquatic Toxicology 5 (2), 143–154.

Hermens, J., Leeuwangh, P., Musch, A., 1984b. Quantitative structure-activity relationships and mixture toxicity studies of chloro-and alkylanilines at an acute lethal toxicity level to the guppy (*Poecilia reticulata*). Ecotoxicology and Environmental Safety 8 (4), 388–394.

Hermens, J., Leeuwangh, P., 1982. Joint toxicity of mixtures of 8 and 24 chemicals to the guppy (*Poecilia reticulata*). Ecotoxicology and Environmental Safety 6 (3), 302–310.

Hill, B., Herlihy, A., Kaufmann, P., Stevenson, R., McCormick, F., Johnson, C.B., 2000a. Use of periphyton assemblage data as an index of biotic integrity. Journal of the North American Benthological Society 19 (1), 50–67.

Hill, B.H., Willingham, W.T., Parrish, L.P., McFarland, B.H., 2000b. Periphyton community responses to elevated metal concentrations in a Rocky Mountain stream. Hydrobiologia 428 (1), 161–169.

Hill, B.H., Herlihy, A.T., Kaufmann, P.R., DeCelles, S.J., Vander Borgh, M.A., 2003. Assessment of streams of the eastern United States using a periphyton index of biotic integrity. Ecological Indicators 2 (4), 325–338.

Hirst, H., Chaud, F., Delabie, C., Jüttner, I., Ormerod, S., 2004. Assessing the short term response of stream diatoms to acidity using inter basin transplantations and chemical diffusing substrates. Freshwater Biology 49 (8), 1072–1088.

Hoagland, K.D., Roemer, S.C., Rosowski, J.R., 1982. Colonization and community structure of two periphyton assemblages, with emphasis on the diatoms (*Bacillariophyceae*). American Journal of Botany 69 (2), 188–213.

Hofmann, G., 1996. Recent Developments in the Use of Benthic Diatoms for Monitoring Eutrophication and Organic Pollution in Germany and Austria. In: Whitton, B.A., Rott, E. (Eds.), Use of Algae for Monitoring Rivers II. Institut für Botanik, Universität Innsbruck, pp. 73–77.

Holding, K., Gill, R., Carter, J., 2003. The relationship between epilithic periphyton (biofilm) bound metals and metals bound to sediments in freshwater systems. Environmental Geochemistry and Health 25 (1), 87–93.

Horner, R.R., Welch, E.B., 1981. Stream periphyton development in relation to current velocity and nutrients. Canadian Journal of Fisheries and Aquatic Sciences 38 (4), 449–457.

Hudnell, H.K., 2008. Cyanobacterial Harmful Algal Blooms: State of the Science and Research Needs: State of the Science and Research Needs. Springer.

Hudon, C., Bourget, E., 1983. The effect of light on the vertical structure of epibenthic diatom communities. In: Botanica Marina, 26, p. 317.

Hyenstrand, P., Blomqvist, P., Pettersson, A., 1998. Factors determining cyanobacterial success in aquatic systems: a literature review. In: Arch. Hydrobiol., 15, pp. 41–62.

Ivorra, N., Barranguet, C., Jonker, M., Kraak, M.H., Admiraal, W., 2002. Metal-induced tolerance in the freshwater microbenthic diatom *Gomphonema parvulum*. Environmental Pollution 116 (1), 147–157.

Ivorra, N., Bremer, S., Guasch, H., Kraak, M.H.S., Admiraal, W., 2000. Differences in the sensitivity of benthic microalgae to ZN and CD regarding biofilm development and exposure history. Environmental Toxicology and Chemistry 19 (5), 1332–1339.

Jan Stevenson, R., Peterson, C.G., 1991. Emigration and immigration can be important determinants of benthic diatom assemblages in streams. Freshwater Biology 26 (2), 279–294.

Johnson, M., Shivkumar, S., Berlowitz-Tarrant, L., 1996. Structure and properties of filamentous green algae. Materials Science and Engineering: B 38 (1), 103–108.

Jones, V.J., Juggins, S., 1995. The construction of a diatom-based chlorophyll a transfer function and its application at three lakes on Signy Island (maritime Antarctic) subject to differing degrees of nutrient enrichment. Freshwater Biology 34 (3), 433–445.

Könemann, H., 1981. Fish toxicity tests with mixtures of more than two chemicals: a proposal for a quantitative approach and experimental results. Toxicology 19 (3), 229–238.

Könemann, H., 1980. Structure—activity relationships and additivity in fish toxicities of environmental pollutants. Ecotoxicology and Environmental Safety 4 (4), 415–421.

Kamaya, Y., Takada, T., Suzuki, K., 2004. Effect of medium phosphate levels on the sensitivity of Selenastrum capricornutum to chemicals. Bulletin of Environmental Contamination and Toxicology 73 (6), 995–1000.

Karr, J.R., 1981. Assessment of biotic integrity using fish communities. Fisheries 6 (6), 21–27.

Karr, J.R., Chu, E.W., 1997. Biological Monitoring and Assessment: Using Multimetric Indexes Effectively.

Kelly, M., Whitton, B., 1998. Biological monitoring of eutrophication in rivers. Hydrobiologia 384 (1–3), 55–67.

Kelly, M.G., Penny, C.J., Whitton, B.A., 1995. Comparative performance of benthic diatom indices used to assess river water quality. Hydrobiologia 302 (3), 179–188.

Kelly, M.G., Whitton, B.A., 1995. The Trophic Diatom Index: a new index for monitoring eutrophication in rivers. Journal of Applied Phycology 7 (4), 433–444.

Kireta, A.R., Reavie, E.D., Sgro, G.V., Angradi, T.R., Bolgrien, D.W., Hill, B.H., Jicha, T.M., 2012. Planktonic and periphytic diatoms as indicators of stress on great rivers of the United States: testing water quality and disturbance models. Ecological Indicators 13 (1), 222–231.

Knauer, K., Behra, R., Sigg, L., 1997. Effects of free Cu^{2+} and Zn^{2+} ions on growth and metal accumulation in freshwater algae. Environmental Toxicology and Chemistry 16 (2), 220–229.

Kostel, J.A., Wang, H., St Amand, A.L., Gray, K.A., 1999. 1. Use of a novel laboratory stream system to study the ecological impact of PCB exposure in a periphytic biolayer. Water Research 33 (18), 3735–3748.

Kwandrans, J., Eloranta, P., Kawecka, B., Wojtan, K., 1998. Use of benthic diatom communities to evaluate water quality in rivers of southern Poland. Journal of Applied Phycology 10 (2), 193–201.

Lavoie, I., Hamilton, P., Poulin, M., 2011. Phytoplankton community metrics based on absolute and relative abundance and biomass: implications for multivariate analyses. Journal of Applied Phycology 23 (4), 735–743.

Lawrence, J., Kopf, G., Headley, J., Neu, T., 2001. Sorption and metabolism of selected herbicides in river biofilm communities. Canadian Journal of Microbiology 47 (7), 634–641.

Ledford, H.K., Niyogi, K.K., 2005. Singlet oxygen and photo-oxidative stress management in plants and algae. Plant, Cell & Environment 28 (8), 1037–1045.

Lessard, J., Hicks, D.M., Snelder, T.H., Arscott, D.B., Larned, S.T., Booker, D., Suren, A.M., 2013. Dam design can impede adaptive management of environmental flows: a case study from the Opuha Dam, New Zealand. Environmental Management 51 (2), 459–473.

Lockert, C., Hoagland, K.D., Siegfried, B.D., 2006. Comparative sensitivity of freshwater algae to atrazine. Bulletin of Environmental Contamination and Toxicology 76 (1), 73–79.

Lombardi, A.T., Hidalgo, T.M.R., Vieira, A.A.H., Sartori, A.L., 2007. Toxicity of ionic copper to the freshwater microalga *Scenedesmus acuminatus* (Chlorophyceae, Chlorococcales). Phycologia 46 (1), 74–78.

Lombardi, A.T., Vieira, A.A.H., Sartori, L.A., 2002. Mucilaginous capsule adsorption and intracellular uptake of copper by Kircheneriella aperta (Chlorococcales). Journal of Phycology 38 (2), 332–337.

Lowe, R.L., Pan, Y., 1996. Benthic algal communities as biological indicators. In: Stevenson, R.J., Bothwell, M.L., Lowe, R.L., Thorp, J.H. (Eds.), Algal Ecology: Freshwater Benthic Ecosystem. Academic press, San Diego.

Ludwig, A., Matlock, M., Haggard, B.E., Matlock, M., Cummings, E., 2008. Identification and evaluation of nutrient limitation on periphyton growth in headwater streams in the Pawnee Nation, Oklahoma. Ecological Engineering 32 (2), 178–186.

Luoma, S.N., 1977. Detection of trace contaminant effects in aquatic ecosystems. Journal of the Fisheries Research Board of Canada 34 (3), 436–439.

Luttenton, M.R., Baisden, C., 2006. The relationships among disturbance, substratum size and periphyton community structure. Hydrobiologia 561 (1), 111–117.

Magnusson, M., Heimann, K., Ridd, M., Negri, A.P., 2012. Chronic herbicide exposures affect the sensitivity and community structure of tropical benthic microalgae. Marine Pollution Bulletin 65 (4–9), 363–372.

Mankiewicz, J., Tarczynska, M., Walter, Z., Zalewski, M., 2003. Natural toxins from cyanobacteria. Acta Biologica Cracoviensia Series Botanica 45, 9–20.

Margoum, C., Guillemain, C., Rabiet, M., Gouy, V., Coquery, M., del Re, A., Capri, E., Fragoulis, G., Trevisan, M., 2007. Dissipation of pesticides in surface water and biofilms in a small agricultural catchment: development of a methodology for studying environmental impact of pesticides. In: 13th Symposium Pesticide Chemistry, Piacenza, Italy, 3–6 September 2007. La Goliardica Pavese SRL, pp. 812–818.

McCann, K.S., 2000. The diversity-stability debate. Nature 405 (6783), 228–233.

McComb, A., Davis, J., 1993. Eutrophic waters of southwestern Australia. Fertilizer Research 36 (2), 105–114.

McKnight, D.M., Morel, F.M., 1979. Release of weak and strong copper-complexing agents by algae. Limnology & Oceanography 24 (5), 823–837.

Melandri, B.A., Baccarini-Melandri, A., San Pietro, A., Gest, H., 1970. Role of phosphorylation coupling factor in light-dependent proton translocation by *Rhodopseudomonas capsulata* membrane preparations. Proceedings of the National Academy of Sciences 67 (2), 477–484.

Meng, W., Zhang, N., Zhang, Y., Zheng, B., 2009. Integrated assessment of river health based on water quality, aquatic life and physical habitat. Journal of Environmental Sciences 21 (8), 1017–1027.

Mieczan, T., 2010. Periphytic ciliates in three shallow lakes in eastern Poland: a comparative study between a phytoplankton-dominated lake, a phytoplanktonmacrophyte lake and a macrophyte-dominated lake. Zoological Studies 49 (5), 589–600.

Mihranyan, A., 2011. Cellulose from cladophorales green algae: from environmental problem to high tech composite materials. Journal of Applied Polymer Science 119 (4), 2449–2460.

Moe, S.J., De Schamphelaere, K., Clements, W.H., Sorensen, M.T., Van den Brink, P.J., Liess, M., 2013. Combined and interactive effects of global climate change and toxicants on populations and communities. Environmental Toxicology and Chemistry 32 (1), 49–61.

Molander, S., Blanck, H., 1992. Detection of pollution-induced community tolerance (PICT) in marine periphyton communities established under diuron exposure. Aquatic Toxicology 22 (2), 129–143.

Moore, M.J.C., Langrehr, H.A., Angradi, T.R., 2012. A submersed macrophyte index of condition for the Upper Mississippi River. Ecological Indicators 13 (1), 196–205.

Morin, S., Bottin, M., Mazzella, N., Macary, F., Delmas, F., Winterton, P., Coste, M., 2009. Linking diatom community structure to pesticide input as evaluated through a spatial contamination potential (Phytopixal): a case study in the Neste river system (South-West France). Aquatic Toxicology 94 (1), 28–39.

Morin, S., Duong, T., Herlory, O., Feurtet-Mazel, A., Coste, M., 2008a. Cadmium toxicity and bioaccumulation in freshwater biofilms. Archives of Environmental Contamination and Toxicology 54 (2), 173–186.

Morin, S., Duong, T.T., Dabrin, A., Coynel, A., Herlory, O., Baudrimont, M., Delmas, F., Durrieu, G., Schäfer, J., Winterton, P., Blanc, G., Coste, M., 2008b. Long-term survey of heavy-metal pollution, biofilm contamination and diatom community structure in the Riou Mort watershed, South-West France. Environmental Pollution 151 (3), 532–542.

Morin, S., Pesce, S., Tlili, A., Coste, M., Montuelle, B., 2010. Recovery potential of periphytic communities in a river impacted by a vineyard watershed. Ecological Indicators 10 (2), 419–426.

Norf, H., Arndt, H., Weitere, M., 2009. Effects of resource supplements on mature ciliate biofilms: an empirical test using a new type of flow cell. Biofouling 25 (8), 769–778.

Norkko, A., Bonsdorff, E., 1996. Population responses of coastal zoobenthos to stress induced by drifting algal mats. Marine Ecology Progress Series 140, 141–151.

Norris, R.H., 1995. Biological monitoring: the dilemma of data analysis. Journal of the North American Benthological Society 440–450.

Nunes, M.L., Ferreira Da Silva, E., De Almeida, S.F.P., 2003. Assessment of water quality in the Caima and Mau River Basins (Portugal) using gochemical and biological indices. Water, Air, and Soil Pollution 149 (1–4), 227–250.

Nyström, B., Björnsäter, B., Blanck, H., 1999. Effects of sulfonylurea herbicides on non-target aquatic micro-organisms: growth inhibition of micro-algae and short-term inhibition of adenine and thymidine incorporation in periphyton communities. Aquatic toxicology 47 (1), 9–22.

Nyström, B., Paulsson, M., Almgren, K., Blank, H., 2000. Evaluation of the capacity for development of atrazine tolerance in periphyton from a swedish freshwater site as determined by inhibition of photosynthesis and sulfolipid synthesis. Environmental Toxicology and Chemistry 19 (5), 1324–1331.

Patrick, R., Rhyne, C.F., Richardson, R.W., Larson, R.A., Bott, T.L., Rogenmuser, K., 1983. The Potential for Biological Controls of *Cladophora glomerata*. EPA US Publication.

Paulsson, M., Månsson, V., Blanck, H., 2002. Effects of zinc on the phosphorus availability to periphyton communities from the river Göta Älv. Aquatic Toxicology 56 (2), 103–113.

Paulsson, M., Nyström, B., Blanck, H., 2000. Long-term toxicity of zinc to bacteria and algae in periphyton communities from the river Göta Älv, based on a microcosm study. Aquatic Toxicology 47 (3–4), 243–257.

Pennanen, T., Frostegard, A., Fritze, H., Baath, E., 1996. Phospholipid fatty acid composition and heavy metal tolerance of soil microbial communities along two heavy metal-polluted gradients in coniferous forests. Applied and Environmental Microbiology 62 (2), 420–428.

Peres, F., Florin, D., Grollier, T., Feurtet-Mazel, A., Coste, M., Ribeyre, F., Ricard, M., Boudou, A., 1996. Effects of the phenylurea herbicide isoproturon on periphytic diatom communities in freshwater indoor microcosms. Environmental Pollution 94 (2), 141–152.

Peretyatko, A., Teissier, S., Backer, S.D., Triest, L., 2010. Assessment of the risk of cyanobacterial bloom occurrence in urban ponds: probabilistic approach. Annales de Limnologie – International Journal of Limnology 46 (02), 121–133.

Persson, P.-E., 1983. Off-flavours in aquatic ecosystems—an introduction. Water Science & Technology 15 (6–7), 1–11.

Pesce, S., Fajon, C., Bardot, C., Bonnemoy, F., Portelli, C., Bohatier, J., 2006. Effects of the phenylurea herbicide diuron on natural riverine microbial communities in an experimental study. Aquatic Toxicology 78 (4), 303–314.

Pesce, S., Fajon, C., Bardot, C., Bonnemoy, F., Portelli, C., Bohatier, J., 2008. Longitudinal changes in microbial planktonic communities of a French river in relation to pesticide and nutrient inputs. Aquatic Toxicology 86 (3), 352–360.

Pesce, S., Lissalde, S., Lavieille, D., Margoum, C., Mazzella, N., Roubeix, V., Montuelle, B., 2010a. Evaluation of single and joint toxic effects of diuron and its main metabolites on natural phototrophic biofilms using a pollution-induced community tolerance (PICT) approach. Aquatic Toxicology 99 (4), 492–499.

Pesce, S., Margoum, C., Montuelle, B., 2010b. In situ relationships between spatio-temporal variations in diuron concentrations and phototrophic biofilm tolerance in a contaminated river. Water Research 44 (6), 1941–1949.

Pesce, S., Margoum, C., Montuelle, B., 2010c. *In situ* relationships between spatio-temporal variations in diuron concentrations and phototrophic biofilm tolerance in a contaminated river. Water Research 44 (6), 1941–1949.

Petchey, O.L., Morin, P.J., Hulot, F.D., Loreau, M., McGrady-Steed, J., Naeem, S., 2002. Contributions of aquatic model systems to our understanding of biodiversity and ecosystem functioning. In: Loreau, M., Naeem, S., Inchausti, P. (Eds.), Biodiversity and Ecosystem Functioning: Synthesis and Perspectives. Oxford University Press, pp. 127–154.

Pflugmacher, S., 2004. Promotion of oxidative stress in the aquatic macrophyte *Ceratophyllum demersum* during biotransformation of the cyanobacterial toxin microcystin-LR. Aquatic Toxicology 70 (3), 169–178.

Pienitz, R., Smol, J., Birks, H.J., 1995. Assessment of freshwater diatoms as quantitative indicators of past climatic change in the Yukon and Northwest Territories, Canada. Journal of Paleolimnology 13 (1), 21–49.

Pilotto, L.S., Douglas, R.M., Burch, M.D., Cameron, S., Beers, M., Rouch, G.J., Robinson, P., Kirk, M., Cowie, C.T., Hardiman, S., Moore, C., Attewell, R.G., 1997. Health effects of exposure to cyanobacteria (blue-green algae) during recreational water-related activities. Australian and New Zealand Journal of Public Health 21 (6), 562–566.

Pizzolon, L., Tracanna, B., Prósperi, C., Guerrero, J.M., 1999. Cyanobacterial blooms in Argentinean inland waters. Lakes & Reservoirs: Research & Management 4 (3–4), 101–105.

Ponader, K.C., Charles, D.F., Belton, T.J., 2007. Diatom-based TP and TN inference models and indices for monitoring nutrient enrichment of New Jersey streams. Ecological Indicators 7 (1), 79–93.

Porter, S.D., Mueller, D.K., Spahr, N.E., Munn, M.D., Dubrovsky, N.M., 2008. Efficacy of algal metrics for assessing nutrient and organic enrichment in flowing waters. Freshwater Biology 53 (5), 1036–1054.

Powers, C.D., Nau-Ritter, G.M., Rowland, R.G., Wurster, C.F., 1982. Field and laboratory studies of the toxicity to phytoplankton of polychlorinated biphenyls (PCBs) desorbed from fine clays and natural suspended particulates. Journal of Great Lakes Research 8 (2), 350–357.

Rai, L.C., Gaur, J.P., Kumar, H.D., 1981. Phycology and heavy metal pollution. Biological Reviews 56 (2), 99–151.

Ramelow, G.J., Biven, S.L., Zhang, Y., Beck, J.N., Young, J.C., Callahan, J.D., Marcon, M.F., 1992. The identification of point sources of heavy metals in an industrially impacted waterway by periphyton and surface sediment monitoring. Water, Air, and Soil Pollution 65 (1–2), 175–190.

Reavie, E.D., Jicha, T.M., Angradi, T.R., Bolgrien, D.W., Hill, B.H., 2010. Algal assemblages for large river monitoring: comparison among biovolume, absolute and relative abundance metrics. Ecological Indicators 10 (2), 167–177.

Ressom, R., San, S.F., Fitzgerald, J., Turczynowicz, L., El, S.O., Roder, D., Maynard, T., Falconer, I., 1994. Health Effects of Toxic Cyanobacteria (Blue – Green Algae). Australian Government Publishing Service, Canberra, pp. 27–69.

Richardson, D.C., Oleksy, I.A., Hoellein, T.J., Arscott, D.B., Gibson, C.A., Root, S.M., 2014. Habitat characteristics, temporal variability, and macroinvertebrate communities associated with a mat-forming nuisance diatom (*Didymosphenia geminata*) in Catskill mountain streams, New York. Aquatic Sciences 76 (4), 553–564.

Rimet, F., Cauchie, H.-M., Hoffmann, L., Ector, L., 2005. Response of diatom indices to simulated water quality improvements in a river. Journal of Applied Phycology 17 (2), 119–128.

Rothrock, P.E., Simon, T.P., Stewart, P.M., 2008. Development, calibration, and validation of a littoral zone plant index of biotic integrity (PIBI) for lacustrine wetlands. Ecological Indicators 8 (1), 79–88.

Rott, E., 1991. Methodological aspects and perspectives in the use of periphyton for monitoring and protecting rivers. In: Whitton, B.A., Rott, E., Friedrich, G., Limnologie, D.G.F. (Eds.), Use of Algae for Monitoring rivers: Proceedings of an International Symposium Held at the Landesamt für Wasser und Abfall Nordrhein-Westfalen, Düsseldorf, Germany, May 26–28, 1991. I.U.O.B.S.B. Commission, Institut für Botanik, Universität Innsbruck, Austria, pp. 9–16.

Rotter, S., Sans-Piché, F., Streck, G., Altenburger, R., Schmitt-Jansen, M., 2011. Active biomonitoring of contamination in aquatic systems—An in situ translocation experiment applying the PICT concept. Aquatic Toxicology 101 (1), 228–236.

Roubeix, V., Mazzella, N., Méchin, B., Coste, M., Delmas, F., 2011. Impact of the herbicide metolachlor on river periphytic diatoms: experimental comparison of descriptors at different biological organization levels. Annales de Limnologie – International Journal of Limnology (47), 239–249.

Ruangsomboon, S., Wongrat, L., 2006. Bioaccumulation of cadmium in an experimental aquatic food chain involving phytoplankton (*Chlorella vulgaris*, zooplankton *Moina macrocopa*, and the predatory catfish *Clarias macrocephalus C. gariepinus*. Aquatic Toxicology 78 (1), 15–20.

Rushforth, S., Brotherson, J., Fungladda, N., Evenson, W., 1981. The effects of dissolved heavy metals on attached diatoms in the Uintah Basin of Utah, U.S.A. Hydrobiologia 83 (2), 313–323.

Sabater, S., 2000. Diatom communities as indicators of environmental stress in the Guadiamar River, S-W. Spain, following a major mine tailings spill. Journal of Applied Phycology 12 (2), 113–124.

Sabater, S., Gregory, S.V., Sedell, J.R., 1998. Community dynamics and metabolism of benthic algae colonizing wood and rock substrata in a forest stream. Journal of Phycology 34 (4), 561–567.

Sabater, S., Navarro, E., Guasch, H., 2002. Effects of copper on algal communities at different current velocities. Journal of Applied Phycology 14 (5), 391–398.

Saenz, M.E., Di Marzio, W.D., 2009. Ecotoxicity of herbicide Glyphosate to four chlorophyceaen freshwater algae. Limnetica 28, 149–158.

Say, P.J., Whitton, B.A., 1977. Influence of zinc on lotic plants. Freshwater Biology 7 (4), 377–384.

Schlosser, I., 1990. Environmental variation, life history attributes, and community structure in stream fishes: implications for environmental management and assessment. Environmental Management 14 (5), 621–628.

Schmidt, J.R., Shaskus, M., Estenik, J.F., Oesch, C., Khidekel, R., Boyer, G.L., 2013. Variations in the microcystin content of different fish species collected from a eutrophic lake. Toxins 5 (5), 992–1009.

Schmitt-Jansen, M., Altenburger, R., 2005. Toxic effects of isoproturon on periphyton communities – a microcosm study. Estuarine, Coastal and Shelf Science 62 (3), 539–545.

Schwarzenbach, R.P., Escher, B.I., Fenner, K., Hofstetter, T.B., Johnson, C.A., Von Gunten, U., Wehrli, B., 2006. The challenge of micropollutants in aquatic systems. Science 313 (5790), 1072–1077.

Seele, J., Mayr, M., Staab, F., Raeder, U., 2000. Combination of two indication systems in prealpine lakes — diatom index and macrophyte index. Ecological Modelling 130 (1–3), 145–149.

Seitz, A., Ratte, H., 1991. Aquatic ecotoxicology: on the problems of extrapolation from laboratory experiments with individuals and populations to community effects in the field. Comparative Biochemistry and Physiology Part C: Comparative Pharmacology 100 (1), 301–304.

Serra, A., Guasch, H., Admiraal, W., Van der Geest, H.G., Van Beusekom, S.A., 2010. Influence of phosphorus on copper sensitivity of fluvial periphyton: the role of chemical, physiological and community-related factors. Ecotoxicology 19 (4), 770–780.

Simon, T.P., 2000. The use of biological criteria as a tool for water resource management. Environmental Science & Policy 3 (Suppl. 1), 43–49.

Slooten, L., Nuyten, A., 1983. Arsenylation of nucleoside diphosphates in *Rhodospirillum rubrum* chromatophores. Biochimica et Biophysica Acta (BBA)-Bioenergetics 725 (1), 49–59.

Soldo, D., Behra, R., 2000. Long-term effects of copper on the structure of freshwater periphyton communities and their tolerance to copper, zinc, nickel and silver. Aquatic Toxicology 47 (3), 181–189.

Solomon, S., 2007. Climate Change 2007-the Physical Science Basis: Working Group I Contribution to the Fourth Assessment Report of the IPCC. Cambridge University Press.

Sorace, A., Formichetti, P., Boano, A., Andreani, P., Gramegna, C., Mancini, L., 2002. The presence of a river bird, the dipper, in relation to water quality and biotic indices in central Italy. Environmental Pollution 118 (1), 89–96.

Spaulding, S.A., Elwell, L., 2007. Increase in Nuisance Blooms and Geographic Expansion of the Freshwater Diatom *Didymosphenia geminata*.

Starodub, M.E., Wong, P.T.S., Mayfield, C.I., Chau, Y.K., 1987. Influence of complexation and pH on individual and combined heavy metal toxicity to a freshwater green alga. Canadian Journal of Fisheries and Aquatic Sciences 44 (6), 1173–1180.

Steffen, M.M., Li, Z., Effler, T.C., Hauser, L.J., Boyer, G.L., Wilhelm, S.W., 2012. Comparative metagenomics of toxic freshwater cyanobacteria bloom communities on two continents. PLoS One 7 (8), e44002.

Steffen, M.M., Zhu, Z., McKay, R.M.L., Wilhelm, S.W., Bullerjahn, G.S., 2014. Taxonomic assessment of a toxic cyanobacteria shift in hypereutrophic Grand Lake St. Marys (Ohio, USA). Harmful Algae 33 (0), 12–18.

Steinaman, A.D., McIntire, C.D., 1986. Effects of current velocity and light energy on the structure of periphyton assemblage in laboratory streams. Journal of Phycology 22 (3), 352–361.

Steinberg, C., Schiefele, S., 1988. Biological indication of trophy and pollution of running waters. Zeitschrift fur Wasser- und Abwasser-Forschung 21 (6), 227–234.

Stevenson, R.J., 1984. Epilithic and epipelic diatoms in the Sandusky River, with emphasis on species diversity and water pollution. Hydrobiologia 114 (3), 161–175.

Stevenson, R.J., Pan, Y., Stevenson, R.J., Pan, Y., 1999. Assessing Environmental Conditions in Rivers and Streams With Diatoms. Cambridge University Press.

Stevenson, R.J., Rier, S., Riseng, C., Schultz, R., Wiley, M., 2006. Comparing effects of nutrients on algal biomass in streams in two regions with different disturbance regimes and with applications for developing nutrient criteria. Hydrobiologia 561 (1), 149–165.

Suter, G.W., 1993. A critique of ecosystem health concepts and indexes. Environmental Toxicology and Chemistry 12 (9), 1533–1539.

Tang, T., Cai, Q., Liu, J., 2006. Using epilithic diatom communities to assess ecological condition of Xiangxi river system. Environmental Monitoring and Assessment 112 (1–3), 347–361.

Tlili, A., Berard, A., Roulier, J.L., Volat, B., Montuelle, B., 2010. PO43- dependence of the tolerance of autotrophic and heterotrophic biofilm communities to copper and diuron. Aquatic Toxicology 98 (2), 165–177.

Tlili, A., Corcoll, N., Bonet, B., Morin, S., Montuelle, B., Berard, A., Guasch, H., 2011a. In situ spatio-temporal changes in pollution-induced community tolerance to zinc in autotrophic and heterotrophic biofilm communities. Ecotoxicology 20 (8), 1823–1839.

Tlili, A., Montuelle, B., Bérard, A., Bouchez, A., 2011b. Impact of chronic and acute pesticide exposures on periphyton communities. Science of the Total Environment 409 (11), 2102–2113.

Tlili, A., Dorigo, U., Montuelle, B., Margoum, C., Carluer, N., Gouy, V., Bouchez, A., Bérard, A., 2008. Responses of chronically contaminated biofilms to short pulses of diuron: an experimental study simulating flooding events in a small river. Aquatic Toxicology 87 (4), 252–263.

Town, R.M., Filella, M., 2000. A comprehensive systematic compilation of complexation parameters reported for trace metals in natural waters. Aquatic Sciences 62 (3), 252–295.

Villeneuve, A., Montuelle, B., Bouchez, A., 2011. Effects of flow regime and pesticides on periphytic communities: evolution and role of biodiversity. Aquatic Toxicology 102 (3–4), 123–133.

Villeneuve, A., Montuelle, B., Bouchez, A., 2010. Influence of slight differences in environmental conditions (light, hydrodynamics) on the structure and function of periphyton. Aquatic Sciences 72 (1), 33–44.

Vis, C., Hudon, C., Cattaneo, A., Pinel-Alloul, B., 1998. Periphyton as an indicator of water quality in the St Lawrence river (Quebec, Canada). Environmental Pollution 101 (1), 13–24.

Wängberg, S.-Å., 1995. Effects of arsenate and copper on the algal communities in polluted lakes in the northern parts of Sweden assayed by PICT (Pollution-Induced Community Tolerance). Hydrobiologia 306 (2), 109–124.

Wängberg, S.-Å., Heyman, U., Blanck, H., 1991. Long-term and short-term arsenate toxicity to freshwater phytoplankton and periphyton in limnocorrals. Canadian Journal of Fisheries and Aquatic Sciences 48 (2), 173–182.

Wallace, J.B., Anderson, N.H., 1996. Habitat, life history, and behavioral adaptations of aquatic insects. In: Merritt, R.W., Cummins, K.W. (Eds.), An Introduction to the Aquatic Insects of North America, third ed. Kendall/Hunt Publishing Company, pp. 41–73.

Walter, H., Consolaro, F., Gramatica, P., Scholze, M., Altenburger, R., 2002. Mixture toxicity of priority pollutants at no observed effect concentrations (NOECs). Ecotoxicology 11 (5), 299–310.

Wang, W.-X., Dei, R.C.H., 2006. Metal stoichiometry in predicting Cd and Cu toxicity to a freshwater green alga Chlamydomonas reinhardtii. Environmental Pollution 142 (2), 303–312.

Wefering, F.M., Danielson, L.E., White, N.M., 2000. Using the AMOEBA approach to measure progress toward ecosystem sustainability within a shellfish restoration project in North Carolina. Ecological Modelling 130 (1–3), 157–166.

Weitzel, R., Bates, J., 1981. Assessment of Effluent Impacts Through Evaluation of Periphyton Diatom Community Structure. Ecological Assessments of Effluent Impacts on Communities of Indigenous Aquatic Organisms. American Society for Testing and Materials, Philadelphia, pp. 142–165.

Welch, E.B., 2002. Ecological Effects of Waste Water: Applied Limnology and Pollutant Effects, second ed. CRC Press.

Whitton, B.A., Rott, E., Friedrich, G., 1991. Use of algae for monitoring rivers. Journal of Applied Phycology 3, 287.

Williams, L.G., Mount, D.I., 1965. Influence of zinc on periphytic communities. American Journal of Botany 52 (1), 26–34.

Xu, F.L., Tao, S., Dawson, R.W., Li, P.G., Cao, J., 2001. Lake ecosystem health assessment: indicators and methods. Water Research 35 (13), 3157–3167.

Xu, H., Min, G.S., Choi, J.K., Jung, J.H., Park, M.H., 2009. An approach to analyses of periphytic ciliate colonization for monitoring water quality using a modified artificial substrate in Korean coastal waters. Marine Pollution Bulletin 58 (9), 1278–1285.

Yachi, S., Loreau, M., 1999. Biodiversity and ecosystem productivity in a fluctuating environment: the insurance hypothesis. Proceedings of the National Academy of Sciences of the United States of America 96 (4), 1463–1468.

Chapter 4

Water and Wastewater Purification Using Periphyton

4.1 INTRODUCTION

Bioremediation refers to the use of microorganisms to eliminate or reduce the concentrations of hazardous wastes at a contaminated site (Boopathy, 2000; de Lorenzo, 2008). One important characteristic of bioremediation is that it is carried out in nonsterile open environments comprising of a variety of microorganisms (Huang et al., 2013; Sivakumar et al., 2012). In this diverse group of microorganisms, the central role of contaminant degradation is carried out by bacteria (Huang et al., 2013). A biological treatment system comprised of these microorganisms has various applications such as the rehabilitation of contaminated sites, such as water, soil, sludge, and waste streams (Boopathy, 2000; de Lorenzo, 2008; Wu et al., 2012).

Boopathy (2000) categorized bioremediation methods into ex situ and in situ bioremediation (Boopathy, 2000; de Lorenzo, 2008) but addresses primarily in situ methods, such as composting (Jørgensen et al., 2000; Peng et al., 2013), bioreactors (Wijekoon et al., 2013; Wu et al., 2011a,b; Yan et al., 2011), biofilters (Gómez-Silván et al., 2010; Jing et al., 2012), bioaugmentation (Mrozik and Piotrowska-Seget, 2010; Schauer-Gimenez et al., 2010), and biostimulation (McGlashan et al., 2012). Although in situ bioremediation methods have been in use for 20–30 years, to date they have not yielded expected results. Their limited success has been attributed to reduced ecological sustainability under environmental conditions. It is therefore necessary that a stable microbial ecosystem, including the balance of structure and composition, should be obtained when implementing bioremediation engineering projects.

de Lorenzo (2008) discusses a systematic biological approach providing an excellent overview of bioremediation involving algal species. More specific focus, however, is needed on periphyton communities, which are an important ecological constituent of surface waters and play a major role in primary productivity and food source biomasses (Larned, 2010). Periphyton is mainly composed of algae, diatoms, fungi, bacteria, and associated protozoa as well as small multicellular animals and organic detritus (Wu et al., 2010a). Periphyton

Periphyton. http://dx.doi.org/10.1016/B978-0-12-801077-8.00004-1

can be found virtually anywhere under water; they play an important role in nutrient cycling and trophic transfer of nutrients and affect the length and/or structure of food chains or foodwebs in aquatic ecosystems (Arnon et al., 2007; Azim, 2009). These factors and the sessile nature of periphyton communities make periphyton a potentially important bioremediation measure for contaminants in aquatic ecosystems (Kanavillil et al., 2012; Small et al., 2008).

Periphyton filtration is a renowned bioremediation method for polluted water (Bradac et al., 2010). Moreover, periphyton communities can be easily contrived and/or incorporated into bioreactors, resulting in efficient operation of bioremediation engineering projects (Wu et al., 2010b). In addition, the microbial composition of the periphyton communities has a relatively robust nature, which tends to resist change under variable conditions (Flynn et al., 2013; Larned, 2010; Stenger-Kovács et al., 2013). Finally, periphyton does not need a long acclimation phase for adapting to local environmental conditions.

4.2 CURRENT BIOLOGICAL WATER AND WASTEWATER TREATMENT METHODS

Many measures have been developed to treat domestic wastewater, groundwater, and specific industrial wastewater and/or contaminants, such as the treatment of domestic wastewater by an activated sludge process (Kassab et al., 2010), the bioremediation of groundwater by aeration (Jechalke et al., 2010), the use of bacterial H_2 reduction via a membrane biofilm reactor to remove oxidized contaminants (Xia et al., 2013), and methylene blue dye removal using bioadsorbents (*Sesamumindicum* L.) (Feng et al., 2011). These measures demonstrate the obvious benefits of using periphyton when purifying wastewater. However, surface waters are often affected by multiple contaminants carried by point and nonpoint source fluxes (Bosch et al., 2013; Wu et al., 2011a,d; Yan et al., 2011). Thus the development of a multifunctional water treatment system is of practical significance for simultaneous removal of the diverse array of contaminants from drinking water and wastewater.

4.3 POTENTIAL OF PERIPHYTON

Biological methods have become noticeably prevalent in water and wastewater treatment due to their environmentally benign and sustainable characteristics. The use of periphyton, comprising of diverse species, will be discussed in this chapter to highlight its potential for remediating surface waters that have been polluted by "heterogeneous water" such as nonpoint source runoff or complex industrial wastewater. Although the use of periphyton in bioremediation is still in its infancy, understanding how it may be used in surface water and wastewater management represents a step forward. Microorganisms in periphyton are considered a community, so this chapter will not discuss the interactions among the different organisms due to the lack of knowledge resulting from currently limited technology needed to examine interactions at this scale.

4.3.1 Removal of Heavy Metals

Metal accumulation by periphyton has been relatively well investigated. Serra et al. (2009) recently reported their studies of the effects of copper exposure on the periphyton community in fluvial ecosystems, albeit under controlled conditions. Bere et al. (2012) found that under field conditions, Cr(III) and Pb(II) influence Cd(II) accumulation and toxicity in tropical periphyton communities. Concentration and speciation were observed to vary dynamically in a small stream during rain events. Analyses revealed that the Cd(II) content in periphyton closely followed Cd(II) concentrations in water, despite being in the presence of higher concentrations of Zn(II) and Mn(II). Decreases in the Cd(II) content of periphyton after rain events were slower than decreases in water and are suggestive of metal accumulation (Bradac et al., 2010). Many photosynthetic species, such as the green alga *Chlamydomonas reinhardtii,* also have a large capacity to absorb metals and therefore are also potentially able to remove metals from wastewaters (Fortin et al., 2007; Mehta and Gaur, 2005).

The ability of microbial aggregates, such as periphyton, to remove heavy metals from water is probably related to their structure. It is well known that the structure of these types of communities ranges from patchy monolayers to filamentous accretions during different phases of biofilm formation (Wu et al., 2010a). The basic structure of periphyton includes at least three conceptual models: (1) heterogeneous mosaic biofilm aggregations, (2) penetrated water-channel biofilms, and (3) dense confluent biofilms. Due to the special porous structure of periphyton, pollutants can adsorb onto or desorb from the active sites of an aggregate surface concomitantly (Wu et al., 2010a); such is the case with Cd, Cu, and Pb ions that are freely shuttled into and out of *Ralstionia* sp. and *Bacillus* sp. aggregates (Choi et al., 2009).

Periphyton accumulate heavy metals via three main mechanisms (Holding et al., 2003): adsorption into extracellular polymeric substances (EPS), cell surface adsorption, and intracellular uptake (or absorption). Metal uptake in periphyton by adsorption and absorption has been evaluated by measuring total and intracellular metal content (Meylan et al., 2003; Serra et al., 2009). Other research has revealed that the inactive/dead microbial biomass also passively binds metal ions via various physicochemical mechanisms (Chien et al., 2013; Wang and Chen, 2009).

Complexation plays an important role in heavy metal removal by periphyton. Many functional groups in the EPS, such as carboxyl, phosphoric, sulfhydryl, phenolic, and hydroxyl groups, can complex with heavy metals (Sheng et al., 2010). To date, thousands of studies have shown that there is a significantly practical potential to remove heavy metals from aqueous solutions using microorganisms and periphyton (Joo et al., 2010; Kao et al., 2006, 2008; Tsuruta, 2004). For example, it was found that *Pseudomonas aeruginosa ASU 6a* (gram-negative) aggregates and *Bacillus cereus AUMC B52* (gram-positive) aggregates are inexpensive and efficient biosorbents for Zn(II) removal from aqueous solutions (Joo et al., 2010). Some gram-negative bacterial strains, such as *Acinetobactercalcoaceticus*, *Erwiniaherbicola*, *P. aeruginosa,* and *P. maltophilia*, have a high affinity for gold biosorption, as do *P. maltophilia* cells immobilized with polyacrylamide gel (Tsuruta, 2004). *Escherichia coli* is an effective bacterial biosorbent used for the removal of multiple heavy metals, such as Pb, Cu, Cd, and Zn (Kao et al., 2006). These studies indicate that the removal of heavy metals by periphyton is related to the composition of the microorganisms and/or periphyton used.

The ion exchange mechanism is the main mode of interaction between some divalent cations and the EPS (Sheng et al., 2010). It has been reported that the binding between the EPS and divalent cations, such as Ca^{2+} and Mg^{2+}, is one of the main intermolecular interactions supporting microbial aggregate structures. During the removal of metals by microbial aggregates, Ca^{2+} and Mg^{2+} are simultaneously released into solution, indicating that ion exchange is involved (Yuncu et al., 2006).

Solid–liquid separation mechanisms, such as flocculation and/or precipitation, are important processes used by periphyton in removing heavy metal ions from wastewater (Choi et al., 2009). For example, a brewer's yeast strain (*Saccharomyces cerevisiae*) was used to remove heavy metals [Cu(II), Ni^{2+}, Zn^{2+}, Cd^{2+}, and Cr^{3+}] from a synthetic effluent. The solid–liquid separation process was carried out using the flocculation ability of the strain. The results showed that flocculation by yeast strains can be used as an inexpensive and natural separation process to remove heavy metals from a wide range of industrial effluents (Machado et al., 2008).

During the adsorption of heavy metals, ions can be isolated by adsorption onto EPS from microorganisms and periphyton (Choi et al., 2009). Natural and extreme acidic eukaryotic biofilms have a strong binding capacity for heavy metals, such as Hg(II), Zn, Cu, Co, Ni, As, Cd, Cr, and Pb, by releasing colloid materials such as protein or affecting the ion value (e.g., the transformation of Hg^{2+} to Hg^0) (Choi et al., 2009; George et al., 2005; Neu and Lawrence, 2010). This indicates that EPS plays an important role in heavy metal complexation during the removal of heavy metals by microbial aggregate adsorption. EPS characteristics may significantly affect the chemical forms, mobility, bioavailability, and ecotoxicity of heavy metals in aqueous solutions.

The efficiency of heavy metal adsorption by periphyton is affected by many factors such as biological and chemical composition, functional groups, and pH. For example, mammalian and fish metallothionein expression in *E. coli* aggregates has led to a significant increase (5–210%) in the overall biosorption efficiency of Pb, Cu, Cd, and Zn (Kao et al., 2006). The biosorption of gold from a solution containing hydrogen tetrachloroaurate(III) using *P. maltophilia* with a high affinity for gold adsorption was very rapid and affected by the pH of the solution, external gold concentration, and cell amounts (Tsuruta, 2004). The presence of amino, carboxyl, hydroxyl, and carbonyl groups led to greater Zn biosorption by a gram-negative bacterium (*P. aeruginosa*) than a gram-positive bacterium (*Bacillus cereus*) (Joo et al., 2010).

The major advantages of using periphyton for biosorption are their high effectiveness in reducing heavy metal ions and the use of inexpensive biosorbents. Microbial aggregate biosorption processes are particularly suitable for treating dilute heavy metal wastewater (Fu and Wang, 2011) and may be used under a range of conditions due to the complex composition of heterotrophic and photoautotrophic microorganisms.

Microbial biomass heavy metal biosorbents are characteristically environmentally safe, come from a range of sources, are low cost, have short maturation and acclimation periods, and exhibit rapid adsorption. Periphyton has vast potential for heavy metal removal under various conditions due to its hierarchical and self-maintained microecosystem. These microbial biomasses can be cultivated and fostered in wastewater treatment systems such as bioreactors, which cannot only improve the efficiency of heavy metal removal but maintain the stability of the periphyton microecosystem (Wu et al., 2011b, 2012).

4.3.2 Removal of Nutrients

4.3.2.1 Nitrogen and Phosphorus

Nutrients, especially carbon, nitrogen, and phosphorus, are necessary for periphyton growth. Their incorporation into the microbial biomass is very efficient due to photosynthetic microorganisms (Merchant et al., 2012; Wu et al., 2012). These microorganisms have the ability to utilize inorganic forms of nitrogen, such as nitrite, nitrate, or ammonium, as the sole nitrogen source for growth (Yariv, 2001). In the first reactor of an intermittently aerated anaerobic—aerobic activated sludge process, nitrification and phosphorus uptake occur during the aeration period, followed by denitrification and phosphorus release during the agitation period (Sasaki et al., 1996). In the second reactor, nitrification and phosphorus uptake occur during aeration, and denitrification and weak phosphorus uptake occur during agitation (Sasaki et al., 1996; Villaverde, 2004).

Periphyton often provide the primary habitats for many neighboring suspended microorganisms (De Beer and Stoodley, 2006). The organic materials adsorbed and deposited onto periphyton are often the main nutrient sources for these microorganisms. Most of these organic materials are absorbed and converted into cell materials such as cytoplasm, maintaining microorganism growth and fostering the formation of some active materials such as EPS. The rest of the organic materials are transferred into excretion. During the conversion process, high-energy ATP is released. This energy promotes the growth of microorganisms and the formation of periphyton (Adav et al., 2010; Neu and Lawrence, 2010; Sheng et al., 2010).

Organic materials often "envelop" and/or "carry" nutrients that in turn supply the growth of microorganisms. It is well known that nutrient removal by microorganism assimilation is associated with microbial aggregate (periphyton) layers (Laspidou and Rittmann, 2002). It has been reported that there is a very thin water layer that adheres to a microbial aggregate surface. The particulates in this thin water layer move slowly, and the organic materials attached to the particulates are mostly absorbed by local microorganisms. As a result, nutrient concentrations in the thin water layer of periphyton are lower than those in the inner layers of periphyton. Nutrients carried by particulates in water move to the thin water layer of periphyton, resulting in greater nutrient ingestion by microorganisms (Badireddy et al., 2010; Laspidou and Rittmann, 2002; Sheng et al., 2010).

Not all organic materials adhering to periphyton can be converted into cell bioplasm; many are retained by periphyton as "stored materials." These include dissolved and nondissolved organic materials that can be useful for microorganism metabolism during growth. Polysaccharides and polyhydroxybutyrate (PHB) are stored inside cells as easily degraded compounds, providing sources of carbon and energy for growth. They are also useful for the removal of nitrogen. Under anaerobic and anoxic conditions, these materials can be easily degraded during denitrification (Badireddy et al., 2010; Laspidou and Rittmann, 2002; Sheng et al., 2010).

Most easily degraded materials are stored in the extracellular matrix. Once the intracellular concentration of these compounds decreases to a certain degree, the materials in the extracellular matrix (microbial aggregate matrix) will become the carbon source for the denitrification process (Badireddy et al., 2010; Laspidou and Rittmann, 2002; Sheng et al., 2010).

The fate of the nutrients and organic compounds in periphyton communities is highly diverse, both spatially and temporally (Azim, 2009; Baxter et al., 2013; Bowes et al., 2012). This allows periphyton to be used for the treatment of heterogeneous wastewater, especially the removal of nitrogen and phosphorus, safely, economically, and conveniently. For example, a periphyton–fish system was proposed to remove phosphorus and nitrogen from wastewater effluent and was found to be highly effective (Rectenwald and Drenner, 1999; Wu et al., 2014).

The biodegradation of nitrogenous compounds by microorganisms and periphyton has been well described. This process often features two predominant processes: autotrophic nitrification and heterotrophic denitrification. Some minor processes such as heterotrophic nitrification and aerobic denitrification are also involved (van Loosdrecht and Jetten, 1998). Organisms or periphyton that degrade nitrogenous compounds can be divided into three main groups according to the biological nitrogen removal processes they conduct: (1) degradation of nitrogen-organic matter and the release of ammonia by various microorganisms, (2) conversion of ammonia to nitrate by certain autotrophic microorganisms and, (3) conversion of nitrate to nitrogen gas by a mixed culture that uses nitrate as an electron acceptor (as opposed to free oxygen used by the nitrifiers) in the metabolism of organic carbon (Villaverde, 2004; Wu et al., 2011b,c).

Another major route for the removal of biological nitrogen from waste liquids involves the removal of nitrogen by nitrification–denitrification processes (Yariv, 2001). The use of a multilevel bioreactor fosters the coexistence of photoautotrophic and heterotrophic microorganisms, which provide environments for the combination of oxidative and reductive processes (Wu et al., 2011b). dos Santos et al. (1996) investigated a system that used the oxidative and reductive environments within polymer beads to remove nitrogen via nitrification–denitrification processes. Their results showed that high nitrogen removal rates (up to 5.1 mmol $N\ m^{-3}$ polymer s^{-1}) were achieved under continuous flow and aerobic conditions because the nitrifier *Nitrosomona seuropaea* and either of the denitrifiers *Pseudomonas denitrificans* or *Paracoccus denitrificans* were coimmobilized in the system.

It is well known that the introduction of excessive phosphorus accelerates the eutrophication process of closed water areas. Phosphorus is often regarded as the limiting factor for phytoplankton growth, thereby accelerating the growth of harmful algal blooms (Smith et al., 1999). Thus, the removal of phosphorus, especially biological phosphorus, has recently been a subject of great concern. In microbial aggregate systems such as wastewater treatment plants using the activated sludge biofilm method, biological phosphate removal is based on the capacity of some microorganisms to store *ortho*-phosphate intracellularly as polyphosphate. These microorganisms store PHB anaerobically, which is oxidized in a phase with an electron acceptor such as oxygen or nitrate present (Villaverde, 2004).

The microorganisms in periphyton largely responsible for P removal are known as the polyphosphate-accumulating organisms (PAOs) (Oehmen et al., 2007). PAOs take up readily biodegradable chemical oxygen demand substrates and store them as polyhydroxyalkanoates (PHAs). The energy required for this anaerobic process is derived from the hydrolysis of intracellular polyphosphate. In the subsequent aerobic stage, PAOs use PHAs to generate energy for growth and phosphorus uptake. In this process the PAOs take up more phosphorus than that released during the anaerobic stage (luxury uptake) (Zhou et al., 2010). Removing phosphate using nitrate instead of oxygen has the advantages of saving energy (oxygen input) and using less organic carbon. It is possible to accumulate denitrifying P-removing bacteria (DPB), which can simultaneously remove P and N in periphyton (Tsuneda et al., 2006).

4.3.2.2 Sulfate

Sulfate is a common constituent of many industrial wastewaters (Lens and Hulshoff Pol, 2000), and its reduction is dominated by two stages of inhibition. Primary inhibition is due to competition for common organic and inorganic substrates between sulfate-reducing bacteria, which suppress methane production. Secondary inhibition results from the toxicity of sulfide manifested in various bacterial groups (Chen et al., 2008). Two major groups of sulfate-reducing bacteria (incomplete and complete oxidizers) dominate the sulfate reduction process. Incomplete oxidizers reduce compounds such as lactate to acetate and CO_2. Complete oxidizers completely convert acetate to CO_2 and HCO_3^- (Chen et al., 2008).

Microorganisms play an essential role in the sulfur cycle, catalyzing both oxidation and reduction reactions of sulfur compounds (Fig. 4.1). These reactions include (1) dissimilatory sulfate reduction, the reduction of sulfate to sulfide is coupled with energy conservation and growth; (2) dissimilatory sulfur reduction, the electron acceptor is elemental sulfur; (3) assimilatory sulfate reduction, the reduced sulfide is assimilated in biomass, proteins, amino acids, and cofactors by plants, fungi, and microorganisms; (4) mineralization of organic compounds with hydrogen sulfide release; (5) sulfide oxidation by O_2, NO_3^-, Fe^{3+}, or Mn^{4+} as electron acceptors by lithotrophic and phototrophic

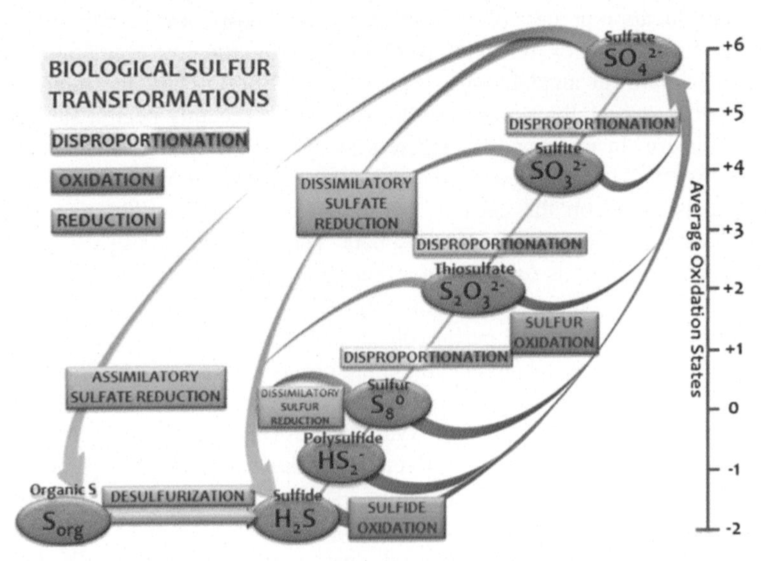

FIGURE 4.1 Biological sulfur transformations. *Taken from Sánchez-Andrea, I., Sanz, J.L., Bijmans, M.F.M., Stams, A.J.M., 2014. Sulfate reduction at low pH to remediate acid mine drainage. In: Research Frontiers in Chalcogen Cycle Based Environmental Technologies Selected Papers Presented at the 3rd International Conference on Research Frontiers in Chalcogen Cycle Science and Technology (G16), UNESCO-IHE, Delft, The Netherlands, May 27th–28th, 2013, vol. 269, pp. 98–109.*

bacteria, producing sulfur and subsequently sulfate; and (6) disproportionation, the coupled oxidation and reduction of sulfur compounds (thiosulfate, sulfite, and sulfur) to sulfate and sulfide. Table 4.1 shows the equations for some of the relevant biological processes with the energy release for each reaction (Sánchez-Andrea et al., 2014).

The microbial conversion of sulfurous compounds involves the metabolism of several different specific groups of bacteria, such as sulfate-reducing bacteria, sulfur- and sulfide-oxidizing bacteria, and phototrophic sulfur bacteria

TABLE 4.1 Stoichiometry and Gibbs Free Energy Changes of Some of the Reactions in the Biological Sulfur Cycle

Reaction Equations	$\Delta G°$ (kJ mol^{-1})
Sulfide Oxidation	
$HS^- + 3/2O_2 + H^+ \rightarrow S^0 + H_2O$	−210
$HS^- + 2O_2 \rightarrow SO_4^{2-} + H^+$	−709
$HS^- + Fe^{3+} \rightarrow S^0 + Fe^{2+} + H^+$	−47
$HS^- + 2/5NO_3^- \rightarrow S^0 + 1/5N_2 + 6/5H_2O$	−214
Sulfur Oxidation	
$S^0 + 1.5O_2 + H_2O \rightarrow SO_4^{2-} + 2H^+$	−499
$S^0 + 6/5NO_3^- + 2/5H_2O \rightarrow SO_4^{2-} + 3/5H_2 + 4/5H^+$	−510
Disproportionation	
$S_2O_3^{2-} + H_2O \rightarrow SO_4^{2-} + HS^- + H^+$	−22
$S^0 + H_2O \rightarrow 1/4SO_4^{2-} + 3/4HS^- + 5/4H^+$	9.5
$SO_3^{2-} + 2/3H^+ \rightarrow 2/3SO_4^{2-} + 1/3S^0 + 1/3H_2O$	−7.6
Sulfate Reduction	
$CH_3COO^{-a} + SO_4^{2-} \rightarrow 2HCO_3^- + HS^-$	−48
$4H_2 + SO_4^{2-} + H^+ \rightarrow HS^- + 4H_2O$	−151.9
Sulfur Reduction	
$1/4C_2H_3O_2^{-a} + H_2O + S^0 \rightarrow 1/2HCO_3^{-2} + 5/4H^+ + HS^-$	−13
$H^2 + S^0 \rightarrow HS^- + H^+$	−27.8

[a]Acetate is used as a representative organic compound, but other organic compounds may be used as well (Rittmann and McCarty, 2001).Gibbs free energy changes were calculated from Thauer et al. (1977).

(Lens and Hulshoff Pol, 2000). Some of these microorganisms can simultaneously use nitrate in autotrophic denitrification by sulfur- and sulfide-oxidizing microorganisms (Villaverde, 2004). In the anaerobic part of microbial aggregates, sulfate reduction contributes considerably to the mineralization process (Zhang et al., 2009). The other important internal cycle in periphyton is sulfate reduction coupled to sulfide oxidation (Villaverde, 2004).

4.3.3 Degradation of Organic Matter

4.3.3.1 Function of Periphyton in Organic Matter Degradation

Biodegradation is the chemical disbanding of organic materials by microorganisms or other biological agents. Microbial degradation of chemicals in the environment is an important route for the removal of these compounds. The types of compounds range from plastics to organic chemicals (both industrial chemicals used in large quantities and trace chemicals such as endocrine disruptors) to organometallics such as methylmercury (Fu and Wang, 2011; Sheng et al., 2010). The biodegradation of these compounds often involves a complex series of biochemical reactions and usually varies with the microorganisms involved. Compounds can be degraded aerobically and/or anaerobically. Biomineralization is the conversion of organic matter to minerals (Diaz, 2008).

Periphyton biofilm is a complex symbiotic system consisting of both photosynthetic and non-photosynthetic microorganisms. A number of species in the periphyton community degrade a range of organic compounds. Most of the organic matter produced by periphyton is degraded microbially within the periphytic community (Azim, 2009).

Currently, most efforts are directed toward the removal of specific contaminants such as nutrients (N and P) and sulfurous compounds due to their significant impact on water quality (Villaverde, 2004; Wu et al., 2011a,b,c). The most important practical use of periphyton is in biological wastewater treatment, while many emerging technologies use periphyton for biodegradation and bioremediation in bioreactors (Liong, 2011). Owing to the inclusion of specific microorganisms with special degradation functions in periphyton (e.g., microcystin-degrading bacteria *Sphingpoyxis* sp. and *Sphingomonas* sp.), periphyton as an ensemble is also often used to remove specific compounds such as microsytin-RR (Wu et al., 2010a), aliphatic homopolyesters, and aliphatic-aromatic copolyesters (Abou-Zeid et al., 2004).

The EPS in periphyton has many available sites for the adsorption of metals and nonbiodegradable and persistent organic matter, such as dyes, aromatics, aliphatics in proteins, and hydrophobic regions in carbohydrates (Sheng et al., 2010). For example, the color removal rate of the azo dye Congo Red by Basidiomycete biosorption in an agitated batch system reached 90% (Tatarko and Bumpus, 1998). Sivasamy and Sundarabal (2011) used

Aspergillus niger and *Trichoderma* sp. as biosorbents for the biosorption of the azo dye *Orange* G. They found that maximum biosorption occurred at pH 2, and the biomass obtained from *Aspergillus niger* was higher than that from *Trichoderma* sp., indicating *Aspergillus niger* was a better biosorbent. Wu et al. (2010a) demonstrated that biosorption by microorganisms and microbial aggregates was the main mechanism for the removal of microcystin-RR from aquatic solutions during the adaption period.

4.3.4 Removal of Phenols

Phenols are an important industrial chemical widely and commonly used as raw materials or intermediates in explosives, medicine, pesticides, dyes, wood preservatives, and rubber production. Phenols are toxic, carcinogenic, mutagenic, and teratogenic and are regarded as priority pollutants in the US Environmental Protection Agency list (Veeresh et al., 2005). The biodegradation rate of phenol in the natural environment is slow. Consequently, phenols accumulate in the environment and persist for a long time, threatening the safety of flora and fauna as well as human beings (Dosta et al., 2011).

Phenols are a common contaminant and their degradation pathways have been extensively investigated. Generally, the metabolic degradation of these organic compounds is associated with oxidative enzymes in microorganisms (Olaniran and Igbinosa, 2011). For example, the degradation pathway of 2,4,6-trichlorophenol (2,4,6-TCP) in the well-known chloroaromatic compound-degrading aerobic bacterium *Cupriavidus necator* JMP134 (pJP4) (formerly *Ralstonia eutropha* JMP134) takes place in the presence of an array of enzymes encoded by the *tcpRXABCYD* gene cluster (Sánchez and González, 2007). Studies have also reported that the *tcpABC* genes from *C. necator* JMP134 (pJP4) (Clément et al., 1995; Olaniran and Igbinosa, 2011; Padilla et al., 2000) encode enzymes that convert 2,4,6-TCP to 2-chloromaleylacetate (2-CMA) (Louie et al., 2002; Olaniran and Igbinosa, 2011). The *tcpABC* genes in *C. necator* are adjacent to four other open reading frames (*tcpY, tcpD, tcpR,* and *tcpX*), thus forming a putative catabolic operon (Olaniran and Igbinosa, 2011; Sánchez and González, 2007). The *tcpR* gene carries a significant identity to the *pcpR* gene, which encodes a LysR-type regulator involved in the degradation of pentachlorophenol (PCP) in *Sphingobium chlorophenolicum* (Cai and Xun, 2002; Olaniran and Igbinosa, 2011). The tcpX protein is supposed to provide $FADH_2$ to *tcpA* because it highly identifies with the TftC protein, which executes this function in the degradation of 2,4,5-TCP in *Burkholderia cepacia* AC1100 (Gisi and Xun, 2003; Olaniran and Igbinosa, 2011). This flavin reductase activity would also be encoded by the *tcpB* gene, which shows sequence similarity to genes coding for nitroreductases. However, a *tcpB* mutant still degrades 2,4,6-TCP (Louie et al., 2002; Olaniran and Igbinosa, 2011). These genes are located in a genetic context, *tcpRXABCYD*, which resembles a putative catabolic operon (Olaniran and Igbinosa, 2011;

Sánchez and González, 2007). Thus, a key point for future explorations of the mechanisms of organic matter degradation by periphyton biofilms, is to investigate the responsible genome. Advanced technologies such as quantitative polymerase chain reaction and metagenomics should be developed as a priority to support the microbial molecular ecology (Wu et al., 2014).

Many phenol-degrading microorganisms, including bacteria, fungi, yeast, and periphyton, have been identified in aqueous solutions (Kang et al., 2006; Kurzbaum et al., 2010; Wang et al., 2007; Yan et al., 2006). For example, the immobilized bacterium *Acinetobacter* sp. has potential for the treatment of phenol-containing wastewater (Wang et al., 2007). Fungi strains (*Graphium* sp. and *Fusarium* sp.) have high percentages of phenol degradation with 75% degradation of 10 mmol L^{-1} phenol in 168 h (Santos and Linardi, 2004).The biodegradation of phenol and *m*-cresol using a pure culture of yeast (*Candida tropicalis*) demonstrated that *C. tropicalis* alone could degrade 2000 mg L^{-1} phenol within 66 h (Yan et al., 2006). Planktonic *Pseudomonas pseudoalcaligenes* cells exhibited a high phenol removal rate in constructed wetland systems, especially those with subsurface flow, suggesting that surface-associated microorganisms (biofilms) can make a much higher contribution to the removal of phenol and other organics due to their greater bacterial biomass (Kurzbaum et al., 2010). Although olive mill wastewater has high polluting power and phenol concentrations as well as high antibacterial activity (Bleve et al., 2011), some bacteria such as *Pleurotus* spp. strains have the ability to remove phenolic compounds from this wastewater stream (Tsioulpas et al., 2002).

The biodegradation of phenol (i.e., bisphenol A or BPA) by microorganisms is mainly carried out by lignin-degrading enzymes such as manganese peroxidase (MnP) and laccase, which are produced by white rot Basidiomycetes microorganisms (Kang et al., 2006). MnP is a heme peroxidase that oxidizes phenolic compounds in the presence of Mn(II) and H_2O_2 while laccase is a multi−copper oxidase that catalyzes the one-electron oxidation of phenolic compounds by reducing oxygen to water (Kang et al., 2006; Reinhammar, 1984). MnP and laccase can degrade BPA and disrupt its estrogenic activity (Kang et al., 2006). In the case of laccase, BPA metabolism is faster in the presence of mediators, such as 1-hydroxybenzotriaxzole (HBT) and 2,2′-azo-bis(3-ethylbenzthiazoline-6-sulfonate), than with laccase alone.

Environmental conditions, that is, whether aerobic or anaerobic, can significantly affect the efficiency of phenol biodegradation. BPA in river waters is biodegraded under aerobic conditions but not under anaerobic conditions (Kang et al., 2006). BpA in spiked samples was rapidly biodegraded under aerobic conditions (>90%), with little decrease in BPA observed under anaerobic conditions (<10%) over 10 days (Kang and Kondo, 2002a). Under anaerobic conditions, such as those in anaerobic marine sediment, BPA was not biodegraded even after 3 months of incubation. These results suggest that anaerobic bacteria have little or no ability to degrade BPA (Kang et al., 2006; Voordeckers et al., 2002).

Phenol biodegradation by microorganisms is also influenced by temperature and microbe counts. The half-lives for phenol biodegradation in 15 river water samples averaged 4 and 7 days at 30 and 20°C, respectively, but only about 20% (0.04 mg L^{-1}) of the spiked phenol was biodegraded at 4°C over a period of 20 days (Kang and Kondo, 2002b). The phenol removal rate is higher in the subsurface flow of constructed wetlands than at the surface due to the greater bacterial biomass (Kurzbaum et al., 2010). It has also been reported that phenol biodegradation does not correlate with bacterial counts (Klecka et al., 2001). These differences may be due to the size of bacterial populations that can execute fast and complete phenol biodegradation or mineralization (Klecka et al., 2001).

4.3.5 Removal of Quinoline

Quinoline is a heterocyclic aromatic organic compound, which is mainly used to synthesize pharmaceuticals, dyes, pesticides, and many chemical additives (Padoley et al., 2008). Due to its toxicity and nauseating odor, discharging quinoline-containing waste greatly damages human health and environmental quality. Studies of quinoline-degrading bacteria not only help reveal the metabolic mechanism of quinoline but also improve the biotreatment of quinoline-containing wastewater (Sun et al., 2009).

The decomposition of quinoline and its derivatives have been enhanced by using either the free or immobilized cells of degrading microorganisms such as *Burkholderia pickettii* (Wang et al., 2002) and *Pseudomonas* sp. BW003 (Sun et al., 2009). In many cases, the biodegradation of quinoline by microorganisms is well described by mathematic models (Wang et al., 2002). For example, *B. pickettii* immobilized on a hybrid carrier could be used to degrade quinoline, with the subsequent degradation process described by a zero-order reaction rate equation when the initial quinoline concentration was in the range of 50–500 mg L^{-1} (Wang et al., 2002).

Although different genera of bacteria may produce different intermediates, almost all transform quinoline into 2-hydroxyquinoline as a first step under anaerobic conditions (Kaiser et al., 1996). During the following transformation step, a new degradation product of quinoline 3,4-dihydro-2-quinoline accumulates and is further transformed into unidentified products (Johansen et al., 1997). The transformation of quinoline by *Pseudomonas* sp. under anaerobic conditions has been reported with the first intermediate metabolite of quinoline catabolism identified as 3-hydroxy coumarin (Padoley et al., 2008). In general, degradation rates, including quinoline degradation, are significantly faster under aerobic conditions (Sun et al., 2009). This is consistent with the accepted view that microbial populations are generally larger and more metabolically active under aerobic conditions.

Because quinoline is one of the most important compounds containing N as a heteroatom (Padoley et al., 2008), this property leads to the retention of quinoline with N transformation during quinoline decomposition. In the aforementioned study, if quinoline was the sole C and N source, N was transformed primarily into ammonia-N, which was then used to synthesize cells. Less than 6% of ammonia-N was transformed into nitrate through heterotrophic nitrification. When glucose was added, more ammonia-N was used by BW003 such that the concentration of ammonia-N was clearly reduced. Therefore, by controlling the C:N ratio, ammonia-N as well as quinoline and its metabolic products can be completely eliminated (Sun et al., 2009).

Endocrine disrupting compounds (EDCs) are chemicals with the potential to elicit negative effects on the endocrine systems of humans and wildlife (Liu et al., 2009). In the past few decades, many methods for EDC elimination from the aquatic ecosystem have been developed, such as activated sludge wastewater treatment systems. During the removal of EDCs by activated sludge wastewater treatment systems and other similar systems, the major mechanisms involved are adsorption and biodegradation (Liu et al., 2009). Considering the EDCs' toxicity to microorganisms, the combination of physical and chemical methods such as activated carbon absorption and advanced oxidation may improve the EDC removal efficiency during the application of technologies based on microbial aggregates.

In summary, periphyton biofilms are capable of entrapping organic detritus, removing metals and nutrients from the water column, and degrading organic compounds in the water column. As a result, periphyton biofilms also help to control the dissolved oxygen concentration and pH of the surrounding water (Crab et al., 2007; Martin and Nerenberg, 2012; Wu et al., 2014).

4.4 CONJOINT ACTION OF ASSIMILATION, ADSORPTION, AND BIODEGRADATION

The diverse metabolic capabilities of microorganisms and their interactions with hazardous organic and inorganic compounds have long been recognized. In many practical cases, assimilation, biodegradation, and biosorption (including complexation, ion exchange, flocculation, and/or precipitation) occur simultaneously during the removal of pollutants. This conjunct action may be described as shown in Fig. 4.2.

Many studies on the conjunct action of assimilation, adsorption, and biodegradation have been conducted in recent years. Some typical cases are summarized as follows.

The biosorption and biodegradation of trichloroethylene (TCE). Experimental results showed that at 25°C the adsorption equilibrium of TCE at concentrations ranging from 10 to 200 mg L^{-1} could be described by the Freundlich isotherm with adsorption completed within 15 min. The results further indicated that glucose could serve as a cosubstrate and enhance TCE

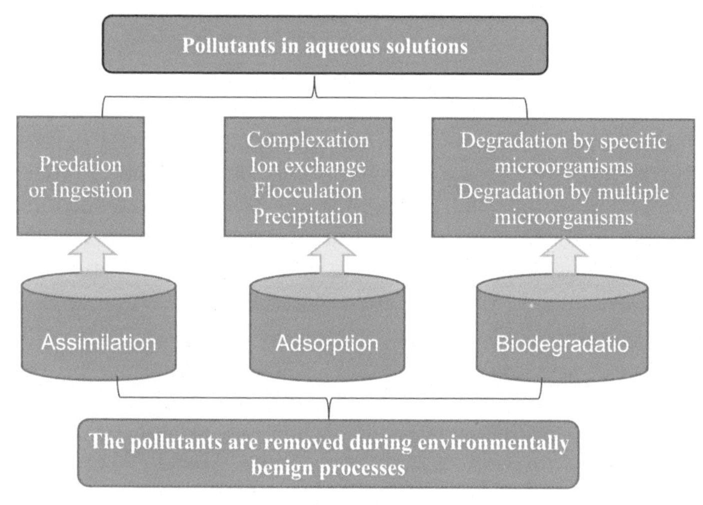

FIGURE 4.2 The process of pollutant removal from aqueous solutions by the conjunct mechanisms of assimilation, adsorption, and biodegradation.

biodegradation through co-metabolism. The TCE biodegradation conformed to first-order reaction kinetics, and the rate constant was 0.3212 d^{-1} at 25°C (Yang et al., 2009). A similar study revealed the effectiveness of microcystin-RR removal by periphyton in surface waters using the dual removal mechanisms of adsorption and biodegradation (Wu et al., 2010a).

The biosorption and biodegradation of PCP. Ye and Li (2007) investigated the biosorption and biodegradation of PCP by anaerobic periphyton to better understand the fate of PCP in an upflow anaerobic sludge blanket reactor. Their results demonstrated that the main mechanism leading to the removal of PCP in anaerobic periphyton was biodegradation, while adsorption was an accessorial process.

The adsorption of organic pollutants by periphyton may be attributed to the fact that there are some hydrophobic regions in EPS (Spath et al., 1998). More than 60% of benzene, toluene, and *m*-xylene were reportedly adsorbed by EPS with only a small fraction of these pollutants adsorbed by cells (Spath et al., 1998). EPS with negative charges is capable of binding with positively charged organic pollutants via electrostatic interaction (Neu and Lawrence, 2010; Sheng et al., 2010). Moreover, proteins have a higher binding strength and better binding capability than humic substances. Soluble EPS has a higher fraction of proteins than bound EPS; thus, it may have a greater binding capacity (Pan et al., 2010; Sheng et al., 2010).

The removal of N, P, and organic pollutants from water using seeding type immobilized microorganisms. Ten strains of dominant heterotrophic bacteria belonging to *Pseudomonas*, *Coccus*, *Aeromonas*, *Bacillus*, and *Enterobateriaceae* were isolated. The rates of TOC, TP, and TN removal were

80.2%, 81.6%, and 86.8%, respectively (Wang et al., 2008). The removal mechanisms simultaneously employed assimilation, adsorption, and biodegradation.

To date, many devices based on the combined mechanisms of assimilation, adsorption, and biodegradation to remove pollutants have been developed in which various robust microorganisms have been cultivated and cultured. These complex microorganisms, which exhibit strong activities, foster the improvement of pollutant removal efficiencies and meet the demand for the treatment of heterogeneous wastewater. Wu et al. (2011b) used a hybrid bioreactor featuring sequential anaerobic, anoxic, and aerobic (A^2/O) processes to treat industrial wastewater and domestic sewage. The removal process included assimilation, adsorption and biodegradation, and the removal efficiencies of the nutrients were 81% for TP, 74% for TDP, 82% for TN, 79% for NO_3-N, and 86% for NH_4-N. In addition, a photobioreactor-wetland (Wu et al., 2011c) and a multilevel bioreactor (Wu et al., 2011b) have been employed to remove UV_{254} matter, metals, and nutrients from non–point-source wastewater.

The individual contributions of assimilation, adsorption, and biodegradation to pollutant removal are not equal. During the interaction between microorganisms and pollutants, the rate of assimilation is typically low. Moreover, most pollutants are adsorbed and biodegraded by the unique metabolic activity of microorganisms and periphyton. Microbial processes are compatible with various environments, which is beneficial to contaminant transform, decompose, and degrade in some degree (Singh and Ward, 2004).

Assimilation, adsorption, and biodegradation may affect each other, thus limiting or stimulating pollutant removal. For example, adsorption might restrict biodegradation, while nutrient limitation and the presence of organic contaminants might stimulate biodegradation (Chen et al., 2010). Adsorption, and not biodegradation, has been observed to be the most common fate of tetracycline entering a biological process (Prado et al., 2009).

4.5 MECHANISMS RESPONSIBLE FOR CONTAMINANT REMOVAL

During the removal of contaminants by periphyton, different processes including uptake, adsorption, and biodegradation are likely to be simultaneously involved. For example, microcystin-RR removal by periphyton in surface waters uses the dual removal mechanisms of adsorption and biodegradation (Wu et al., 2010a). Each process mentioned earlier often involves subprocesses. For example, the conventional subprocesses for removing heavy metals from aqueous solutions by periphyton include complexation (by both chelation and other types of coordination) (Wang and Dei, 2001), adsorption, absorption (Barakat, 2011; Wang and Chen, 2009), filtration, microprecipitation (Bradac et al., 2010; Scinto and Reddy, 2003; Žižek et al., 2011), electrostatic interaction (Pokrovsky et al., 2013), and ion exchange (Yang et al., 2013). Generally, these occur concurrently during the removal of contaminants by periphyton. Due to the conjoint action of these different processes, there is significant potential to remove multiple contaminants synchronously.

Sheng et al. (2010) presented a comprehensive summary describing that EPS can substantially affect the physicochemical properties of periphyton biofilms (i.e., structure, surface charge, flocculation, settling properties, dewatering properties ,and adsorption ability). For example, the composition and spatial distribution of EPS in periphyton biofilm can significantly influence adsorption and thus change the removal efficiencies of contaminants (Sheng et al., 2010). This implies that to enhance contaminant removal efficiency, EPS composition and spatial distribution should be kept in a relatively steady state. Thus, mature phase periphyton with a relatively stable composition and structure are strongly recommended for use in removing multiple contaminants.

When multiple contaminants are being removed, a series of reactions involving complex stoichiometry and microbial energetics take place. These redox processes are carried out by a variety of microorganisms. The empirical cell synthesis reaction, whereby NH_4^+ as a pollutant, is removed from the water body and incorporated into the biomass follows Eq. (4.1).

$$0.2CO_2 + 0.05NH_4^+ + 0.05HCO_3 = 0.05C_5H_7O_2N + 0.45H_2O \tag{4.1}$$

Microorganisms carry out the majority of these chemical reactions to capture energy for cell synthesis and maintain cellular activity (Rittmann and McCarty, 2001; Schnurr et al., 2013).

When organic matter, represented as $C_6H_{12}O_6$, is removed, it acts as the electron donor and is oxidized to produce CO_2, HCO_3^-, and H_2O. Molecular oxygen is the electron acceptor. The electron-donor half reaction, electron-acceptor half-reaction, and overall reaction are shown in Eq. (4.2) through Eq. (4.4). The CO_2 produced, in turn, supports the growth of photoautotrophic microorganisms such as cyanobacteria and diatoms in the periphyton communities. This ameliorated microorganism matrix provides habitats for additional microorganisms, resulting in the development of the periphyton community in a benign cyclic fashion.

$$0.0417C_6H_{12}O_6 + 0.25N_2O = 0.25CO_2 + H^+ + e^- \tag{4.2}$$

$$0.25O_2 + H^+ + e^- = 0.5H_2O \tag{4.3}$$

Overall, Eq. (4.2) plus Eq. (4.3):

$$0.0417C_6H_{12}O_6 + 0.25O_2 = 0.25CO_2 + 0.25H_2O \tag{4.4}$$

The majority of wastewater applications involve non-nitrogenous organic matter along with reduced nitrogen in the form of NH_4^+ and organic-N (Hasar et al., 2008). It is known that NH_4^+ in aqueous solution is in equilibrium with NH_3 and is thus able to act as an electron donor that is oxidized by bacteria using O_2 as the electron acceptor. Eq. (4.5) through Eq. (4.7) show the donor, acceptor, and overall reactions.

$$0.125NH_4{}^+ + 0.375H_2O^- = 0.125NO_3{}^- + 1.25H^+ + e^- \quad (4.5)$$

$$0.25O_2 + H^+ + e^- = 0.5H_2O \quad (4.6)$$

Overall, Eq. (4.5) plus Eq. (4.6):

$$0.125NH_4{}^+ + 0.25O_2 = 0.125NO_3{}^- + 0.25H^+ + 0.125H_2O \quad (4.7)$$

Eq. (4.7) shows the complete aerobic oxidation of $NH_4{}^+$ to $NO_3{}^-$. However, nitrification actually proceeds in two steps, with $NO_2{}^-$ as an intermediate (Hasar et al., 2008).

In the presence of organic electron donors, such as glucose, denitrification can occur. The reactions for this heterotrophic denitrification process are summarized in Eq. (4.8) through Eq. (4.10).

$$0.0417C_6H_{12}O_6 + 0.25H_2O = 0.25CO_2 + H^+ + e^- \quad (4.8)$$

$$0.2NO_3{}^- + 1.2H^+ + e^- = 0.1N_2 + 0.6H_2O \quad (4.9)$$

Overall, Eq. (4.8) plus Eq. (4.9):

$$0.0417C_6H_{12}O_6 + 0.2NO_3{}^- + 0.2H^+ = 0.25CO_2 + 0.35H_2O + 0.1N_2 \quad (4.10)$$

Eq. (4.10) illustrates the anoxic removal of $NO_3{}^-$ under mildly acidic conditions.

During the removal of contaminants, as illustrated here, nitrogen species—either $NO_3{}^-$ or $NH_4{}^+$ can be synthesized into periphyton biomass. For example, under acidic conditions in the presence of organic matter such as ethanol (as the electron-donor substrate), with $NO_3{}^-$ as the N source and electron acceptor, the biomass will be synthesized by heterotrophs according to Eq. (4.11).

$$\begin{aligned} &NO_3{}^- + 0.6685CH_3CH_2OH + 1.399H^+ \\ &= 0.4245N_2 + 0.1325C_5H_7O_2N + 2.0383H_2O + 0.6780CO_2 \end{aligned} \quad (4.11)$$

Under basic, anoxic conditions, the heterotrophic nitrifier can use organic matter (such as glucose) to synthesize biomass with $NH_4{}^+$ as the N source according Eq. (4.12).

$$\begin{aligned} &0.0417C_6H_{12}O_6 + 0.011NH_4{}^+ + 0.011HCO_3{}^- \\ &= 0.011C_5H_7O_2N + 0.065CH_3CH_2OH + 0.076CO_2 + 0.044H_2O \end{aligned} \quad (4.12)$$

Moreover, under aerobic conditions, the $NH_4{}^+$ acts as the N source. The autotrophic nitrifier can use CO_2 to synthesize biomass as seen in Eq. (4.13).

$$\begin{aligned} &0.13NH_4{}^+ + 0.225O_2 + 0.02CO_2 + 0.005HCO_3{}^- \\ &= 0.005C_5H_7O_2N + 0.125NO_3{}^- + 0.25H^+ + 0.12H_2O \end{aligned} \quad (4.13)$$

4.6 SUCCESS OF CELL IMMOBILIZATION TECHNOLOGY

In many cases, contaminant removal efficiencies (such as NO_3^-, NH_4^+, phosphorus) are directly associated with the periphyton mass, composition, structure, and activity (Rajakumar et al., 2008; Wu et al., 2011b). Although periphyton is present on all underwater solid surfaces, their naturally occurring biomass may be too small to bring about the rapid reactions resulting in substantial simultaneous contaminant removal. Moreover, it is impossible to artificially adjust periphyton composition and structure during periphyton growth on natural soil surfaces under natural conditions. Consequently, there is a need to improve contaminant removal efficiencies under these natural conditions.

To enhance periphyton mass on an industrial scale, many types of substrates (e.g., industrial soft carrier and artificial aquatic mats (Wu et al., 2010b), hollow-fiber (Hasar et al., 2008; Tang et al., 2012), and TiO_2-coated biofilm carrier (Li et al., 2012)), have been produced to concentrate microorganisms and form microbial aggregates on the surface of these carriers. These substrates provide habitats and nutrients required for the growth and maintenance of a high-density microbial population (Rittmann and McCarty, 2001). This process of concentrating microorganisms is referred to as cell immobilization.

Cell immobilization is a relatively new aspect of biotechnology. It occurs via adsorption of the microorganisms onto a support matrix leading to entrapment or embedding in polymer gels or sponges that have been extensively studied as effective means for controlling biomass morphology and achieving higher cell density, production rate, and product yield (Wang et al., 2010). The technology is extensively and efficiently applied to purify wastewater, in particular industrial wastewater, by immobilizing microbial cells in alginate and/or PVA gels (Pramanik et al., 2011; Zamalloa et al., 2013). It is notable that with increased public understanding and requests for environmental protection, this environmentally benign procedure has bright prospects for removing contaminants.

The use of the cell immobilization technique for periphyton biofilm in the treatment of complex wastewater offers many advantages over more conventional biomeasures of free cells including:

1. Prolonged activity and stability of the biocatalyst. The immobilization supports may act as a protective agent against physicochemical effects of pH, temperature, solvents, or heavy metals.
2. Higher cell densities per unit bioreactor volume, which leads to high volumetric productivity, shorter acclimation and maturation phases, and the elimination of nonproductive cell growth phases.
3. Ability to use smaller bioreactors with simplified process designs and, therefore, lower capital costs. This will lead to easier biomass recovery through the reduction of water quality adjustment (or pretreatment), thus reducing equipment costs and energy demands.

4. Feasibility of treating wastewater in a continuous process and in low-temperature conditions.
5. Regeneration and reuse of the biocatalyst for extended periods in batch operations, without removal from the bioreactor.
6. Reduction of risk of microbial contamination because the aggregated periphyton biofilms with high cell densities, are self-maintaining and self-sustainable microecosystems, with more activity and greater stability.

By cell immobilization, native or foreign microorganisms can be obtained. Moreover, the periphyton composition and structure can be designed according to the varied nutrient compositions of the water and substrate surface, as well as the bioreactor hydraulic conditions.

4.7 SAFETY CONSIDERATIONS

When adding a considerable amount of periphyton to surface water, the environmental safety must be considered. The inclusion of "excessive" periphyton can potentially change water chemistry, hydraulic conditions, habitat availability, and foodweb dynamics (Larned, 2010), thus affecting the balance of aquatic ecosystems. Many studies have shown however, that periphyton communities are self-organizing and self-sustaining microsystems (Ksoll et al., 2007; Larned, 2010).

It is likely that periphyton is a buffer system that can counteract the effects of the matter produced or excreted by periphyton itself, maintaining water quality in a relatively steady state. For example, a previous study showed that the enrichment of periphyton by man-made substrates (wet biomass from 1600 to 2650 g m^{-3} water) could adjust reactive phosphorus release from sediments into overlying water, maintaining the experimental water system in a healthy state (Wu et al., 2010b). Moreover, it has been shown that the periphyton biofilm biomass in a healthy aquatic ecosystem can reach 10 g dry matter cm^{-2} (Azim et al., 2003) or 1200 mg chlorophyll a m^{-2} (Welch et al., 1992).

According to previous studies, the periphyton biofilm biomass reaches 1882 g wet matter cm^{-3} water (equal to about 3.25 g dry matter cm^{-2}) when a common biofilm substrate, soft fiber carrier, was used to immobilize periphyton biofilm in simulated ponds where conditions were adequate for periphyton growth (Wu et al., 2011c). This suggests that the introduction of periphyton biofilm via cell immobilization technologies to natural waters will not pose potential risks to aquatic ecosystems.

4.8 FUTURE PROSPECTS

Research fields such as microbiology, water environmental engineering, and aquaculture have verified the universality of periphyton. These ubiquitous microbial communities are being investigated as an important source for bioremediation, and their bioprocesses are being studied in the fields of clinical, food, water, and environmental microbiology from a biofilm perspective. These studies will lead to further understanding of the potential of periphyton for bioremediation. The technologies based on periphyton are both capable of and suited for in situ remediation of surface waters that have been polluted by complex pollution sources such as non–point source wastewater.

There are some aspects that should be considered before an efficient and reliable bioremediation strategy based on periphyton biofilms using cell immobilization technology is developed and applied to large-scale surface waters. Such considerations include (1) the development of an in situ device installed with high-density periphyton biofilms and their substrates, which must be environmentally friendly and easy to operate, and (2) amendment of technology based on periphyton to remediate surface waters, it is necessary to identify the responsible microorganisms and then immobilize these target microorganisms on the substrates to produce functionally efficient periphyton biofilms.

The progress made by recent molecular explorations in microbial technologies has had an enormous impact on our efforts to understand and manage microsocieties with a cleaner "green" concept. In the future, it is possible that highly effective periphyton biofilms will be formed via cloning from the target microorganism genome responsible for the contaminant removal.

4.9 CONCLUSIONS

With the development of various industries, more and more complex wastewater comprising of diverse contaminants will be discharged. The current biological treatment technology based on single microbial communities will be insufficient to effectively treat this heterogeneous wastewater. This chapter demonstrates the significant potential of the application of periphyton comprising heterotrophic and phototrophic microorganisms to remove a range of contaminants. The periphyton communities can be easily contrived and/or incorporated into bioreactors based on cell immobilization technology. Multifunctional water treatment systems based on periphyton have practical significance for bioremediation of contaminated surface waters.

REFERENCES

Abou-Zeid, D.M., Muller, R.J., Deckwer, W.D., 2004. Biodegradation of aliphatic homopolyesters and aliphatic-aromatic copolyesters by anaerobic microorganisms. Biomacromolecules 5 (5), 1687–1697.

Adav, S.S., Lin, J.C.-T., Yang, Z., Whiteley, C.G., Lee, D.-J., Peng, X.-F., Zhang, Z.-P., 2010. Stereological assessment of extracellular polymeric substances, exo-enzymes, and specific bacterial strains in bioaggregates using fluorescence experiments. Biotechnology Advances 28 (2), 255–280.

Arnon, S., Packman, A.I., Peterson, C.G., Gray, K.A., 2007. Effects of overlying velocity on periphyton structure and denitrification. Journal of Geophysical Research 112, G01002.

Azim, M.E., 2009. Photosynthetic periphyton and surfaces. In: Encyclopedia of Inland Waters. Academic Press, Oxford, pp. 184–191.

Azim, M.E., Milstein, A., Wahab, M.A., Verdegam, M.C.J., 2003. Periphyton-water quality relationships in fertilized fishponds with artificial substrates. Aquaculture 228 (1–4), 169–187.

Badireddy, A.R., Chellam, S., Gassman, P.L., Engelhard, M.H., Lea, A.S., Rosso, K.M., 2010. Role of extracellular polymeric substances in bioflocculation of activated sludge microorganisms under glucose-controlled conditions. Water Research 44 (15), 4505–4516.

Barakat, M.A., 2011. New trends in removing heavy metals from industrial wastewater. Arabian Journal of Chemistry 4, 361–377.

Baxter, L.R., Sibley, P.K., Solomon, K.R., Hanson, M.L., 2013. Interactions between atrazine and phosphorus in aquatic systems: effects on phytoplankton and periphyton. Chemosphere 90 (3), 1069–1076.

Bere, T., Chia, M.A., Tundisi, J.G., 2012. Effects of Cr III and Pb on the bioaccumulation and toxicity of Cd in tropical periphyton communities: implications of pulsed metal exposures. Environmental Pollution 163, 184–191.

Boopathy, R., 2000. Factors limiting bioremediation technologies. Bioresource Technology 74, 63–67.

Bosch, N.S., Allan, J.D., Selegean, J.P., Scavia, D., 2013. Scenario-testing of agricultural best management practices in Lake Erie watersheds. Journal of Great Lakes Research 39, 429–436.

Bowes, M.J., Ings, N.L., McCall, S.J., Warwick, A., Barrett, C., Wickham, H.D., Harman, S.A., Armstrong, L.K., Scarlett, P.M., Roberts, C., Lehmann, K., Singer, A.C., 2012. Nutrient and light limitation of periphyton in the River Thames: implications for catchment management. Science of the Total Environment 434, 201–212.

Bradac, P., Wagner, B., Kistler, D., Traber, J., Behra, R., Sigg, L., 2010. Cadmium speciation and accumulation in periphyton in a small stream with dynamic concentration variations. Environmental Pollution 158 (3), 641–648.

Cai, M., Xun, L., 2002. Organization and regulation of pentachlorophenol-degrading genes in *Sphingobium chlorophenolicum* ATCC 39723. Journal of Bacteriology 184, 4672–4680.

Chen, B.L., Wang, Y.S., Hu, D.F., 2010. Biosorption and biodegradation of polycyclic aromatic hydrocarbons in aqueous solutions by a consortium of white-rot fungi. Journal of Hazardous Materials 179 (1–3), 845–851.

Chen, Y., Cheng, J.J., Creamer, K.S., 2008. Inhibition of anaerobic digestion process: a review. Bioresource Technology 99 (10), 4044–4064.

Chien, C.-C., Lin, B.-C., Wu, C.-H., 2013. Biofilm formation and heavy metal resistance by an environmental *Pseudomonas* sp. Biochemical Engineering Journal 78, 132–137.

Choi, A., Wang, S., Lee, M., 2009. Biosorption of cadmium, copper, and lead ions from aqueous solutions by *Ralstonia* sp and *Bacillus* sp isolated from diesel and heavy metal contaminated soil. Geoscience Journal 13 (4), 331–341.

Clément, P., Matus, V., Cárdenas, L., González, B., 1995. Degradation of trichlorophenols by *Alcaligenes eutrophus* JMP134. FEMS Microbiological Letters 127, 51–55.

Crab, R., Avnimelech, Y., Defoirdt, T., Bossier, P., Verstraete, W., 2007. Nitrogen removal techniques in aquaculture for a sustainable production. Aquaculture 270, 1–14.

De Beer, D., Stoodley, P., 2006. Microbial biofilms. Prokaryotes 1, 904–937. Chapter 3.10.

de Lorenzo, V., 2008. Systems biology approaches to bioremediation. Current Opinoin in Biotechnology 19, 579–589.

Diaz, E., 2008. Microbial Biodegradation: Genomics and Molecular Biology. Caister Academic Press, Madrid, Spain.

Dosta, J., Nieto, J.M., Vila, J., Grifoll, M., Mata-Álvarezlvarez, J., 2011. Phenol removal from hypersaline wastewaters in a membrane biological reactor (MBR): operation and microbiological characterisation. Bioresource Technology 102 (5), 4013–4020.

Feng, Y., Yang, F., Wang, Y., Ma, L., Wu, Y., Kerr, P.G., Yang, L., 2011. Basic dye adsorption onto an agro-based waste material – Sesame hull (*Sesamum indicum L.*). Bioresource Technology 102 (22), 10280–10285.

Flynn, K.F., Chapra, S.C., Suplee, M.W., 2013. Modeling the Lateral Variation of Bottom-attached Algae in Rivers, vol. 267, pp. 11–25.

Fortin, C., Denison, F.H., Garnier-Laplace, J., 2007. Metal-phytoplankton interactions: modeling the effect of competing ions (H^+, Ca^{2+}, and Mg^{2+}) on uranium uptake. Environmental Toxicology and Chemistry 26, 242–248.

Fu, F., Wang, Q., 2011. Removal of heavy metal ions from wastewaters: a review. Journal of Environmental Management 92, 407–418.

Gómez-Silván, C., Molina-Muñoz, M., Poyatos, J.M., Ramos, A., Hontoria, E., Rodelas, B., González-López, J., 2010. Structure of archaeal communities in membrane-bioreactor and submerged-biofilter wastewater treatment plants. Bioresource Technology 101 (7), 2096–2105.

George, S., Kishen, A., Song, P., 2005. The role of environmental changes on *Monospecies* biofilm formation on root canal wall by *Enterococcus* faecalis. Journal of Endodontics 31, 867–872.

Gisi, M.R., Xun, L., 2003. Characterization of chlorophenol 4-monooxygenase (TftD) and NADH: flavin adenine dinucleotide oxidoreductase (TftC) of *Burkholderia cepacia* AC1100. Journal of Bacteriology 185, 2786–2792.

Hasar, H., Xia, S., Ahn, C.H., Rittmann, B.E., 2008. Simultaneous removal of organic matter and nitrogen compounds by an aerobic/anoxic membrane biofilm reactor. Water Research 42, 4109–4116.

Holding, K.L., Gill, R.A., Carter, J., 2003. The relationship between epilithic periphyton (biofilm) bound metals and metals bound to sediments in freshwater systems. Environmental Geochemistry and Health 25, 87–93.

Huang, Y., Zhang, J., Zhu, L., 2013. Evaluation of the application potential of bentonites in phenanthrene bioremediation by characterizing the biofilm community. Bioresource Technology 134, 17–23.

Jørgensen, K.S., Puustinen, J., Suortti, A.M., 2000. Bioremediation of petroleum hydrocarbon-contaminated soil by composting in biopiles. Environmental Pollution 107 (2), 245–254.

Jechalke, S., Vogt, C., Reiche, N., Franchini, A.G., Borsdorf, H., Neu, T.G., Richnow, H.H., 2010. Aerated treatment pond technology with biofilm promoting mats for the bioremediation of benzene, MTBE and ammonium contaminated groundwater. Water Research 44, 1785–1796.

Jing, Z., Li, Y.-Y., Cao, S., Liu, Y., 2012. Performance of double-layer biofilter packed with coal fly ash ceramic granules in treating highly polluted river water. Bioresource Technology 120, 212–217.

Johansen, S.S., Licht, D., Arvin, E., Masbaek, H., Hansen, A.B., 1997. Metabolic pathways of quinoline, indole and their methylated analogs by *Desulphobacterium indolicum* (DSM 3383). Applied and Environmental Biotechnology 47, 292–300.

Joo, J.H., Hassan, S.H.A., Oh, S.E., 2010. Comparative study of biosorption of Zn^{2+} by *Pseudomonas aeruginosa* and *Bacillus cereus*. International Biodeterioration & Biodegradation 64 (8), 734–741.

Kaiser, J.P., Feng, Y.C., Bollag, J.M., 1996. Microbial metabolism of pyridine, quinoline, acridine, and their derivatives under aerobic and anaerobic conditions. Microbiological Review 60, 483–498.

Kanavillil, N., Thorn, M., Kurissery, S., 2012. Characterization of natural biofilms in temperate inland waters. Journal of Great Lakes Research 38 (3), 429–438.

Kang, J.-H., Katayama, Y., Kondo, F., 2006. Biodegradation or metabolism of bisphenol A: from microorganisms to mammals. Toxicology 217 (2–3), 81–90.

Kang, J.H., Kondo, F., 2002a. Bisphenol A degradation by bacteria isolated from river water. Archive of Environmental Contamination and Toxicology 43, 265–269.

Kang, J.H., Kondo, F., 2002b. Effects of bacterial counts and temperature on the biodegradation of bisphenol A in river water. Chemosphere 49, 493–498.

Kao, W.C., Chiu, Y.P., Chang, C.C., Chang, J.S., 2006. Localization effect on the metal biosorption capability of recombinant mammalian and fish metallothioneins in *Escherichia coli*. Biotechnology Progress 22 (5), 1256–1264.

Kao, W.C., Huang, C.C., Chang, J.S., 2008. Biosorption of nickel, chromium and zinc by MerP-expressing recombinant *Escherichia coli*. Journal of Hazardous Materials 158 (1), 100–106.

Kassab, G., Halalsheh, M., Klapwijk, A., Fayyad, M., van Lier, J.B., 2010. Sequential anaerobic-aerobic treatment for domestic wastewater – a review. Bioresource Technology 101 (10), 3299–3310.

Klecka, G.M., Gonsior, S.J., West, R.J., Goodwin, P.A., Markham, D.A., 2001. Biodegradation of bisphenol A in aquatic environments: river die-away. Environmental Toxicology and Chemistry 20, 2725–2735.

Ksoll, W.B., Ishii, S., Sadowsky, M.J., Hicks, R.E., 2007. Presence and sources of fecal coliform bacteria in epilithic periphyton communities of Lake Superior. Applied and Environmental Microbiology 73 (12), 3771–3778.

Kurzbaum, E., Kirzhner, F., Sela, S., Zimmels, Y., Armon, R., 2010. Efficiency of phenol biodegradation by planktonic *Pseudomonas pseudoalcaligenes* (a constructed wetland isolate) vs. root and gravel biofilm. Water Research 44 (17), 5021–5031.

Larned, S.T., 2010. A prospectus for periphyton: recent and future ecological research. Journal of the North American Benthological Society 29 (1), 182–206.

Laspidou, C.S., Rittmann, B.E., 2002. A unified theory for extracellular polymeric substances, soluble microbial products, and active and inert biomass. Water Research 36 (11), 2711–2720.

Lens, P.N.L., Hulshoff Pol, L., 2000. Environmental Technologies to Treat Sulfur Pollution. IWA Publishing.

Li, G., Park, S., Rittmann, B.E., 2012. Developing an efficient TiO_2-coated biofilm carrier for intimate coupling of photocatalysis and biodegradation. Water Research 46 (19), 6489–6496.

Liong, M.-T., 2011. Bioprocess Sciences and Technology. Nova Science Publishers, Hauppauge, NY.

Liu, Z.-H., Kanjo, Y., Mizutani, S., 2009. Removal mechanisms for endocrine disrupting compounds (EDCs) in wastewater treatment-physical means, biodegradation, and chemical advanced oxidation: a review. Science of the Total Environment 407 (2), 731–748.

Louie, T.M., Webster, C.M., Xun, L., 2002. Genetic and biochemical characterization of a 2,4,6-trichlorophenol degradation pathway in *Ralstonia eutropha* JMP134. Journal of Bacteriology 184, 3492–3500.

Machado, M.D., Santos, M.S., Gouveia, C., Soares, H.M., Soares, E.V., 2008. Removal of heavy metals using a brewer's yeast strain of *Saccharomyces cerevisiae*: the flocculation as a separation process. Bioresource Technology 99 (7), 2107–2115.

Martin, K.J., Nerenberg, R., 2012. The membrane biofilm reactor (MBfR) for water and wastewater treatment: principles, applications, and recent developments. Bioresource Technology 122, 83–94.

McGlashan, M.A., Tsoflias, G.P., Schillig, P.C., Devlin, J.F., Roberts, J.A., 2012. Field GPR monitoring of biostimulation in saturated porous media. Journal of Applied Geophysics 78, 102–112.

Mehta, S.K., Gaur, J.P., 2005. Use of algae for removing heavy metal ions from wastewater: progress and prospects. Critical Review in Biotechnology 25 (3), 113–152.

Merchant, S.S., Helmann, J.D., Robert, K.P., 2012. Elemental economy: microbial strategies for optimizing growth in the face of nutrient limitation. In: Advances in Microbial Physiology, vol. 60. Academic Press, pp. 91–210. Chapter 2.

Meylan, S., Behra, R., Sigg, L., 2003. Accumulation of copper and zinc in periphyton in response to dynamic variations of metal speciation in freshwater. Environmental Science & Technology 37, 5204–5212.

Mrozik, A., Piotrowska-Seget, Z., 2010. Bioaugmentation as a strategy for cleaning up of soils contaminated with aromatic compounds. Microbiological Research 165 (5), 363–375.

Neu, T.R., Lawrence, J.R., 2010. Extracellular polymeric substances in microbial biofilms. In: Holst, O., Brennan, P.J., Itzstein, M. (Eds.), Microbial Glycobiology. Academic Press, San Diego, pp. 733–758. Chapter 37.

Oehmen, A., Lemos, P.C., Carvalho, G., Yuan, Z., Keller, J., Blackall, L.L., Reis, M.A.M., 2007. Advances in enhanced biological phosphorus removal: from micro to macro scale. Water Research 41 (11), 2271–2300.

Olaniran, A.O., Igbinosa, E.O., 2011. Chlorophenols and other related derivatives of environmental concern: properties, distribution and microbial degradation processes. Chemosphere 83 (10), 1297–1306.

Padilla, L., Matus, V., Zenteno, P., González, B., 2000. Degradation of 2,4,6-trichlorophenol via chlorohydroxyquinol in *Ralstonia eutropha* JMP134 and JMP222. Journal of Basic Microbiology 40 (4), 243–249.

Padoley, K.V., Mudliar, S.N., Pandey, R.A., 2008. Heterocyclic nitrogenous pollutants in the environment and their treatment options – an overview. Bioresource Technology 99, 4029–4043.

Pan, X.L., Liu, J., Zhang, D.Y., Chen, X., Song, W.J., Wu, F.C., 2010. Binding of dicamba to soluble and bound extracellular polymeric substances (EPS) from aerobic activated sludge: a fluorescence quenching study. Journal of Colloid Interface and Science 345, 442–447.

Peng, J., Zhang, Y., Su, J., Qiu, Q., Jia, Z., Zhu, Y.-G., 2013. Bacterial communities predominant in the degradation of 13C4-4,5,9,10-pyrene during composting. Bioresource Technology 143, 608–614.

Pokrovsky, O.S., Martinez, R.E., Kompantseva, E.I., Shirokova, L.S., 2013. Interaction of metals and protons with anoxygenic phototrophic bacteria *Rhodobacter blasticus*. Chemical Geology 335, 75–86.

Prado, N., Ochoa, J., Amrane, A., 2009. Biodegradation and biosorption of tetracycline and tylosin antibiotics in activated sludge system. Process in Biochemistry 44 (11), 1302–1306.

Pramanik, S., McEvoy, J., Siripattanakul, S., Khan, E., 2011. Effects of cell entrapment on nucleic acid content and microbial diversity of mixed cultures in biological wastewater treatment. Bioresource Technology 102 (3), 3176–3183.

Rajakumar, S., Ayyasamy, P.M., Shanthi, K., Thavamani, P., Velmurugan, P., Song, Y.C., Lakshmanaperumalsamy, P., 2008. Nitrate removal efficiency of bacterial consortium (*Pseudomonas* sp. KW1 and *Bacillus* sp. YW4) in synthetic nitrate-rich water. Journal of Hazardous Materials 157, 553–563.

Rectenwald, L.L., Drenner, R.W., 1999. Nutrient removal from wastewater effluent using an ecological water treatment system. Environmental Science & Technology 34 (3), 522–526.

Reinhammar, B., 1984. Laccase. In: Lontie, R. (Ed.), Copper Proteins and Copper Enzymes, vol. 3. CRC Press, Boca Raton, FL, pp. 1–35.

Rittmann, B., McCarty, P., 2001. Environmental Biotechnology: Principles and Applications. The McGraw-Hill Companies, Inc., New York.

Sánchez-Andrea, I., Sanz, J.L., Bijmans, M.F.M., Stams, A.J.M., 2014. Sulfate reduction at low pH to remediate acid mine drainage. In: Research Frontiers in Chalcogen Cycle Based Environmental Technologies Selected Papers Presented at the 3rd International Conference on Research Frontiers in Chalcogen Cycle Science and Technology (G16), UNESCO-IHE, Delft, The Netherlands, May 27th–28th, 2013, vol. 269, pp. 98–109.

Sánchez, M.A., González, B., 2007. Genetic characterization of 2,4,6-trichlorophenol degradation in *Cupriavidus necator* JMP134. Applied and Environmental Microbiology 73, 2769–2776.

dos Santos,, V.A.P.M., Bruijnse, M., Tramper, J., Wijffels, R.H., 1996. The magic-bead concept: an integrated approach to nitrogen removal with co-immobilized micro-organisms. Applied Microbiology and Biotechnology 45, 447–453.

Santos, V.L., Linardi, V.R., 2004. Biodegradation of phenol by a filamentous fungi isolated from industrial effluents-identification and degradation potential. Process in Biochemistry 39 (8), 1001–1006.

Sasaki, K., Yamamoto, Y., Tsumura, K., Ouchi, S., Mori, Y., 1996. Development of 2-reactor intermittent-aeration activated sludge process for simultaneous removal of nitrogen and phosphorus. Water Science & Technology 34 (1–2), 111–118.

Schauer-Gimenez, A.E., Zitomer, D.H., Maki, J.S., Struble, C.A., 2010. Bioaugmentation for improved recovery of anaerobic digesters after toxicant exposure. Water Research 44 (12), 3555–3564.

Schnurr, P.J., Espie, G.S., Allen, D.G., 2013. Algae biofilm growth and the potential to stimulate lipid accumulation through nutrient starvation. Bioresource Technology 136, 337–344.

Scinto, L.J., Reddy, K.R., 2003. Biotic and abiotic uptake of phosphorus by periphyton in a subtropical freshwater wetland. Aquatic Botany 77 (3), 203–222.

Serra, A., Corcoll, N., Guasch, H., 2009. Copper accumulation and toxicity in fluvial periphyton: the influence of exposure history. Chemosphere 74 (5), 633–641.

Sheng, G.-P., Yu, H.-Q., Li, X.-Y., 2010. Extracellular polymeric substances (EPS) of microbial aggregates in biological wastewater treatment systems: a review. Biotechnology Advances 28 (6), 882–894.

Singh, A., Ward, O.P., 2004. Biodegradation and Bioremediation. Springer, Berlin, New York.

Sivakumar, G., Xu, J., Thompson, R.W., Yang, Y., Randol-Smith, P., Weathers, P.J., 2012. Integrated green algal technology for bioremediation and biofuel. Bioresource Technology 107, 1–9.

Sivasamy, A., Sundarabal, N., 2011. Biosorption of an azo dye by *Aspergillus niger* and *Trichoderma* sp. fungal biomasses. Current Microbiology 62 (2), 351–357.

Small, J.A., Bunn, A., McKinstry, C., Peacock, A., Miracle, A.L., 2008. Investigating freshwater periphyton community response to uranium with phospholipid fatty acid and denaturing gradient gel electrophoresis analyses. Journal of Environmental Radioactivity 99 (4), 730–738.

Smith, V.H., Tilman, G.D., Nekola, J.C., 1999. Eutrophication: impacts of excess nutrient inputs on freshwater, marine, and terrestrial ecosystems. Environmental Pollution 100 (1–3), 179–196.

Spath, R., Flemming, H.C., Wuertz, S., 1998. Sorption properties of biofilms. Water Science & Technology 37 (4–5), 207–210.

Stenger-Kovács, C., Lengyel, E., Crossetti, L.O., Üveges, V., Padisák, J., 2013. Diatom ecological guilds as indicators of temporally changing stressors and disturbances in the small Torna-stream, Hungary. Ecological Indicators 24, 138–147.

Sun, Q., Bai, Y., Zhao, C., Xiao, Y., Wen, D., Tang, X., 2009. Aerobic biodegradation characteristics and metabolic products of quinoline by a *Pseudomonas* strain. Bioresource Technology 100 (21), 5030–5036.

Tang, Y., Zhou, C., Van Ginkel, S.W., Ontiveros-Valencia, A., Shin, J., Rittmann, B.E., 2012. Hydrogen permeability of the hollow fibers used in H_2-based membrane biofilm reactors. Journal of Membrane Science 407–408, 176–183.

Tatarko, M., Bumpus, J.A., 1998. Biodegradation of Congo Red by *Phanerochaete chrysosporium*. Water Research 32 (5), 1713–1717.

Thauer, R.K., Jungermann, K., Decker, K., 1977. Energy conservation in chemotrophic anaerobic bacteria. Bacteriological Review 41, 100.

Tsioulpas, A., Dimou, D., Iconomou, D., Aggelis, G., 2002. Phenolic removal in olive oil mill wastewater by strains of *Pleurotus* spp. in respect to their phenol oxidase (laccase) activity. Bioresource Technology 84 (3), 251–257.

Tsuneda, S., Ohno, T., Soejima, K., Hirata, A., 2006. Simultaneous nitrogen and phosphorus removal using denitrifying phosphate-accumulating organisms in a sequencing batch reactor. Biochemical Engineering Journal 27 (3), 191–196.

Tsuruta, T., 2004. Biosorption and recycling of gold using various microorganisms. The Journal of General and Applied Microbiology 50 (4), 221–228.

van Loosdrecht, M.C.M., Jetten, M.S.M., 1998. Microbiological conversions in nitrogen removal. Water Science & Technology 38 (1), 1–7.

Veeresh, G.S., Kumar, P., Mehrotra, I., 2005. Treatment of phenol and cresols in upflow anaerobic sludge blanket (UASB) process: a review. Water Research 39 (1), 154–170.

Villaverde, S., 2004. Recent development on biological nutrient removal processes for wastewater treatment. Reviews in Environmental Science and Bio/Technology 3, 171–183.

Voordeckers, J.W., Fennell, D.E., Jones, K., Haggblom, M.M., 2002. Anaerobic biotransformation of tetrabromobisphenol A, tetrachlorbisphenol A, and bisphenol A in estuarine sediments. Environmental Science & Technology 36, 696–701.

Wang, I., Huang, L., Yun, L., Tang, F., Zhao, J., Liu, Y., Zeng, X., Luo, Q., 2008. Removal of nitrogen, phosphorus, and organic pollutants from water using seeding type immobilized microorganisms. Biomedince and Environmental Science 21, 150–156.

Wang, J., Chen, C., 2009. Biosorbents for heavy metals removal and their future. Biotechnology Advances 27, 195–226.

Wang, J., Quan, X., Han, L., Qian, Y., Werner, H., 2002. Microbial degradation of quinoline by immobilized cells of *Burkholderia pickettii*. Water Research 36 (9), 2288–2296.

Wang, W.-X., Dei, R.C.H., 2001. Effects of major nutrient additions on metal uptake in phytoplankton. Environmental Pollution 111 (2), 233–240.

Wang, Y., Tian, Y., Han, B., Zhao, H.-B., Bi, J.-N., Cai, B.-L., 2007. Biodegradation of phenol by free and immobilized *Acinetobacter* sp. strain PD12. Journal of Environmental Sciences 19 (2), 222–225.

Wang, Z., Wang, Y., Yang, S.-T., Wang, R., Ren, H., 2010. A novel honeycomb matrix for cell immobilization to enhance lactic acid production by *Rhizopus oryzae*. Bioresource Technology 101 (14), 5557–5564.

Welch, E.B., Quinn, J.M., Hickey, C.W., 1992. Periphyton biomass related to point-source nutrient enrichment in seven new Zealand streams. Water Research 26 (5), 669–675.

Wijekoon, K.C., Fujioka, T., McDonald, J.A., Khan, S.J., Hai, F.I., Price, W.E., Nghiem, L.D., 2013. Removal of N-nitrosamines by an aerobic membrane bioreactor. Bioresource Technology 141, 41–45.

Wu, Y., He, J., Hu, Z., Yang, L., Zhang, N., 2011c. Removal of UV_{254} nm matter and nutrients from a photobioreactor-wetland system. Journal of Hazardous Materials 194, 1–6.

Wu, Y., He, J., Yang, L., 2010a. Evaluating adsorption and biodegradation mechanisms during the removal of microcystin-RR by periphyton. Environmental Science & Technology 44, 6319–6324.

Wu, Y., Hu, Z., Kerr, P.G., Yang, L., 2011b. A multi-level bioreactor to remove organic matter and metals, together with its associated bacterial diversity. Bioresource Technology 102, 736–741.

Wu, Y., Hu, Z., Yang, L., Graham, B., Kerr, P., 2011a. The removal of nutrients from non-point source wastewater by a hybrid bioreactor. Bioresource Technology 102 (3), 2419–2426.

Wu, Y., Li, T., Yang, L., 2012. Mechanisms of removing pollutants from aqueous solutions by microorganisms and their aggregates: a review. Bioresource Technology 107, 10–18.

Wu, Y., Liu, J., Yang, L., Chen, H., Zhang, S., Zhao, H., Zhang, N., 2011c. Allelopathic control of cyanobacterial blooms by periphyton biofilms. Environmental Microbiology 13 (3), 604–615.

Wu, Y., Xia, L., Hu, Z., Liu, S., Liu, H., Nath, B., Zhang, N., Yang, L., 2011d. The application of zero-water discharge system in treating diffuse village wastewater and its benefits in community afforestation. Environmental Pollution 159 (10), 2968–2973.

Wu, Y., Xia, L., Yu, Z., Shabbir, S., Kerr, P.G., 2014. In situ bioremediation of surface waters by periphytons. Bioresource Technology 151, 367–372.

Wu, Y., Zhang, S., Zhao, H., Yang, L., 2010b. Environmentally benign periphyton bioreactors for controlling cyanobacterial growth. Bioresource Technology 101 (24), 9681–9687.

Xia, S., Liang, J., Xu, X., Shen, S., 2013. Simultaneous removal of selected oxidized contaminants in groundwater using a continuously stirred hydrogen-based membrane biofilm reactor. Journal of Environmental Sciences 25 (1), 96–104.

Yan, J., Jianping, W., Jing, B., Daoquan, W., Zongding, H., 2006. Phenol biodegradation by the yeast *Candida tropicalis* in the presence of m-cresol. Biochemical Engineering Journal 29 (3), 227–234.

Yan, R., Yang, F., Wu, Y., Hu, Z., Nath, B., Yang, L., Fang, Y., 2011. Cadmium and mercury removal from non-point source wastewater by a hybrid bioreactor. Bioresource Technology 102 (21), 9927–9932.

Yang, Q., Shang, H.T., Wang, J.L., 2009. Biosorption and biodegradation of trichloroethylene by acclimated activated sludge. International Journal of Environmental Pollution 38 (3), 289–298.

Yang, S., Ngwenya, B.T., Butler, I.B., Kurlanda, H., Elphick, S.C., 2013. Coupled Interactions Between Metals and Bacterial Biofilms in Porous Media: Implications for Biofilm Stability, Fluid Flow and Metal Transport, vol. 337–338, pp. 20–29.

Yariv, C., 2001. Biofiltration – the treatment of fluids by microorganisms immobilized into the filter bedding material: a review. Bioresource Technology 77, 257–274.

Ye, F.X., Li, Y., 2007. Biosorption and biodegradation of pentachlorophenol (PCP) in an upflow anaerobic sludge blanket (UASB) reactor. Biodegradation 18 (5), 617–624.

Yuncu, B., Sanin, F.D., Yetis, U., 2006. An investigation of heavy metal biosorption in relation to C/N ratio of activated sludge. Journal of Hazardous Materials 137, 990–997.

Zamalloa, C., Boon, N., Verstraete, W., 2013. Decentralized two-stage sewage treatment by chemical biological flocculation combined with microalgae biofilm for nutrient immobilization in a roof installed parallel plate reactor. Bioresource Technology 130, 152–160.

Zhang, L., Keller, J., Yuan, Z., 2009. Inhibition of sulfate-reducing and methanogenic activities of anaerobic sewer biofilms by ferric iron dosing. Water Research 43 (17), 4123–4132.

Zhou, S., Zhang, X., Feng, L., 2010. Effect of different types of electron acceptors on the anoxic phosphorus uptake activity of denitrifying phosphorus removing bacteria. Bioresource Technology 101, 1603–1610.

Žižek, S., Milačič, R., Kovač, N., Jaćimović, R., Toman, M.J., Horvat, M., 2011. Periphyton as a bioindicator of mercury pollution in a temperate torrential river ecosystem. Chemosphere 85 (5), 883–891.

Chapter 5

Periphytic Biofilm and Its Functions in Aquatic Nutrient Cycling

5.1 INTRODUCTION

Nutrient cycling is of critical importance in aquatic ecosystems and subsequently has been the focus of much ecological research. Since the introduction of the European Water Framework Directive (WFD) regulatory requirements to achieve "good" or "high" ecological status in a range of water bodies by 2015, an understanding of (1) the nutrient delivery process, (2) the multiple sources of nutrients entering waters, and (3) how nutrients circulate in aquatic ecosystems, has been urgently needed (Neal and Jarvie, 2005). Before the mid-1900s, most research concentrated on the role of biotic components such as macrophytes, plankton (zooplankton and phytoplankton) and invertebrates (benthos, nekton, and neuston) in aquatic nutrient cycling. At that time no attention was given to the role of periphytic biofilm or periphyton in aquatic ecosystems. In 1963, Wetzel highlighted the importance of periphyton as primary producers in aquatic ecosystems in his revolutionary review paper (Wetzel, 1963). Essentially, periphyton communities are solar-powered biogeochemical reactors with assemblages of photoautotrophic algae, heterotrophic and chemoautotrophic bacteria, fungi, protozoans, metazoans, and viruses (Larned, 2010). They are commonly attached to submerged surfaces in most aquatic ecosystems and play a significant role in natural aquatic ecosystems through their influence on primary production, food chains, organic matter, and nutrient recycling (Battin et al., 2003; Cantonati and Lowe, 2014; Saikia, 2011). Despite this recognized importance, periphytic biofilm is still ignored as a major contributor of most nutrients to aquatic ecological cycles (Saikia et al., 2013). With this background, certain basic questions must be examined to help understand the roles and functions of periphytic biofilm on nutrient cycling in aquatic ecosystems. The role of this chapter is to (1) characterize the periphyton biofilm, (2) state the factors that control the growth and death of periphyton, (3) evaluate the important functions of periphytic biofilm in aquatic nutrient (C, N, and P) cycling, and (4) provide some valuable directions for future research on periphytic biofilm in relation to nutrient cycling.

Periphyton. http://dx.doi.org/10.1016/B978-0-12-801077-8.00005-3

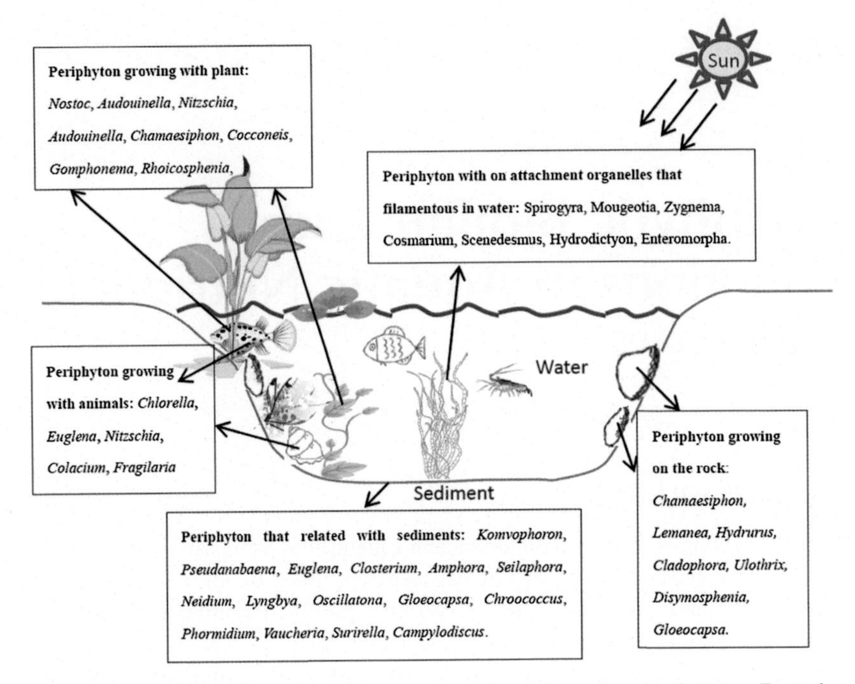

FIGURE 5.1 Periphyton communities categorized by their preferred substrates. Examples of organisms within each community are given. *Reproduced from Poulíčková, A., Hašler, P., Lysáková, M., Spears, B., 2008. The ecology of freshwater epipelic algae: an update. Phycologia 47 (5), 437–450.*

From the perspective of taxonomic composition, periphytic biofilm communities are highly diverse, containing mainly algae, bacteria, fungi, and protozoans (Fig. 5.1). This diversity is dependent on a range of factors such as habitat and surface types, light intensity, grazing pressure, seasonality, nutrient availability, and physical disturbances. The algae are the most fundamental and frequently encountered groups in natural periphyton. This is because algae are very diverse and cosmopolitan in nature, and they are normally attached to submerged substrate surfaces throughout or partially during their life cycles. The most frequently encountered algal groups are the diatoms (*Bacillariophyceae*), cyanobacteria (*Cyanophyceae*), red algae (*Rhodophyceae*), chrysophytes (*Chrysophyceae*), and green algae (*Chlorophyceae*). Some taxa of brown (*Phaeophyceae*) algae are also common in periphyton, especially in brackish/saline waters (Azim, 2009).

5.2 FUNCTIONS OF PERIPHYTIC BIOFILM IN NUTRIENT CYCLING

Research has demonstrated that periphytic biofilm plays important roles in the nutrient cycling of water bodies, not only by being important primary producers (Liboriussen and Jeppesen, 2003; Vadeboncoeur et al., 2001) and serving as an energy and food source for higher trophic levels (Saikia and Das, 2009; Saikia et al., 2013; Saikia, 2011), but also by affecting nutrient turnover (Wetzel, 1993) and the transfer of nutrients between the water and sediment interface (Battin et al., 2003; Vander Zanden and Vadeboncoeur, 2002; Woodruff et al., 1999; Wu et al., 2010). Compared to suspended phytoplankton, periphyton contributes more substantially to aquatic nutrient cycling through regulation of nutrient condition, maintenance of stable nutrient resource flows, and performing as a nutrient retention tool and excess nutrient removal agent (Saikia, 2011). Consequently, the knowledge of "nutrient cycling" in aquatic ecosystems will remain incomplete if studies on periphyton are not conducted.

5.2.1 Source and Sink of Nutrients

Many previous studies have suggested that periphyton was generally a sink and sometimes a source for water nutrients such as C, N, and P, depending on many factors like light intensity, taxonomic composition, and the stoichiometry of periphyton (Baldwin et al., 2006; Drake et al., 2012; Matheson et al., 2012; McCormick et al., 2006; Saikia, 2011). The role of periphyton as a nutrient sink is often ignored in comparison to other components such as plants. Periphyton, however, can concentrate nutrients (especially P) from water through luxury nutrient uptake. Luxury P-uptake is the storage of P within the biomass in the form of polyphosphate that can be present as acid-soluble or acid-insoluble polyphosphate (Saikia, 2011). The P content of periphyton has been reported to range from 1 to 2 mg Pg^{-1} DM, which represented a 30,000–60,000-fold nutrient bioconcentration relative to the average P content of the ambient water (McCormick et al., 2006). This result suggests that periphyton can be an important short-term sink for P in wetlands, which are strongly affected by light availability and the taxonomic composition of the periphyton assemblage (Drake et al., 2012; McCormick et al., 2006).

Periphytic biofilms also have the potential to be active sites for N transformation in aquatic ecosystems. Periphytic community members exhibit close spatial relationships and synergistic metabolisms that can result in anaerobic zones within the biofilm, even though the surrounding water column may be aerobic (Decho, 2000; Paerl and Pinckney, 1996). Proximate aerobic and anaerobic environments can facilitate coupled nitrification/denitrification. Thus, denitrification within biofilms may be a quantitatively important sink for N (mainly nitrate) in rivers with sufficient habitat for periphyton and could account for up to 50% of the NO_3^- taken up (Baldwin et al., 2006). Baldwin et al. (2006) also suggested that other processes, particularly biofilm assimilation, may be more important than denitrification as a nitrate sink, particularly in the lower reaches of the river. Sobczak et al. (2003) showed that microbial assimilation of NO_3^- accounted for approximately 60% of NO_3^- loss in mesocosm studies of simulated hyporheic flow paths of the Neversink River, while denitrification accounted for less than 5%.

Periphyton can be a sink for S in aquatic systems. In the oxic water column, sulfate (SO_4^{2-}) is the dominant form of inorganic sulfur. While many microbes in periphyton can reduce SO_4^{2-} to S^{-2} intracellularly (with subsequent synthesis of S^{-2}-containing amino acids), larger scale, biogeochemically important SO_4^{2-} reduction must occur in anoxic microenvironments. Microbial mats, such as biofilms, could provide such conditions (Revsbech et al., 1989). These microenvironments also provide habitats for photosynthetic and chemosynthetic bacteria that reoxidize S^{-2} to elemental sulfur (S), thiosulfate ($S_2O_3^{2-}$), or SO_4^{2-} (Paerl and Pinckney, 1996).

As previously mentioned, periphyton consists of various nutrient materials like extracellular polymeric substance (EPS), polyunsaturated fatty acid (PUFA), sterols, amino acids, vitamins, and pigments (Thompson et al., 2002). Consequently, it could provide a source of polysaccharides, proteins, nucleic acid, and other polymers in nutrient cycling of aquatic ecosystems under certain conditions. Aquatic periphyton could remove soluble nutrients from the ambient water during their growth phase. Nutrients acquired by periphyton may be released back into the water by sloughing, scouring, and dislodgement of periphytic communities (Saikia, 2011).

Many microorganisms, such as *Microcystis aeruginosa*, can excrete EPS into water. The high cell density of *Microcystis* frequently results in high EPS concentrations in the water column. The EPS in cultures of *Microcystis aeruginosa* reached 130 μg per 10^7 cells. The main component of EPS in *Microcystis* cultures is polysaccharides followed by proteins (Xu et al., 2013). EPS enrich hydroxyl groups, carboxylic groups, acetylated amino acids, and also contain some noncarbohydrate constituents, e.g., phosphate and sulfate (De Philippis et al., 2011). These chemical groups in EPS can effectively bind with heavy metal ions through ion exchange or complexation (Fang et al., 2011; Gong et al., 2005).

Senescent and periphytic detritus (as dead particulate organic matter) is mineralized either within the periphyton or in the water column to release soluble nutrients. For example, C and N sources in periphyton establish from the time of initiation of bacterial biofilm formation. The bacteria, as initial substrate colonizers, can secrete and form EPS outside the periphyton. This EPS mainly consists of polysaccharides and proteins, which provide a significant source of C and N to the periphytic community and higher trophic levels. It therefore represents a trophic link between dissolved organic and inorganic substrates in the water column and the higher trophic levels of the ecosystem (Hynes and Hynes, 1970). The bacterial EPS from early biofilm exists as a part of dissolved organic matter (Lignell, 1990) as well as particulate matter (Decho, 2000). It acts both as rich organic C and N storage (Freeman and Lock, 1995) and a chief supplier of C and N demand for organisms that feed on periphytic aggregates (Decho and Moriarty, 1990; Hoskins et al., 2003).

Being polyanionic in nature, EPS permits inorganic nutrient entrapment through ion exchange processes leading to storage of organic C and N in the biofilm (Freeman et al., 1995). In addition, periphyton could actively fix atmospheric N_2 via the nitrogenase enzyme complex (N_2 fixation) (Inglett et al., 2004, 2009). Drake et al. (2012) found that periphyton were sometimes a source of P when light availability and periphyton C:P ratios were both low (Drake et al., 2012). Periphyton could also be an important source of oxygen in streams due to photosynthesis by attached algae in natural substrates. This periphyton photosynthesis could supply higher amounts of oxygen to the water column (Pratiwi et al., 2012).

5.2.2 Ingestion and Retention of Nutrients

Periphytic biofilm can play significant roles in the cycling of ambient nutrients as they can trap particulate material and assimilate or uptake the nutrients (such as C, N, and P) from the water column (Adey et al., 1993; Matheson et al., 2012). Biofilm theory holds that the uptake of nutrients from water is diffusion-limited. Before assimilation, solutes must first pass from the bulk liquid to the biofilm surface (external mass transfer) and then through the biofilm matrix to the cells (internal mass transfer) (Gantzer et al., 1988). Such nutrient uptake is controlled by three principal processes: (1) diffusion from the ambience into the viscous sublayer of the periphytic boundary layer; (2) slower transport, dominated by molecular diffusion, through the inner portion of the viscous sublayer (the diffusive boundary layer, or DBL) to periphyton cell surfaces; and (3) membrane transport from cell surfaces into cells. Nutrients, especially C, N, and P, are necessary for periphyton growth (Larned et al., 2004). Their incorporation into the microbial biomass is very efficient due to photosynthetic microorganisms (Wu et al., 2012). Periphyton has a high capacity for photosynthesis due to its large proportion of algae. It is well known that the overall chemical reaction involved in photosynthesis is:

$$6CO_2 + 6H_2O(+\text{ light energy}) \rightarrow C_6H_{12}O_6 + 6O_2$$

Through the above process, CO_2 in the water column is immobilized as organic compounds, like sugar, and stored in the biomass of periphyton. However, the fate of the nutrients and organic compounds in periphyton communities is spatially and temporally diverse (Azim, 2009; Baxter et al., 2013). Most of the organic matter produced by periphyton is degraded to CO_2 microbially by bacteria and protists within the periphytic community. Many of the nutrients released during decomposition are actively sequestered and retained by the viable components of the periphyton community.

Moreover, nutrient uptake is further determined by the activity of enzymes prevalent in periphytic bodies called intracellular enzymes. For example, inorganic N assimilation by algal groups follows the following pathway:

$$NO_3^-(+\text{ Nitrate Reductase}) \rightarrow NO_2^- \rightarrow NH_4^+$$

Nitrate reductase (NR) catalyzes the initial reduction of NO_3^- to NO_2^-, mostly available from decomposition of bacterial components, which is believed to be the rate-limiting step in the uptake and assimilation of NO_3^- into amino acids and proteins. The activity of NR is regulated in response to available NO_3^-, NO_2^-, and NH_4^-. The expression of NR is dependent on NO_3^- and light and is suppressed in most algae by high ambient concentrations of NH_4^- (Berges et al., 1995; Young et al., 2005).

Under aerobic conditions, NH_4^- could act as the N source. The autotrophic nitrifier in periphyton can use CO_2 to synthesize amino acids and proteins as follows (Wu et al., 2014):

$$0.13NH_4^+ + 0.225O_2 + 0.02CO_2 + 0.005HCO_3^- \\ = 0.005C_5H_7O_2N + 0.125NO_3^- + 0.25H^+ + 0.12H_2O$$

Similar to N, P is also an element essential for periphyton growth. Many studies have verified that periphyton has a high affinity for P and can effectively remove or uptake P from waters through assimilation, adsorption, and coprecipitation mechanisms (Crispim et al., 2009; De Godos et al., 2009; Dodds, 2003; Drake et al., 2012; Havens et al., 1999; Hayashi et al., 2012; Jarvie et al., 2002; Lu et al., 2014; Rejmánková and Komárková, 2000; Scinto and Reddy, 2003; Woodruff et al., 1999; Wu et al., 2014). Luxury P-uptake is the storage of P within the periphyton biomass in the form of polyphosphate (Saikia, 2011). Much of the P-fraction available in the aquatic environment, however, is not directly available for uptake by periphyton because it is bound to organic chelators. A widely distributed enzyme which helps cleave orthophosphate from organic chelators is alkaline phosphatase (AP). The periphyton matrix can form a diffusion barrier that promotes retention of extracellular enzymes such as alkaline phosphatases and proteases along with their substrates and products (Ellwood et al., 2012; Flemming and Wingender, 2010). The extracellular hydrolytic enzyme is retained and accumulates over time, contributing to the overall biofilm metabolism, acting as an external digestive system and rendering the products more available for uptake (Flemming and Wingender, 2010).The overall process is:

$$\text{Organically-bound } PO_4^{3-}(+\text{ AP}) \rightarrow \text{free } PO_4^{3-} \rightarrow \text{P available for uptake}$$

Generally, periphyton could assimilate P from the water column by several processes, including P uptake and biodegradation/enzymolysis, filtering particulate P from the water, and attenuating flow, which it decreases advective transport of particulate and dissolved P from sediments. Furthermore, periphyton photosynthesis locally increases pH, which can lead to increased precipitation of calcium phosphate, concurrent deposition of carbonate–phosphate complexes, and long-term burial of P. Actively photosynthesizing periphyton can cause supersaturated O_2 concentrations near the sediment surface encouraging deposition of metal phosphates.

All the above information indicates that periphyton could significantly influence nutrient retention in aquatic ecosystems. It has been proposed that much of the metabolism within the periphyton community was mutualistically coupled. Because of this intimate physiological interdependence, internal nutrient and energetic recycling within the relatively closed community is intense. Metabolism, growth, and productivity within periphyton communities rely heavily on internal recycling and conservation of nutrients—resulting in unusually high efficiency of utilization and retention of captured external nutrients (Azim, 2009). Dodds (2003) summarized the role of periphyton assemblages in increasing the retention of nutrients in waters, especially in shallow waters. First, they can remove nutrients from the water column and cause a net flux of nutrients toward the sediments. Second, they can slow water exchange across the sediment–water column boundary thus decreasing advective transport of P away from sediments. Third, they can intercept nutrients diffusing from the benthic sediments or senescent macrophytes. Fourth, they can cause biochemical conditions that favor P deposition. Finally, they can trap particulate material from the water column (Dodds, 2003).

5.3 MEDIATOR, REGULATOR, AND INDICATOR OF NUTRIENTS

5.3.1 Mediator

As noted earlier, periphyton is the primary transducer of the sun's rays into biologically based energy for stream ecosystems. Thus, this community is the "grass" of streams for aquatic grazing animals. Indeed, periphyton is a mediator of nutrient transfer in the water column. The periphytic nutrient transfer pathway mainly involves ambient nutrient entrapment, storage and transfer to the immediate higher trophic level, which can be summarized to two possible pathways. First, by direct nutrient uptake from the environment and transfer to immediate grazer. Second, through de novo synthesis of metabolic products as raw materials and trophic upgrading to immediate predators (Saikia and Das, 2009; Saikia and Nandi, 2010; Saikia et al., 2013; Saikia, 2011). For example, periphyton can capture the energy of sunlight via their chlorophyll molecules, absorb CO_2 and other nutrients such as phosphorus and nitrogen from the surrounding water, and then synthesize organic C in the form of new or enlarged cells. Periphyton could then secrete a portion of this C and a host of other organisms such as communities of bacteria, fungi, and protozoa can live off the secreted C. This is a normal pathway for C transfer in periphyton communities. In addition, the bacterial C of periphyton may enter the next trophic group as complex C-rich compounds, such as fatty acids (FA), saturated fatty acids (SAFA), and monounsaturated fatty acids (MUFA). The dominance of algae in periphytic biofilm provides a rich source of C in the FA form to periphyton-grazing animals. As the food chain proceeds, organic C is transferred from periphyton to grazers through predation (Saikia, 2011). It has been suggested that periphyton is often the dominant C source for grazers in shallow lakes with low planktonic productivity (Hecky and Hesslein, 1995; James et al., 2000). Some invertebrate taxa (such as aquatic insects, *Hemiptera* and *Trichoptera*, and the freshwater shrimp, *Caridina*) that generally use allochthonous C sources, attain up to 65% of their dietary C from periphyton (Mihuc and Toetz, 1994). However, the utilization of periphytic C by predators might be disproportionate, even when standing consumer biomass is low (Hecky and Hesslein, 1995; James et al., 2000).

Periphytic bacteria are important nutrient sources at the base of periphytic food webs. For example, during the initial colonization of periphytic communities, the bacterial C transfers to the immediate neighbors of periphyton consortia through bacterial metabolizable substrates. In a growing periphyton, cynobacteria and other early colonizing algae share this bacterial C. In aged periphyton, such C transfers to later successional algae groups, such as filamentous *Spirogyra*, from bacterial decomposition. In addition, the concentration of nutrients in periphtyon groups, such as the luxury P-uptake by periphytic bacteria, adds additional dimensions for nutrient storage in bacterial cells and influences nutrient uptake and transfer.

5.3.2 Regulator

Periphytic biofilms play important simultaneous roles in controlling aquatic biogeochemical cycling and regulating nutrient exchange between the water column and the hyporheic zone (Paerl and Pinckney, 1996). For instance, the periphyton can play a major role in regulating nutrient transfer across the sediment–water interface, including (1) the buffering of physicochemical nutrient release via direct sequestration of nutrients, (2) reducing hypoxic nutrient release and denitrification–nitrification processes via the maintenance of oxygenated surface sediment, and (3) decreasing nutrient release across the sediment–water interface via flow attenuation (Poulíčková et al., 2008). Woodruff et al. (1999) argued that the presence of periphytic biofilm could control chemical fluxes of calcium, alkalinity, and phosphorus from the sediment to the overlying water. Microbial biofilms have also changed the physical and chemical microhabitat and contributed to ecosystem processes in 30-m-long stream mesocosms. Biofilm growth increased hydrodynamic transient storage by 300% and the retention of suspended particles by 120%. In addition, by enhancing the relative uptake of organic molecules of lower bioavailability, the interplay of biofilm microarchitecture and mass transfer changed their downstream linkage. As living zones of transient storage, biofilms bring hydrodynamic retention and biochemical processing into close spatial proximity and influence biogeochemical processes and patterns in streams (Battin et al., 2003). In addition, the fine particulate organic matter (FPOM) in water has a high nutrient content and lower C:nutrient ratio (Bonin et al., 2000; Cross et al., 2003), and is an important source of nutrient input to peripyton. Detritivory on periphyton at an optimum accumulation period of FPOM could ensure a higher rate of nutrient transfer to consumers (Saikia, 2011). When periphyton die and dehydrate, their biomass, such as algal material (mainly charophytes), is directly incorporated into the sediment or soil through desiccation and pulverization. Nutrients like N (mainly as NH_3) and P then dramatically increase in the sediment or soil (Gómez-Pompa et al., 2003; Novelo and Tavera, 2003). Several studies have shown that periphyton regulated water quality by quickly and efficiently removing excess P from the water column (Gaiser et al., 2006; Thomas et al., 2006). Therefore, the accumulation of nutrients by periphyton grown on substrates is an important way to regulate nutrient condition in aquatic ecosystems.

Another way for periphytic biofilm to regulate nutrient conditions in water columns is through the activity of extracellular enzymes that are secreted or immobilized by periphyton. Extracellular enzymes may become attached to EPS through processes similar to the formation of enzyme humus complexes in soil (Lock et al., 1984). It has also been suggested that the binding of enzymes protects them from degradation and increases their exposure to substrates (Espeland and Wetzel, 2001). The resulting efficient recycling of organic matter and increasing concentration of nutrients within this matrix-bound hydrolytic system buffers the community metabolism to changes in ambient nutrient dynamics (Sekar et al., 2002). Such high effectiveness in removing or uptaking nutrients from water columns helps periphyton act as a bioeliminator to improve water quality (Sabater et al., 2002).

5.3.3 Indicator

Periphyton responds predictably and quickly (days to weeks) to changes in environmental conditions at a large range of spatial scales (meters to tens of kilometers). It therefore serves as an excellent indicator, early responder, and can be used as a warning of impending change in aquatic ecosystems (Gaiser, 2009; Gaiser et al., 2005; McCormick et al., 2001; Noe et al., 2002; Reavie et al., 2010). The desirable features of periphyton making it a reliable ecological indicator include: (1) it is ubiquitously distributed throughout the study system, (2) ability to rapidly respond to environmental change in ways that are, (3) quantifiable at several levels of biological organization (individual, species, population, and community) with (4) consequences for levels above and below its placement in the food web (Karr, 1999). Studies of variation in Everglades periphyton along naturally existing and experimentally created gradients found strong relationships between structural (i.e., species composition, growth form) and functional (e.g., nutrient uptake and productivity) properties and the nutrient states in water (Biggs, 1995; Gaiser et al., 2006; McCormick et al., 1996; Pan et al., 2000). Reliance on periphyton to indicate environmental change is well justified by scientific research conducted in the Everglades ecosystem (Gaiser et al., 2006; McCormick and Stevenson, 1998). For instance, nutrient enrichment in water elevates periphyton nutrient content, reduces the proportion of calcareous floating and epiphytic periphyton mats, and replaces native species with nonmat-forming filamentous species (Gaiser, 2009). The total phosphorus (TP) content of periphyton tissue is one of the best indicators of P load history in water systems (Gaiser et al., 2005, 2006). Using periphyton P content, rather than water or soil P, to indicate P-enrichment history, is recommended because periphyton P content has repeatedly been shown to more reliably indicate P load history, and has been adopted in most large-scale monitoring programs in the Everglades (Gaiser, 2009; Gaiser et al., 2005). Research also shows that the response of periphyton is rapid and easily detected, as even low-level nutrient enhancements (>10 ppb TP) lead to the complete demise of the periphyton community (Gaiser et al., 2005; Thomas et al., 2006).

5.4 FUTURE WORK NEEDED

Periphytic biofilm have important roles in aquatic nutrient cycling, mainly as potential sources and sinks in nutrient dynamics, the buffering (ingestion and retention) of nutrient transport, and as mediators, regulators, and indicators of nutrient states. Despite the state of current knowledge, more work is needed on the role and importance of periphyton in nutrient cycling.

5.4.1 From Macro/Mesoscales to Microscales

Most research aimed at elucidating periphytic consortial structure, function, production, and biogeochemical nutrient dynamics has underscored the importance of characterizing and quantifying key processes (e.g., photosynthesis, respiration, redox-dependent C, N, and other nutrient transformations) at appropriate spatial and temporal scales. There has been substantial progress on how periphyton affects nutrient dynamics at macro/mesoscales like lakes, rivers, and streams, however, the study of nutrient dynamics in periphyton at the microscale, especially in the microzone of periphyton, needs further evaluation. Recent rapid development of molecularly and genetically advanced techniques such as microelectrode system, confocal scanning laser microscopy, nuclear magnetic resonance, synchrotron-based X-ray absorption near-edge structure spectroscopy, and high-throughput sequencing, have made it possible to study not only the dynamics of nutrient transportation, transformation, and interaction in the microzone in which periphytic consortia thrive, but also explain the molecular and genetic mechanisms of nutrient dynamics by periphyton. Representative microzones or microhabitats including EPS, detrital aggregates, biofilms, microbial mats, endosymbioses, and more loosely knit plant–microbe and animal–microbe associations, consist of not only the organisms themselves, but also complex organic and inorganic matrices. A common property of periphytic biofilm and similar aggregates is the occurrence of mass transfer resistance, which is due to the limited water flow inside the matrix and the presence of a hydrodynamic boundary layer between the matrix and the surrounding water. Matrices include mucilaginous slimes, gels, and partially lignified and/or lithified laminae produced by consortia members. These matrices serve as highly selective diffusional and structural modifiers of molecular and larger particle transport processes, and as such can be formidable barriers to the entry and movement of gases, dissolved nutrients, and macromolecules. Matrix constituents may bind physically, chemically, structurally, and functionally, subsequently altering molecules. To describe transport inside matrices, transport in the voids (advection and diffusion), the "layering" of different periphytic groups, and in the base film and cell clusters (diffusion only) must be distinguished through measuring techniques of high spatial resolution.

Periphyton is a microbial assemblage that consists of various species, mainly algae, bacteria, fungi, and protozoans and the contribution of these different species to nutrient dynamics may differ. Thus, the functional microbial species contributing to nutrient transportation and transformation need further identification and investigation. The role of periphyton microbial biodiversity in regulating nutrient cycling (both diffuse and turbulence related) should also be assessed over a range of habitat types.

5.4.2 From Waters to Soils/Sediments

Most work conducted on periphytic biofilm to date has generally focused on periphyton on "hard" surfaces such as stone, bamboo, wood, and plant surfaces. Few studies have examined periphyton on "soft" surfaces such as sediments or soil surfaces. In aquatic ecosystems, especially in marshes or wetlands, the sediment surface is covered by a layer of periphytic biofilm. Essentially, the main input or output of periphyton in soil systems is through the increase or decrease of biological nutrients (e.g. C, N, P). The role of periphyton in nutrient cycling between water and soils might be a "bridge" to exploring the interactions of nutrient cycling between waters and soils. This not only contributes to in-depth understanding of biogeochemical process in aquatic ecosystems, but also to nutrient transport and management in soil systems (particularly in paddy soil). Therefore, further work is needed on periphyton on soil surfaces and how it influences soil nutrient transfer and uptake by plants (especially crops).

Periphyton is often widely distributed on the surface of soils, especially in paddy fields. Soil nutrient properties, such as nutrient retention and runoff, are important for grain production and food security, however, little attention has focused on the influence of soil periphyton on nutrient properties of soil (Wu, 2013). Similar emphasis should also focus on sediments. Some types of periphyton, like motile epipelic diatoms, could increase sediment stability through the maintenance of periphytic biomass or a matrix composed of EPSs in the upper microns of the sediment surface (Sheng et al., 2010; Wu et al., 2010). Previous studies have also suggested that periphytic biodiversity may be critical to the biostabilization and nutrient transport of sediments (Wu et al., 2010). For instance, epipelic diatoms may reduce sensitivity to erosion through biochemical processes like EPS production, while cyanobacteria may reduce sensitivity to erosion by the maintenance of physical barriers like the mat itself (Sheng et al., 2010).

5.4.3 From Theory to Application

As mentioned above, substantial theoretical progress has been made in the study of periphyton–nutrient relationships, suggesting that periphyton could be regarded as a potentially promising strategy to indicate, regulate, and alleviate excessive nutrient inputs to aquatic ecosystems. However, it is necessary to determine longer-term effects for realistic and feasible applications of periphyton. Consequently, the application of periphyton-based technologies in the monitoring, indicating, and restoration of aquatic ecosystems will be a hot topic for future work.

Periphyton application is of growing interest worldwide. Currently, the most common applications of periphyton are in wastewater treatment systems, especially for nutrient removal. Increasing human development has led to an increase in the complexity of the composition of wastewater that enters downstream surface waters and increasing inputs of nonpoint-source wastewater. It is postulated that periphyton can remove a variety of water contaminants such as N, P, organic material, and heavy metals. As it is necessary and practical to develop an integrated technology based on periphyton to remove multiple pollutants simultaneously, future studies should focus on developing pollutant-removing microbial aggregates with multiple compositions.

As nutrients such as phosphorus becomes increasingly scarce, nutrient recovery and reuse will increase. Previous studies have already indicated that periphyton has the potential to capture nutrients from eutrophic waters and reuse them as biofertilizers for animal or plant growth (Wu, 2013; Wu and Yang, 2012). For example, the phosphorus content of periphyton is relatively high, at 100- to1000-fold that of surrounding waters. Periphytic algae have also been identified as good supplementary livestock feed (such as for fishes and chickens) due to their high protein content (Wu, 2013; Wu and Yang, 2012). These issues, from harvest to application, however, are currently quite poorly covered by the literature. Further research is needed on periphyton harvesting, transportation, stability, biofertilizer techniques, and the proportion of periphytic nutrient availability for crops.

REFERENCES

Adey, W., Luckett, C., Jensen, K., 1993. Phosphorus removal from natural waters using controlled algal production. Restoration Ecology 1 (1), 29–39.

Azim, M.E., 2009. Photosynthetic periphyton and surfaces. In: Likens, G.E. (Ed.), Encyclopedia of Inland Waters. Academic Press, Oxford, pp. 184–191.

Baldwin, D.S., Mitchell, A., Rees, G., Watson, G., Williams, J., 2006. Nitrogen processing by biofilms along a lowland river continuum. River Research and Applications 22 (3), 319–326.

Battin, T.J., Kaplan, L.A., Denis Newbold, J., Hansen, C.M.E., 2003. Contributions of microbial biofilms to ecosystem processes in stream mesocosms. Nature 426 (27), 439–442.

Baxter, L.R., Sibley, P.K., Solomon, K.R., Hanson, M.L., 2013. Interactions between atrazine and phosphorus in aquatic systems: effects on phytoplankton and periphyton. Chemosphere 90 (3), 1069–1076.

Berges, J.A., Cochlan, W.P., Harrison, P.J., 1995. Laboratory and field responses of algal nitrate reductase to diel periodicity in irradiance, nitrate exhaustion, and the presence of ammonium. Marine Ecology Progress Series. Oldendorf 124 (1), 259–269.

Biggs, B.J., 1995. The contribution of flood disturbance, catchment geology and land use to the habitat template of periphyton in stream ecosystems. Freshwater Biology 33 (3), 419–438.

Bonin, H., Griffiths, R., Caldwell, B., 2000. Nutrient and microbiological characteristics of fine benthic organic matter in mountain streams. Journal of the North American Benthological Society 19 (2), 235–249.

Cantonati, M., Lowe, R.L., 2014. Lake benthic algae: toward an understanding of their ecology. Freshwater Science 33 (2), 475–486.

Crispim, M., Vieira, A., Coelho, S., Medeiros, A., 2009. Nutrient uptake efficiency by macrophyte and biofilm: practical strategies for small-scale fish farming. Acta Limnologica Brasiliensia 21 (4), 387–391.

Cross, W.F., Benstead, J.P., Rosemond, A.D., Bruce Wallace, J., 2003. Consumer-resource stoichiometry in detritus-based streams. Ecology Letters 6 (8), 721–732.

De Godos, I., González, C., Becares, E., García-Encina, P.A., Muñoz, R., 2009. Simultaneous nutrients and carbon removal during pretreated swine slurry degradation in a tubular biofilm photobioreactor. Applied Microbiology and Biotechnology 82 (1), 187–194.

De Philippis, R., Colica, G., Micheletti, E., 2011. Exopolysaccharide-producing cyanobacteria in heavy metal removal from water: molecular basis and practical applicability of the biosorption process. Applied Microbiology and Biotechnology 92 (4), 697–708.

Decho, A.W., Moriarty, D.J.W., 1990. Bacterial exopolymer utilization by a harpacticoid copepod: a methodology and results. Limnology and Oceanography 35 (5), 1039–1104.

Decho, A.W., 2000. Microbial biofilms in intertidal systems: an overview. Continental Shelf Research 20 (10), 1257–1273.

Dodds, W.K., 2003. The role of periphyton in phosphorus retention in shallow freshwater aquatic systems. Journal of Phycology 39 (5), 840–849.

Drake, W., Scott, J.T., Evans-White, M., Haggard, B., Sharpley, A., Rogers, C.W., Grantz, E.M., 2012. The effect of periphyton stoichiometry and light on biological phosphorus immobilization and release in streams. Limnology 13 (1), 97–106.

Ellwood, N.T.W., Di Pippo, F., Albertano, P., 2012. Phosphatase activities of cultured phototrophic biofilms. Water Research 46 (2), 378–386.

Espeland, E., Wetzel, R., 2001. Complexation, stabilization, and UV photolysis of extracellular and surface-bound glucosidase and alkaline phosphatase: implications for biofilm microbiota. Microbial Ecology 42 (4), 572–585.

Fang, L., Wei, X., Cai, P., Huang, Q., Chen, H., Liang, W.R., ong, X., 2011. Role of extracellular polymeric substances in Cu(II) adsorption on *Bacillus subtilis* and *Pseudomonas putida*. Bioresource Technology 102 (2), 1137–1141.

Flemming, H.-C., Wingender, J., 2010. The biofilm matrix. Nature Reviews. Microbiology 8 (9), 623–633.

Freeman, C., Lock, M.A., 1995. The biofilm polysaccharide matrix: a buffer against changing organic substrate supply? Limnology and Oceanography 40 (2), 273–278.

Freeman, C., Chapman, P., Gilman, K., Lock, M., Reynolds, B., Wheater, H., 1995. Ion exchange mechanisms and the entrapment of nutrients by river biofilms. Hydrobiologia 297 (1), 61–65.

Gaiser, E.E., Trexler, J.C., Richards, J.H., Childers, D.L., Lee, D., Edwards, A.L., Scinto, L.J., Jayachandran, K., Noe, G.B., Jones, R.D., 2005. Cascading ecological effects of low-level phosphorus enrichment in the Florida Everglades. Journal of Environmental Quality 34 (2), 717–723.

Gaiser, E.E., Trexler, J.C., Jones, R.D., Childers, D.L., Richards, J.H., Scinto, L.J., 2006. Periphyton responses to eutrophication in the Florida Everglades: cross-system patterns of structural and compositional change. Limnology and Oceanography 51, 617–630.

Gaiser, E., 2009. Periphyton as an indicator of restoration in the Florida Everglades. Ecological Indicators 9 (6 Suppl.), S37–S45.

Gantzer, C.J., Rittmann, B.E., Herricks, E.E., 1988. Mass transport to streambed biofilms. Water Research 22 (6), 709–722.

Gómez-Pompa, A., Allen, M., Fedick, S., Jiménez-Osornio, J., 2003. The Lowland Maya Area: Three Millennia at the Human-Wildland Interface. Food Products Press, New York.

Gong, R., Ding, Y.D., Liu, H., Chen, Q., Liu, Z., 2005. Lead biosorption by intact and pretreated *Spirulina maxima* biomass. Chemosphere 58 (1), 125–130.

Havens, K.E., East, T.L., Hwang, S.J., Rodusky, A.J., Sharfstein, B., Steinman, A.D., 1999. Algal responses to experimental nutrient addition in the littoral community of a subtropical lake. Freshwater Biology 42 (2), 329–344.

Hayashi, M., Vogt, T., Mächler, L., Schirmer, M., 2012. Diurnal fluctuations of electrical conductivity in a pre-alpine river: effects of photosynthesis and groundwater exchange. Journal of Hydrology 450–451 (0), 93–104.

Hecky, R., Hesslein, R., 1995. Contributions of benthic algae to lake food webs as revealed by stable isotope analysis. Journal of the North American Benthological Society 631–653.

Hoskins, D.L., Stancyk, S.E., Decho, A.W., 2003. Utilization of algal and bacterial extracellular polymeric secretions (EPS) by the deposit-feeding brittlestar *Amphipholis gracillima* (Echinodermata). Marine Ecology Progress Series 247, 93–101.

Hynes, H.B.N., Hynes, H., 1970. The Ecology of Running Waters. Liverpool University Press Liverpool.

Inglett, P., Reddy, K., McCormick, P., 2004. Periphyton chemistry and nitrogenase activity in a northern Everglades ecosystem. Biogeochemistry 67 (2), 213–233.

Inglett, P.W., D'Angelo, E.M., Reddy, K.R., McCormick, P.V., Hagerthey, S.E., 2009. Periphyton nitrogenase activity as an indicator of wetland eutrophication: spatial patterns and response to phosphorus dosing in a northern Everglades ecosystem. Wetlands Ecology and Management 17 (2), 131–144.

James, M., Hawes, I., Weatherhead, M., 2000. Removal of settled sediments and periphyton from macrophytes by grazing invertebrates in the littoral zone of a large oligotrophic lake. Freshwater Biology 44 (2), 311–326.

Jarvie, H.P., Neal, C., Warwick, A., White, J., Neal, M., Wickham, H.D., Hill, L.K., Andrews, M.C., 2002. Phosphorus uptake into algal biofilms in a lowland chalk river. Science of the Total Environment 282–283 (0), 353–373.

Karr, J.R., 1999. Defining and measuring river health. Freshwater Biology 41 (2), 221–234.

Larned, S.T., Nikora, V.I., Biggs, B.J., 2004. Mass-transfer-limited nitrogen and phosphorus uptake by stream periphyton: a conceptual model and experimental evidence. Limnology and Oceanography 49 (6), 1992–2000.

Larned, S.T., 2010. A prospectus for periphyton: recent and future ecological research. Journal of the North American Benthological Society 29 (1), 182–206.

Liboriussen, L., Jeppesen, E., 2003. Temporal dynamics in epipelic, pelagic and epiphytic algal production in a clear and a turbid shallow lake. Freshwater Biology 48 (3), 418–431.

Lignell, R., 1990. Excretion of organic carbon by phytoplankton: its relation to algal biomass, primary productivity and bacterial secondary productivity in the Baltic Sea. Marine Ecology Progress Series MESEDT 68 (1/2).

Lock, M., Wallace, R., Costerton, J., Ventullo, R., Charlton, S., 1984. River epilithon: toward a structural-functional model. Oikos 10–22.

Lu, H., Yang, L., Shabbir, S., Wu, Y., 2014. The adsorption process during inorganic phosphorus removal by cultured periphyton. Environmental Science and Pollution Research 21 (14), 8782–8791.

Matheson, F.E., Quinn, J.M., Martin, M.L., 2012. Effects of irradiance on diel and seasonal patterns of nutrient uptake by stream periphyton. Freshwater Biology 57 (8), 1617–1630.

McCormick, P.V., Stevenson, R.J., 1998. Periphyton as a tool for ecological assessment and management in the Florida Everglades. Journal of Phycology 34 (5), 726–733.

McCormick, P.V., Rawlik, P.S., Lurding, K., Smith, E.P., Sklar, F.H., 1996. Periphyton-water quality relationships along a nutrient gradient in the northern Florida Everglades. Journal of the North American Benthological Society 433–449.

McCormick, P.V., O'Dell, M.B., Shuford III, R.B., Backus, J.G., Kennedy, W.C., 2001. Periphyton responses to experimental phosphorus enrichment in a subtropical wetland. Aquatic Botany 71 (2), 119–139.

McCormick, P.V., Shuford III, R.B., Chimney, M.J., 2006. Periphyton as a potential phosphorus sink in the Everglades Nutrient Removal Project. Ecological Engineering 27 (4), 279–289.

Mihuc, T., Toetz, D., 1994. Determination of diets of alpine aquatic insects using stable isotopes and gut analysis. The American Midland Naturalist 146–155.

Neal, C., Jarvie, H.P., 2005. Agriculture, community, river eutrophication and the water Framework directive. Hydrological Processes 19 (9), 1895–1901.

Noe, G.B., Childers, D.L., Edwards, A.L., Gaiser, E., Jayachandran, K., Lee, D., Meeder, J., Richards, J., Scinto, L.J., Trexler, J.C., 2002. Short-term changes in phosphorus storage in an oligotrophic Everglades wetland ecosystem receiving experimental nutrient enrichment. Biogeochemistry 59 (3), 239–267.

Novelo, E., Tavera, R., 2003. The role of periphyton in the regulation and supply of nutrients in a wetland at El Edén, Quintana Roo. The Lowland Maya Area: Three Millennia at the Human-Wildland Interface 217–239.

Paerl, H., Pinckney, J., 1996. A mini-review of microbial consortia: their roles in aquatic production and biogeochemical cycling. Microbial Ecology 31 (3), 225–247.

Pan, Y., Stevenson, R.J., Vaithiyanathan, P., Slate, J., Richardson, C.J., 2000. Changes in algal assemblages along observed and experimental phosphorus gradients in a subtropical wetland, USA. Freshwater Biology 44 (2), 339–353.

Poulíčková, A., Hašler, P., Lysáková, M., Spears, B., 2008. The ecology of freshwater epipelic algae: an update. Phycologia 47 (5), 437–450.

Pratiwi, N.T.M., Hariyadi, S., Tajudin, R., 2012. Photosynthesis of periphyton and diffusion process as source of oxygen in rich-riffle upstream waters. Microbiology Indonesia 5 (4), 182.

Reavie, E.D., Jicha, T.M., Angradi, T.R., Bolgrien, D.W., Hill, B.H., 2010. Algal assemblages for large river monitoring: comparison among biovolume, absolute and relative abundance metrics. Ecological Indicators 10 (2), 167–177.

Rejmánková, E., Komárková, J., 2000. A function of cyanobacterial mats in phosphorus-limited tropical wetlands. Hydrobiologia 431 (2–3), 135–153.

Revsbech, N.P., Christensen, P.B., Nielsen, L.P., Sørensen, J., 1989. Denitrification in a trickling filter biofilm studied by a microsensor for oxygen and nitrous oxide. Water Research 23 (7), 867–871.

Sabater, S., Guasch, H., Roman, A., Muñoz, I., 2002. The effect of biological factors on the efficiency of river biofilms in improving water quality. Hydrobiologia 469, 149–156.

Saikia, S., Das, D., 2009. Potentiality of periphyton-based aquaculture technology in rice-fish environment. Journal of Scientific Research 1 (3), 624–634.

Saikia, S., Nandi, S., 2010. C and P in aquatic food chain: a review on C: P stoichiometry and PUFA regulation. Knowledge and Management of Aquatic Ecosystems 398 (03).

Saikia, S., Nandi, S., Majumder, S., 2013. A review on the role of nutrients in development and organization of periphyton. Journal of Research in Biology 3 (1), 780–788.

Saikia, S.K., 2011. Review on periphyton as mediator of nutrient transfer in aquatic ecosystems. Ecologia Balkanica 3 (2), 65–78.

Scinto, L.J., Reddy, K.R., 2003. Biotic and abiotic uptake of phosphorus by periphyton in a subtropical freshwater wetland. Aquatic Botany 77 (3), 203–222.

Sekar, R., Nair, K., Rao, V., Venugopalan, V., 2002. Nutrient dynamics and successional changes in a lentic freshwater biofilm. Freshwater Biology 47 (10), 1893–1907.

Sheng, G.P., Yu, H.Q., Li, X.Y., 2010. Extracellular polymeric substances (EPS) of microbial aggregates in biological wastewater treatment systems: a review. Biotechnology Advances 28, 882–892.

Sobczak, W.V., Findlay, S., Dye, S., 2003. Relationships between DOC bioavailability and nitrate removal in an upland stream: an experimental approach. Biogeochemistry 62 (3), 309–327.

Thomas, S., Gaiser, E.E., Gantar, M., Scinto, L.J., 2006. Quantifying the responses of calcareous periphyton crusts to rehydration: a microcosm study (Florida Everglades). Aquatic Botany 84 (4), 317–323.

Thompson, F.L., Abreu, P.C., Wasielesky, W., 2002. Importance of biofilm for water quality and nourishment in intensive shrimp culture. Aquaculture 203 (3), 263–278.

Vadeboncoeur, Y., Lodge, D.M., Carpenter, S.R., 2001. Whole-lake fertilization effects on distribution of primary production between benthic and pelagic habitats. Ecology 82 (4), 1065–1077.

Vander Zanden, M.J., Vadeboncoeur, Y., 2002. Fishes as integrators of benthic and pelagic food webs in lakes. Ecology 83 (8), 2152–2161.

Wetzel, R.G., 1963. Primary productivity of periphyton. Nature 197 (4871), 1026–1027.

Wetzel, R.G., 1993. Microcommunities and microgradients: linking nutrient regeneration, microbial mutualism, and high sustained aquatic primary production. Netherland Journal of Aquatic Ecology 27 (1), 3–9.

Woodruff, S., House, W., Callow, M., Leadbetter, B., 1999. The effects of a developing biofilm on chemical changes across the sediment-water interface in a freshwater environment. International Review of Hydrobiology 84 (5), 509–532.

Wu, Y., Yang, L., 2012 3rd Quarter. Chapter 8. "A prospectus for bio-organic fertilizer based on microorganisms: recent and future research in agricultural ecosystem". In: Organic Fertilizers: Types, Production and Environmental Impact. Series: Agriculture Issues and Policies. Nova Science Publisher, New York, ISBN 978-1-62081-422-2.

Wu, Y., Zhang, S., Zhao, H., Yang, L., 2010. Environmentally benign periphyton bioreactors for controlling cyanobacterial growth. Bioresource Technology 101 (24), 9681–9687.

Wu, Y., Li, T., Yang, L., 2012. Mechanisms of removing pollutants from aqueous solutions by microorganisms and their aggregates: a review. Bioresource Technology 107 (0), 10–18.

Wu, Y., Xia, L., Yu, Z., Shabbir, S., Kerr, P.G., 2014. In situ bioremediation of surface waters by periphytons. Bioresource Technology 151 (0), 367–372.

Wu, Y., 2013. The studies of periphytons: from waters to soils. Hydrology: Current Research 4, 2.

Xu, H., Cai, H., Yu, G., Jiang, H., 2013. Insights into extracellular polymeric substances of cyanobacterium *Microcystis aeruginosa* using fractionation procedure and parallel factor analysis. Water Research 47 (6), 2005–2014.

Young, E., Lavery, P., Van Elven, B., Dring, M., Berges, J., 2005. Dissolved inorganic nitrogen profiles and nitrate reductase activity in macroalgal epiphytes within seagrass meadows. Marine Ecology Progress Series 288, 103–114.

Chapter 6

Periphyton Functions in Adjusting P Sinks in Sediments

6.1 INTRODUCTION

Sediments play an important role in overall phosphorus (P) cycling in aquatic ecosystems, acting both as a sink and a source of P due to continuous transport of chemical species across the sediment and water interface (Jöbgen et al., 2004; Jørgensen et al., 2000; Jarvie et al., 2002; Jorgensen and Des Marais, 1990). Many previous studies have indicated that sediment could continuously release P into the overlying water, which may result in continued eutrophication of these waters (Koski-Vähälä and Hartikainen, 2001; Monbet et al., 2007; Palmer-Felgate et al., 2011; Qin, 2009; Søndergaard et al., 2003; Smith et al., 2006; Wang et al., 2008). Generally, sediment P can be divided into labile P (NH_4Cl-P), reductant P (BD-P), metal-bound P (NaOH-P), calcium-bound P (HCl-P), and residual P (Res-P) using chemical extraction methods (Fytianos and Kotzakioti, 2005; Kaiser et al., 1996; Kaiserli et al., 2002; Rydin, 2000). These P fractions in sediment, however, may have different P exchangeability and bioavailability (Wang et al., 2006, 2009). It has been documented that only part of the P fractions in sediment are easily exchangeable and biologically available (such as NH_4Cl-P), as most tend to adsorb on the surface or interior of various metal oxides and hydroxides (especially those of Fe, Mn, Ca, and Al) (Christophoridis and Fytianos, 2006; Renjith et al., 2011; Rzepecki, 2010). Therefore, the evaluation of P fractions in sediment not only increases the understanding of P cycling in aquatic ecosystems, but also contributes to the management of eutrophication caused by the release of internal P loads (Kaiser et al., 1996; Kaiserli et al., 2002; Solomon, 2007).

The release of P from sediments is a highly complex process, involving a number of physical processes (i.e., adsorption and desorption), chemical processes (i.e., ligand exchange and precipitation), and biological processes (i.e., release from living cells and autolysis of cells) (Christophoridis and Fytianos, 2006). Many factors can influence the release process of P between the sediment and water interface, including redox potential (Eh), pH, temperature, dissolved oxygen, salinity, and sediment resuspension

Periphyton. http://dx.doi.org/10.1016/B978-0-12-801077-8.00006-5

(Christophoridis and Fytianos, 2006; Jin et al., 2006a,b; Kaiserli et al., 2002; Koski-Vähälä and Hartikainen, 2001; Perkins and Underwood, 2001; Wang et al., 2013). Most research, however, has been based on the two phases (water and sediment), ignoring or underestimating the importance of the biota between them.

Periphytic biofilm, also called periphyton, is biota ubiquitously distributed between overlying water and sediments, especially on the surface of sediments and suspended particles in shallow aquatic ecosystems (Poulíčková et al., 2008; Writer et al., 2011). It is often a complex microbial consortium of algae, bacteria, and other micro- and mesoorganisms (Wu et al., 2012). Many previous studies have demonstrated that periphytic biofilm plays a significant role in natural aquatic ecosystems by affecting primary production, food chains, and the migration of nutrients or contaminants in the sediment–water interface (Battin et al., 2003; Paerl and Pinckney, 1996; Saikia, 2013; Writer et al., 2011). In particular, periphytic biofilm is essential for P cycling between water and the sediment interface due to its high affinity for P (Drake et al., 2012; McCormick et al., 2006; Scinto and Reddy, 2003).

Generally, the high affinity of periphytic biofilm to P is due to its assimilation (Guzzon et al., 2008), adsorption (Lu et al., 2014a,b; Scinto and Reddy, 2003), coprecipitation (Dodds, 2003; Hill and Fanta, 2008), and interception or entrapment (Adey et al., 1993) of P from water. Consequently, periphytic biofilm can act as a potential sink of P in water (McCormick et al., 2006). Some previous studies have found that P migration between water and sediment interfaces was controlled more by periphytic biofilm than diffusion (Gainswin et al., 2006; Pietro et al., 2006; Woodruff et al., 1999b). It has also been recently suggested that the presence of periphytic biofilm in the water–sediment interface could not only decrease the P content in overlying water, but also reduce the release of sediment P to overlying waters (Wu et al., 2010a; Zhang et al., 2013b). The changes in P concentrations and fractions between a sediment–periphytic biofilm-overlying water system however, have not been systematically investigated. Therefore, how P migrates in this "three-phase" system and whether the presence of periphytic biofilm can alter P migration and transformation remain unclear.

The objectives of this study were to (1) systematically evaluate the change in P concentration and species between overlying water and sediments in the presence of periphytic biofilms; (2) assess the role of periphytic biofilms in P migration between overlying water and sediments; and (3) clarify P content and form in periphytic biofilms. This study will help clarify the role of periphytic biofilm in P migration between overlying water and sediments. It will also provide valuable information for better understanding P cycling in wetlands that contain periphytic biofilms or similar microbial assemblages.

6.2 MATERIAL AND METHODS

6.2.1 Cultivation of Periphytic Biofilm

Periphytic biofilms were incubated in a glass tank (50 cm length, 20 cm width, and 60 cm height). The water ($\sim$50 L) used for the periphytic biofilm gathering was collected from Xuanwu Lake in Nanjing, Jiangsu Province of China. Industrial soft carriers (diameter 12 cm and length 55 cm, Jineng Environmental Protection Company of YiXing, China) were immersed into the water for microorganism gathering. The modified BG-11 medium (5 L, composing of 0.1 g $NaCO_3$, 0.75 g $NaNO_3$, 0.2 g K_2HPO_4, 0.375 g $MgSO_4 \cdot 7H_2O$, 0.18 g $CaCl_2 \cdot 2H_2O$, 14.3 mg H_3BO_4, 9.05 mg $MnCl_2 \cdot 4H_2O$, 1.1 mg $ZnSO_4$, 1.95 mg Na_2MoO_4, 0.395 mg $CuSO_4 \cdot 5H_2O$, 24.7 mg $Co(NO_3)_2 \cdot 6H_2O$, 30 mg citric acid and ammonium ferric citrate) was added periodically (weekly) to sustain periphytic biofilm growth. To reduce the influence of environmental conditions on periphyton growth, the tank was kept in a greenhouse with air temperature maintained between 25 and 35°C. By day 60 many native microorganisms had grown on the carriers and formed dense and stable periphytic biofilm (green in color). The periphytic biofilms were collected and used in the following experiments.

6.2.2 Experimental Design

The sediments used in the experiments were collected from the shallow area of Xuanwu Lake. After air-drying, the large rocks and wood pieces in the sediment were carefully picked out. Then, simulation of the water–sediment interface was conducted in a 5.0-L beaker. Firstly, 1.0 kg of dry sediment (TP: 12.29 $\pm$ 0.75 mg kg^{-1}, Labile-P:1.02 $\pm$ 0.07 mg kg^{-1}, Fe/Al-P:3.67 $\pm$ 0.36 mg kg^{-1}, Ca-P:6.34 $\pm$ 0.24 mg kg^{-1}) was added to the beaker. Secondly, 3.0 L water (20 mg L^{-1} $NaCO_3$, 150 mg L^{-1} $NaNO_3$, 10 mg L^{-1} K_2HPO_4, 75 mg L^{-1} $MgSO_4 \cdot 7H_2O$, 36 mg L^{-1} $CaCl_2 \cdot 2H_2O$, pH: 7.8) was slowly poured into the beaker along the inside wall to avoid resuspending the sediments. Finally, the beaker was kept static for 24 h to settle any suspended sediments. The periphytic biofilms and their substrates were carefully taken out of the tank, divided into similar-sized aliquots (5 cm each, about 20 g wet weight), and placed into the beakers at the sediment–over water interface. In the control, substrates with no periphytic biofilm growth were added. The control was also covered with black cardboard to avoid periphytic biofilm formation at the water–sediment interface. A total of 10 beakers were prepared, five treatment and five controls. To avoid the influence of environmental conditions on the experiment, the beakers were kept in a greenhouse with air temperature between 25 and 30°C.

6.2.3 Samples and Analyses

The experiment started on September 10, 2013, with water and sediments sampled after 0 d, 7 d, 15 d, 30 d, and 60 d. Total P (TP), total dissolved P (TDP), and dissolved inorganic P (DIP) in water were determined by a flow injection analyzer (SEAL AA3, Germany). Particular P (PP) and dissolved organic P (DOP) were calculated as TP minus TDP, and TDP minus DIP, respectively. The P fractionation in sediment was based on the method of Rydin (2000). The successive fractionation steps for extracting different forms of P are described in Table 6.1. Briefly, the sediment was subjected to sequential chemical extractions with 1.0 M NH_4Cl, 0.1 M NaOH, and 0.5 M HCl with different extraction times. The extracts were then centrifuged and the supernatants filtered through a 0.45-μm phosphorus-free membrane. The TP of sediment was determined by the persulfate digestion–molybdophosphate reaction method after strong acid digestion ($H_2SO_4 + HClO_4$). Residual-P was calculated as TP minus total extracted P. 1 g (DW) of sediment was extracted by 25 mL of extractant. Each extraction step terminated with a short wash (10 mL) by the extractant. The soluble reactive phosphorus (SRP) in each fraction was determined by the molybdenum blue/ascorbic acid method. Changes in P loadings were calculated based on the difference between P loadings in water, periphytic biofilms, and sediments at the beginning and end of the experiment.

After the experiments, the P fractions in the periphytic biofilms were subjected to sequential P fractionation according to the method of Borovec et al. (2010). The P was divided into three groups: immediately available phosphorus (IAP, $MgCl_2$-P), potentially available phosphorus (PAP, NaOH/HCl-P), and no available phosphorus (NAP, $H_2SO_4 + HNO_3 + H_2O_2$-P). Briefly, a fixed amount of sample (approximately 5 g wet weight) was extracted using 15 mL of chemical extractant (1.0 M NH_4Cl, 0.1 M NaOH and 0.5 M HCl). The residual P of periphytic biofilm was determined by the persulfate digestion–molybdophosphate reaction method after strong acid digestion ($H_2SO_4 + HNO_3 + H_2O_2$).

TABLE 6.1 Successive Steps of P Fractionation in Sediments

Step/Reagent	Process of Extraction	P Species
NH_4Cl	1.0 M NH_4Cl at pH 7 for 2 h, 25°C + rinse	Labile-P
NaOH	0.1 M NaOH for 16 h, 25°C + rinse	Fe/Al-P
HCl	0.5 M HCl for 16 h, 25°C + rinse	Ca-P
$H_2SO_4 + HClO_4$	18.4 M H_2SO_4 + 70–72% $HClO_4$ + 180°C	Residual-P

Each experiment in this study was conducted in triplicate, and the mean results (±SD) are presented. Statistical analyses were performed using SPSS 19.0, and $p < 0.05$ indicates statistical significance.

6.3 RESULTS

6.3.1 pH Changes in the Water Column

The pH was dramatically enhanced from about 7.5 to 10 in the presence of periphytic biofilm over 60 days and did not change in the control ($p > 0.05$, Fig. 6.1). This implies that the presence of periphyton in waters can substantially change the pH of the water column.

6.3.2 P Changes in the Water Column

The presence of a periphytic biofilm clearly altered the P content in the water column (Fig. 6.2A–E). TP decreased continuously from 1.4 to 0.04 mg L^{-1} over 60 days in the presence of periphytic biofilm. In the control, TP initially increased from 1.2 to 2.1 mg L^{-1} then plateaued at about 1.8 mg L^{-1}. TDP, DIP, PP, and DOP exhibited similar patterns to TP under both periphytic biofilm and control treatments. These results imply that periphytic biofilm plays an important role in P removal from the water column.

6.3.3 P Changes in Sediment

TP of sediment in the control decreased from about 11.8 to 8.8 mg kg^{-1} after 14 days and then plateaued. TP of sediment increased slightly from about 11.8 to 12.1 mg kg^{-1} in the initial 7 days and then decreased dramatically to about

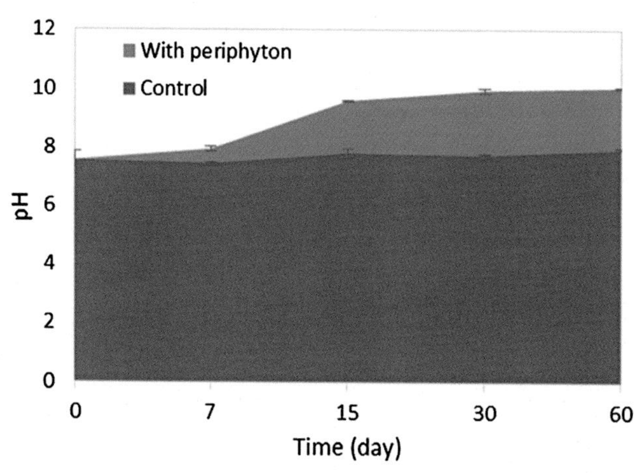

FIGURE 6.1 Temporal change in water column pH in the presence or absence (control) of periphytic biofilm.

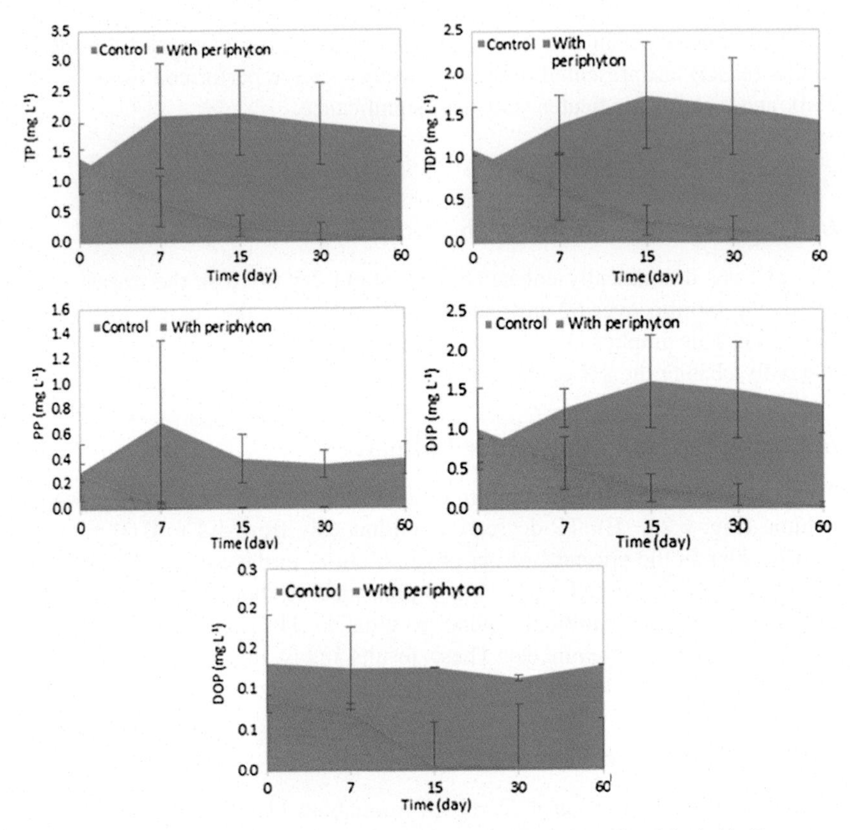

FIGURE 6.2 Temporal changes in water column P in the presence of periphytic biofilm.

8.7 mg kg^{-1} in the following 53 days in the presence of periphytic biofilm between water and sediment (Fig. 6.3A). Fe/Al-P in sediment decreased from 3.7 to 2.5 mg kg^{-1} after 60 days in the control, but rose from 3.7 to 4.0 mg kg^{-1} in the initial 7 days and then continuously dropped to 2.0 mg kg^{-1} at 60 days in the presence of periphytic biofilm (Fig. 6.3B). Ca-P in sediment declined from 6.3 to 5.7 mg kg^{-1} in 60 days in the control and increased slightly from 6.3 to 6.6 mg g^{-1} in 15 days and then reduced to 6.0 mg g^{-1} in the presence of periphytic biofilm (Fig. 6.3C). For Labile-P and Res-P in sediment, similar patterns were observed for both periphytic biofilm and control treatments, with decreases from about 1.0 to 0.1 mg kg^{-1} and 0.8 to 0.5 mg kg^{-1}, respectively (Fig. 6.3D and E). These results were consistent with the P changes in the water column (Fig. 6.2).

The presence of periphytic biofilm did not influence the P fraction in sediments (Fig. 6.4). Fe/Al-P and Ca-P dominated the sediment P, accounting for ~84–92% combined. After 60 days, the percentage of Labile-P decreased from about 8.6% to 1.1% in the control and to 1.6% in the presence of

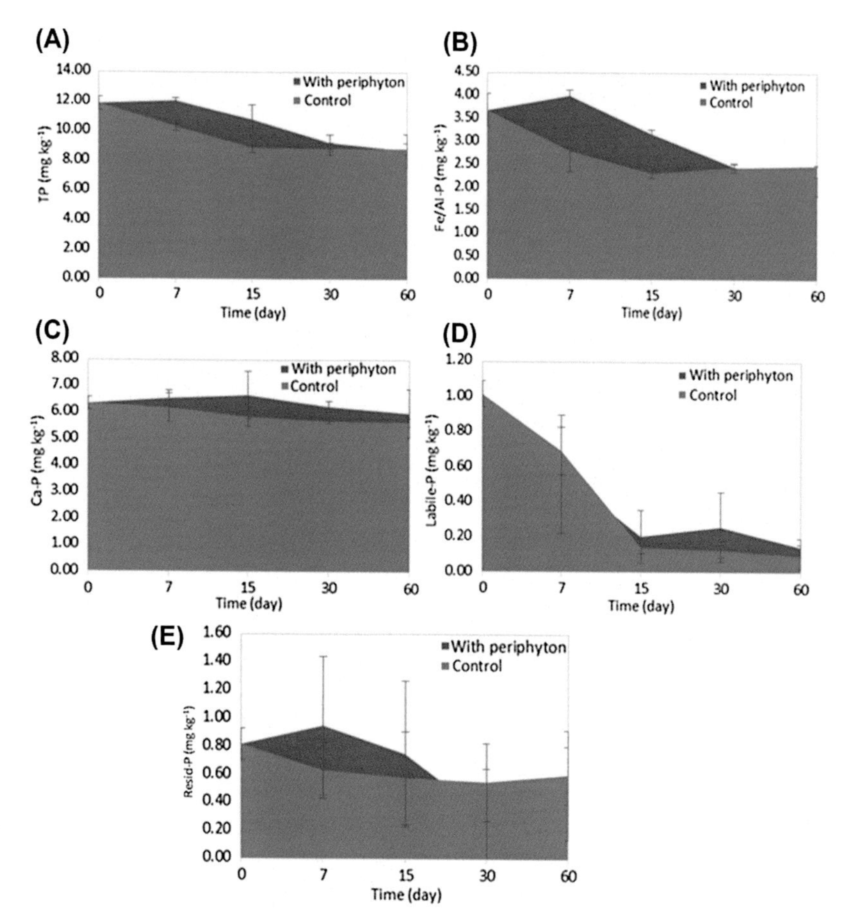

FIGURE 6.3 Temporal changes in sediment P in the presence of periphytic biofilm and controls (without periphyton).

periphytic biofilm. Similar patterns were observed for Fe/Al-P (Fig. 6.4). In contrast, the percentage of Ca-P increased from 53.6% to 64.0% in the control and to 68.9% in the presence of periphytic biofilm. The percentage of Res-P showed no significant variation ($p > 0.05$). These results indicate that the sediment in our experiments could release P into overlying water, mainly in the forms of Labile-P and Fe/Al-P. Compared to the control, the presence of periphytic biofilm could promote the release of Fe/Al-P while attenuating the release of Labile-P and Ca-P after 60 days.

P loadings in this stimulated system were continuously estimated during the experiment. It was observed that the total P amounts of water and sediment in the control were generally maintained at about 15 mg while significantly decreasing from about 16 to 9 mg in the presence of periphytic biofilm (Fig. 6.5). Results also showed that the increased amount of water P

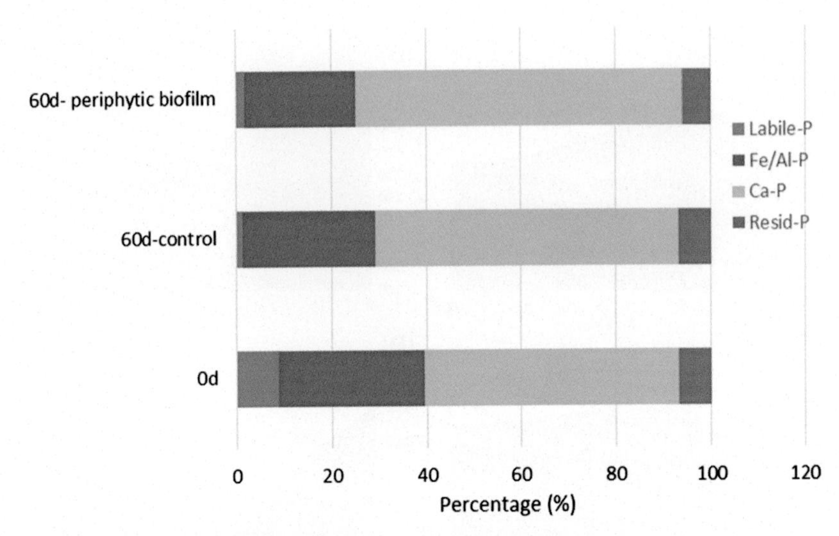

FIGURE 6.4 The percentage of P fraction in sediment after 60 days.

(maximal at about 3 mg) was consistent with the decreased amount of sediment P in the control, indicating that the additional P in water came from the release of sediment P. The amount of P in water and sediment in the presence of periphytic biofilm both exhibited a downward trend, decreasing by about 4 and 3 mg, respectively. This suggests that the total amount of lost P (about 7 mg) from the water and sediment was trapped in the periphytic biofilm system. Furthermore, compared with the control, the presence of periphytic biofilm could attenuate P release from sediment. For example, at day 14, the P amount released from sediment was about 3 mg in the control and about 1 mg in the presence of periphytic biofilm. This implies that periphytic biofilm could decrease the release of sediment P by 67%. The percentage of P attenuation by periphytic biofilm, however, decreased consistently over time.

6.3.4 P Changes in Periphytic Biofilm

The P of periphytic biofilms was divided into four fractions (Labile-P, Fe/Al-P, Ca-P, and Res-P) based on successive chemical extractions and determined at the start (0 d) and end (60 d) of the experiment. After 60 days, the total P content of periphytic biofilm increased significantly from about 0.4–0.8 mg g^{-1}, suggesting periphytic biofilm was capable of removing P from water and storing P (Fig. 6.6). The Fe/Al-P and Ca-P of periphytic biofilm increased from 0.09 and 0.11 mg g^{-1} to 0.28 and 0.31 mg g^{-1}, respectively. This further implies that periphytic biofilm was capable of storing P, mainly in the forms of Fe/Al-P and Ca-P (72% of periphytic TP combined), on its macrosurface.

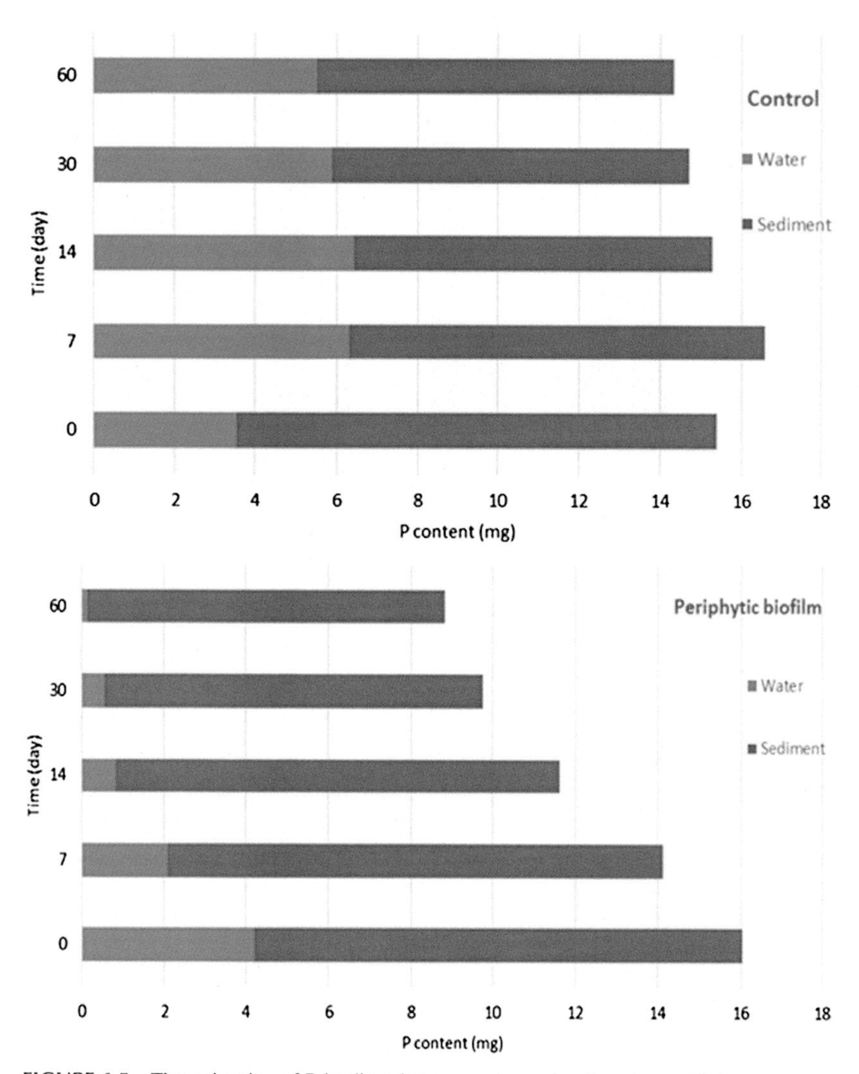

FIGURE 6.5 The estimation of P loadings between water and sediment over 60 days.

6.4 DISCUSSION

The study of P migration and transformation between water and sediment is always one of the most important topics for P cycling in aquatic ecosystems. Most current work, however, is based on the sediment−water interface, ignoring the presence of periphytic biofilm between them. Periphyton makes a substantial contribution (42−97%) to the total annual productivity of ecosystems, especially in shallow waters (Azim, 2009; Pratiwi et al., 2012). As a result, periphytic biofilm significantly affects the migration of nutrients (especially P) between the water and sediment interface (Battin et al., 2003;

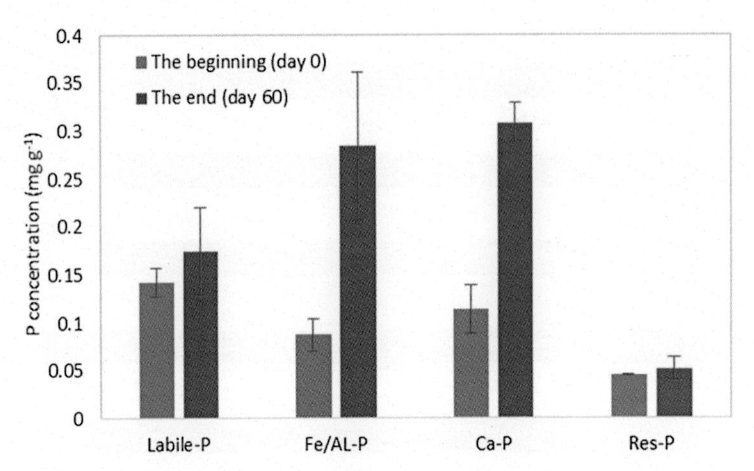

FIGURE 6.6 The content of various P fractions in periphytic biofilm.

Woodruff et al., 1999c). The evaluation of P migration and transformation based on a sediment–periphytic biofilm–water system will therefore improve the understanding of P cycling in aquatic ecosystems that contain periphytic biofilm or similar microbial aggregates.

6.4.1 P Removal From the Water Column by Periphytic Biofilm

It has been postulated by many previous studies that the periphytic biofilm plays an important role in P biogeochemistry of aquatic ecosystems (Dodds, 2003; McCormick et al., 2006) as periphytic biofilm can uptake or remove P from the water column (Boelee et al., 2011; Guzzon et al., 2008; Jöbgen et al., 2004; Lu et al., 2014b; Pietro et al., 2006). Our study showed clear evidence that the introduced periphytic biofilm could significantly remove P from water, including all P species in the water column. This suggests that periphytic biofilm has vast potential as an efficient biomaterial for different methods of P removal from wastewater.

The mechanisms used by periphytic biofilm to remove P from water are mainly assimilation (Guzzon et al., 2008), adsorption (Lu et al., 2014a,b; Scinto and Reddy, 2003), coprecipitation (Dodds, 2003; Hill and Fanta, 2008), and interception or trapping of P (Adey et al., 1993). In this study, it was notable that the pH of water in the presence of periphytic biofilm was substantially higher (about 10) than in the control. This was probably because the periphytic biofilm consisted mainly of phototrophic algae whose intense photosynthesis would consume CO_2 from water and release O_2, resulting in the increase in water pH (Hayashi et al., 2012; Pratiwi et al., 2012; Zhang et al., 2013a). Such water column conditions are favorable for the coprecipitation of P and metal salts such as $CaCO_3$ (Dodds, 2003; Jarvie et al., 2002; Neal and Jarvie, 2005; Scinto and Reddy, 2003), which greatly contribute to the removal of P from water. The P concentration of periphytic biofilm also increased significantly from about 0.4–0.8 mg g^{-1} after 60 days, which means that the periphytic biofilm itself was able to capture P from water. Therefore, we suggest that the P removal from water by periphytic biofilm was controlled by several mechanisms in combination, rather than one single mechanism like assimilation or coprecipitation.

The relatively high pH of water in this experiment was surprising. A pH value of 8.0 in the water column and 9.0 and 7.0 in periphytic biofilm and sediment, respectively, had been previously documented (Borchardt, 1996; Carlton and Wetzel, 1988). One possible reason was that the water–sediment system in our experiment was a closed artificial system that far from simulated the natural conditions, as the water volume was small and water depth was shallow, with limited diffusion and exchange. Moreover, we measured the pH of water at about 14:00 each day when periphytic photosynthesis reaches the highest level (data not shown).

6.4.2 The Influence of Periphytic Biofilm on Sediment P

Previous studies have already indicated that periphytic biofilm can control chemical fluxes of calcium, alkalinity, and P from the sediment to the overlying water (Woodruff et al., 1999a,b). Some studies have also indicated that the presence of periphytic biofilm in the sediment–water interface could reduce the release of sediment P to overlying water (Wu et al., 2010b; Zhang et al., 2013a). Our experiments showed similar results. For example, the sediment under control conditions was a P source that released P, especially Labile-P and Fe/Al-P, into the overlying water. In the presence of periphytic biofilm, the sediment could be a short-term P sink that accumulated P (like Fe/Al-P and Ca-P) in the initial 7 days. After 7 days, all types of P fractions in the sediment were released into the water column, but generally less was released by sediment with periphytic biofilm. This further indicated that periphytic biofilm could reduce the release of sediment P to overlying waters.

It was interesting to find that more Fe/Al-P of sediment was released in the presence of periphytic biofilm than in the control after 30 days. This implied that periphytic biofilm could release more Fe/Al-P over a relatively long time. One possible explanation is that the long-term presence of periphytic biofilm between water and sediment may induce a rise in sediment pH, favoring the release of Fe/Al-P from sediment (Jin et al., 2006a,b; Rydin, 2000). As a result, more Fe/Al-P was released from the sediment to the overlying water in the presence of periphytic biofilm than in the control after 30 days.

The release of P from surface sediments is an important process for understanding aquatic P cycling. In sediment, only some P fractions are easily exchangeable and biologically available, while most of them (about 8–82%) bind with cations like Ca, Fe, and Al (Renjith et al., 2011; Rzepecki, 2010). It is well known that Labile-P, extracted with NH_4Cl, often represents pore water-P, loosely sorbed P, and $CaCO_3$-associated P in hard waters, while Fe/Al-P, extracted with NaOH-P, is a type of P exchangeable with OH^-, mainly aluminum and ferric. Ca-P, extracted with HCl, is a P form sensitive to low pH, assumed to consist mainly of apatite, and Residual-P consists mainly of refractory organic P as well as the inert inorganic P fraction (Rydin, 2000). In our experiment, the presence of periphytic biofilm generally decreased the release of Labile-P, Fe/Al-P, and Ca-P, and did not alter the trend of sediment P in 60 days. Among the various P fractions, it is worth noting the change of Ca-P in sediment. Ca-P is thermodynamically regarded as a more stable mineral-bound form than Fe/Al-P (Ann et al., 1999). Normally, Ca-P accounts for about 1–52% of total sediment P in natural aquatic ecosystems like wetlands and lakes (Qian et al., 2010; Renjith et al., 2011; Rzepecki, 2010), and has not been highlighted as an exchangeable P unless there is a rapid change in sediment pH and dissolution of metal oxides (Renjith et al., 2011). In our study, however, Ca-P content generally decreased over 60 days, whether in the presence of periphytic biofilm or not, meaning that Ca-P in our study was solubilized and released to water. It has been reported that one type of microbe called phosphate solubilizing bacteria (PSB) is abundantly present in waters and sediments and is very effective in solubilizing Ca-P (Fankem et al., 2006; Maitra et al., 2015). Although the exact reasons responsible for the declines in Ca-P content are unknown, in our study dissolution by PSB might contribute to such variation. Thus, investigations of PSB should be included in future work when evaluating P cycling between water and sediment (Maitra et al., 2015).

In the presence of periphytic biofilm, the concentration of Ca-P in sediment increased at the beginning of the experiment (0–14 days). This outcome differs from the change in Ca-P content in the control. Ca-P could be mobilized naturally with the aging of sediment (Froelich, 1988), resulting in an increase of Ca-P content over time. In this study, however, the increase in Ca-P concentration is mainly due to the presence of periphytic biofilm. One possible explanation is that the rise in pH of water resulting from active periphytic photosynthesis favors Ca-P deposition to sediment. Many previous studies have already confirmed that the presence of periphytic biofilm between water and sediment could cause concurrent deposition of carbonate–phosphate complexes, namely the coprecipitation phenomenon, resulting in Ca-P deposited onto the sediment (Dodds, 2003; Jarvie et al., 2002; Noe et al., 2003; Woodruff et al., 1999b).

6.4.3 P Reserved in Periphytic Biofilm

The presence of periphytic biofilm in shallow waters could lead to the coprecipitation of P with $CaCO_3$ or its adsorption onto $CaCO_3$ crystals (Dodds, 2003; Scinto and Reddy, 2003), thus making P less available or limited for other ecosystem components such as macrophytes or planktonic communities. The P content of the water column after the inclusion of periphytic biofilm in our experiments was direct evidence of this phenomenon. This phenomenon is highly consistent with our finding that Ca-P increased and accounted for nearly 38% of periphytic TP after 60 days. Similar proportions of Ca-P were found by Borovec et al. (2010). Furthermore, the majority of P in periphytic biofilm was found in exchangeable and loosely bound forms such as Labile-P and Fe/Al-P, which together could account for about 56% of total periphytic P after 60 days. This also implied that the growth of periphytic microorganisms was primarily limited by factors other than the scarcity of P. As the partitioning of periphytic P species and bioavailability were influenced by features of the periphytic biofilm itself, such as EPS and environmental factors such as pH, light (Borovec et al., 2010), the processes and mechanisms of P release from periphytic biofilms need further investigation.

6.5 CONCLUSIONS

Periphytic biofilm is an important component of shallow aquatic ecosystems and can significantly influence the change of P content and species. This study demonstrates that periphytic biofilm can markedly decrease the content of all P species (TP, TDP, DIP, PP, DOP) in the water column, reduce the release of some sedimentary P such as Fe/Al-P and Ca-P to overlying water, and increase the content of periphytic P that occurs mainly in forms of exchangeable P. It also showed that the presence of periphytic biofilm can alter physicochemical features between the water and sediment interface such as water pH, resulting in the occurrence of coprecipitation between P and metal salt. These findings suggest that periphytic biofilm can be regarded as an important P buffer between the water and sediment interface that is capable of not only capturing and storing P from waters, but attenuating P release from sediment. This has substantial ecological implications that not only contribute to better understanding of P cycling in shallow water ecosystems with periphytic biofilm or similar microbial assemblages, but also benefits ecological engineering and restoring of periphytic biofilm-based techniques. More importantly, as P resources become dramatically scarce, P recovery and reuse will undoubtedly increase. Capturing P from waters, especially from wastewater, by periphytic biofilm will become an important technique and requires further investigation.

REFERENCES

Adey, W., Luckett, C., Jensen, K., 1993. Phosphorus removal from natural waters using controlled algal production. Restoration Ecology 1 (1), 29−39.

Ann, Y., Reddy, K.R., Delfino, J.J., 1999. Influence of redox potential on phosphorus solubility in chemically amended wetland organic soils. Ecological Engineering 14, 169−180.

Azim, M.E., 2009. Photosynthetic periphyton and surfaces. In: Likens, G.E. (Ed.), Encyclopedia of Inland Waters. Academic Press, Oxford, pp. 184−191.

Battin, T.J., Kaplan, L.A., Denis Newbold, J., Hansen, C.M.E., 2003. Contributions of microbial biofilms to ecosystem processes in stream mesocosms. Nature 426 (6965), 439−442.

Boelee, N., Temmink, H., Janssen, M., Buisman, C., Wijffels, R., 2011. Nitrogen and phosphorus removal from municipal wastewater effluent using microalgal biofilms. Water Research 45 (18), 5925–5933.

Borchardt, M.A., 1996. Nutrients. In: Lowe, R.L. (Ed.), Algal Ecology. Academic Press, San Diego, pp. 183–227.

Borovec, J., Sirová, D., Mošnerová, P., Rejmánková, E., Vrba, J., 2010. Spatial and temporal changes in phosphorus partitioning within a freshwater cyanobacterial mat community. Biogeochemistry 101 (1–3), 323–333.

Carlton, R.G., Wetzel, R.G., 1988. Phosphorus flux from lake sediments: effect of epipelic algal oxygen production. Limnology and Oceanography 33, 562–570.

Christophoridis, C., Fytianos, K., 2006. Conditions affecting the release of phosphorus from surface lake sediments. Journal of Environmental Quality 35, 1181–1192.

Dodds, W.K., 2003. The role of periphyton in phosphorus retention in shallow freshwater aquatic systems. Journal of Phycology 39 (5), 840–849.

Drake, W., Scott, J.T., Evans-White, M., Haggard, B., Sharpley, A., Rogers, C.W., Grantz, E.M., 2012. The effect of periphyton stoichiometry and light on biological phosphorus immobilization and release in streams. Limnology 13 (1), 97–106.

Fankem, H., Nwaga, D., Deubel, A., Dieng, L., Merbach, W., Etoa, F.X., 2006. Occurrence and functioning of phosphate solubilizing microorganisms from oil palm tree (*Elaeis guineensis*) rhizosphere in Cameroon. African Journal of Biotechnology 5, 1–9.

Froelich, P.N., 1988. Kinetic control of dissolved phosphate in natural rivers and estuaries: a primer on the phosphate buffer mechanism. Limnology and Oceanography 33, 649–668.

Fytianos, K., Kotzakioti, A., 2005. Sequential fractionation of phosphorus in lake sediments of Northern Greece. Environmental Monitoring and Assessment 100, 191–200.

Gainswin, B.E., House, W.A., Leadbeater, B.S.C., Armitage, P.D., 2006. Kinetics of phosphorus release from a natural mixed grain-size sediment with associated algal biofilms. Science of the Total Environment 360, 127–141.

Guzzon, A., Bohn, A., Diociaiuti, M., Albertano, P., 2008. Cultured phototrophic biofilms for phosphorus removal in wastewater treatment. Water Research 42 (16), 4357–4367.

Hayashi, M., Vogt, T., Mächler, L., Schirmer, M., 2012. Diurnal fluctuations of electrical conductivity in a pre-alpine river: effects of photosynthesis and groundwater exchange. Journal of Hydrology 450–451 (0), 93–104.

Hill, W.R., Fanta, S.E., 2008. Phosphorus and light colimit periphyton growth at subsaturating irradiances. Freshwater Biology 53 (2), 215–225.

Jöbgen, A., Palm, A., Melkonian, M., 2004. Phosphorus removal from eutrophic lakes using periphyton on submerged artificial substrata. Hydrobiologia 528 (1–3), 123–142.

Jørgensen, K.S., Puustinen, J., Suortti, A.M., 2000. Bioremediation of petroleum hydrocarbon-contaminated soil by composting in biopiles. Environmental Pollution 107 (2), 245–254.

Jarvie, H.P., Neal, C., Warwick, A., White, J., Neal, M., Wickham, H.D., Hill, L.K., Andrews, M.C., 2002. Phosphorus uptake into algal biofilms in a lowland chalk river. Science of the Total Environment 282–283 (0), 353–373.

Jin, X., Jiang, X., Yao, Y., Li, L., Wu, F., 2006a. Effects of light and oxygen on the uptake and distribution of phosphorus at the sediment-water interface. Science of the Total Environment 357, 231–236.

Jin, X., Wang, S., Pang, Y., Chang Wu, F., 2006b. Phosphorus fractions and the effect of pH on the phosphorus release of the sediments from different trophic areas in Taihu Lake, China. Environmental Pollution 139, 288–295.

Jorgensen, B.B., Des Marais, D.J., 1990. The diffusive boundary layer of sediments: oxygen microgradients over a microbial mat. Limnology and Oceanography 35 (6), 1343–1355.

Kaiser, J.P., Feng, Y.C., Bollag, J.M., 1996. Microbial metabolism of pyridine, quinoline, acridine, and their derivatives under aerobic and anaerobic conditions. Microbiological Review 60, 483–498.

Kaiserli, A., Voutsa, D., Samara, C., 2002. Phosphorus fractionation in lake sediments-Lakes Volvi and Koronia, N. Greece. Chemosphere 46, 1147–1155.

Koski-Vähälä, J., Hartikainen, H., 2001. Assessment of the risk of phosphorus loading due to resuspended sediment. Journal of Environmental Quality 30, 960–966.

Lu, H., Yang, L., Shabbir, S., Wu, Y., 2014a. The adsorption process during inorganic phosphorus removal by cultured periphyton. Environmental Science and Pollution Research 21 (14), 8782–8791.

Lu, H., Yang, L., Zhang, S., Wu, Y., 2014b. The behavior of organic phosphorus under non-point source wastewater in the presence of phototrophic periphyton. PLoS One 9 (1), e85910.

Maitra, N., Manna, S., Samanta, S., Sarkar, K., Debnath, D., Bandopadhyay, C., Sahu, S., Sharma, A., 2015. Ecological significance and phosphorus release potential of phosphate solubilizing bacteria in freshwater ecosystems. Hydrobiologia 745, 69–83.

McCormick, P.V., Shuford III, R.B., Chimney, M.J., 2006. Periphyton as a potential phosphorus sink in the Everglades Nutrient Removal Project. Ecological Engineering 27 (4), 279–289.

Monbet, P., McKelvie, I.D., Saefumillah, A., Worsfold, P.J., 2007. A protocol to assess the enzymatic release of dissolved organic phosphorus species in waters under environmentally relevant conditions. Environmental Science & Technology 41, 7479–7485.

Neal, C., Jarvie, H.P., 2005. Agriculture, community, river eutrophication and the Water Framework Directive. Hydrological Processes 19 (9), 1895–1901.

Noe, G., Scinto, L., Taylor, J., 2003. Phosphorus cycling and partitioning in an oligotrophic Everglades wetland ecosystem: a radioisotope tracing study. Freshwater Biology 48, 1993–2008.

Paerl, H., Pinckney, J., 1996. A mini-review of microbial consortia: their roles in aquatic production and biogeochemical cycling. Microbial Ecology 31, 225–247.

Palmer-Felgate, E.J., Bowes, M.J., Stratford, C., Neal, C., MacKenzie, S., 2011. Phosphorus release from sediments in a treatment wetland: contrast between DET and EPC0 methodologies. Ecological Engineering 37, 826–832.

Perkins, R.G., Underwood, G.J.C., 2001. The potential for phosphorus release across the sediment-water interface in an eutrophic reservoir dosed with ferric sulphate. Water Research 35, 1399–1406.

Pietro, K.C., Chimney, M.J., Steinman, A.D., 2006. Phosphorus removal by the *Ceratophyllum*/periphyton complex in a south Florida (USA) freshwater marsh. Ecological Engineering 27, 290–300.

Poulíčková, A., Hašler, P., Lysáková, M., Spears, B., 2008. The ecology of freshwater epipelic algae: an update. Phycologia 47 (5), 437–450.

Pratiwi, N.T.M., Hariyadi, S., Tajudin, R., 2012. Photosynthesis of periphyton and diffusion process as source of oxygen in rich-riffle upstream waters. Microbiology. Indonesia 5 (4), 182.

Qian, Y., Shi, J., Chen, Y., Lou, L., Cui, X., Cao, R., Li, P., Tang, J., 2010. Characterization of phosphate solubilizing bacteria in sediments from a shallow eutrophic lake and a wetland: isolation, molecular identification and phosphorus release ability determination. Molecules 15, 8518–8533.

Qin, B., 2009. Lake eutrophication: control countermeasures and recycling exploitation. Ecological Engineering 35, 1569–1573.

Renjith, K., Chandramohanakumar, N., Joseph, M.M., 2011. Fractionation and bioavailability of phosphorus in a tropical estuary, Southwest India. Environmental Monitoring and Assessment 174, 299–312.

Rydin, E., 2000. Potentially mobile phosphorus in Lake Erken sediment. Water Research 34, 2037–2042.

Rzepecki, M., 2010. The dynamics of phosphorus in lacustrine sediments. Polish Journal of Ecology 58, 409–427.

Søndergaard, M., Jensen, J.P., Jeppesen, E., 2003. Role of sediment and internal loading of phosphorus in shallow lakes. Hydrobiologia 506, 135–145.

Saikia, S.N., Majumder, S., 2013. A review on the role of nutrients in development and organization of periphyton. Journal of Research in Biology 3 (1), 780–788.

Scinto, L.J., Reddy, K.R., 2003. Biotic and abiotic uptake of phosphorus by periphyton in a subtropical freshwater wetland. Aquatic Botany 77 (3), 203–222.

Smith, D.R., Warnemuende, E.A., Haggard, B.E., Huang, C., 2006. Changes in sediment-water column phosphorus interactions following sediment disturbance. Ecological Engineering 27, 71–78.

Solomon, S., 2007. Climate Change 2007. The Physical Science Basis: Working Group I Contribution to the Fourth Assessment Report of the IPCC. Cambridge University Press.

Wang, H., Holden, J., Spera, K., Xu, X., Wang, Z., Luan, J., Xu, X., Zhang, Z., 2013. Phosphorus fluxes at the sediment-water interface in subtropical wetlands subjected to experimental warming: a microcosm study. Chemosphere 90, 1794–1804.

Wang, S., Jin, X., Bu, Q., Jiao, L., Wu, F., 2008. Effects of dissolved oxygen supply level on phosphorus release from lake sediments. Colloids and Surfaces A. Physicochemical and Engineering Aspects 316, 245–252.

Wang, S., Jin, X., Zhao, H., Wu, F., 2006. Phosphorus fractions and its release in the sediments from the shallow lakes in the middle and lower reaches of Yangtze River area in China. Colloids and Surfaces A: Physicochemical and Engineering Aspects 273, 109–116.

Wang, S., Jin, X., Zhao, H., Wu, F., 2009. Phosphorus release characteristics of different trophic lake sediments under simulative disturbing conditions. Journal of Hazardous Materials 161, 1551–1559.

Woodruff, S., House, W., Callow, M., Leadbetter, B., 1999a. The effects of a developing biofilm on chemical changes across the sediment-water interface in a freshwater environment. International Review of Hydrobiology 84, 509–532.

Woodruff, S., House, W., Callow, M., Leadbetter, B., 1999b. The effects of biofilms on chemical processes in surficial sediments. Freshwater Biology 41, 73–89.

Woodruff, S., House, W., Callow, M., Leadbetter, B., 1999c. The effects of a developing biofilm on chemical changes across the sediment-water interface in a freshwater environment. International Review of Hydrobiology 84 (5), 509–532.

Writer, J.H., Ryan, J.N., Barber, L.B., 2011. Role of biofilms in sorptive removal of steroidal hormones and 4-nonylphenol compounds from streams. Environmental Science & Technology 45, 7275–7283.

Wu, Y., Kerr, P.G., Hu, Z., Yang, L., 2010a. Removal of cyanobacterial bloom from a biopond–wetland system and the associated response of zoobenthic diversity. Bioresource Technology 101 (11), 3903–3908.

Wu, Y., Li, T., Yang, L., 2012. Mechanisms of removing pollutants from aqueous solutions by microorganisms and their aggregates: a review. Bioresource Technology 107 (0), 10–18.

Wu, Y., Zhang, S., Zhao, H., Yang, L., 2010b. Environmentally benign periphyton bioreactors for controlling cyanobacterial growth. Bioresource Technology 101 (24), 9681–9687.

Zhang, H., Li, G., Song, X., Yang, D., Li, Y., Qiao, J., Zhang, J., Zhao, S., 2013a. Changes in soil microbial functional diversity under different vegetation restoration patterns for Hulunbeier Sandy Land. Acta Ecologica Sinica 33 (1), 38–44.

Zhang, X., Liu, Z., Gulati, R., Jeppesen, E., 2013b. The effect of benthic algae on phosphorus exchange between sediment and overlying water in shallow lakes: a microcosm study using ^{32}P as a tracer. Hydrobiologia 710 (1), 109–116.

Chapter 7

The Evaluation of Phosphorus Removal Processes and Mechanisms From Surface Water by Periphyton

7.1 INTRODUCTION

Excessive P input is one of the main causes of eutrophication and harmful algal blooms that lead to deterioration of water quality and aquatic ecosystems (Smith and Schindler, 2009; Smith et al., 1999; Wu et al., 2010b). Many measures have been proposed to remove P from water including physical (e.g., reverse osmosis membranes), chemical (e.g., aluminum chloride application), and biological (e.g., biofilm bioreactor) methods (de-Bashan and Bashan, 2004; Pratt et al., 2012; Wu et al., 2005, 2010c). Recently, the biological method based on periphyton biofilms has been considered a promising measure for P removal due to its cost-effectiveness, easy-harvesting, high-effectiveness and environment-friendly advantages (Guzzon et al., 2008; Roeselers et al., 2008).

Periphyton is a complex microbial assemblage of algae, bacteria, protozoa, metazoa, epiphytes, and detritus (Wu et al., 2012). It is commonly attached to submerged surfaces in most aquatic ecosystems and plays a significant role in natural aquatic ecosystems through its influence on primary production, food chains, organic matter, and nutrient cycling (Battin et al., 2003). It has been postulated that periphyton biofilms are capable of absorbing and/or adsorbing water contaminants (Wicke et al., 2008), such as microcystins (Li et al., 2012), hormones (Writer et al., 2011), and toxic metals (Dong et al., 2003). Consequently, periphyton biofilms have great potential for complicated surface water removal of multiple contaminants.

It has been reported that periphyton biofilms play a key role in removing P in wetlands and can act as an important potential sink for P (Drake et al., 2012; McCormick et al., 2006). They have subsequently been developed to remove P from wastewater (Guzzon et al., 2008; Li et al., 2003).

Periphyton. http://dx.doi.org/10.1016/B978-0-12-801077-8.00007-7

Currently, activated sludge biofilm systems are the most common method for P removal, and it is clear that the P removal process is dominated by two independent mechanisms: the direct absorption of P by suspended cells, and the storage of P as polyphosphate by microorganisms (de-Bashan and Bashan, 2004). Periphyton biofilms differ from activated sludge biofilms in wastewater treatment plants in a number of ways. For example, periphyton is ubiquitous in aquatic environments and performs numerous important environmental functions such as nutrient cycling and self-purification of aquatic ecosystems (Sabater et al., 2002). Moreover, the components of periphyton biofilms in this study are dominated by more photoautotrophic microorganisms (e.g., cyanobacteria and microalgae). These multilayer constructions are embedded in a common extracellular polymeric substance (EPS) secreted by the community, that mediates the adhesion of phototrophs and heterotrophs as well as gas and nutrients fluxes (Donlan, 2002). These special components distinguish periphyton and sludge biofilm structures and their subsequent capacity to capture or remove contaminants such as P from wastewaters.

Most current studies on P removal methods or technologies have focused on purely inorganic P or a specific type of P-based contaminant such as organophosphorus pesticides (Gatidou and Iatrou, 2011). P_o commonly includes nucleic acids, phospholipids, inositol phosphates, phosphoamides, phosphoproteins, sugar phosphates, amino phosphoric acids, and organic condensed P species. It is often as abundant as (sometimes greatly in excess of) P_i in natural water bodies and sediments (McKelvie, 2005). Previous studies show that soluble P_o in aquatic systems, especially in lakes, often exceeded that of orthophosphate and accounted for 50–90% of total P (Herbes et al., 1975; Minear, 1972). In aquatic systems, the role of P_o has largely been underestimated not only because of its compositional and structural complexity (Turner et al., 2005), but also due to the limitation in analytical methods and techniques. As a result, P_o has usually been grouped in the "nonreactive" and "nonbioavailable" component of total phosphorus (P_t) (McKelvie, 2005). However, there is strong evidence that some organisms such as algae and bacteria are adapted to P_o via enzymatic hydrolysis and/or bacterial decomposition (Cotner and Wetzel, 1992; Dyhrman et al., 2006; Sanudo-Wilhelmy, 2006). As a result, the importance of P_o is not widely recognized as a potentially large pool of bioavailable P, and its influence on P cycling and eutrophication of aquatic ecosystems is ignored in comparison to organic P species. Most importantly, P_o is typically not susceptible to traditional removal technologies used for inorganic phosphorus (Rittmann et al., 2011), which may be due to its complex species and chemical dynamics. Therefore, a comprehensive investigation of P_o conversion and removal processes by periphyton ubiquitously distributed in natural waters will not only help development of potential technology for P_o removal from high-organic waters such as animal wastes, but also provide strong evidence to fully understand P biogeochemical cycling of aquatic ecosystems that contain periphyton or similar microbial aggregates.

Previous research has shown that the basic structures of periphyton biofilm can be described by three conceptual models: (1) heterogeneous mosaic biofilm aggregations, (2) penetrated water-channel biofilms, and (3) dense confluent biofilms (Wimpenny and Colasanti, 1997). Due to these special porous structures, nutrients such as N and P can be freely transported into periphyton (Wetzel, 2001). Moreover, it has been reported that adsorption of P was suspected to account for P not recovered (48–72%) in periphyton-dominated marshes in Belize (Rejmánková and Komárková, 2000). Similarly, Scinto and Reddy (2003) revealed that adsorption mechanisms were responsible for water column P removal by periphyton. Adsorption has also been identified as the main mechanism involved in microcystin removal by periphyton (Wu et al., 2010a).

The main objectives of this study were to: (1) determine the efficiency of P removal by periphyton biofilms, (2) determine whether P removal by periphyton biofilms is dominated by adsorption mechanisms, and (3) describe the process of P removal by periphyton biofilms using kinetic models (pseudo-first-order, pseudo-second-order, intraparticle diffusion, and Boyd's model), isotherm models (Langmuir and Dubinin–Radushkevich) and mathematical models of biosorption behavior (adsorption intensity, adsorption energy, and Gibbs free energy). The findings will: (1) evaluate a promising environmentally benign biomeasure for removing P from nonpoint resource wastewater; (2) assist in improving the P removal efficiency of periphyton through distinguishing (or modulating) adsorption mechanisms; and (3) provide insight into the P removal processes by periphyton or similar microbial aggregates.

7.2 MATERIALS AND METHODS

7.2.1 Periphyton Culture

The periphyton substrate – Industrial Soft Carriers (diameter 12 cm and length 55 cm, Jineng Environmental Protection Company of Yixing, China) – was used for in situ collecting and culturing of periphyton from Xuanwu Lake, East China. During the experiment, the substrates were submerged into the lake water (TN: 1.90 mg L^{-1}, TP: 0.1 mg L^{-1}, pH: 7.8, ammonia: 0.53 mg L^{-1}; nitrate: 0.73 mg L^{-1}); and the microorganisms in the hypereutrophic water attached to the substrate surfaces and formed periphyton. Once the thickness of periphyton on the substrate surface was >5 mm, the periphyton with their substrates were removed for indoor culture.

The indoor culture of periphyton was conducted in glass tanks (each tank: 100 cm length, 100 cm width, and 60 cm height). Firstly, the tanks were sterilized using a 95% alcohol solution and rinsed with water. Then, the collected periphyton and their substrates were submerged into the glass tanks filled with simulated artificial wastewater (modified BG-11 medium composed of macronutrients [$NaCO_3$, $NaNO_3$, K_2HPO_4, $MgSO_4 \cdot 7H_2O$, $CaCl_2 \cdot 2H_2O$], micronutrients [H_3BO_4, $MnCl_2 \cdot 4H_2O$, $ZnSO_4$, Na_2MoO_4, $CuSO_4 \cdot 5H_2O$, $Co(NO_3)_2 \cdot 6H_2O$] and organic nutrients [citric acid and ammonium ferric citrate]). The glass tanks were kept in a greenhouse with air temperature at 25–30°C. When dense periphyton was formed (periphyton thickness >5 mm) after 60 days, it was collected for the following experiments.

7.2.2 P Removal Rate

P_i stock solution (1000 mg P L^{-1}) and P_o (100 mg P L^{-1}) were prepared by dissolving 4.387 g K_2HPO_4 and 0.6505 g ATP (disodium adenosine triphosphate, $C_{10}H_{14}O_{13}N_5P_3Na_2 \cdot 3H_2O$, sigma) into 1 L distilled water. All P concentrations used in experiments were dilutions of P stock. The ratio of dry mass of periphyton to wet mass of periphyton is 0.0532 ± 0.0085 ($n = 10$). A 5% ratio was selected as a standard for dry weight conversion of periphyton. The periphyton mass in the whole study was expressed by dry weight. Five different treatment levels were tested in this experiment, each with different biomass (0–0.2 g) of periphyton. The periphyton was placed into 250-mL flasks with simulated artificial wastewater of an initial Pi concentration of 13 mg P L^{-1}. The P concentrations in solution were determined after 1, 4, 8, 12, 24, 36, and 48 h.

To distinguish which mechanisms (adsorption and assimilation) are responsible for the P removal process, 0.25 g NaN_3 was added to the solution to inhibit microbial activity. The phosphorus concentrations in solution were determined after the specific intervals listed above (Saisho et al., 2001). The role of EPS in the P removal process by periphyton was simultaneously explored by a comparative assay of periphyton with EPS (intact periphyton) and without EPS (the EPS was extracted from periphyton before the experiment).

To explore the influence of environmental conditions on P removal processes by periphyton, light intensity (12,000, 480, and 0 Lux), temperature (45, 25, and 5°C), and initial P concentration (13, 26, 33, 41, and 49 mg P L^{-1}) were investigated independently.

7.2.3 P Removal Process

For the P_i removal process, batch kinetic experiments were conducted in 250-mL flasks containing different concentrations of adsorbent (0.2, 0.4, and 0.6 g L^{-1} periphyton) and adsorbate (13, 32, and 60 mg P L^{-1}). The isotherm experiments were conducted in 50-mL flasks with 0.6 g L^{-1} periphyton and different phosphorus concentrations ranging from 13 to 60 mg P L^{-1} (conditions: light intensity = 12,000 Lux, temperature = 25°C). An equilibration time of 48 h was used.

For the P_o removal process, the experiment comprised of two parts: conversion experiments and removal experiments. The conversion kinetic experiments were conducted in 250-mL flasks that contained 0 (control), 0.05, 0.1, and 0.2 g of the periphyton biomass in an incubator with an initial P_o concentration (C_0) of about 20 mg P L^{-1} (conditions: light intensity = 12,000 Lux, temperature = 25°C). The total phosphorus (P_t) and inorganic phosphorus (P_i) concentrations in solution were determined after 1, 4, 8, 12, 24, 36, and 48 h, respectively. The removal batch kinetic assay was conducted in 250-mL flasks with different biomass of periphyton (0.2, 0.4, and 0.6 g) and temperature (10, 20, and 30°C). P_t and P_i concentrations in solution were determined after 1, 4, 8, 12, 24, 36, and 48 h.

7.2.4 Sample Analysis

To estimate periphyton mass, dry weight (DW) was determined by oven drying samples at 80°C for 72 h. The morphology of the periphyton was characterized by optical microscopy (OM), scanning electron microscopy (SEM), and Zeiss confocal laser scanning microscopy (CLSM). The thickness of the periphyton was measured (estimated) by calipers. The removal of the EPS of periphyton and the maintenance of cell integrity followed the procedure of Ellwood et al. (2012). Briefly, periphyton was incubated in distilled water at 50°C for 1 h followed by centrifugation (3500 rpm, 20 min, 15°C).

All P_i concentrations in solution were determined using a flow injection analyzer (SEAL AA3, Germany). The morphology of the phototrophic periphyton was characterized by OM, SEM, and CLSM. The phosphatase assay procedure broadly follows that of Ellwood et al. (2012). Briefly, the periphyton biomasses cultured in different P_o concentration (from 10 to 50 mg P L^{-1}) were carefully prepared, divided into similar-sized aliquots and placed into 15-mL tubes containing 9.5 mL of artificial surface water (without N or P), while the control contained no phototrophic periphyton. The tubes were then placed in a shaking incubator at 25°C for 20 min before the addition of 0.5 mL substrate (final concentration of 0.25 mM). The samples were then incubated for 3 h, after which the assay reaction was terminated by the addition 0.5 mL of 0.5 M NaOH. Finally, the phosphatase was determined using the absorbance at 405 nm wavelength. Periphyton biomasses were removed from the solution, rinsed and dried and weighed to an accuracy of 1.0 mg. The P_t and P_i concentrations in solution were determined simultaneously using a flow injection analyzer (SEAL AA3, Germany).

The detailed methods for determining microbial activity and diversity are based on Biolog ECO Microplates. The BiologECO Microplates (Hayward, CA, USA) were applied to investigate the microbial respiratory activity in periphyton. The plate comprised of an array of 96 wells and 31 types of carbon sources. The wells are filled with a redox-sensitive tetrazolium dye (oxidation indicator) which turns purple as a result of respiratory electron transport in metabolically active cells (Balser and Wixon, 2009). Therefore, plate color was directly proportional to respiratory activity. For all samples, 50-mL aliquots were used for each Biolog and 150-μL aliquots were added into each well of every Biolog ECO Microplate, which was incubated at 25°C. Color development (590 nm) was evaluated using a Biolog Microplate Reader every 12 h for 7 days (168 h).

7.2.5 Data Analyses

Each experiment was conducted in triplicate. The mean results (±SD) are presented. Statistical analysis (*T*-test) was performed among samples and $P < 0.05$ indicated statistical significance.

For the Biolog trial, average well color development (AWCD) was calculated according to the following equation:

$$\text{AWCD} = \frac{\sum (C - R)}{n} \tag{7.1}$$

where C is color production within each well, R is the absorbance value of the plate's control well, and n is the number of substrates ($n = 31$).

The Shannon index (H) is commonly used to characterize species diversity in a community, which was obtained using the following equation:

$$H = -\sum pi \ln pi \tag{7.2}$$

where pi is the proportion of the relative absorbance value of well i to the total plate's wells.

For the enzyme activity assays, phosphatase activity (PA) was calculated by calibration curves constructed from *p*-nitrophenol (pNP) standards (0–0.2 mM) in assay medium, which was expressed as mmol pNP released g^{-1} DW (dry weight) h^{-1} and obtained by the following equation

$$\text{PA} = \frac{C_i}{mt} V \tag{7.3}$$

where C_i is the concentration of pNP (mmol L^{-1}), m is the dry weight of phototrophic periphyton (g), t is the reaction time (h), and V is the volume of the solution.

For the conversion kinetic study, the amount of P_o transformed to P_i (q_c) at time t (0, 1, 2, 4, 8, 12, 24, and 48 h) was obtained based on Eq. (7.4):

$$q_c = \frac{(C_t - C_0)V}{m} \tag{7.4}$$

where C_0 is the initial P_i concentration (mg L^{-1}), C_t is the concentration of P_i at time t, V is the volume of solution (L), and m is the dry weight of the periphyton (g).

Pseudo-first-order kinetic (Eq. 7.5) and pseudo-second-order kinetic (Eq. 7.6) models were used to evaluate the conversion process (Dawood and Sen, 2012):

$$\log(q_e - q_c) = \log q_e - \frac{k_1}{2.303}t \tag{7.5}$$

where q_c and q_e represent the amount of P_o transformed to P_{inorg} (mg g^{-1}) at time t and at equilibrium time, respectively. Parameter k_1 represents the adsorption first-order rate constant (min^{-1}) that is calculated from the plot of log ($q_e - q_c$) against time (t).

$$\frac{t}{q_c} = \frac{1}{k_2 {q_e}^2} + \frac{t}{q_e} \tag{7.6}$$

where k_2 is the pseudo-second-order rate constant (g mg^{-1} h^{-1}). A plot between t/q_c versus t gives the value of the constants k_2 and allows the calculation of q_e (mg g^{-1}).

For removal kinetics, the amount of P_i adsorbed onto adsorbents at time t (0, 1, 2, 4, 8, 12, 24, and 48 h) was obtained by means of the following equation:

$$q_t = \frac{(C_0 - C_t)V}{m} \tag{7.7}$$

where C_0 is the initial phosphate concentration (mg L^{-1}), C_t is the concentration of phosphorus at any time t, V is the volume of solution (L), and m is the dry weight of periphyton (g).

To further evaluate the adsorption mechanism, the intraparticle diffusion model was used. It is useful for identifying the adsorption mechanism and is commonly used for design purposes. It is expressed by the following equation:

$$q_t = K_{id} t^{0.5} + C \tag{7.8}$$

where q_t is the amount adsorbed at time t, $t^{0.5}$ is the square root of the time and K_{id} (mg g^{-1} min$^{-0.5}$) is the rate constant of intraparticle diffusion.

The Arrhenius equation was used to further analyze the removal process, and is calculated as follows:

$$\ln k = \frac{-E_a}{RT} + \ln A \tag{7.9}$$

where E_a is activation energy, T is the temperature in Kelvin, R is the gas constant (8.314 J mol^{-1} K^{-1}), and A is a constant called the frequency factor. Values of E_a can be determined from the slope of ln k versus T^{-1} plot.

Boyd's model was applied to predict the actual slow step involved in the adsorption process, and is expressed as follows (Tang et al., 2012):

$$B_t = -0.4977 - \ln\left(1 - \frac{q_t}{q_e}\right) \tag{7.10}$$

The adsorption isotherm is obtained using the following equation:

$$q_e = \frac{(C_0 - C_e)V}{m} \tag{7.11}$$

where C_e is the equilibrium phosphate concentration (mg L^{-1}), V is the volume of solution (L), and m is the dry weight of periphyton (g).

To determine the phosphorus adsorption characteristics of periphyton and to evaluate the applicability of this adsorption process, adsorption thermodynamic experiments were carried out at 298.15 K. The Langmuir isotherms model was applied to analyze the experimental data. Based on further analysis of the Langmuir equation, the dimensionless parameter of the equilibrium or adsorption intensity (R_L) can be expressed by following equation:

$$R_L = \frac{1}{1 + K_L C_0} \tag{7.12}$$

The R_L parameter is considered to be one of the more reliable indicators of the adsorption intensity of a surface. There are four possibilities for the R_L value: (1) favorable adsorption, $0 < R_L < 1$, (2) unfavorable adsorption, $R_L > 1$, (3) linear adsorption, $R_L = 1$, and (4) irreversible adsorption, $R_L = 0$ (Ho et al., 2002).

The Dubinin–Radushkevich (D–R) isotherm model was also used to determine the adsorption type (physical or chemical). The linear form of the D–R model is expressed by the following equation (Wu et al., 2010a):

$$\ln q_e = \ln q_m - \beta\varepsilon^2 \tag{7.13}$$

$$\varepsilon = RT \ln\left(1 + \frac{1}{C_e}\right) \tag{7.14}$$

where ε is the Polanyi potential:

Finally, thermodynamic parameters such as adsorption energy (E) and Gibbs free energy (ΔG°) were evaluated to study the feasibility of the process, and are expressed by the following equations:

$$E = \frac{1}{\sqrt{2\beta}} \tag{7.15}$$

$$\Delta G^\circ = -RT \ln(K_L) \tag{7.16}$$

7.3 RESULTS AND DISCUSSION

7.3.1 Characteristics of the Periphyton

The periphyton was mainly composed of green algae, diatoms, bacteria, and protozoa, which was dominated by phototrophic algae (Fig. 7.1). These algae

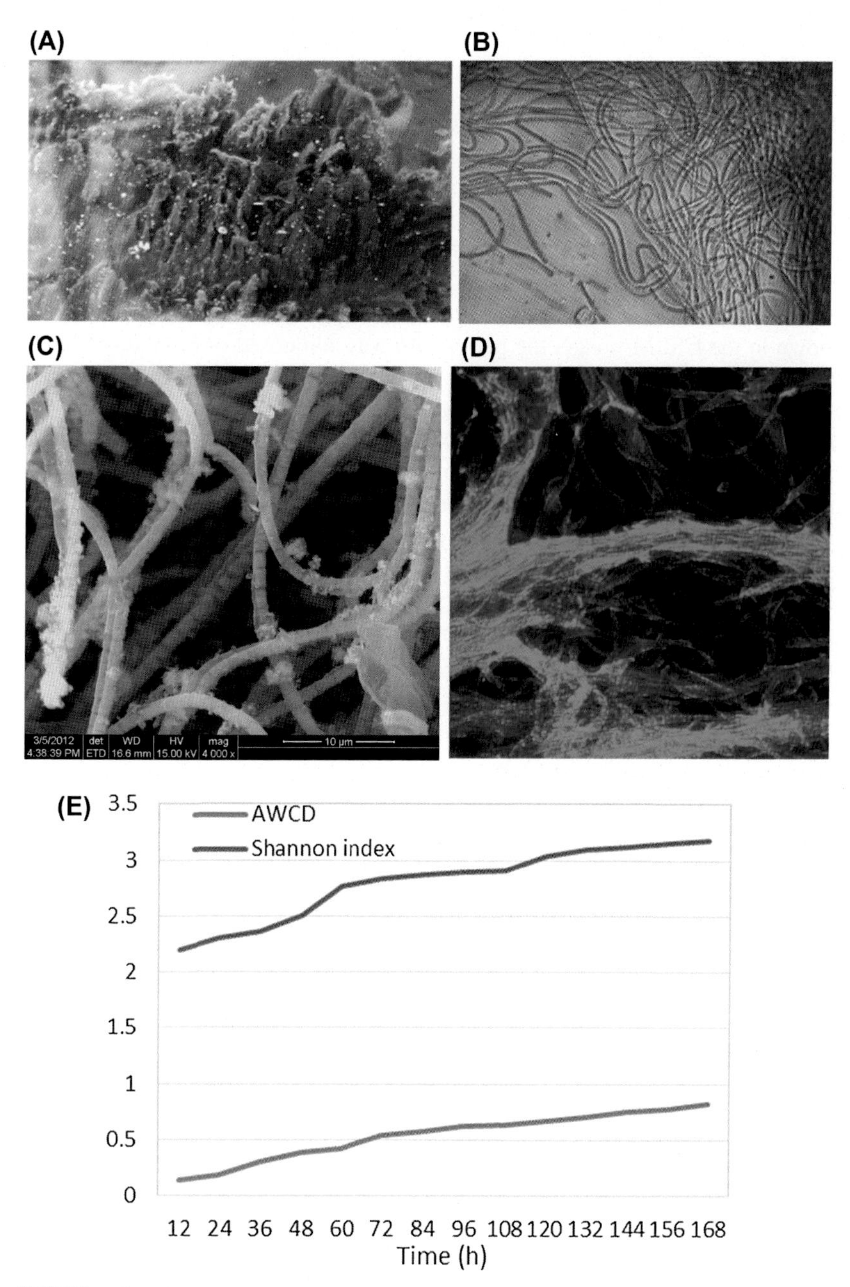

FIGURE 7.1 The characteristics of periphyton observed in water (A) and under different microscopes: OM (B), SEM (C), CLSM (D), and based on diversity indices (E).

with a diameter of about 1.5 μm intertwined each other, forming the base matrix for other microorganisms such as bacteria. The Shannon index of the periphyton based on the Biolog analyses was about 3.1 after 7 days, indicating that there were many types of microorganisms living in the periphyton (de Beer and Stoodley, 2006).

Microstructure is a significant determinant of the activity of the biofilm since it plays an important role in the transportation of nutrients and water (Sabater et al., 2002). Previous studies indicated biofilm structure was heterogeneous and complex, containing voids, channels, cavities, pores, and filaments, and with cells arranged in clusters or layers (De Beer et al., 1994). As shown in the CSLM image, the periphyton was composed of biomass clusters separated by interstitial voids, which might have considerable consequences for mass transfer inside the biofilms and exchange of substrates and products with the water phase (Fig. 7.1C). These periphyton microvoids could play many important roles such as interception of nutrient transportation, especially granular nutrients, between sediment and water interface. These structural characteristics are, however, obviously influenced by the species arrangement of organisms composing the biofilms (James et al., 1995). For example, the microvoids constructed among complex cells such as algae, bacteria, and protozoa may be larger than those constructed by single species. These voids might also provide more microspaces or adsorption sites for capturing nutrients, especially for particulate nutrients such as polyphosphate particles.

7.3.2 P_i Removal by Periphyton

The rates of P_i removal by periphyton ranged from 6% to 100%, while the control ranged from 0.1% to 1.1% after 1–48 h (Fig. 7.2A). The removal rate of P_i by periphyton could reach 90% after 12 h when the initial phosphorus concentration was 13 mg P L^{-1} and initial periphyton mass was over 0.2 g L^{-1}. These results demonstrate that periphyton is effective P_i removal material. This outcome was further evidence that periphyton has a high affinity for P and plays an important role in the uptake or storage of P in waters (Dodds, 2003; Noe et al., 2002).

The P removal rate under the four dosage levels (0.1, 0.2, 0.4, and 0.6 g L^{-1}) demonstrated a consistent trend over time within 48 h, and increased from 13%, 6%, 9%, and 20% to 87%, 100%, 100%, and 100%, respectively (Fig. 7.2A). This implies that the higher the content of periphyton available for P removal, the greater the amount of phosphorus removed. In addition, the dynamic curves calculated for P_i removal rate showed that the slope in the period from t_0 to t_{12} was larger than that from t_{12} to t_{48}.

To determine whether adsorption dominated the removal process of P_i by periphyton, NaN_3 was used to impede microbial activity by restraining microbial respiration and inhibiting assimilation (Saisho et al., 2001). AWCD values (representing microbial activity) in the NaN_3 treatment and the control

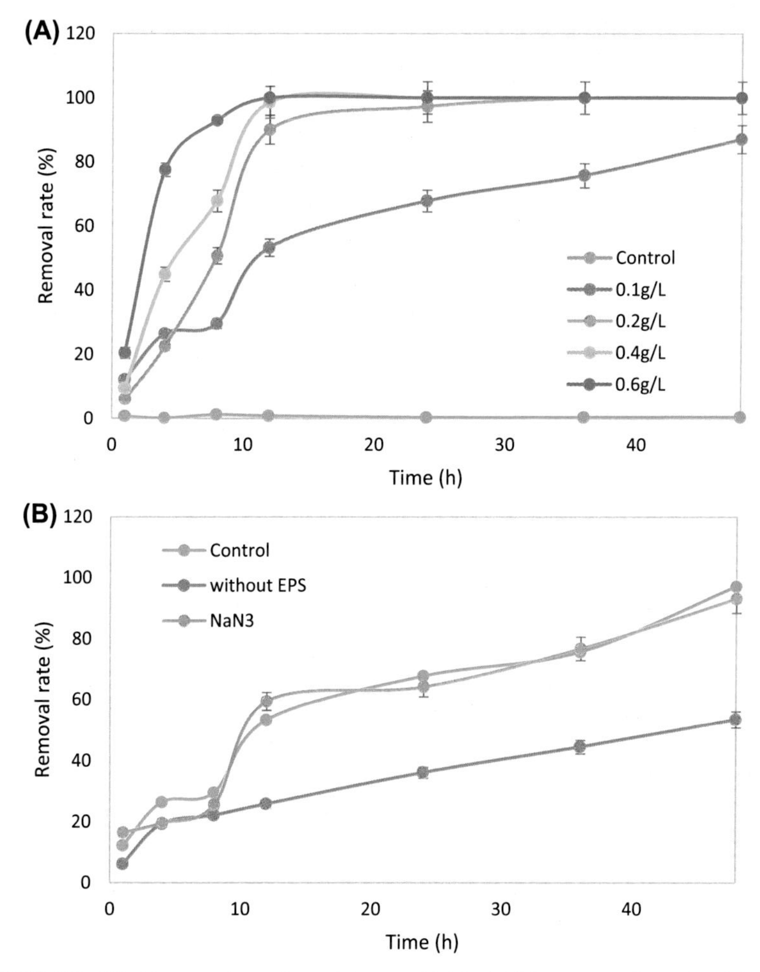

FIGURE 7.2 The P_i removal rate by periphyton. (A) The initial concentration of P_i (C_0) was 13 mg P L^{-1}, periphyton dosage was 0.1, 0.2, 0.4, 0.6 g L^{-1} (dry weight) and no periphyton was in the control. (B) The control was intact periphyton and no NaN_3 was add (experimental conditions: illumination = 12,000 Lux; temperature = 25°C).

were significantly different ($P < 0.05$). The AWCD values in the NaN_3 treatment were zero over time while the AWCD values in the control increased with time (Fig. 7.3). This implies that microbial respiration in periphyton was completely inhibited and the effect of microorganism ingestion on phosphorus removal was trivial. The removal rates of P_i by periphyton with NaN_3 treatment within 48 h were slightly decreased but not significantly different from the controls ($P > 0.05$; Fig. 7.2B). This further suggested that the assimilation of phosphorus by microbes was minimal during the removal of P_i by periphyton in 48 h.

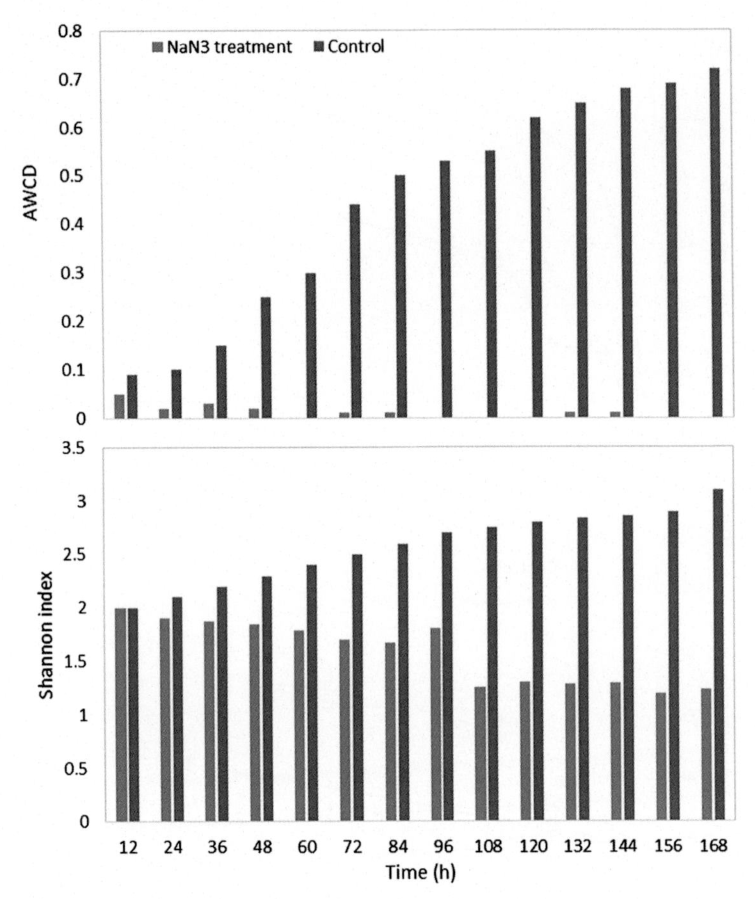

FIGURE 7.3 Microbial community activity (represented by AWCD) and functional diversity (represented by Shannon index) based on Biolog analyses.

In addition, the EPS significantly affected P_i removal rate by periphyton in 48 h ($P < 0.05$; Fig. 7.2B). After 48 h, the maximum phosphorus removal rates were 54% and 100% under cells (without EPS) and intact periphyton (control), respectively. EPS may account for 50–90% of the total organic carbon of periphyton and can be considered the primary matrix material and sorption site (Flemming et al., 2007). This implies that EPS played an important role in P_i removal by periphyton and that adsorption due to EPS might be a major process during P_i removal. These outcomes corresponded with the result that removal rates from t_0 to t_{12} were higher than those from t_{12} to t_{48} in the phosphorus removal experiments. This further suggests that the P_i removal process by periphyton was dominated by adsorption and provides the foundation for the kinetic and isotherm investigations.

7.3.3 P_o Removal by the Periphyton

The P_t concentration in solution decreased over time from 13 to 6, 2, 0, and 0 mg P L^{-1}, respectively, under the treatments with periphyton masses of 0.4, 0.8, 1.6, and 2.4 g L^{-1}, respectively (Fig. 7.4). The P_t content in the control showed a slight reduction. This indicates that periphyton could effectively remove P_o from artificial surface water. Simultaneously, the P_i concentration in solution showed the same change under varied biomass content treatments. This indicates that the conversion of P_o by periphyton occurred throughout the whole removal process.

The transformation process of P_o (ATP) by periphyton was studied by monitoring the P_t, P_i, and q_c over time (Fig. 7.5). When the initial P_o concentration was about 20 mg P L^{-1}, the P_i concentrations in solution increased over time from about 0.7 to 6.4, 10.2, and 14.3 mg P L^{-1} under 0.2, 0.4, and

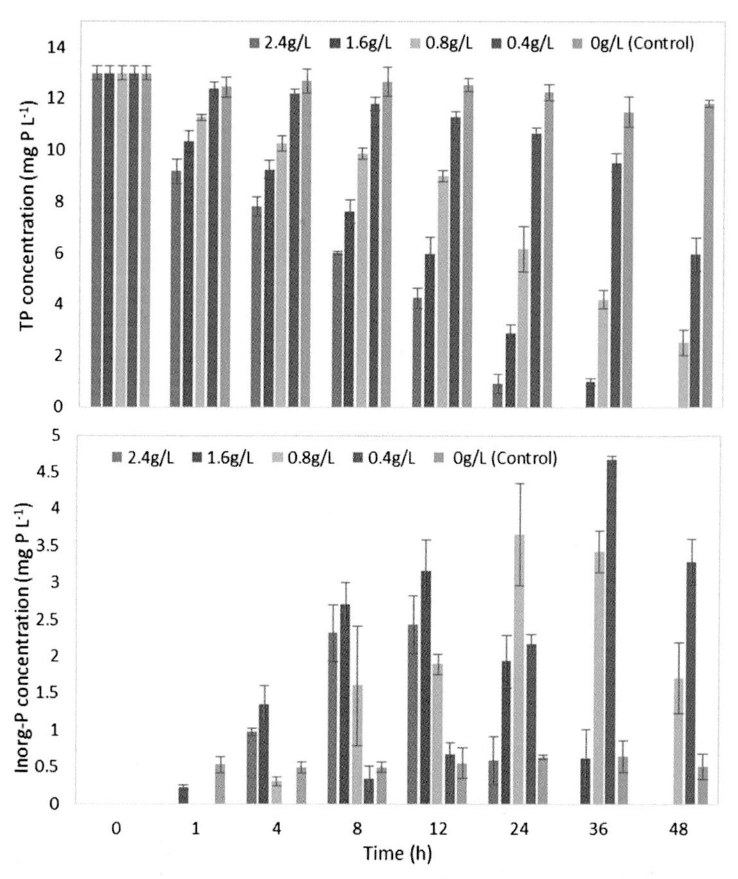

FIGURE 7.4 The removal process of P_o by periphyton. *Inorg-P*, inorganic phosphorus content; *TP*, total phosphorus content.

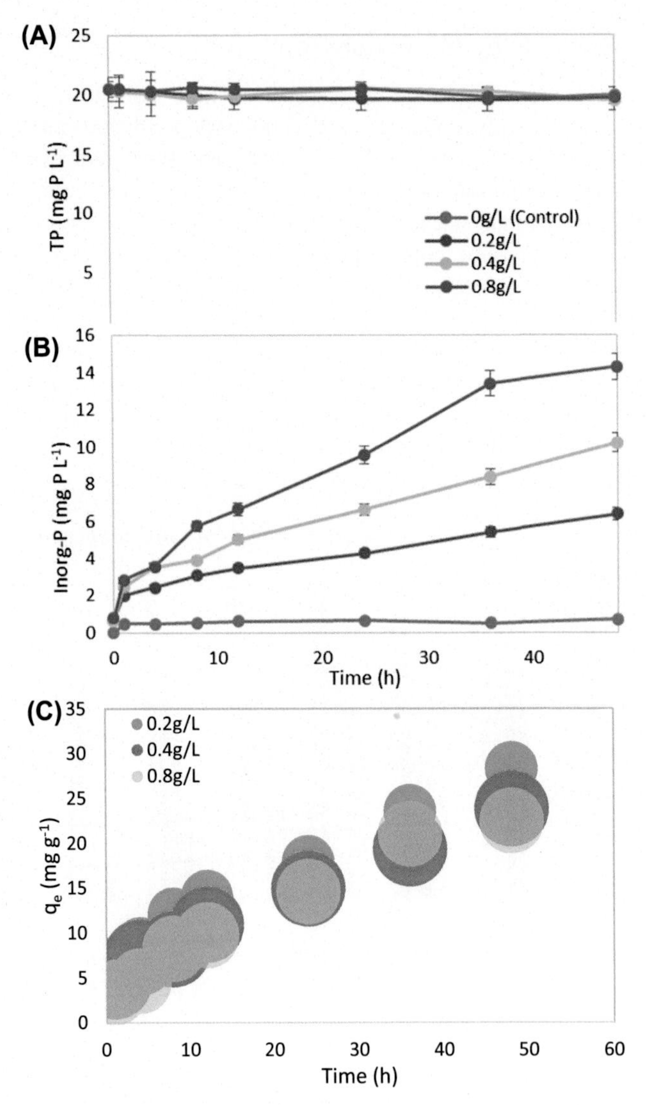

FIGURE 7.5 The conversion process of P_o (A) the change in P_t and P_i over time, (B) the change in q_c over time (experimental conditions: light intensity = 12,000 Lux, temperature = 25°C).

0.6 g L^{-1} of periphyton, respectively. The control (no periphyton) showed no significant change in 48 h ($P > 0.05$). These results indicated that the periphyton had a relatively substantial ability to convert P_o to P_i, which became stronger with the increasing biomass of the periphyton. It is well established that the reaction rate of the conversion of P_o to P_i was directly associated with phosphatase based on the reaction equation (ATP + enzyme → ADP + Pi + energy). This means that the periphyton produced phosphates, which are

TABLE 7.1 The Kinetic Parameters of P_o Transformation by the Periphyton

Periphyton Biofilm Content	Pseudo-First-Order Kinetic Model			Pseudo-Second-Order Kinetic Model		
	k_1 (h^{-1})	q_1 (mg g^{-1})	R^2	k_2 (g mg^{-1} h^{-1})	q_2 (mg g^{-1})	R^2
0.2 g L^{-1}	0.0432	24.632	0.964	0.0028	31.847	0.922
0.4 g L^{-1}	0.0416	21.463	0.977	0.0026	27.778	0.897
0.6 g L^{-1}	0.0412	20.888	0.984	0.0017	30.303	0.888

beneficial to the P_o conversion reaction. Moreover, the content of phosphates increased with the increasing periphyton biomass.

To quantify the transformation capacity of P_o by periphyton, the P_o transformation data were described using kinetic models (pseudo-first-order and pseudo-second-order kinetic equations). According to Fig. 7.5B, the amounts of P_o converted to P_i (q_c) after 48 h by the periphyton were 28.3, 23.9, and 22.5 mg g^{-1} under 0.2, 0.4, and 0.6 g L^{-1}, respectively. The pseudo-first-order and pseudo-second-order kinetic constant k and q values determined from the plots, decreased with increasing biomass of the periphyton (Table 7.1). This implies that the treatment with higher periphyton biomass contains a dense layer, resulting in smaller contribution to P_o transformation. This result is consistent with observations in previous studies (de los Ríos et al., 2004; Roeselers et al., 2008). It is notable that the correlation coefficient (R^2) of the pseudo-first-order model is higher than the pseudo-second-order coefficient, suggesting that the pseudo-first-order kinetic model is more suitable for describing the P_o transformation process.

To identify the role of phosphatase in P_o transformation by periphyton, phosphatase activity was evaluated under various P_o concentrations. Phosphatase activity increased initially but then decreased as P_o concentration increased (Fig. 7.6). The maximal phosphatase activity was about 22 μmol pNP g^{-1} h^{-1} when the P_o concentration was 20 mg P L^{-1}. Phosphatase plays an important role in the biochemical cycles of P in aquatic systems by hydrolyzing dissolved organic P to phosphates that are available for cellular uptake. Similar enzymatic responses to P limitation have been previously demonstrated in both planktonic communities and biofilms (Bentzen et al., 1992; Espeland and Wetzel, 2001; Huang et al., 1998).

The phototrophic biofilm cultured under limited phosphate and organic P supply exhibited higher phosphatase activities than that cultured with sufficient phosphate supply (Ellwood et al., 2012), suggesting that end-product repression and derepression of phosphatase activity was a main limiting

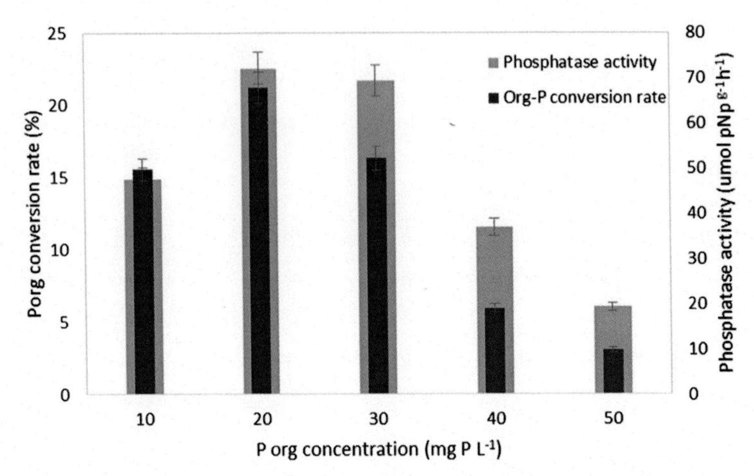

FIGURE 7.6 Phosphatase activity of the periphyton under different P_o concentrations.

factor of phosphatase activity. In this study, the concentration of P_i transformed from P_o increased slowly when the periphyton mass was under $0.6\ g\ L^{-1}$ in the later stages (after the 36 h) (Fig. 7.5). This result is similar to Ellwood et al. (2012) study. However, the P_o concentration was also an important factor determining the phosphatase activity of phototrophic periphyton (Fig. 7.3). There was a significant reduction in phosphatase activity under high P_o concentration ($P < 0.05$). One possible explanation could be growth repression under high P_o concentration, which showed that the periphyton were not functioning after being cultured at high P_o concentration (i.e., over $30\ mg\ P\ L^{-1}$).

The conversion of P_o to P_i is an important process for P removal and recovery since P_i is the removable and recoverable form of P in wastewater-treatment systems. Many measures such as the advanced-oxidation processes (AOPs) have been used to convert P_o to P_i, and are regarded as promising means for the transformation of P_o in low-concentration streams (Rittmann et al., 2011). AOPs rely on nonspecific free-radical species, such as hydroxyl radicals, to quickly attack the structure of organic compounds. However, this method is mostly used for the destruction of specific P-based and trace contaminants, such as organophosphorus pesticides (Badawy et al., 2006). Therefore, AOP methods might be impractical for use in surface waters such as streams and lakes, which often have high organic pollution loadings. Periphyton-based conversion systems have many advantages over AOP methods, such as low capital cost, easy-harvest, and powerfully converting ability for high content and nonspecific P_o (ATP), suggesting that is a potential promising technology for P_o removal and recovery.

The removal rate under the four periphyton biomass levels (0.4, 0.8, 1.6, and $2.4\ g\ L^{-1}$) demonstrated a consistent trend over time within 48 h, and

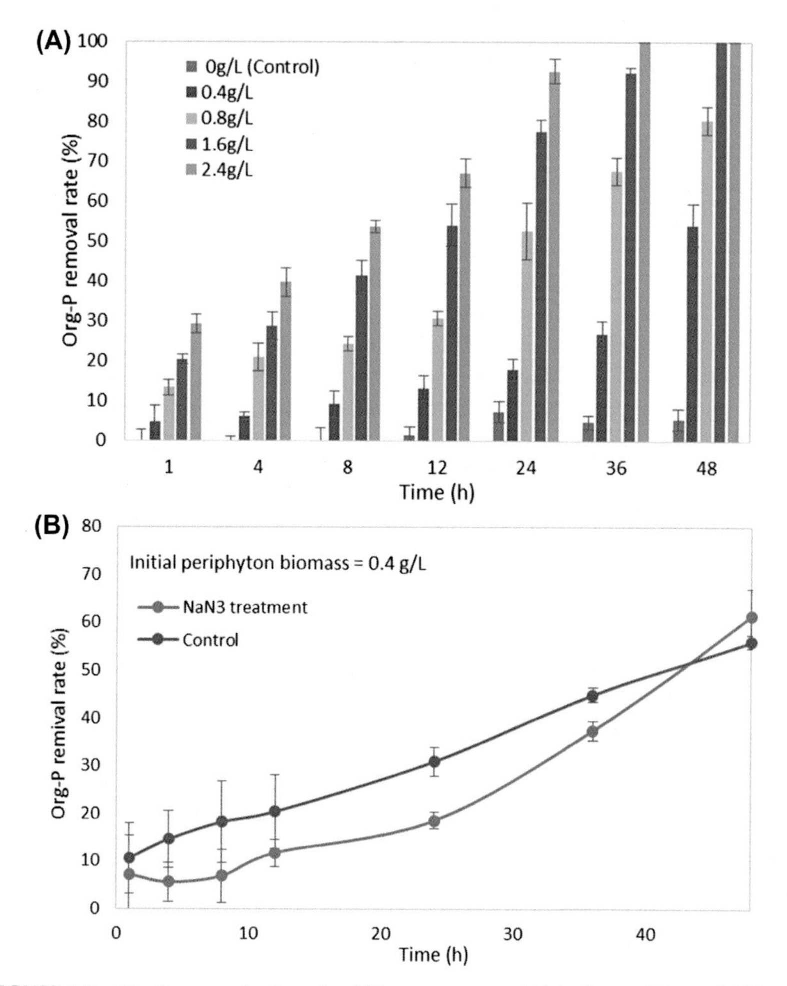

FIGURE 7.7 The P_o removal rate under different treatments (A) in the conditions of different periphytic mass initial concentrations, (B) with the addition of NaN_3 (for inhibiting microbial activity), and the control (without the addition of NaN_3).

increased from 5%, 13%, 20%, and 29% to 54%, 80%, 100%, and 100%, respectively (Fig. 7.7A). This implies that the higher the periphyton content available for P_o removal, the greater the amount of P_o removed. Furthermore, to determine whether assimilation mechanisms dominated the removal of P_o by periphyton, NaN_3 was used to impede periphyton microbial activity by restraining microbial respiration and inhibiting assimilation (Saisho et al., 2001; Wu et al., 2010a). The removal rates of P_o by periphyton under the NaN_3 treatment within 48 h were not significantly different from the control ($P > 0.05$; Fig. 7.7B), which indicates that the assimilation of P by microbes was minimal during the removing of P_o by the periphyton in 48 h. This further suggests that the P_o removal by periphyton is dominated by adsorption within 48 h.

7.3.4 P_i adsorption by Periphyton

To describe the adsorption process, the widely used pseudo-first-order and pseudo-second-order kinetic models were selected to fit the data collected based on three different adsorbent and adsorbate dosages. All the regression coefficients (R^2) of the pseudo-second-order kinetic model were very high ($R^2 > 0.99$; Table 7.2) under the treatments of different adsorbent content, and were higher than that of the pseudo-first-order kinetic model. This showed that the adsorption of phosphorus onto periphyton closely follows the pseudo-second-order kinetic model. This result was similar to a previous adsorption study (Ozacar, 2003). Moreover, it is obvious that the q_2 values were very close to the experimental q_e values, which decreased with increasing adsorbent. In addition, models fit relatively well under low initial phosphorus concentrations. This implies that the adsorption capacity may be easily saturated and the adsorption process complicated when the concentration of the adsorbate (P_i) was high.

In a solid–liquid adsorption system, solute molecule transfer is usually characterized by intraparticle diffusion, boundary layer diffusion or both. This determines whether the controlling step of the adsorption is an intraparticle process, an external diffusion process, or both (Dawood and Sen, 2012). An intraparticle diffusion model and Boyd plots were used to analyze the process of P_i adsorption onto periphyton (Table 7.1). The determination coefficients (R^2) were between 0.53 and 0.89 in all treatments, indicating that intraparticle diffusion was not the only rate-limiting step and that some other processes might be controlling the rate of adsorption.

The plots are multilinear, containing at least three linear segments (Fig. 7.8A). According to intraparticle diffusion models, if the plot of q_t versus $t^{0.5}$ presents a multilinearity correlation, it indicates that at least three steps occur during the adsorption process: the first is the transport of molecules from the bulk solution to the adsorbent external surface by diffusion through the boundary layer (film diffusion). The second portion is the diffusion of the molecules from the external surface into the pores of the adsorbent. The third portion is the final equilibrium stage, where the molecules are adsorbed on the active sites on the internal surface of the pores and the intraparticle diffusion starts to slow down due to the solute concentration becoming lower (Hameed and El-Khaiary, 2008; Kiran et al., 2006; Sun and Yang, 2003). Therefore, the P_i adsorption process by periphyton could be divided into three steps: external mass transfer at the beginning (0–4 h), intraparticle diffusion in the middle period (4–8 h), and equilibrium after 8 h.

The linearity of the plot of B_t against time (Fig. 7.8B) can provide information to demonstrate how intraparticle diffusion and the boundary layer

TABLE 7.2 Kinetic Model Parameters for P_i Adsorption Under Different Treatments

Treatments		Pseudo-First-Order Kinetic Model			Pseudo-Two-Order Kinetic Model			Intraparticle Diffusion Model		
		k_1 (h^{-1})	q_1 ($mg\ g^{-1}$)	R^2	k_2 ($g\ mg^{-1}\ h^{-1}$)	q_2 ($mg\ g^{-1}$)	R^2	k_{id} ($mg\ g^{-1}\ h^{-0.5}$)	C ($mg\ g^{-1}$)	R^2
Periphyton content ($g\ L^{-1}$)	0.2	0.165	104.400	0.863	0.007	70.922	0.999	9.993	14.926	0.750
	0.4	0.098	56.195	0.869	0.012	56.180	0.997	8.352	11.650	0.676
	0.6	0.392	52.312	0.971	0.099	36.232	0.999	4.695	13.350	0.529
P_i Content ($mg\ P\ L^{-1}$)	13	0.084	32.137	0.943	0.003	44.643	0.994	5.665	4.234	0.891
	32	0.006	22.444	0.839	0.001	29.326	0.355	2.466	1.511	0.739
	60	0.009	13.483	0.682	0.005	18.116	0.738	2.064	1.162	0.846

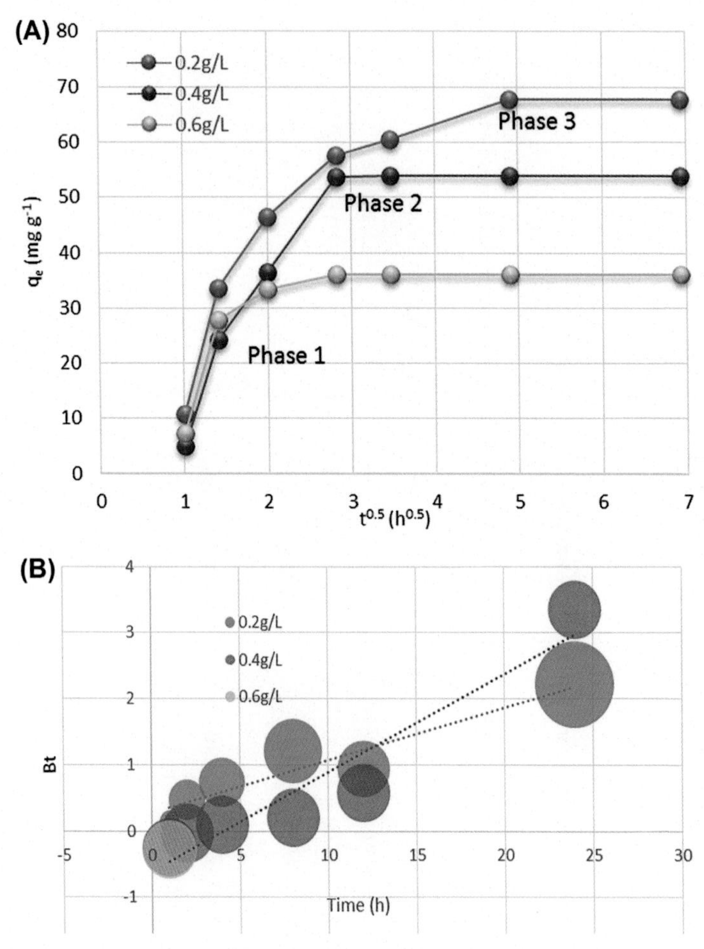

FIGURE 7.8 Intraparticle diffusion and Boyd plots for P_i adsorption at different periphyton biomass (experimental conditions: illumination = 12,000 Lux, temperature = 25°C).

can affect the rates of adsorption. If a plot of B_t versus t is a straight line passing through the origin, then adsorption reflects the layer effect. As the plots were linear only in the initial period of adsorption and passed through the origin in our study, the external mass transfer was the rate-limiting process at the beginning of adsorption potentially followed by intraparticle diffusion. The C value in Eq. (7.3) represents the boundary layer effect (film diffusion). Higher C values indicate greater contributions of the surface sorption in the rate-controlling step (Özcan et al., 2005). The C values decreased with increases in adsorbent and adsorbate concentrations (Table 7.2), reflecting the larger boundary layer thickness at lower concentrations of adsorbent and adsorbate.

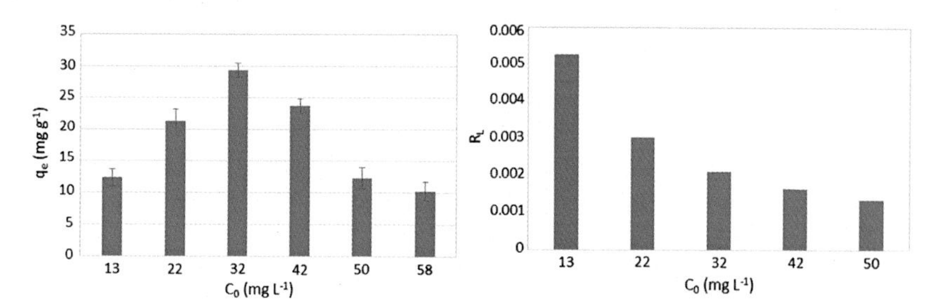

FIGURE 7.9 Variation in adsorption intensity with different initial P_i concentrations (experimental conditions: illumination = 12,000 Lux, temperature = 25°C).

Surface models such as Langmuir and D–R equations are often used to examine adsorption data collected from heterogeneous surfaces (García-Calzón and Díaz-García, 2007). Therefore, Langmuir and D–R isotherms were selected for the heterogeneous surface of periphyton. The maximum q_e was about 29.3 mg g^{-1} when initial P_i content was 32 mg L^{-1} (Fig. 7.9). The linear plots of C_e/q_e versus C_e were examined to determine the q_m and K_L values. The high determination coefficient ($R^2 = 0.997$, Table 7.2) showed that the P_i adsorption isotherm by periphyton can be described using the Langmuir model. During P_i adsorption by periphyton, the R_L values decreased from 0.0055 to 0.0011 when the initial phosphate concentration increased (Fig. 7.9). This implies that the P_i adsorption by periphyton was favorable (Ho et al., 2002).

The equilibrium data could be described using a linear D–R equation ($R^2 = 0.835$) (Table 7.3). The β value was 1×10^{-7} mol^2 J^{-2} under 0.6 g L^{-1} periphyton mass. Generally, Langmuir and D–R equations have been regarded as empirical, rather than mechanistic (e.g., surface models) adsorption isotherms models. In this study, data collected from the adsorption process simultaneously fit these two empirical equations. This implies that the adsorption process of periphyton had mechanistic relevance (or surface adsorption properties) and that the model parameters included in these

TABLE 7.3 Isotherm Parameters for Langmuir and D–R Models of P_i Adsorption by Periphyton

Periphyton Content	Langmuir				D–R			
	q_m	K_L	R^2	$\Delta G°$	β	Lnq_m	R^2	E
0.6 g L^{-1}	22.779	14.6	0.997	−6.651	1E-07	3.246	0.835	4.083

empirical equations had specific thermodynamic significance. Moreover, the mean sorption energy (E) can be used to distinguish chemical and physical adsorption (Wu et al., 2010a). If the E value is between 8 and 16 kJ mol^{-1}, the adsorption process follows chemical exchange. If E is less than 8 kJ mol^{-1}, then the adsorption is physical in nature (Wu et al., 2010a). The E value was 4.08 kJ mol^{-1}, implying that the adsorption process was physical in nature (Table 7.3). The thermodynamic results found that ΔG° was −6.6 kJ mol^{-1}, showing that P_i adsorption by periphyton was a spontaneous process (Wu et al., 2010a).

Previous studies have revealed that the adsorption of P by periphyton could be an effective mechanism for the removal of P from agriculturally enriched Everglades water, the removed P without recovery accounting for 48−72% (Scinto and Reddy, 2003). This finding also implies that natural wetlands with periphyton are a promising method to immobilize P. Compared to other measures like one-side technologies, the application of ubiquitous periphyton in natural wetlands can collect and immobilize P of fluctuating load and discharge times from wastewater flux in situ on a large scale. The artificial introduction of periphyton into natural wetlands and the modulation of adsorption is a promising option for P removal from wastewater. Moreover, as P resources have become increasingly scarce recently, P recovery from wastewaters by algae and macrophytes are regarded as a promising strategy and are already widely used (Shilton et al., 2012). Periphyton is also easier to acquire and harvest than algae and macrophytes. Therefore, P recovery and reuse technologies based on periphyton have vast potential application prospects. However, the fact that adsorption might be reversible and periphyton might detach under high-flow conditions (de Beer and Stoodley, 2006) needs to be considered during the long-term application of periphyton.

7.3.5 P_o adsorption by Periphyton

For a pseudo-second-order model, the correlation coefficient (R^2) is generally less than the pseudo-first-order coefficient (Table 7.4). Accordingly, kinetic parameters k_1 and q_1 showed the same trend and increased from 0.019 and 5.84 to 0.041 and 11.97, respectively, with temperature rises, while k_1 increased from 0.047 to 0.102 and q_1 decreased from 12.57 to 4.99 when the periphyton content increased. In view of these results, it can be safely concluded that the pseudo-first-order kinetic model provides a better correlation and description for the adsorption process of P_o by the periphyton at different temperatures and biomass.

In a solid−liquid system, most adsorption reactions take place through multistep mechanisms, which contain external film diffusion, intraparticle diffusion, and interactions between adsorbates and active sites. Thus, an intraparticle diffusion model was chosen to analyze the process of P_o adsorption onto the periphyton. The determination coefficients (R^2) increased

TABLE 7.4 Kinetic Parameters for P_o Removal by the Periphyton

Treatments	Pseudo-First-Order Kinetic			Pseudo-Second-Order Kinetic			Intraparticle Diffusion		
	k_1 (h^{-1})	q_1 (mg g^{-1})	R^2	k_2 (g mg^{-1} h^{-1})	q_2 (mg g^{-1})	R^2	k_{id} (mg g^{-1} $h^{-0.5}$)	C	R^2
10°C	0.019	5.84	0.944	0.005	7.15	0.670	0.810	−0.71	0.876
20°C	0.025	11.51	0.991	0.003	15.34	0.707	1.707	−0.91	0.901
30°C	0.041	11.97	0.994	0.006	16.95	0.953	1.955	1.31	0.991
0.8 g L^{-1}	0.047	12.57	0.972	0.003	16.86	0.843	1.908	−0.66	0.973
1.6 g L^{-1}	0.067	7.74	0.986	0.009	9.85	0.961	1.173	0.27	0.990
2.4 g L^{-1}	0.102	4.99	0.973	0.024	6.22	0.966	0.719	0.95	0.953

from 0.88 to 0.99 as temperature rose (Table 7.4), which suggested intraparticle diffusion may be the rate-controlling step under high temperature (30°C). The relatively high R^2 under different biomass contents (Table 7.4) implies that intraparticle diffusion in the adsorption process of P_o by the periphyton was influenced by biomass. According to the intraparticle diffusion model (Fig. 7.10), if the plot of q_t versus $t^{0.5}$ presents a multilinearity correlation, it indicates that three steps occur during the adsorption process: the first is the transport of molecules from the bulk solution to the adsorbent external surface by diffusion through the boundary layer (film diffusion).

The second portion is the diffusion of the molecules from the external surface into the pores of the adsorbent. The third portion is the final equilibrium stage, where the molecules are adsorbed on the active sites on the internal surface of the pores and the intraparticle diffusion starts to slow down due to the solute concentration becoming lower (Hameed and El-Khaiary, 2008; Sun and Yang, 2003). It was shown that the plot of q_t versus $t^{0.5}$ presents a multilinearity correlation and does not pass through the origin under low temperature (Fig. 7.10), which indicates the adsorption of P_o by the periphyton was controlled by processes other than intraparticle diffusion under relatively low temperature. The large intercept (C) suggests that the process is largely of surface adsorption, this implied that the adsorption processes of P_o by the periphyton at temperature of 30°C and biomass of 2.4 g L^{-1} were more inclined to surface adsorption (Table 7.5).

To further reveal the types of P_o adsorption (physical and chemical) by the periphyton, the Arrhenius equation was chosen to calculate the activation energy (E_a) based on kinetic parameters. The magnitude of E_a may indicate the type of adsorption. Two main types of adsorption may occur, physical and chemical. In physical adsorption, the E_a value is usually low between 5 and 40 kJ mol^{-1} since the equilibrium is usually rapidly attained and the energy requirements are weak (Aksakal and Ucun, 2010). Chemical adsorption is specific and involves forces much stronger than physical adsorption, with E_a values commonly high between 40 and 800 kJ mol^{-1} according to the Arrhenius equation (Doğan et al., 2006). However, in some systems the chemical adsorption occurs very rapidly and E_a was relatively low, which is termed as a nonactivated chemisorption (Kay et al., 1995).

The correlation coefficient of the corresponding linear plot of ln k against $1/T$ is 0.96 (Fig. 7.11). The E_a value for the adsorption of P_o onto the periphyton was 27.082 kJ mol^{-1}, which suggests that the adsorption of P_o in the presence of the periphyton exhibited the characteristics of physical adsorption.

Organic P can be found commonly in municipal, agricultural, and animal wastewaters, but there is scant information on its removal and recovery as current P removal techniques are typically for inorganic P. Furthermore, as P resources are becoming increasingly scarce, P recovery from wastewaters by algae and macrophytes is regarded as a promising strategy and already in

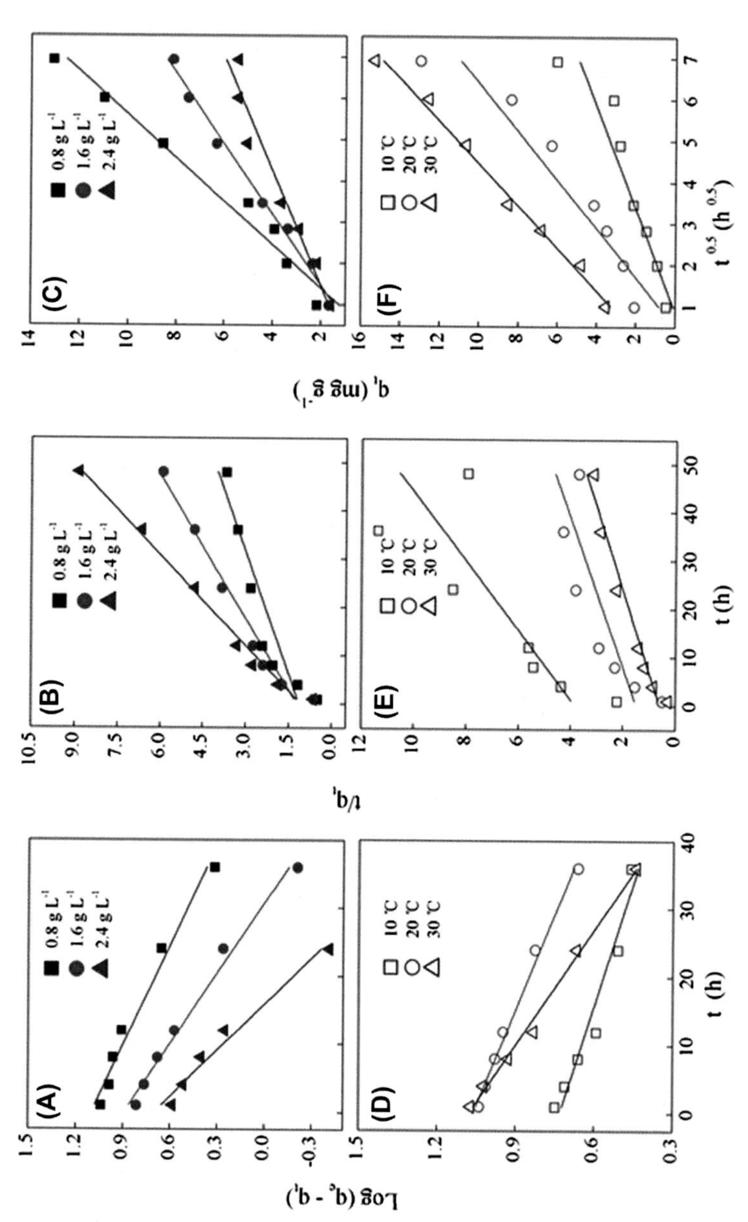

FIGURE 7.10 Adsorption kinetic analysis, (A, D) the pseudo first-order kinetic, (B, E) the pseudo second-order kinetic, and (C, F) the intraparticle diffusion kinetic of the periphyton biofilm for the P_{org}. with different biomass content at different temperatures (Lu et al., 2014).

TABLE 7.5 Adsorption Kinetic Analysis, log $(q_e - q_t)$ for the Pseudo-First-Order Kinetic and t/q_t for the Pseudo-Second-Order Kinetic and q_t (mg g^{-1}) for the Intraparticle Diffusion Kinetic of the Periphyton Biofilm for the P_o With Different Biomass Content at Different Temperatures

	Pseudo First-Order Kinetics			Pseudo First-Order Kinetics		
Time (*d*)	0.8 g L^{-1}	1.6 g L^{-1}	2.4 g L^{-1}	10°C	20°C	30°C
1	1.04	0.81	0.58	0.75	1.04	1.07
4	0.98	0.76	0.51	0.71	1.01	1.02
8	0.96	0.68	0.40	0.66	0.98	0.93
12	0.91	0.57	0.25	0.59	0.95	0.83
24	0.65	0.26	−0.41	0.51	0.82	0.67
36	0.32	−0.21	0	0.46	0.66	0.44
	Pseudo Second-Order Kinetic			Pseudo Second-Order Kinetic		
Time (*d*)	0.8 g L^{-1}	1.6 g L^{-1}	2.4 g L^{-1}	10°C	20°C	30°C
1	0.46	0.60	0.63	2.22	0.49	0.28
4	1.17	1.71	1.86	4.36	1.52	0.83
8	2.04	2.38	2.75	5.43	2.30	1.17
12	2.40	2.73	3.30	5.60	2.91	1.41
24	2.81	3.80	4.77	8.51	3.81	2.25
36	3.27	4.80	6.64	11.37	4.30	2.87
48	3.67	5.91	8.86	7.94	3.70	3.14
	Intraparticle Diffusion Kinetics			Intraparticle Diffusion Kinetics		
Time (*d*)	0.8 g L^{-1}	1.6 g L^{-1}	2.4 g L^{-1}	10°C	20°C	30°C
1	2.16	1.66	1.59	0.45	2.05	3.52
4	3.41	2.34	2.16	0.92	2.63	4.79
8	3.92	3.36	2.91	1.47	3.48	6.82
12	4.99	4.40	3.64	2.14	4.13	8.53
24	8.55	6.31	5.03	2.82	6.30	10.67
36	11.00	7.50	5.42	3.17	8.38	12.56
48	13.06	8.12	5.42	6.04	12.96	15.30

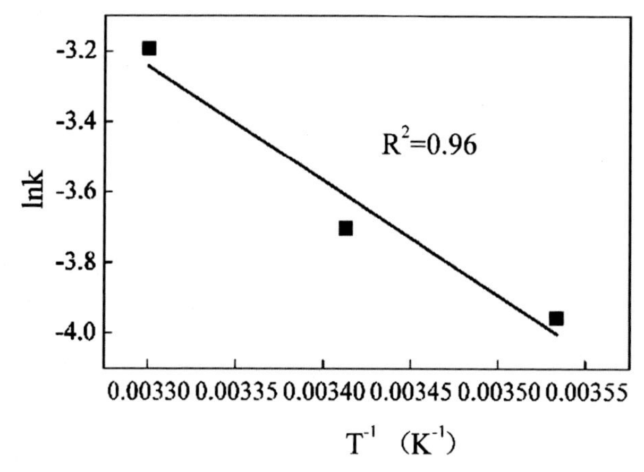

FIGURE 7.11 Arrhenius plot.

widespread use (Shilton et al., 2012). As periphyton are easily acquired and harvested, the development of methods for removing and capturing P from nonpoint source wastewaters for reuse based on periphyton will be of primary importance to agriculture in the near future. In this study, our experimental results reveal that periphyton not only possesses substantial capacity for effective organic P removal, but also has the ability to convert organic P to inorganic P that is readily captured. There are many advantages of the periphyton itself – it is environmentally friendly, economically viable, and operationally simple. Given the above advantages, P removal, recovery, and reuse technologies based on periphyton have vast practical potentials despite its dependence on numerous factors such as light, temperature, water column P concentration, water flow velocity, and thickness of periphyton (Matheson et al., 2012; McCormick et al., 2001). Most importantly, the native conditions of wastewaters in natural systems (especially in agricultural wastewaters) are more complicated and the adsorption process may be reversible under high-flow conditions. In such conditions, whether the adsorbed P will be released back into aquatic ecosystems from the periphyton needs further investigation.

7.3.6 Influence of Environmental Conditions on P Removal Rate

Previous research suggests that the uptake or removal of P in water depends on environmental factors such as light, temperature, water column P concentration, and water flow velocity (DeNicola, 1996; Matheson et al., 2012; Panswad et al., 2003). In this study, the P_i removal rates by periphyton under various temperatures (96%, 36%, 22% under 25, 45, and 5°C), and light levels

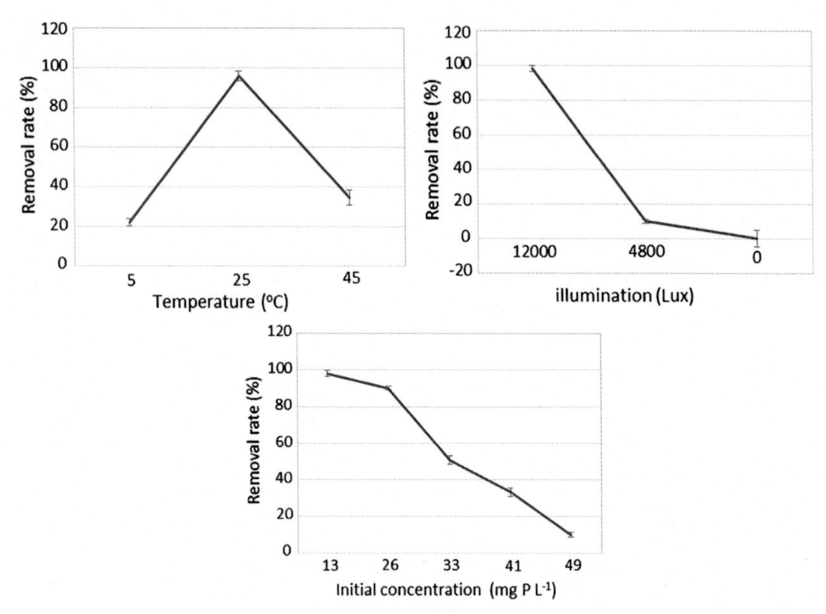

FIGURE 7.12 The effect of environmental conditions on P_i removal rate by periphyton (experimental conditions: periphyton mass was 0.2 g L^{-1}, initial concentration of P_i was 13 mg P L^{-1}).

(98%, 10%, 0.1% under 12,000, 4000, 0 Lux) were significantly different ($P < 0.05$; Fig. 7.12). Moreover, the P_i removal rates decreased from 98% to 10% while C_0 increased from 13 to 49 mg P L^{-1} ($P < 0.05$). These results mean that the P_i removal process by periphyton was significantly influenced by environmental conditions. One reasonable explanation for these findings could be the restraint of phototrophic organism growth or activities in periphyton. It was expected that the light, temperature, and P_i concentration could alter species composition and architecture characteristics, and the release of extracellular organic carbon (EOC) and EPS from living algal cells in periphyton (Espeland and Wetzel, 2001; McCormick et al., 2001; Van der Grinten et al., 2004), which could subsequently strongly affect P_i removal or adsorption processes.

7.4 CONCLUSIONS

The ubiquitous and environmentally benign periphyton is a promising and available biomaterial for removing both organic and inorganic P from wastewater. Also, periphyton can produce large amounts of phosphatase that can facilitate the transformation of organic P to inorganic P. This conversion process is influenced by the concentration of organic P and periphyton biomass in solution. This study firstly demonstrated that the P removal by periphyton was dominated by adsorption. It was also suggested that the adsorption of P by periphyton has mechanistic relevance and exhibited physical characteristics. However, this bioadsorption process is distinct from nonbioadsorption, which is more influenced by environmental conditions such as temperature. These results also imply that the inclusion of a periphyton-based P adsorption system in natural wetlands will be of major benefit when protecting surface waters from wastewater pollution. This work also gives an insight into organic P conversion and removal processes in the presence of periphyton or other similar microbial aggregates, contributes to the full understanding of P biogeochemical circulation in aquatic systems, and provides kinetic data for the design of P removal, recovery, and reuse technologies based on the periphyton.

REFERENCES

Rejmánková, E., Komárková, J., 2000. A function of cyanobacterial mats in phosphorus-limited tropical wetlands. Hydrobiologia 431 (2–3), 135–153.

Aksakal, O., Ucun, H., 2010. Equilibrium, kinetic and thermodynamic studies of the biosorption of textile dye (Reactive Red 195) onto *Pinus sylvestris* L. Journal of Hazardous Materials 181 (1), 666–672.

Badawy, M.I., Ghaly, M.Y., Gad-Allah, T.A., 2006. Advanced oxidation processes for the removal of organophosphorus pesticides from wastewater. Desalination 194 (1–3), 166–175.

Balser, T.C., Wixon, D.L., 2009. Investigating biological control over soil carbon temperature sensitivity. Global Change Biology 15 (12), 2935–2949.

Battin, T.J., Kaplan, L.A., Denis Newbold, J., Hansen, C.M.E., 2003. Contributions of microbial biofilms to ecosystem processes in stream mesocosms. Nature 426 (27), 439–442.

Bentzen, E., Taylor, W., Millard, E., 1992. The importance of dissolved organic phosphorus to phosphorus uptake by limnetic plankton. Limnology and Oceanography 37 (2), 217–231.

de Beer, D., Stoodley, P., 2006. Microbial biofilms. Prokaryotes 1, 904–937.

De Beer, D., Stoodley, P., Roe, F., Lewandowski, Z., 1994. Effects of biofilm structures on oxygen distribution and mass transport. Biotechnology and Bioengineering 43 (11), 1131–1138.

Cotner Jr., J.B., Wetzel, R.G., 1992. Uptake of dissolved inorganic and organic phosphorus compounds by phytoplankton and bacterioplankton. Limnology and Oceanography 37, 232–243.

Dawood, S., Sen, T.K., 2012. Removal of anionic dye congo red from aqueous solution by raw pine and acid-treated pine cone powder as adsorbent: equilibrium, thermodynamic, kinetics, mechanism and process design. Water Research 46, 1933–1946.

DeNicola, D.M., 1996. Periphyton responses to temperature at different ecological levels. In: Stevenson, R.J. (Ed.), Algal Ecology: Freshwater Benthic Ecosystem. Academic Press, San Diego, pp. 149–181.

de-Bashan, L.E., Bashan, Y., 2004. Recent advances in removing phosphorus from wastewater and its future use as fertilizer (1997–2003). Water Research 38 (19), 4222–4246.

Dodds, W.K., 2003. The role of periphyton in phosphorus retention in shallow freshwater aquatic systems. Journal of Phycology 39 (5), 840–849.

Doğan, M., Alkan, M., Demirbas, Ö., Özdemir, Y., Özmetin, C., 2006. Adsorption kinetics of maxilon blue GRL onto sepiolite from aqueous solutions. Chemical Engineering Journal 124 (1–3), 89–101.

Dong, D., Li, Y., Zhang, J., Hua, X., 2003. Comparison of the adsorption of lead, cadmium, copper, zinc and barium to freshwater surface coatings. Chemosphere 51 (5), 369–373.

Donlan, R.M., 2002. Biofilms: microbial life on surfaces. Emerging Infectious Diseases 8, 881–890.

Drake, W.M., Scott, J.T., Evans-White, M., Haggard, B., Sharpley, A., Rogers, C.W., Grantz, E.M., 2012. The effect of periphyton stoichiometry and light on biological phosphorus immobilization and release in streams. Limnology 13 (1), 97–106.

Dyhrman, S.T., Chappell, P.D., Haley, S.T., Moffett, J.W., Orchard, E.D., Waterbury, J.B., Webb, E.A., 2006. Phosphonate utilization by the globally important marine diazotroph *Trichodesmium*. Nature 439 (7072), 68–71.

Ellwood, N.T.W., Di Pippo, F., Albertano, P., 2012. Phosphatase activities of cultured phototrophic biofilms. Water Research 46 (2), 378–386.

Espeland, E.M., Wetzel, R.G., 2001. Effects of photosynthesis on bacterial phosphatase production in biofilms. Microbial Ecology 42, 328–337.

Flemming, H.C., Neu, T.R., Wozniak, D.J., 2007. The EPS matrix: the "house of biofilm cells". Journal of Bacteriology 189 (22), 7945–7947.

Gatidou, G., Iatrou, E., 2011. Investigation of photodegradation and hydrolysis of selected substituted urea and organophosphate pesticides in water. Environmental Science and Pollution Research 18 (6), 949–957.

García-Calzón, J.A., Díaz-García, M.E., 2007. Characterization of binding sites in molecularly imprinted polymers. Sensors and Actuators B: Chemical 123 (2), 1180–1194.

Guzzon, A., Bohn, A., Diociaiuti, M., Albertano, P., 2008. Cultured phototrophic biofilms for phosphorus removal in wastewater treatment. Water Research 42 (16), 4357–4367.

Hameed, B.H., El-Khaiary, M.I., 2008. Kinetics and equilibrium studies of malachite green adsorption on rice straw-derived char. Journal of Hazardous Materials 153 (1–2), 701–708.

Herbes, S.E., Allen, H.E., Mancy, K.H., 1975. Enzymatic characterization of soluble organic phosphorus in lake water. Science 187 (4175), 432–434.

Ho, Y.S., Huang, C.T., Huang, H.W., 2002. Equilibrium sorption isotherm for metal ions on tree fern. Process Biochemistry 37 (12), 1421–1430.

Huang, C.-T., Xu, K.D., McFeters, G.A., Stewart, P.S., 1998. Spatial patterns of alkaline phosphatase expression within bacterial colonies and biofilms in response to phosphate starvation. Applied and Environmental Microbiology 64 (4), 1526–1531.

James, G.A., Beaudette, L., Costerton, J.W., 1995. Interspecies bacterial interactions in biofilms. Journal of Industrial Microbiology 15 (4), 257–262.

Kay, M., Darling, G.R., Holloway, S., White, J.A., Bird, D.M., 1995. Steering effects in non-activated adsorption. Chemical Physics Letters 245 (2–3), 311–318.

Kiran, I., Akar, T., Ozcan, A.S., Ozcan, A., Tunali, S., 2006. Biosorption kinetics and isotherm studies of Acid Red 57 by dried Cephalosporium aphidicola cells from aqueous solutions. Biochemical Engineering Journal 31, 197–203.

Li, J., Shimizu, K., Maseda, H., Lu, Z., Utsumi, M., Zhang, Z., Sugiura, N., 2012. Investigations into the biodegradation of microcystin-LR mediated by the biofilm in wintertime from a biological treatment facility in a drinking-water treatment plant. Bioresource Technology 106 (0), 27–35.

Li, J., Xing, X.-H., Wang, B.-Z., 2003. Characteristics of phosphorus removal from wastewater by biofilm sequencing batch reactor (SBR). Biochemical Engineering Journal 16 (3), 279–285.

Lu, H., Yang, L., Zhang, S., Wu, Y., 2014. The behavior of organic phosphorus under non-point source wastewater in the presence of phototrophic periphyton. PLoS One 9 (1), e85910.

Matheson, F.E., Quinn, J.M., Martin, M.L., 2012. Effects of irradiance on diel and seasonal patterns of nutrient uptake by stream periphyton. Freshwater Biology 57, 1617–1630.

McCormick, P.V., O'Dell, M.B., Shuford Iii, R.B.E., Backus, J.G., Kennedy, W.C., 2001. Periphyton responses to experimental phosphorus enrichment in a subtropical wetland. Aquatic Botany 71 (2), 119–139.

McCormick, P.V., Shuford, R.B.E., Chimney, M.J., 2006. Periphyton as a potential phosphorus sink in the everglades nutrient removal project. Ecological Engineering 27 (4), 279–289.

McKelvie, I.D., 2005. Separation, Preconcentration and Speciation of Organic Phosphorus in Environmental Samples. CAB International, London.

Minear, R.A., 1972. Characterization of naturally occurring dissolved organophosphorus compounds. Environmental Science and Technology 6 (5), 431–437.

Noe, G.B., Childers, D.L., Edwards, A.L., Gaiser, E., Jayachandran, K., Lee, D., Jones, R.D., 2002. Short-term changes in phosphorus storage in an oligotrophic Everglades wetland ecosystem receiving experimental nutrient enrichment. Biogeochemistry 59 (3), 239–267.

Ozacar, M., 2003. Equilibrium and kinetic modelling of adsorption of phosphorus on calcined alunite. Adsorption 9 (2), 125–132.

Özcan, A.S., Erdem, B., Özcan, A., 2005. Adsorption of Acid Blue 193 from aqueous solutions onto BTMA-bentonite. Colloids and Surfaces A: Physicochemical and Engineering Aspects 266 (1), 73–81.

Panswad, T., Doungchai, A., Anotai, J., 2003. Temperature effect on microbial community of enhanced biological phosphorus removal system. Water Research 37 (2), 409–415.

Pratt, C., Parsons, S.A., Soares, A., Martin, B.D., 2012. Biologically and chemically mediated adsorption and precipitation of phosphorus from wastewater. Current Opinion in Biotechnology 23, 1–7.

de los Ríos, A., Ascaso, C., Wierzchos, J., Fernández-Valiente, E., Quesada, A., 2004. Microstructural characterization of cyanobacterial mats from the McMurdo Ice Shelf, Antarctica. Applied and Environmental Microbiology 70 (1), 569–580.

Rittmann, B.E., Mayer, B., Westerhoff, P., Edwards, M., 2011. Capturing the lost phosphorus. Chemosphere 84, 846–853.

Roeselers, G., Loosdrecht, M., Muyzer, G., 2008. Phototrophic biofilms and their potential applications. Journal of Applied Phycology 20 (3), 227–235.

Sabater, S., Guasch, H., Roman, A., Muñoz, I., 2002. The effect of biological factors on the efficiency of river biofilms in improving water quality. Hydrobiologia 469, 149–156.

Saisho, D., Nakazono, M., Tsutsumi, N., Hirai, A., 2001. ATP synthesis inhibitors as well as respiratory inhibitors increase steady-state level of alternative oxidase mRNA in *Arabidopsis thaliana*. Journal of Plant Physiology 158 (2), 241–245.

Sanudo-Wilhelmy, S.A., 2006. Oceanography: a phosphate alternative. Nature 439 (7072), 25–26.

Scinto, L.J., Reddy, K.R., 2003. Biotic and abiotic uptake of phosphorus by periphyton in a subtropical freshwater wetland. Aquatic Botany 77 (3), 203–222.

Shilton, A.N., Powell, N., Guieysse, B., 2012. Plant based phosphorus recovery from wastewater via algae and macrophytes. Current Opinion in Biotechnology 23 (6), 884–889.

Smith, V.H., Schindler, D.W., 2009. Eutrophication science: where do we go from here? Trends in Ecology & Evolution 24 (4), 201–207.

Smith, V.H., Tilman, G.D., Nekola, J.C., 1999. Eutrophication: impacts of excess nutrient inputs on freshwater, marine, and terrestrial ecosystems. Environmental Pollution 100 (1–3), 179–196.

Sun, Q., Yang, L., 2003. The adsorption of basic dyes from aqueous solution on modified peat–resin particle. Water Research 37 (7), 1535–1544.

Tang, H., Zhou, W., Zhang, L., 2012. Adsorption isotherms and kinetics studies of malachite green on chitin hydrogels. Journal of Hazardous Materials 209–210, 218–225.

Turner, B.L., Cade-Menun, B.J., Condron, L.M., Newman, S., 2005. Extraction of soil organic phosphorus. Talanta 66 (2), 294–306.

Van der Grinten, E., Janssen, M., Simis, S.G.H., Barranguet, C., Admiraal, W., 2004. Phosphate regime structures species composition in cultured phototrophic biofilms. Freshwater Biology 49 (4), 369–381.

Wetzel, R.G., 2001. Limnology: Lake and River Ecosystems. Academic Press, San Diego.

Wicke, D., Böckelmann, U., Reemtsma, T., 2008. Environmental influences on the partitioning and diffusion of hydrophobic organic contaminants in microbial biofilms. Environmental Science and Technology 42 (6), 1990–1996.

Wimpenny, J.W.T., Colasanti, R., 1997. A unifying hypothesis for the structure of microbial biofilms based on cellular automaton models. FEMS Microbiology Ecology 22 (1), 1–16.

Writer, J.H., Ryan, J.N., Barber, L.B., 2011. Role of biofilms in sorptive removal of steroidal hormones and 4-nonylphenol compounds from streams. Environmental Science and Technology 45, 7275–7283.

Wu, Y., Feng, M., Liu, J., Zhao, Y., 2005. Effects of polyaluminium chloride and copper sulfate on phosphorus and UV254 under different anoxic levels. Fresenius Environmental Bulletin 13 (5), 406–412.

Wu, Y., He, J., Yang, L., 2010a. Evaluating adsorption and biodegradation mechanisms during the removal of microcystin-RR by periphyton. Environmental Science and Technology 44 (16), 6319–6324.

Wu, Y., Kerr, P.G., Hu, Z., Yang, L., 2010b. Removal of cyanobacterial bloom from a biopond–wetland system and the associated response of zoobenthic diversity. Bioresource Technology 101 (11), 3903–3908.

Wu, Y., Li, T., Yang, L., 2012. Mechanisms of removing pollutants from aqueous solutions by microorganisms and their aggregates: a review. Bioresource Technology 107 (0), 10–18.

Wu, Y., Zhang, S., Zhao, H., Yang, L., 2010c. Environmentally benign periphyton bioreactors for controlling cyanobacterial growth. Bioresource Technology 101 (24), 9681–9687.

Chapter 8

Periphyton: An Interface Between Sediments and Overlying Water

8.1 INTRODUCTION

The quality of drinking water has deteriorated with increasing inputs of external pollution loads (Rectenwald and Drenner, 2000). As a result, eutrophication has become the primary problem in surface waters around the world (Chen et al., 2016; Zhao et al., 2013). Eutrophic waters experience nuisance algal blooms, which can result in loss of water clarity, and taste and odor problems (Chen et al., 2016; Loubet et al., 2016; Schindler, 1977, 1998). The primary causative agent of eutrophication and algal blooms is excessive loading of nutrients (Bothwell, 1989; Dodds et al., 2002; El-Shafai et al., 2007), in particular phosphorus (P) (Schindler, 1977, 1998). The P cycle differs significantly from the nitrogen cycle since it exhibits a multiple compound sedimentary phase (Adey et al., 1993; Carlton and Wetzel, 1988; Ding et al., 2006). Phosphate is present in the sediment matrix in the form of calcium, iron, or aluminum precipitates or it is adsorbed on the surface of minerals (Lu et al., 2016). Despite reduced external P loading, many waters have failed to recover from eutrophication due to P release from sediment during summer (Søndergaard et al., 2003) and is especially pronounced in waters with a long residence time (Søndergaard et al., 2003). In addition, P is often a limiting nutrient for bacterioplankton growth (Cotner and Wetzel, 1992; Hörnström, 2002) and dominates whether the harmful algal blooms occur.

Many technologies have been used in situ to reduce P loads in water via controlling P release from sediments. These include the application of lake marl (Stüben et al., 1998) and gypsum-based techniques (Eila et al., 2003) for reducing P loadings in situ. Coprecipitation of nutrients with calcite, aluminum sulfate, or iron chloride have also been applied to immobilize P in sediments (Jarvie et al., 2002; Plant and House, 2002; Woodruff et al., 1999). These chemical treatments are often associated with changes in pH values or salinity, which may threaten life in the waters (Agostinho et al., 2004; Chen et al., 2016; Dillon, 2000; Steinman, 1996), and often involve expensive

Periphyton. http://dx.doi.org/10.1016/B978-0-12-801077-8.00008-9

chemicals. The use of microorganisms (Azim et al., 2003; Cooke, 1956; Cross et al., 2005) and ecological water treatment systems, such as the periphyton–fish system (Rectenwald and Drenner, 2000), for removal of P has been studied. Results demonstrated that the effects of microorganism use were not steady and the periphyton–fish system could not be used in drinking waters due to the feces excreted by the fish.

Periphyton is a complex mixture of algae, cyanobacteria, heterotrophic microbes, and detritus that is attached to submerged surfaces in most aquatic ecosystems, serving as an important food source for invertebrates, tadpoles, and fish (Azim et al., 2005). Periphyton, as a type of microbial assemblages, is very similar to the composition and structure of biofilm (Gottlieb et al., 2005; Sladeckova, 1962). The organic and inorganic P in the periphyton varied with the changes in periphyton biomass. This indicates that the P cycle in water and sediments can be affected by the periphyton growing on any substrate through P transfer processes including adsorption, assimilation, ligand exchange, and enzymatic hydrolysis (Bushong and Bachmann, 1989; Sainto and Reddy, 2003). However, the present studies regarding P mitigation pay more attention to the release of P from sediments (or sediment porewater) into overlying water (Monbet et al., 2007; Reitzel et al., 2005), ignoring the presence of periphyton attached to the surfaces of sediments in natural waters.

One of the objectives of this chapter is to examine whether the P release between sediments and overlying water is affected by the introduction of periphyton. As the P released from sediments is directly corrected with the growth of algae, the second objective of this chapter is to investigate whether the algal growth was inhibited by periphyton through affecting the P release and to what extent.

8.2 MATERIAL AND METHODS

8.2.1 Experimental Processes

The parameters of the water and sediments used in the experiment are presented in Table 8.1. The water was hypereutrophic and the sediments in the lake demonstrated high P loading.

The experimental microcosm was made of glass and its dimensions were as follows: length $\times$ width $\times$ height $=$ 120 cm $\times$ 100 cm $\times$ 100 cm. The practical use volume was 1.0×10^6 cm^3 with sediments of 15 cm thickness at the bottom and 85 cm deep overlying water. The microcosms were wrapped with shade cloth to avoid direct exposure to the sun and put outside. On rainy days, the tops of these microcosms were not covered.

Three microcosms were used for this experiment. One was used as the control and the other two for different treatments. The collected lake water was added into the microcosms after the sediments were laid on the bottoms of the microcosms. After the microcosms were static for 1 day, the periphyton that

TABLE 8.1 The Parameters of Water and Sediments Used for Experimentation (Means, $n = 3$)

Items	Parameters	Units	Values
Sediments	Labile-P	$mg\ g^{-1}$	0.27
	Fe/Al-P	$mg\ g^{-1}$	1.83
	Ca-P	$mg\ g^{-1}$	3.94
	Residual-P	$mg\ g^{-1}$	4.71
	Total phosphorus (TP)	$mg\ g^{-1}$	10.45
Water	Transparency	cm	21.00
	pH		8.15
	Dissolved oxygen (DO)	$mg\ L^{-1}$	7.14
	Total phosphorus (TP)	$mg\ L^{-1}$	0.64
	Total nitrogen (TN)	$mg\ L^{-1}$	24.58

had been cultured for 270 days was fixed to the water surface under 30 cm in the microcosms with their substrates, artificial aquatic mats (AAMs) and industrial soft carriers (ISCs). AAMs are made from the inert material polyethylene and ISCs are made from inert polypropylene. The periphyton growing on AAM and ISC surfaces were designated as AAM- and ISC-periphyton, respectively. Each microcosm (AAM and ISC periphyton) contained 1600 g (filtration weight at 25–30°C for 2 h, moisture 85 ± 5%) fresh periphyton. Nothing was added into the third microcosm as it was used as the control. Aeration was applied at the interface between sediments and water by a common piscicultural pump (power 3.2 W) to keep dissolved oxygen in the range 8.5–9.5 mg L^{-1} during the whole experimental period. In the following days of the experiment, nothing was done except data collection. During the whole experimental period, the temperature coincided with the ambient temperature ranging from 22–38°C. The control and the AAM and ISC periphyton treatments were performed in triplicate.

Two control experiments were conducted. One experiment was designed to determine whether the two periphyton substrates, AAMs and ISCs, had affected P movement and cyanobacterial growth. A 1000-mL measuring cylinder with 150 mL sediments at the bottom and 850 mL of overlying water acted as the experimental microcosm. Then the AAMs and ISCs were fixed 300 mL from the water surface. To inhibit spontaneous periphyton formation 2.0 g Na_3N was added directly into each measuring cylinder. No substrates were added to the control.

The other control experiment was designed to determine the extent of cyanobacterial growth inhibition by periphyton. A 1000-mL measuring cylinder with 150-mL sediments at the bottom and 850 mL of overlying BG-11 medium (Rippka et al., 1979) containing *M. aeruginosa* cells acted as the experimental microcosm. Then 1.6 g (filtration weight at 25–30°C for 2 h, moisture 85 $\pm$ 5%) of fresh periphyton with ISCs was directly fixed 3 cm under the BG-11 media surface. No periphyton was added to this control. The BG-11 media were prepared as below:

(A) Stock solutions for BG-11:

Stock 1:
Na_2MG EDTA 0.1 g L^{-1}
Ferric ammonium citrate 0.6 g L^{-1}
Citric acid $\cdot 1H_2O$ 0.6 g L^{-1}
$CaCl_2 \cdot 2H_2O$ 3.6 g L^{-1}
Filter into a sterile bottle or autoclave

Stock 2:
$MgSO_4 \cdot 7H_2O$ 75 g L^{-1}
Filter into a sterile bottle or autoclave

Stock 3:
$K_2HPO_4 \cdot 3H_2O$ 4.0 g L^{-1} or K_2HPO_4—3.05 g L^{-1}
Filter into a sterile bottle or autoclave

Stock 5 (microelements):
$H_3BO_3 \cdot$ 2.86 g L^{-1}
$MnCl_2 \cdot 4H_2O$ 1.81 g L^{-1}
$ZnSO_4 \cdot 7H_2O$ 0.222 g L^{-1}
$CuSO_4 \cdot 5H_2O$ 0.079 g L^{-1}
$COCl_2 \cdot 6H_2O$ 0.050 g L^{-1}
$NaMoO_4 \cdot 2H_2O$ 0.391 g L^{-1} or MoO_4 (85%)—0.018 g L^{-1}

(B) For basic BG11 medium the following stock solutions were combined:

Stock solution per L of medium
Stock 1—10 mL
Stock 2—10 mL
Stock 3—10 mL
Na_2CO_3—0.02 g
Stock 5—1.0 mL
$NaNO_3$—1.5 g

The stocks were combined and pH adjusted to 7.5 (using 1.0 N HCl). Aliquots of 50 mL/125 mL flask were added to flasks with cotton stoppers and autoclaved. After autoclaving and cooling the pH was about 7.1.

The sediments and overlying water used in the control experiments were as in Table 8.1. The control experiments were conducted in triplicate. All measuring cylinders were incubated at 28 ± 1°C with a light intensity of 2500 lux under 12/12 h of a light/dark cycle.

8.2.2 Sample and Analyses

Sediment samples: Total P (TP) and P fractions were determined for the freeze-dried sediment collected from the bottom of the microcosm using a self-made core sampler. Total P was measured after 1.0 g of freeze-dried sediments was washed at 500°C and acid dissolution ($HF + HNO_3$) according to Rao and Reddi (1990). Sequential phosphate extraction of 1.0 g freeze-dried sediments was undertaken following the procedure used by Williams et al. (1976), as modified by Hieltjes and Lijklema (1980), to divide the major reservoirs of sedimentary P into five pools (Fig. 8.1). (1) The labile-P was extracted by 1 mol L^{-1} NH_4Cl solution at pH 7 for 2.2 h; (2) the phosphate adsorbed to metal oxides (mainly the Fe, Al compound), was extracted by 0.1 M NaOH solution for 16 h; (3) the calcium-bound-phosphate (Ca-P) fraction, composed of apatites (hence the current term, AP) and of carbonate-associated P, was extracted by 0.5 M HCl solution for 16 h; (4) residual P (Residual-P), which cannot be extracted by common chemicals, was extracted by 1 M HCl solution after ashing at 550°C; and (5) the nonapatite inorganic phosphate (NAIP) fraction, which represents all inorganic forms of P other than those bound to calcium (including apatite), i.e., Fe- and Al-bound P (Fe/Al-P) and labile P (labile P), so-called exchange phosphate (exch-P). All analyses were performed in triplicate.

8.2.2.1 Water Samples

The water samples were collected from 0.5 m under the pond water surface. The transparency of pond water was determined by using a Secchi disk

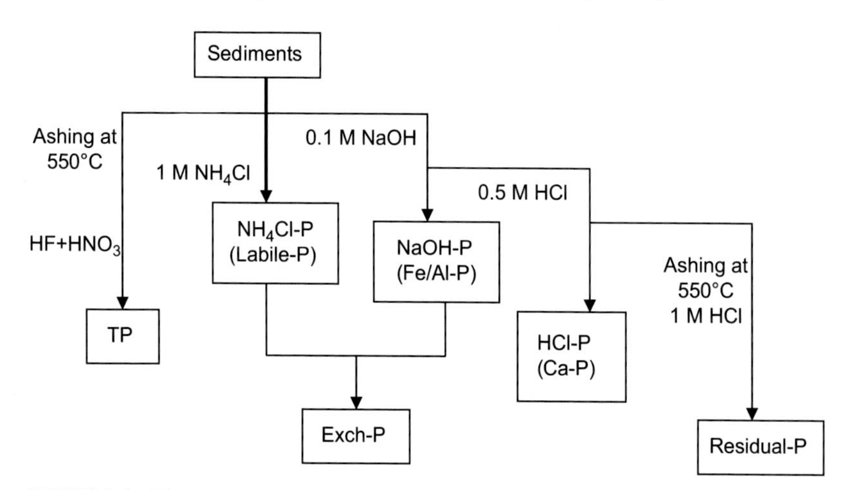

FIGURE 8.1 The extraction and analytical processes of P fraction in sediments.

(APHA et al., 1998). The chlorophyll-a content was determined after extraction of the filtered algal mat with 90% acetone as per APHA (APHA et al., 1998). Total phosphorus (TP) was measured calorimetrically by the persulfate digestion–molybdophosphate reaction method (APHA et al., 1998). The dissolved total phosphorus (DTP) was measured in the same way after the water was filtered through a 0.45-μm GF/C filter (APHA et al., 1998). Dissolved oxygen (DO) in the overlying water was measured in situ by a meter (YSI 52 dissolved oxygen meter). The illuminances in the interfaces of the overlying water and sediments in the microcosms were determined using an illuminometer (ZDS-10W-2D, Shanghai Yueci Electron Sci-Tech Ltd., Co.). The species of phytoplankton in the overlying water were determined using a Phyto-PAM phytoplankton analyzer (WALZ, Germany).

8.2.2.2 Periphyton Samples

The periphyton was peeled off by ultrasonication rather than a knife to avoid breaking microorganism cell walls. The periphyton biomass was weighed after the periphyton was filtered at 25–30°C for 2 h, moisture about 85 ± 5%. The total periphyton biomasses growing on the surface of AAM and ISC substrates were calculated according to the periphyton biomass of every square surface area and the total surface area of AAM and ISC substrates. The total P content in periphyton (Periphyton-TP) was measured after 1.0 g freeze-dried periphyton was ashed at 500°C and treated by acid dissolution ($HF + HNO_3$) according to Rao and Reddi (1990). The P uptake by periphyton (Uptake-P) was measured using the same method after the periphyton was dipped in double-distilled water until the dissolved P in the double-distilled water arrived at the balance point of adsorption and desorption. These analyses were performed in triplicate.

The bacteria in periphyton were characterized by ERIC-PCR as follows. One gram of periphyton (filtration weight at 25–30°C for 2 h, moisture 85 ± 5%) was centrifuged for 10 min. DNA was extracted from periphyton by the method of Giovanni et al. (1999). Briefly, a buffer solution that contained 10 mM Tris–HCl, 1 mM EDTA (pH 8.0), 10% sodium dodecyl sulfate (Calbiochem), a 20 mg mL^{-1} proteinase K solution (Promega), and 0.25 mg mL^{-1} of lysozyme (Fluka, BioChemika) were added to the pellets. The mixture was incubated for 1 h at 37°C, and the enzymes were then inactivated by heating at 100°C for 10 min. The DNA was pelleted by centrifugation at 9000 × g for 30 min, precipitated overnight at −20°C with 2.5 mL of absolute ethanol, and then air-dried. The pellets were resuspended in 50 μL of sterile distilled water, and the DNA was quantified by measuring the optical density at 260 nm.

PCR amplification of the flaA gene was performed by using the forward oligonucleotide primer (5′AGCTCTTAGCTCCATGAGTT3′) and the reverse primer (5′ACATTGTAGCTAAGGCGACT 3′) (Dons et al., 1992; Gray and Kroll,

1995). For ERIC-PCR, the forward primer (5′-ATGTAAGCTCCTGGGGA-TTCAC-3′) and the reverse primer (5′-AAGTAAGTGACTGGGG-TGAGCG) were used (Jersek et al., 1999). Reactions were performed in a 50-μL volume in a Perkin Elmer thermocycler (Gene Amp PCR System 2400). Reaction mixtures contained 1 μM of each primer, 50−100 ng of genomic DNA, 1.5 mM $MgCl_2$, each deoxynucleoside triphosphate (dATP, dTTP, dCTP, dGTP; Promega, USA) at a concentration of 0.2 mM, and 1 U of Taq DNA polymerase (Promega, USA). For flaA amplification, cycle conditions of 1 cycle at 94°C for 2 min; 35 cycles of 94°C for 1 min, 65°C for 30 s, and 72°C for 30 s; and 1 cycle at 72°C for 10 min were used. For ERIC-PCR, the amplification was accomplished by running 30 cycles of denaturing at 90°C for 30 s, annealing at 50°C for 30 s and extending at 50°C for 30 s, initial denaturing at 95°C for 5 min, denaturing extending at 72°C for 8 min. All amplification products were electrophoresized in agarose gels, stained with ethidium bromide, detected under a short-wavelength UV light source, and photographed with a Polaroid 667 camera. A 100-bp DNA Ladder (Promega) was used as a molecular size marker.

The Shannon−Weaver diversity index was calculated to assess the bacterial diversity in the periphyton.

8.2.3 Statistics

Statistical analysis (one-way ANOVA and regression analysis) using the software package SPSS 12.0 was performed. The level of statistical significance was accepted as $P < 0.05$.

8.3 RESULT

8.3.1 Periphyton Characteristics

To investigate the dominant species composition, the periphyton was observed under optional microscope and scanning electron microscope. The results showed that the heterotrophs in both AAM and ISC periphytons were dominated by bacteria. The algal communities in both AAM and ISC periphyton were dominated by diatoms (Fig. 8.2A). The lane quantities (AAM periphyton and ISC periphyton) were similar at the same period of incubation (Fig. 8.2B), which means that the microbial communities of both periphyton were similar. The average Shannon−Weaver diversity indices of bacteria were 2.03 and 1.97 for the AAM and ISC periphyton, respectively. Both indices were always greater than 1.90, implying that the periphyton communities were abundant and tended to form stable microecosystems (Giovanni et al., 1999).

The total P content in the periphyton (Periphyton-TP) was 0.59 mg g^{-1} and the Uptake-P content was 0.25 mg g^{-1}. According to the specific surface area of AAMs and ISCs, the total periphyton biomasses in AAM and ISC

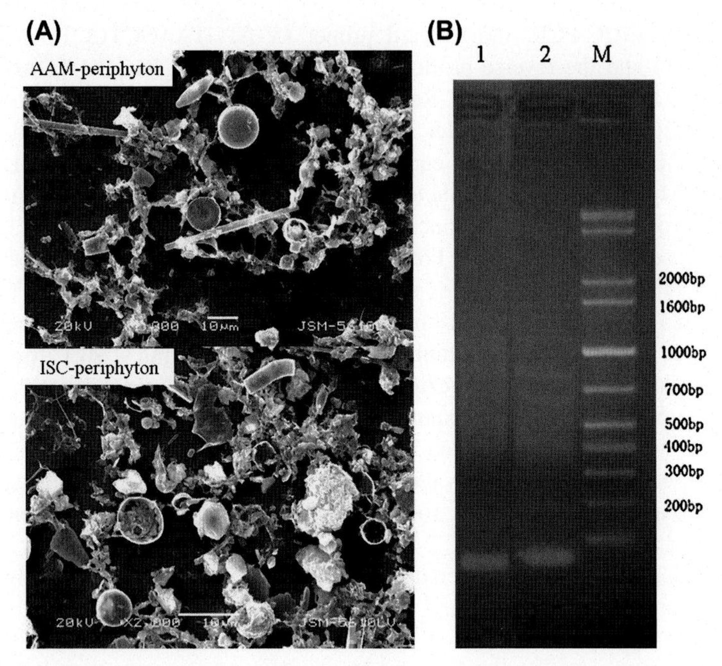

FIGURE 8.2 (A) The structure of the periphyton recorded by scanning electric microscope (SEM); (B) ERIC-PCR fingerprints of periphyton (lane 1: AAM periphyton; lane 2: ISC periphyton; M: kb plus DNA ladder).

periphyton microcosms were estimated, increasing from 1600 to 2650 g and 2410 g between the beginning and the end of the experiment for AAM and ISC periphyton microcosms, respectively.

8.3.2 Changes in Dissolved Total Phosphorus and Total Phosphorus in Overlying Water

The DTP TP in overlying water are directly associated with P release in sediments (Cao et al., 2011; Lin et al., 2016). Thus, DTP and TP contents in overlying water were determined during the experiment. The results showed that the average DTP concentrations in the water of AAM and ISC periphyton treatments decreased with time, and arrived at a stable range from 0.12 to 0.14 mg L^{-1} after 23 days. The average concentrations of TP in the water of periphyton treatments (AAM and ISC periphyton) also decreased with time, while large changes in TP concentration occurred after day 23 (Fig. 8.3). The average DTP and TP concentrations in the water in the control all increased markedly after day 23, corresponding with the decrease in DTP and TP in the sediments from the beginning of the experiment to day 23 day (Fig. 8.3). This implied that the P released from sediments had increased the DTP and TP concentrations in water in the control. Statistics showed that the average TD

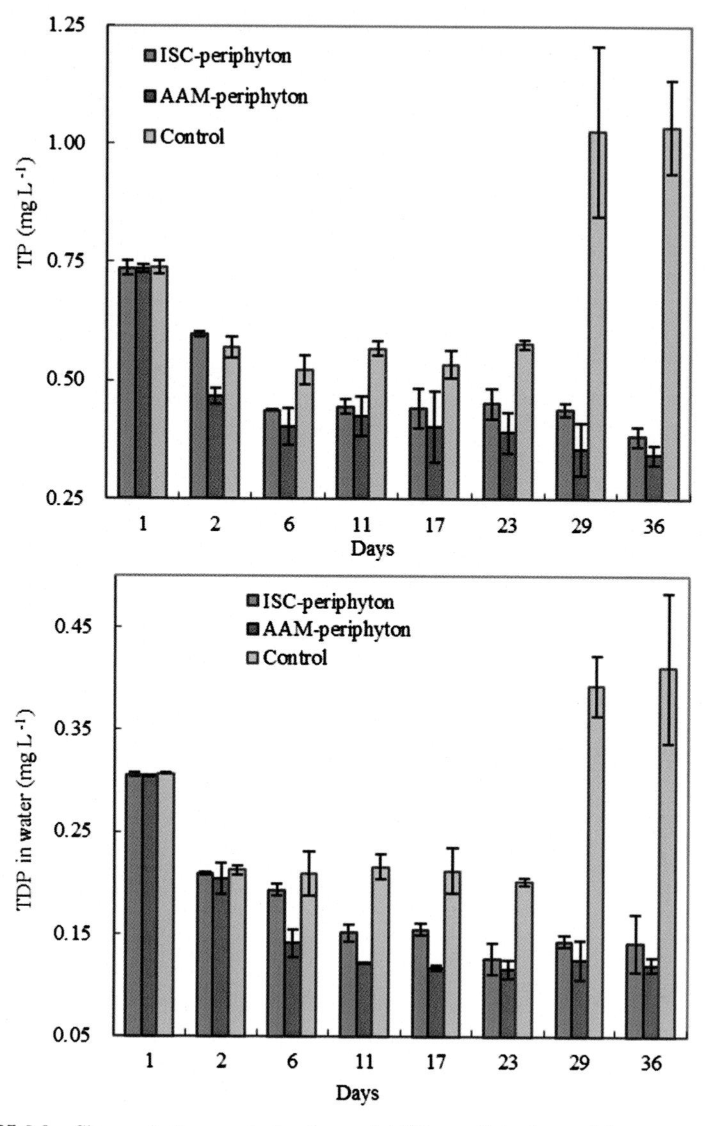

FIGURE 8.3 Changes in P concentration (mean ± 1 SD, $n = 3$) in the overlying water.

and DTP in the overlying water of each periphyton treatment were significantly different from that of the control ($P < 0.05$).

8.3.3 Phosphorus Release From Sediments

The largest proportion of the TP in the experiment sediments was Ca-P, followed by Residual-P. Their concentrations were maintained in the range

6.5–7.2 and 2.2–2.9 mg g^{-1} for Ca-P and Residual-P, respectively. There were no significant differences in the average concentrations of Ca-P and Residual-P between treatments (AAM and ISC periphyton) and the control.

The average TP concentration in the control sediments decreased after periphyton was introduced for 25 days. The average TP concentrations in the sediments decreased from 11.8 to 10.2 and 10.4 mg g^{-1} after 25 and 38 days, respectively (Fig. 8.4). The TP concentrations in the AAM and ISC periphyton treatments also decreased over time (Fig. 8.4), but the magnitude of the decreases were lower than those of the control.

The average Fe/Al-P concentration in the control sediments markedly decreased, from 1.9 to 0.38 mg g^{-1} between the beginning and the end of the experiment. In addition, the overall average Fe/Al-P concentrations in the sediments of each treatment (AAM and ISC periphyton) were significantly different from the overall average Fe/Al-P concentrations in the control sediments control ($P < 0.05$).

The Fe/Al-P was the largest portion of exch-P, accounting for 86.4% at the beginning of the experiment. The Fe/Al-P concentrations decreased with time not only in the control but also in the periphyton microcosms; in particular, the average exch-P concentration in the controls between day 25 and day 38 decreased by more than half of the original concentration (2.2 mg g^{-1}) (Fig. 8.4). The overall average concentrations of exch-P in the sediments

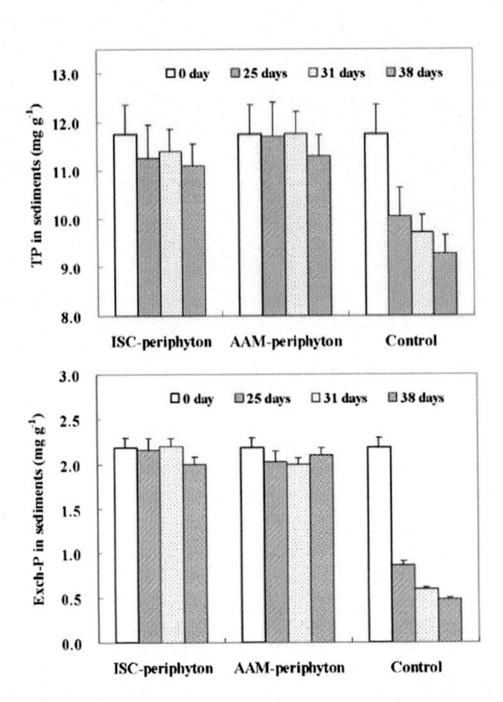

FIGURE 8.4 Changes (mean ± 1 SD, $n = 3$) in total P and exch-P in the sediments.

between each treatment (AAM and ISC periphyton) were significantly different from the control ($P < 0.05$).

8.3.4 Changes in Chlorophyll-a Concentrations

Cyanobacterial growth can be represented by chlorophyll-a concentration in overlying water, thus, the chlorophyll-a levels in the overlying water were determined to investigate whether cyanobacterial blooms occurred. During the first 23 days, the transparencies increased from 6 to 65 and 80 cm under the treatments of ISCs and AAMs, respectively (Fig. 8.5). However, the initial microcosm experiment showed that the algae, cyanobacteria accounting for 100% of the algal species, grew well in the control but grew poorly in AAM and ISC periphyton treatments (Fig. 8.5). Significant differences in growth were observed between periphyton and nonperiphyton treatments after 5 days of experiment ($P < 0.05$). Investigation showed that a cyanobacterial bloom had formed in the control microcosm, a dense layer of cyanobacterial biomass floated at the surface of the overlying water.

There were two good relationships between the average concentration of TP and chlorophyll-a in the control during the whole experimental period ($n = 8$, $R^2 = 0.80$, $P < 0.05$), and between the average concentration of DTP and chlorophyll-a during the period from the day 6 to day 38 ($n = 6$, $R^2 = 0.63$, $P < 0.05$). This implies that the higher the TDP concentration in the overlying water, the better the cyanobacterial growth. Furthermore, there were negative linear relationships between the exch-P in the sediments and chlorophyll-a (or DTP or TP) in the overlying water in the control microcosm ($n = 5$, $R^2 > 0.91$, $P < 0.05$), implying that P release from sediment occurred once the P concentration in overlying water could not meet the demands of cyanobacterial growth.

8.3.5 Control Experiment

Some middle- and large-sized protozoa are capable of capturing young cyanobacteria. Thus, a control experiment without the presence of protozoans was conducted to exclude the effects of protozoa on the growth of cyanobacteria. Microscope observation ($\times 1000$) did not find any microorganisms (i.e., protozoans) growing on the AAM and ISC surfaces at the end of the experiment, indicating no periphyton spontaneously formed during the experiment. The change trend lines of the average TDP and chlorophyll-a concentrations in the overlying water of both control and AAM and ISC treatments were very similar (Fig. 8.6). Statistical analysis showed that the average TDP and chlorophyll-a concentrations of the control were not significantly different from those of the AAM or ISC treatments ($P > 0.05$), which indicated that AAMs or ISCs could not directly affect P release and cyanobacterial growth.

When the original chlorophyll-a concentration of *M. aeruginosa* in the overlying water was more than 119.32 $\mu g\ L^{-1}$, *M. aeruginosa* growth increased with time (Fig. 8.7), which showed that the inhibitory effect of

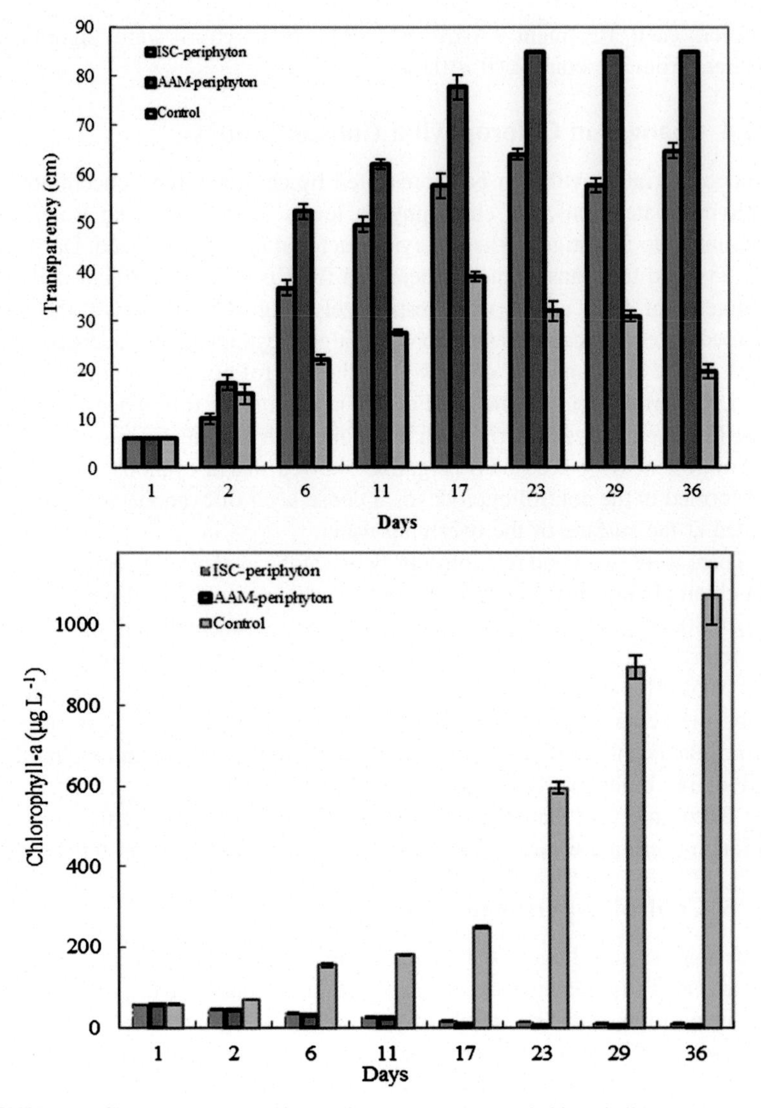

FIGURE 8.5 Changes (mean ± 1 SD, $n = 3$) in transparency and chlorophyll-a content.

periphyton on cyanobacterial growth was affected by the original cyanobacterial concentration.

8.4 DISCUSSION

Phosphorus is an essential element for the synthesis of nucleic acids, ATP, and proteins and is also the necessary element for cell division and growth of cyanobacteria (Conley et al., 2009). Thus, it is important to investigate the release of P

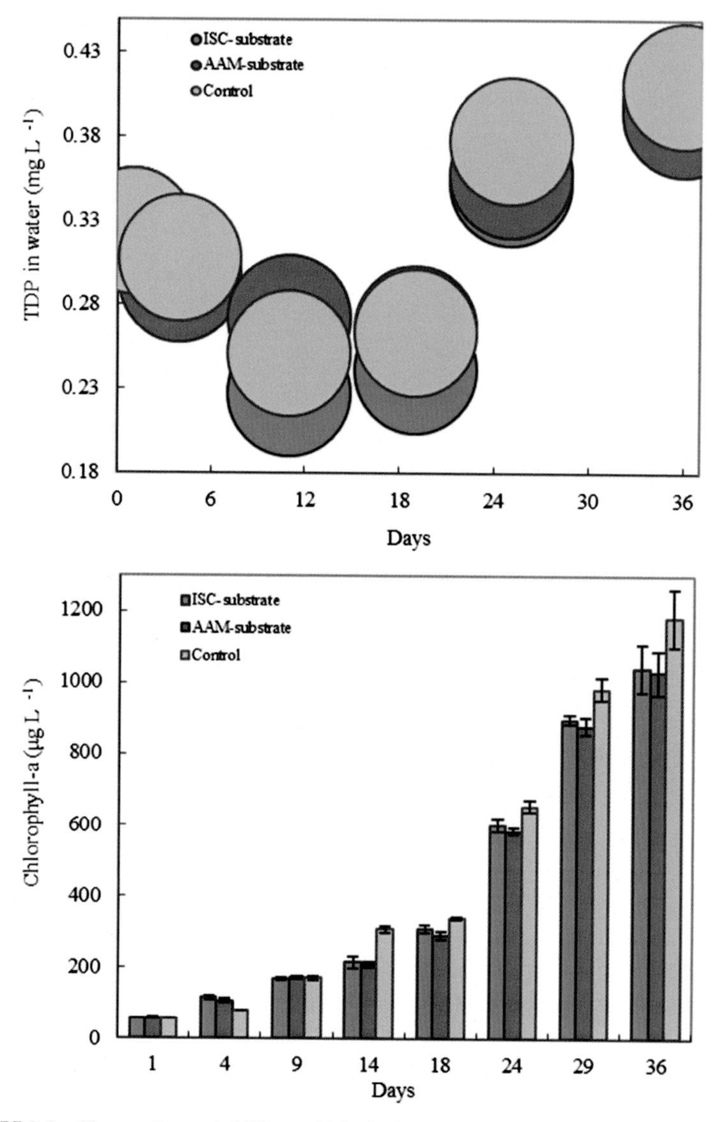

FIGURE 8.6 Changes (mean ± 1 SD, $n = 3$) in the DTP and chlorophyll-a concentrations in the overlying water of the control experiment in the absence of periphyton.

from sediments to overlying water in the presence of periphyton. Investigation showed that the periphyton plays an important role in regulating the P transportation between sediments and overlying water, and directly affects cyanobacterial biomass.

The increase of transparency in the control microcosm from day 0 to day 16 of the experiment was due to coprecipitation (Leandrini and Rodrigues,

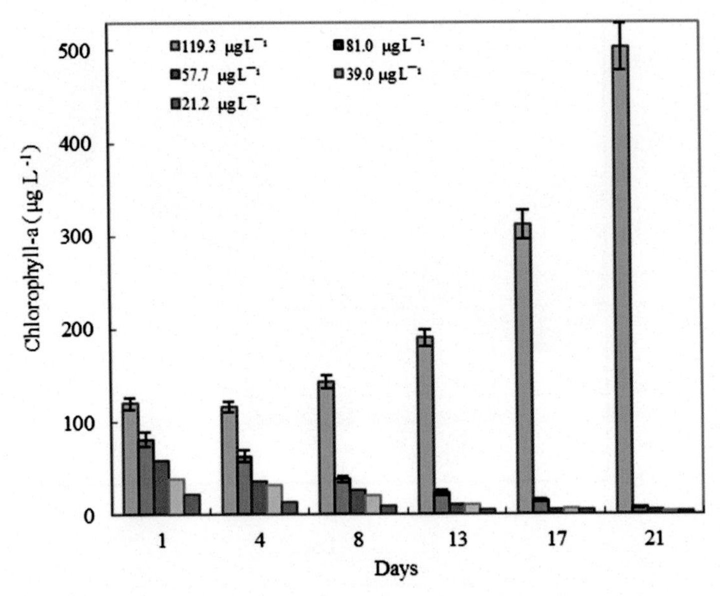

FIGURE 8.7 Differences (mean ± 1 SD, $n = 3$) in the inhibition of *M. aeruginosa* growth by periphyton with the original *M. aeruginosa* concentrations.

2008; Plant and House, 2002), while the decrease in the transparency after day 16 was due to the cyanobacterial bloom, in which the chlorophyll-a content in control microcosm increased from 57.5 to 1075.2 $\mu g\ L^{-1}$. The sustained decrease of chlorophyll-a concentrations in the overlying water in both treatments (AAM and ISC periphyton) indicated that the cyanobacterial growth was inhibited by the introduction of periphyton.

The decreases in the TP and DTP concentrations in the overlying water in the periphyton microcosms were partly attributed to the uptake of P by periphyton, which resulted in phosphorus-limited cyanobacterial growth to a certain extent. According to the conservation law of materials, the amount of P released from sediment in the AAM and ISC periphyton microcosms was not equal to the total amount of the TP in the corresponding overlying water and Uptake-P in the periphyton. Periphyton often acts as a role of short-term sink for P (McCormick, 1994; McCormick et al., 1997, 2001, 2006) and plays an important role in removal, uptake, or transformation of biologically available solute P (Bushong and Bachmann, 1989; Sainto and Reddy, 2003). During the experimental period, the TP loadings accumulated in periphyton increased from 940.8 to 1558.2 and 1417.1 mg for AAM and ISC periphyton treatments, respectively. The other reason for the decrease in TP and DTP content in the overlying water in the periphyton microcosms was P adsorption by periphyton (periphyton aggregates). The structure of the periphyton aggregates was very similar to that of biofilms, which ranged from patchy monolayers to filamentous accretions during different phases of biofilm formation (Gottlieb et al., 2005, 2006; Wolf et al., 2007), with many adsorption points (cavities) for nutrients.

The increases in the TP and DTP contents in the overlying water in the control microcosm were supplied by P released from sediments during the algal bloom from day 23 to day 38 of the experiment. It has been reported that some algal bacteria, like cyanobacteria, could induce the production of alkaline phosphatase (Agostinho et al., 2004; Carey et al., 2007; Roelke et al., 1997; Sterner et al., 1997), which led to more organic and polyinorganic P in the water (Jones, 1972; Lu et al., 2014). Increasing release of P from sediment leads to increasing TP and DTP contents in the overlying water and results in larger differences in P concentrations between overlying water and sediment pore water (Gonsiorczyk et al., 2001). The linear relationship between TP (or DTP) in the overlying water and chlorophyll-a concentration indicated that cyanobacterial blooms could lead to (even accelerate) P release from the sediments in the control microcosm.

The amount of P released in the control microcosm was dependent on exch-P release from sediments, accounting for 68.4%. This is inconsistent with the driving process for the rapid P release being the remineralization of organic P (Hupfer et al., 2004). This is because the experimental sediments were fresh and the dissolved oxygen contents in overlying water in the control microcosm were kept at 8.5–9.5 mg L^{-1} during the experiment, levels which resisted the mineralization process of organic P (Gonsiorczyk et al., 2001).

In the AAM and ISC periphyton treatments the periphyton acted as a P sink that could maintain the balance in overlying water of P concentration (Bushong and Bachmann, 1989; McCormick et al., 2006), resulting in the balance of P concentrations between sediments and overlying water. Therefore, no P release occurred from sediments in periphyton treatments.

The negative linear relationship between the exch-P concentration in the sediments and the chlorophyll-a, TP, and DTP in the overlying water in the control microcosm showed that the released exch-P from the sediments was the main source of P for cyanobacterial growth in the absence of periphyton. In addition, the significant difference of exch-P in the sediments between each periphyton treatment and the control indicated that the introduction of the periphyton could positively retard the exch-P release from sediments into overlying water. Furthermore, the significant differences in the TP and the DTP in the overlying water between the periphyton treatments and the control showed that the application of periphyton decreased the P content in the overlying water through retarding P release as well as uptaking and adsorbing P from the water.

However, the DTP contents of the overlying water of the periphyton microcosms from day 25 to day 38, ranging from 0.12–0.14 mg L^{-1}, were sufficient to meet the demand of cyanobacterial growth (Lehtiniemi et al.,

2002; Stewart and Falconer, 2008; Wu et al., 2010, 2011). This suggested that the inhibition of cyanobacterial growth was not directly P-limited in the treatment microcosms (AAM and ISC periphyton) and that, based on available data, light availability was likely to be a factor inhibiting cyanobacterial growth. Indeed, other studies (e.g., Bushong and Bachmann, 1989; Hepinstall and Fuller, 1994). found that nutrient increases alone may not excessively limit algal growth in situations where light is the limiting factor. During the experimental period the illuminances in the interface of the overlying water and sediments in the AAM and ISC periphyton microcosms were kept between 600–5500 lux, which met the demand of cyanobacterial growth (Lehtiniemi et al., 2002; Stewart and Falconer, 2008; Wu et al., 2010, 2011).

In addition, the control experiment excluded the effects of AAM and ISC substrates on P release and cyanobacterial growth. It implies that cyanobacterial growth in the periphyton treatment microcosms could not contribute to the adsorption of cyanobacterial cells by periphyton. One reason is that the increase of the average chlorophyll-a contents in the overlying water in the periphyton treatment microcosms was very trivial, only from 48.9 to 38.2 and 39.3 mg between day 0 and day 2 of the experiment for AAM and ISC periphyton treatments, respectively. This was inconsistent with observations that physical adsorption was a rapid process, often in a short time (less than 24 h) when material similar to periphyton biofilms acted as absorbent (Babica et al., 2005). The other reason is that the excessive cyanobacterial growth in the overlying water of the periphyton microcosms occurred once the cyanobacteria desorbed from the periphyton, if adsorption had occurred between periphyton and cyanobacteria.

Another mechanism inhibiting cyanobacterial growth in the periphyton microcosms should be considered after excluding other possible mechanisms. Allelopathy, defined as the production and release of biochemical substances by one organism that influences others positively or negatively (Rice, 1974), has been frequently reported in aquatic ecosystems in various contexts such as: between bacteria of different species (Chao and Levin, 1981), between bacteria and phytoplankton (Cole, 1982; Hulot and Huisman, 2004), between phytoplankton and zooplankton (Turner and Tester, 1997), and between calanoid copepods (Folt and Goldman, 1981). The periphyton in this study was a type of bacterial–microalgal coimmobilized system, in which some bacteria may release certain compounds to inhibit algal growth, such as *H. akashiwo*-killing bacteria playing a dominant role in the termination of *Heterosigma akashiwa* (Raphidophyceae) blooms in 1994 and 1995 in Hiroshima Bay (Kim and Richardson, 1999; Kim et al., 1998). We assume that allelopathy of the periphyton inhibited the cyanobacterial growth in the AAM and ISC periphyton treatment microcosms, but further targeted tests are needed to confirm this hypothesis.

This study showed that the effects of periphyton should not be ignored during studies of P release between sediment and overlying water. The introduction of periphyton could delay (retard) the release of exchange phosphorus (Exch-P) from sediments, which also provides a promising biomeasure to reduce P release from sediments in eutrophic waters. Most importantly, the inhibitory effect of periphyton on cyanobacterial growth provides an environmentally benign biomeasure to control cyanobacterial blooms that frequently occur in centralized drinking waters worldwide. In addition, the potential inhibition of cyanobacteria growth through periphyton allelopathy provides a potential pathway for cyanobacteria growth control.

REFERENCES

Adey, W., Luckett, C., Jensen, K., 1993. Phosphorus removal from natural waters using controlled algal production. Restoration Ecology 1 (1), 29–39.

Agostinho, A., Gomes, L., Veríssimo, S., Okada, E.K., 2004. Flood regime, dam regulation and fish in the Upper Paraná River: effects on assemblage attributes, reproduction and recruitment. Reviews in Fish Biology and Fisheries 14 (1), 11–19.

APHA, AWWA, WPCF, 1998. Standard Methods for the Examination of Water and Wastewater, twentieth ed. American Public Health Association, Washington, DC.

Azim, M.E., Milstein, A., Wahab, M.A., Verdegam, M.C.J., 2003. Periphyton–water quality relationships in fertilized fishponds with artificial substrates. Aquaculture 228 (1–4), 169–187.

Azim, M.E., Verdegem, M.C., van Dam, A.A., Beveridge, M.C., 2005. Periphyton: Ecology, Exploitation and Management. CABI.

Babica, P., Bláha, L., Maršálek, B., 2005. Removal of microcystins by phototrophic biofilms: a microcosm study. Environmental Science and Pollution Research 12, 369–374.

Bothwell, M.L., 1989. Phosphorus–limited growth dynamics of lotic periphytic diatom communities: areal biomass and cellular growth rate responses. Canadian Journal of Fisheries and Aquatic Sciences 46 (8), 1293–1301.

Bushong, S.J., Bachmann, R.W., 1989. In situ nutrient enrichment experiments with periphyton in agricultural streams. Hydrobiology 178, 1–10.

Cao, Z., Zhang, X., Ai, N., 2011. Effect of sediment on concentration of dissolved phosphorus in the three gorges reservoir. International Journal of Sediment Research 26 (1), 87–95.

Carey, R.O., Vellidis, G., Lowrance, R., Pringle, C.M., 2007. Do nutrients limit algal periphyton in small blackwater coastal plain streams? Journal of the American Water Resources Association 43 (5), 1183–1193.

Carlton, R.G., Wetzel, R.G., 1988. Phosphorus flux from lake sediments: effect of epipelic algal oxygen production. Limnology and Oceanography 33, 562–570.

Chao, L., Levin, B.R., 1981. Structured habitats and the evolution of anticompetitor toxins in bacteria. Proceedings of the National Academy of Sciences of the United States of America 78, 6324–6328.

Chen, M., Li, X.-H., He, Y.-H., Song, N., Cai, H.-Y., Wang, C., Li, Y.-T., Chu, H.-Y., Krumholz, L.R., Jiang, H.-L., 2016. Increasing sulfate concentrations result in higher sulfide production and phosphorous mobilization in a shallow eutrophic freshwater lake. Water Research 96, 94–104.

Cole, J.J., 1982. Interactions between bacteria and algae in aquatic ecosystems. Annual Review of Ecology and Systematics 13, 291–314.

Conley, J.D., Paerl, H.W., Howarth, R.W., Boesch, D.F., Seitzinger, S.P., Havens, K.E., 2009. Controlling eutrophication: nitrogen and phosphorus. Science 323, 1014–1015.

Cooke, W.B., 1956. Colonization of artificial bare areas by microorganisms. The Botanical Review 22 (9), 613–638.

Cotner Jr., J.B., Wetzel, R.G., 1992. Uptake of dissolved inorganic and organic phosphorus compounds by phytoplankton and bacterioplankton. Limnology and Oceanography 37, 232–243.

Cross, W.F., Benstead, J.P., Frost, P.C., Thomas, S.A., 2005. Ecological stoichiometry in freshwater benthic systems: recent progress and perspectives. Freshwater Biology 50 (11), 1895–1912.

Dillon, R.T., 2000. The Ecology of Freshwater Molluscs. Cambridge University Press.

Ding, Y.W., Wang, L., Wang, B.Z., Wang, Z., 2006. Removal of nitrogen and phosphorus in a combined A2/O-BAF system with a short aerobic SRT. Journal of Environmental Sciences 18, 1082–1087.

Dodds, W.K., Smith, V.H., Lohman, K., 2002. Nitrogen and phosphorus relationships to benthic algal biomass in temperate streams. Canadian Journal of Fisheries and Aquatic Sciences 59 (5), 865–874.

Dons, L., Rasmussen, O.F., Olsen, J.E., 1992. Cloning and characterization of a gene encoding flagellin of Listeria monocytogenes. Molecular Microbiology 6, 2919–2929.

Eila, V., Anu, L., Salonen, V.P., 2003. A new gypsum-based technique to reduce methane and phosphorus release from sediments of eutrophied lakes. Water Research 10, 1–10.

El-Shafai, S.A., El-Gohary, F.A., Nasr, F.A., Van der Steen, N.P., Gijzen, H.J., 2007. Nutrient recovery from domestic wastewater using a UASB-duckweed ponds system. Bioresource Technology 98, 798–807.

Folt, C., Goldman, C.R., 1981. Allelopathy between zooplankton: a mechanism for interference competition. Science 213, 1133–1135.

Giovanni, D.I., Lidia, S., Watrud, R.J., 1999. Fingerprinting of mixed bacterial strains and biological gram-negative (GN) substrate communities by enterobacterial repetitive intergenic consensus sequence-PCR (ERIC-PCR). Current Microbiology 38, 217–223.

Gonsiorczyk, T., Casper, P., Koschel, R., 2001. Mechanisms of phosphorus release from the bottom sediment of the oligotrophic Lake Stechlin: importance of the permanently oxic sediment surface. Archiv fur Hydrobiologie 151, 203–219.

Gottlieb, A., Richards, J., Gaiser, E., 2006. Comparative study of periphyton community structure in long and short-hydroperiod Everglades marshes. Hydrobiologia 569 (1), 195–207.

Gottlieb, A., Richards, J., Gaiser, E.E., 2005. Effects of desiccation duration on the community structure and nutrient retention of short and long-hydroperiod Everglades periphyton mats. Aquatic Botany 82, 99–112.

Gray, D.I., Kroll, R.G., 1995. Polymerase chain reaction amplification of the flaA gene for the rapid identification of Listeria spp. Letters of Applied Microbiology 20, 65–68.

Hörnström, E., 2002. Phytoplankton in 63 limed lakes in comparison with the distribution in 500 untreated lakes with varying pH. Hydrobiologia 470 (1–3), 115–126.

Hepinstall, J.A., Fuller, R.L., 1994. Periphyton reactions to different light and nutrient levels and response of bacteria to those manipulations. Archiv fur Hydrobiologie 2131, 161–173.

Hieltjes, A.H.M., Lijklema, L., 1980. Fractionation of inorganic phosphates in calcareous sediments. Journal of Environmental Quality 9, 405–407.

Hulot, F.D., Huisman, J., 2004. Allelopathic interactions between phytoplankton species: the roles of heterotrophic bacteria and mixing intensity. Limnology and Oceanography 49, 1424–1434.

Hupfer, M., Herzog, C., Lewandowski, J., 2004. Is a large sedimentary phosphorus surplus a necessary pre-requisite for a high phosphorus release rate?. In: Annual Report 2004, Leibniz-Institute of Freshwater Ecology and Inland Fisheries, pp. 59–67.

Jarvie, H.P., Neal, C., Warwick, A., White, J., Neal, M., Wickham, H.D., Hill, L.K., Andrews, M.C., 2002. Phosphorus uptake into algal biofilms in a lowland chalk river. Science of the Total Environment 282–283 (0), 353–373.

Jersek, B., Giltot, P., Gubina, M., Klun, N., Mehle, J., Tcherneva, E., 1999. Typing of Listeria monocytogenes strains by repetitive element sequence-based PCR. Journal of Clinic Microbiology 37, 103–109.

Jones, J.G., 1972. Studies on freshwater bacteria: association with algae and alkaline phosphatase activity. Journal of Ecology 216, 1345–1347.

Kim, M.A., Richardson, J.S., 1999. Effects of light and nutrients on Grazer–Periphyton interactions. In: Biology and Management of Species and Habitats at Risk, 15–19 Feb., 1999, Kamloops, B.C., pp. 497–502.

Kim, M.C., Yoshinaga, I., Imai, I., Nagasaki, K., Itakura, S., Ishida, Y., 1998. A close relationship between algicidal bacteria and termination of *Heterosigma akashiwo (Raphidophyceae)* blooms in Hiroshima Bay, Japan. Marine Ecology Progress Series 170, 25–32.

Leandrini, J., Rodrigues, L., 2008. Temporal variation of periphyton biomass in semilotic environments of the upper Paraná River floodplain. Acta Limnologica Brasiliensia 20 (1), 21–28.

Lehtiniemi, M., Engström-Öst, J., Karjalainen, M., Kozlowsky-Suzuki, B., Viitasalo, M., 2002. Fate of cyanobacterial toxins in the pelagic food web: transfer to copepods or to faecal pellets? Marine Ecology Progress Series 241, 13–21.

Lin, P., Klump, J.V., Guo, L., 2016. Dynamics of dissolved and particulate phosphorus influenced by seasonal hypoxia in Green Bay, Lake Michigan. Science of the Total Environment 541, 1070–1082.

Loubet, P., Roux, P., Guérin-Schneider, L., Bellon-Maurel, V., 2016. Life cycle assessment of forecasting scenarios for urban water management: a first implementation of the WaLA model on Paris suburban area. Water Research 90, 128–140.

Lu, H., Wan, J., Li, J., Shao, H., Wu, Y., 2016. Periphytic biofilm: a buffer for phosphorus precipitation and release between sediments and water. Chemosphere 144, 2058–2064.

Lu, H., Yang, L., Zhang, S., Wu, Y., 2014. The behavior of organic phosphorus under non-point source wastewater in the presence of phototrophic periphyton. PLoS One 9 (1), e85910.

McCormick, P., 1994. Evaluating the multiple mechanisms underlying herbivore-algal interactions in streams. Hydrobiologia 291 (1), 47–59.

McCormick, P., Shuford Iii, R.E., Backus, J., Kennedy, W., 1997. Spatial and seasonal patterns of periphyton biomass and productivity in the northern Everglades, Florida, U.S.A. Hydrobiologia 362 (1–3), 185–210.

McCormick, P.V., O'Dell, M.B., Shuford III, R.B., Backus, J.G., Kennedy, W.C., 2001. Periphyton responses to experimental phosphorus enrichment in a subtropical wetland. Aquatic Botany 71 (2), 119–139.

McCormick, P.V., Shuford III, R.B., Chimney, M.J., 2006. Periphyton as a potential phosphorus sink in the everglades nutrient removal project. Ecological Engineering 27 (4), 279–289.

Monbet, P., McKelvie, I.D., Saefumillah, A., Worsfold, P.J., 2007. A protocol to assess the enzymatic release of dissolved organic phosphorus species in waters under environmentally relevant conditions. Environmental Science and Technology 41, 7479–7485.

Plant, L., House, W., 2002. Precipitation of calcite in the presence of inorganic phosphate. Colloids and Surfaces A: Physicochemical and Engineering Aspects 203 (1), 143–153.

Rao, C.R.M., Reddi, G.S., 1990. Decomposition procedure with aqua regia and hydrofluoric acid at room temperature for the spectrophotometric determination of phosphorus in rock and minerals. Analytica Chimica Acta 237, 251–252.

Rectenwald, L.L., Drenner, R.W., 2000. Nutrient removal from wastewater effluent using an ecological water treatment system. Environmental Science and Technology 34, 522–526.

Reitzel, K., Hansen, J., Andersen, F.O., Hansen, K.S., Jensen, H.S., 2005. Lake restoration by dosing aluminum relative to mobile phosphorus in the sediment. Environmental Science and Technology 39, 4134–4140.

Rice, E.L., 1974. Allelopathy. Academic Press, New York.

Rippka, R., Deruelles, J., Waterbury, J.B., Herdman, M., Stanier, R.Y., 1979. Generic assignments, strain histories and properties of pure cultures of cyanobacteria. Journal of Genetic Microbiology 111, 1–61.

Roelke, D.I., Cifuentes, L.A., Eldridge, P.M., 1997. Nutrient and phytoplankton dynamics in a sewage-impacted Gulf coast estuary: a field test of the PEG-model and equilibrium resource competition theory. Estuaries 20 (4), 725–742.

Søndergaard, M., Jensen, J.P., Jeppesen, E., 2003. Role of sediment and internal loading of phosphorus in shallow lakes. Hydrobiologia 506, 135–145.

Sainto, L.J., Reddy, K.R., 2003. Biotic and abiotic uptake of phosphorus by periphyton in a subtropical freshwater wetland. Aquatic Botany 77, 203–222.

Schindler, D., 1977. Evolution of phosphorus limitation in lakes. Science 195 (4275), 260–262.

Schindler, D.W., 1998. Whole-Ecosystem experiments: replication versus realism: the need for ecosystem-scale experiments. Ecosystems 1 (4), 323–334.

Sladeckova, A., 1962. Limnological investigation methods for the periphyton ('Aufwuchs') community. Botany Review 28, 286–350.

Stüben, D., Walpersdorf, E., Voss, K., Rönicke, H., 1998. Application of lake marl at Lake Arendsee, NE Germany: first results of a geochemical monitoring during the restoration experiment. The Science of the Total Environment 218, 33–44.

Steinman, A.D., 1996. 12-Effects of grazers on freshwater benthic algae. In: Lowe, R.L. (Ed.), Algal Ecology. Academic Press, San Diego, pp. 341–373.

Sterner, R.W., Elser, J.J., Fee, E.J., Guildford, S.J., Chrzanowski, T.H., 1997. The light: nutrient ratio in lakes: the balance of energy and materials affects ecosystem structure and process. The American Naturalist 150 (6), 663–684.

Stewart, I., Falconer, I.R., 2008. Cyanobacteria and cyanobacterial toxins. In: Walsh, P.J., Smith, S.L., Fleming, L.E. (Eds.), Oceans and Human Health: Risks and Remedies From the Seas. Academic Press.

Turner, J.T., Tester, P.A., 1997. Toxic marine phytoplankton, zooplankton grazers, and pelagic food webs. Limnology and Oceanography 42, 1203–1214.

Williams, J.D.H., Jaquet, J.M., Thomas, R.L., 1976. Forms of phosphorus in the surficial sediments of Lake Erie. Journal of Fisheries Research and Board Canada 33, 413–429.

Wolf, G., Picioreanu, C., van Loosdrecht, M.C.M., 2007. Kinetic model of phototrophic biofilms-the PHOBIA model. Biotechnology and Bioengineering 97, 1064–1079.

Woodruff, S.L., House, W.A., Callow, M.E., Leadbeater, B.S.C., 1999. The effects of biofilms on chemical processes in surficial sediments. Freshwater Biology 41 (1), 73–89.

Wu, Y., Liu, J., Yang, L., Chen, H., Zhang, S., Zhao, H., Zhang, N., 2011. Allelopathic control of cyanobacterial blooms by periphyton biofilms. Environmental Microbiology 13 (3), 604–615.

Wu, Y., Zhang, S., Zhao, H., Yang, L., 2010. Environmentally benign periphyton bioreactors for controlling cyanobacterial growth. Bioresource Technology 101 (24), 9681–9687.
Zhao, L., Li, Y., Zou, R., He, B., Zhu, X., Liu, Y., Wang, J., Zhu, Y., 2013. A three-dimensional water quality modeling approach for exploring the eutrophication responses to load reduction scenarios in Lake Yilong (China). Environmental Pollution 177, 13–21.

Chapter 9

Periphyton: A Promising Bio-Organic Fertilizer Source in Agricultural Ecosystems

9.1 INTRODUCTION

Periphyton is capable of entrapping nutrients such as phosphorus and can be a potential source of biofertilizer (Lu et al., 2014a,b, 2016). The disposal of the used periphyton carrier however, poses a difficult problem. The development of an environmentally benign and biodegradable carrier will allow periphyton loaded with nutrients and its carrier to be recycled together.

To maintain a stable habitat for periphytic formation, environmental conditions, such as carriers for periphytic growth, have been investigated (Khatoon et al., 2007). Usually, periphytic growth is limited by the surface roughness and surface charge of carriers in the early stages of microorganism attachment and periphyton development (Azim et al., 2002a; Azim, 2005). The physicochemical properties, and spatial and temporal variations of carriers can restrain periphyton growth (Renner and Weibel, 2011). Subsequently, studies of the carriers (both artificial and natural) are needed to promote the biomass and functions of periphyton.

Periphytic carriers can be classified into three main groups: (1) natural surfaces, including but not limited to, wood (epixylon) (Vadeboncoeur, 2009), sediment (epipelon), rocks (epilithon) (Graba et al., 2013), sand (epipsammon), macrophytes (epiphyton), and animal bodies (epizoon) (Azim, 2009); (2) artificial, such as those used in constructed wetlands or biological aerated filters, including quartz sand, anthracite, shale, biological ceramsite (Azim et al., 2002a; Azim, 2005); and (3) synthetic, such as plastic, PVC piping (Burkholder, 1996; Putz, 1997), glass tubing (Moschini-Carlos et al., 2000), and fiber bundles (Dodds, 1989; Graham et al., 1996; Lowe, 1986).

Among the above-mentioned carriers, ceramsite, plastic, and fiber carriers have been widely used in water/wastewater treatment systems. The use of ceramsite in constructed wetlands often causes plugging of the pores and channels to the inner periphyton, thereby decreasing its pollution-removal efficiency (Wu et al., 2014b). There are some concerns with synthetic

Periphyton. http://dx.doi.org/10.1016/B978-0-12-801077-8.00009-0

carriers such as PVC pipes which might cause secondary contamination due to leaching of plasticizers from the material (Briassoulis et al., 2013). Moreover, synthetic carriers such as plastic and fiber bundles do not have a carbon-rich surface to maintain the development and nutrient requirements of periphyton (Abdul Khalil et al., 2012; Zhang and Mei, 2013). It is possible that the periphyton attached to plastic or fiber bundle carriers can be easily washed off by water due to their smooth surfaces. Additionally, the disposal of synthetic plastic materials is difficult (Ivar do Sul et al., 2014). Therefore, it is important to use natural carriers that better support the growth of periphyton.

Carriers made from agrowastes have shown promising advantages in water/wastewater treatment systems based on microbial aggregates, which are abundant, inexpensive, renewable, and fully biodegradable (Abdul Khalil et al., 2012; Zhang and Mei, 2013). The agrowaste is a good source of carbon and nutrients such as nitrogen and phosphorus. Although carriers based on agrowastes such as wood, bamboo, and bagasse in wastewater/water treatment systems have been widely used (Azim, 2009; Azim and Asaeda, 2005; Azim et al., 2003), some natural carriers, such as leaves and stems, might be easily degraded, releasing colored substances such as humic acids. Thus, it is necessary to pretreat the agrowastes before they are used as periphyton carriers.

The aims of this study were to (1) obtain an environmentally benign measure to develop biofertilizer based on periphyton loaded with nutrients and its carrier; and (2) summarize the development of biofertilizer based on microbial aggregates such as periphyton. The findings will present a useful measure to develop biofertilizer based on nutrient recycling by periphyton. They will also provide valuable insight into the fundamentals of biofertilizer development and thereby extend the application of microbial aggregates such as periphyton from waters to soils.

9.2 MATERIALS AND METHODS

9.2.1 Modified Agrowastes

Peanut shell (PS), rice husks (RH), and ceramsite (C) were procured in Lianyungang City, Jiangsu Province. The raw materials were washed with distilled water until the water was clear, dried in an oven at 105°C for 2 h, then at 85°C for 24 h. The raw materials were cut into pieces of length × width × thickness (1.5 cm × 0.4 cm × 0.15 cm). Then, the materials were treated as follows: (1) decomposed peanut shell (DPS) was fermented in a humid environment (86–88% humidity) at 35 ± 1°C for 1 year; (2) acidified peanut shell (APS) was treated in 18.4 mol L^{-1} sulfuric acid for 24 h; (3) acidified rice husk (ARH) was treated in 18.4 mol L^{-1} sulfuric acid for 24 h; (4) ceramsite (C) was not treated.

9.2.2 The Culture of Periphyton on Modified Agrowastes

The formula of the periphyton cultivating medium is $NaNO_3$ (85.1 mg L^{-1}), $CaCl_2 \cdot 2H_2O$ (36.76 mg L^{-1}), $MgSO_4 \cdot 7H_2O$ (36.97 mg L^{-1}), $NaHCO_3$ (12.6 mg L^{-1}), $Na_2SiO_3 \cdot 9H_2O$ (28.42 mg L^{-1}), K_2HPO_4 (8.71 mg L^{-1}), H_3BO_3 (24 mg L^{-1}), 1 mL trace Elements Solution ($Na_2EDTA \cdot 2H_2O$ [4.36 mg L^{-1}], $FeCl_3 \cdot 6H_2O$ [3.15 mg L^{-1}], $CuSO_4 \cdot 5H_2O$ [2.5 μg L^{-1}], $ZnSO_4 \cdot 7H_2O$ [22 μg L^{-1}], $CoCl_2 \cdot 6H_2O$ [10 μg L^{-1}], $MnCl_2 \cdot 4H_2O$ [180 μg L^{-1}], $Na_2MoO_4 \cdot 2H_2O$ [6.3 g L^{-1}], Na_3VO_4 [18 g L^{-1}]), vitamin B_{12} (0.135 mg L^{-1}), thiamine (0.335 mg L^{-1}), biotin (0.025 g L^{-1}). One milliliter of each composition of the medium was added into 1000 mL of water collected from a eutrophic lake, Lake Xuanwu, Nanjing City, eastern China, which is defined as mixed liquid. Water chemistry characteristics were pH 8.02, total nitrogen (TN) 1.9 mg L^{-1}, NO_3^--N 0.43 mg L^{-1}, NH_4^+-N 0.07 mg L^{-1}, total phosphorus (TP) 0.1 mg L^{-1}, and dissolved phosphorus (DP) 0.035 mg L^{-1}.

The mixed liquid was used for the culture of periphyton. The periphyton culture process was as follows. First, the modified agrowastes (DPS, APS, and ARH) and C were individually added into a tank of volume (length × width × height = 15 cm× 15 cm× 10 cm). Secondly, the mixed liquid was added into each tank until the liquid/solid volume ratios (the ratio of the volume of the mixed liquid to the volume of the modified agrowaste) was 7:3. Then, the tanks were placed in an incubator with illumination of 2500−2800 Lux and air temperature 28 ± 1°C. The microorganisms in the mixed liquid automatically concentrated on the surfaces of the modified agrowastes (DPS, APS, and ARH) and C, forming periphyton. On day 60, the periphyton became brown with a thickness of about 0.2−0.3 cm. The periphyton attached on different carriers was peeled off using a knife sterilized in 0.1 M HCl solution. One part of the collected periphyton was used for investigating the periphyton properties. The other part of the periphyton was employed in the following experiments involving nitrogen and phosphorous capture.

9.2.3 The Removal of Nutrients by Periphyton Attached on Modified Agrowastes

To test whether the self-purification ability of surface waters was improved during the application of periphyton attached on the modified agrowaste carriers, the periphyton attached to DPS was chosen to purify artificial water. The experiment was conducted in six 150 mL simulated surface waters. The sediment was laid at the bottom of the simulated surface waters and then the water sample was added. The sediment thickness was c.2.0 cm while the overlying water depth was c.10 cm. The simulated surface waters were placed in the open air to obtain stable light (from 2800 to 300 Lux). The periphyton immobilized DPS that had been cultured for 60 days in the cultivating medium were fixed 10 cm under the water surface (on the top surface of sediments).

Three simulated surface waters had 2.4 g of fresh periphyton (weight at 25–30°C, wetness 85 ± 5%). The control did not contain any periphyton. The dissolved oxygen (DO) concentration was maintained in the range 8.5–9.5 mg L^{-1} via aeration throughout the whole experiment. The temperature was set at 25°C. The water samples were collected for determining the nutrient concentrations.

9.2.4 Methods and Analyses

The amended carriers were collected and dried in an oven at 105°C for 2 h, then at 85°C for 24 h, then sent for Fourier transform infrared (FTIR) spectroscopy (KBr pellet) using a Nicolet IS10 (Thermo Electron Co., USA) with spectra recorded in the range 4000–450 cm^{-1}, at a resolution of 4 cm^{-1} (Zhao et al., 2014). The periphytic assemblage grown on amended carriers after 60 days cultivation was collected and also dried in an oven at 105°C for 2 h, then at 85°C for 24 h. It was then used for the determination of nutrients.

For TP and TN measurement, the periphyton samples were immersed in 10 mL concentrated sulfuric acid for 24 h then digested at 160°C until the sample turned to a brownish black color. The sample was cooled and eight drops of hydrogen peroxide were added. The drops were repeated 3–5 times until the sample became transparent at 360°C digestion. Digested samples were cooled, 25 mL of water was added, and the samples then vortexed to dissolve precipitated salts and diluted to 250 mL for analysis (Maher et al., 2002). TP was measured colorimetrically by an Auto Analyzer 3 (SEALAA3, Holland). TN was measured by a Kjeldahl machine (KjeltecTM8400, Denmark). The periphyton nutrient-capturing ability was defined as the potential of periphyton to remove N and P from water. The N and P in water were directly determined using an Auto Analyzer 3 after the water was filtered through a 0.45-μm filter. The nutrient removed from water by periphyton was defined as the difference between the initial load and the determined load at a specific time.

Biolog analysis was employed to investigate the functional potential of periphyton. The Biolog EcoPlates (Biolog, Hayward, USA) are comprised of three replicate wells of 31 sole carbon substrates, including amino acids ($n = 6$), carbohydrates ($n = 10$), carboxylic acids ($n = 7$), polymers ($n = 4$), amine polymers ($n = 4$), phenolic compounds ($n = 2$), and a water blank. Temporal changes in physiological characteristics were described according to the microbial metabolic capacity to distinguish microbial communities (Choi et al., 1999).

Fresh periphyton (0.5 g), with constant moisture, was placed into a conical flask filled with 50 mL sterile buffer solution and some glass beads under sterile conditions. The suspended periphyton was serially diluted to 1:1000 with sterile inoculating solution (0.85% NaCl) for periphyton incubation. After shaking for 30 min, 150 μL of bacterial inoculation was added to each of the

96 wells in each Biolog EcoPlate using a pipette. The plates were incubated in the dark at 25°C, and the absorbance reading was measured every 24 h at 590 nm for 7 days (Zhang et al., 2013).

The color development in each plate is expressed as average well color development (AWCD), which indicates the overall microbial metabolic capacity based on the redox dye tetrazolium violet color change. The following equation is used to calculate AWCD.

$$\mathrm{AWCD} = \sum\left(\frac{C_i - R}{n}\right)$$

where C_i is the optical density value from each well and R is the optical density value from the water blank.

The Shannon index (H) was used to assess the richness of microbial metabolic response. The Simpson index (D) was used to describe the dominance of the response. The Mclntosh index (U) describes the evenness of the response, and reflects the similarity of the microbial activity.

After approximately 60 days, the periphyton became stable and was collected for dry mass measurement (DM drying at 70°C, 8 h), ash free dry mass (at 445°C, 1 h), and chlorophyll-a (Chl-a) determination (Moschini-Carlos et al., 2000).

All experiments were performed in triplicate. The results are presented as means ±standard deviation (SD). All data were analyzed using Excel (Microsoft). The principal component analysis (PCA) was performed using SPSS for Windows ver. 19.0 (SPSS Company).

9.3 RESULTS AND DISCUSSION

9.3.1 Amended Carriers With More Hydrophilic Groups

Fig. 9.1 presents FTIR spectra for decomposed peanut shell, acidified peanut shell, and acidified rice husks in the region 4000–450 cm^{-1}. The spectra were in the 1800–450 cm^{-1} region. DPS had more functional groups, the band represented the signal for C=O of COOH and aromatic stretching vibration in 1700–1500 cm^{-1} region. The peaks around 1450 cm^{-1} were mainly due to the $-CH_3$ shearing and bending mode. The absorbance bands in the region from 1200 to 1000 cm^{-1} were assigned to C–O–C deformation vibrations of cellulosic compounds. In the byproduct composite the peak was assigned to the OeH stretching vibration around 3200 cm^{-1}. However, there was no significant difference between the frequency value of the CeH stretching vibration about CH_3 and CH_2 asymmetry around 2910 cm^{-1} (Dick et al., 2003; Ge et al., 2014).

Hydrophobic rough mutant bacterial cells were found to adhere more readily onto carrier surfaces possessing more hydrophilic groups than hydrophobic carriers with smooth surfaces (Mazumder et al., 2011). The FTIR

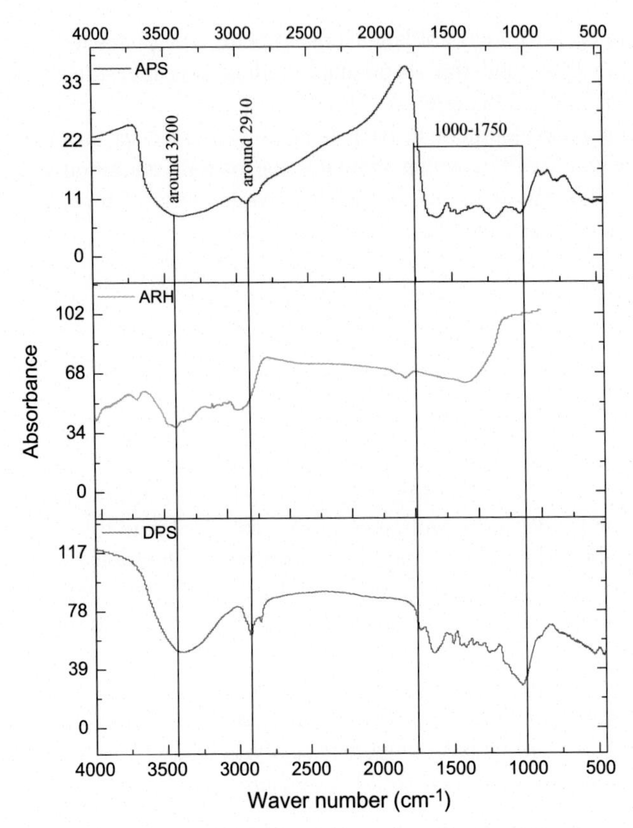

FIGURE 9.1 FTIR spectra of amended agricultural carriers (*APS*, acidified peanut shell; *ARH*, acidified rice husks; *DPS*, decomposed peanut shell.) of periphyton.

results revealed that the amended carriers, especially DPS, possess more hydrophilic groups.

9.3.2 The Capture of Nitrogen and Phosphorus by Periphyton

To test the ability of periphyton attached to different carriers to capture nutrients, the removal of nutrients was investigated (Fig. 9.2). The capture of total nitrogen (TN) and total phosphorus (TP) by periphyton were enhanced by 3255.00% (TP) and 600.88% (TN) for DPS, 624.29% (TP) and 1156.17% (TN) for APS, 833.57% (TP) and 659.95% (TN) for ARH in comparison to ceramsite (C). Substrate surface loading is a key parameter in determining periphyton structure (e.g., porous structures) and function (e.g., activity and nitrification) (Wijeyekoon et al., 2004). In this study, the substrate nutrient loadings and carbon contents differed between carriers, resulting in differences in periphytic structure and activity and leading to improved nutrient entrapment from water.

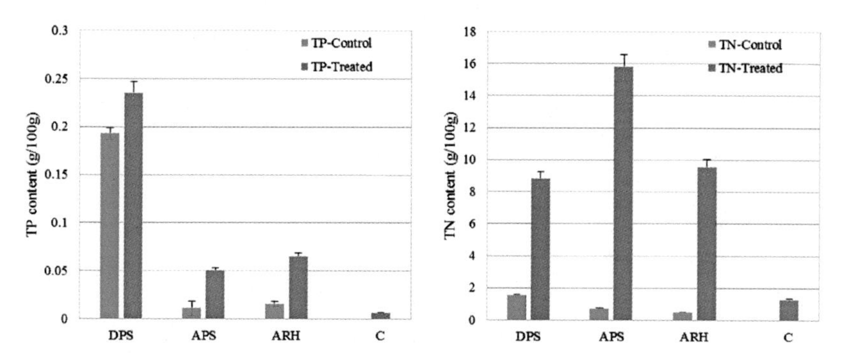

FIGURE 9.2 The concentration of total phosphorus (TP) and total nitrogen (TN) in different carriers (TP-CK, TN-CK), and periphyton (TP, TN).

9.3.3 The Removal of Nitrogen and Phosphorus From Surface Waters

As periphyton has great potential for purifying nonpoint source wastewater (Wu et al., 2014a), the periphyton attached to amended carriers were employed to in situ purify eutrophic water in simulated surface waters. DPS was chosen to concentrate periphyton in this experiment because DPS possesses more hydrophilic groups. The results showed that the TP removal rates in the simulated eutrophic waters with periphyton were 43.8−56.1% when the DPS carrier was used, which were 24.9−47.5% higher than those in the control. The TN removal rates from eutrophic water were 53.3−69.2% when DPS carrier was used, which were 12.6−48.7% higher than those in the control (Fig. 9.3). The nitrate removal rates by periphyton attached on DPS were 88.2−96.5% while the nitrate removal rates in the control were 10.4−77.8% (Fig. 9.3). The TN in the simulated eutrophic waters was dominated by nitrate, accounting for 87−92% of the TN. The ratio of the removed nitrate to the removed TN by periphyton attached on DPS was 77.5−84.7%. Combined with the previous results the use of agricultural wastes (i.e., rice straw) as carriers demonstrates good denitrification potential (Yang et al., 2015). This implies that the amended carriers are capable of improving denitrification.

The TN removal rate in this study is lower than that in a similar study where TN removal increased from 43.44% to 82.34% in membrane bioreactor (MBR) during the use of agricultural wastes (i.e., rice straw) as carriers (Yang et al., 2015). This is because the MBR was employed to remove high loading influent in a relatively closed system while in this study the periphyton was used for improving the self-purification ability of surface waters. Moreover, both experiments were conducted under different conditions, such as initial TN concentrations and running patterns.

Periphyton, an aggregation containing multiple autotrophs and heterotrophs (Azim, 2005), plays a significant role in nutrient cycling in aquatic

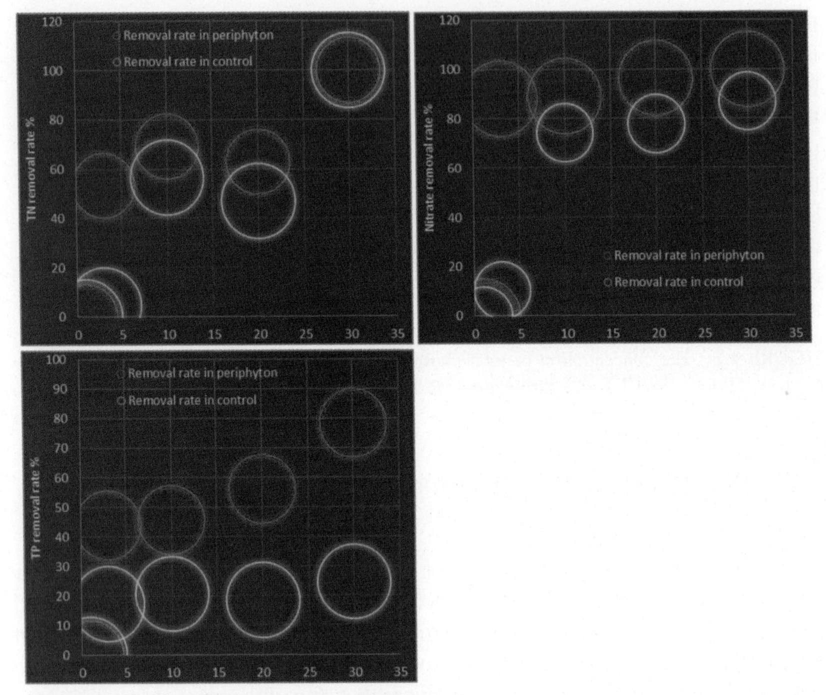

FIGURE 9.3 The removal of TN, nitrates, and TP by periphyton in eutrophic simulated surface waters.

ecosystems. Biodegradation is one of the main ways currently used for decomposing a wide range of polluting substances (Abdul Khalil et al., 2012; Zhang and Mei, 2013), implying that detoxification may have occurred when DPS was used as the periphyton carrier when the peanut was employed to be carrier of periphyton. Indeed, the results of our previous study showed that periphyton is capable of removing microcystin-RR via adsorption and biodegradation mechanisms (Wu et al., 2010).

9.3.4 Beneficial Biological Property of Periphyton Attached to Modified Carriers

The periphytic biomass calculated as fresh weight (FW), chlorophyll-a, and autotrophic index (AI) are presented in Table 9.1. The biomass of periphyton decreased in the order DPS > APS > ARH > C. The biomasses of periphyton attached to DPS, APS, and ARH were higher than those grown on C. This is because the release of P and N by the agro-carrier provided extra nutrient sources for the periphyton community. Moreover, the high surface areas of the agro-carriers such as DPS, APS, and ARH offer more suitable habitats for microorganisms of periphyton (Yang et al., 2015). The degradation of organic matter on the carrier surfaces by microorganisms attached to surfaces might

TABLE 9.1 Comparison of the Biomass and Chlorophyll-A Concentration of Periphyton Attached to the Amended Carriers and Other Frequently Used Carriers

	Biomass (mg cm^{-2})	Chlorophyll-A (μg g^{-1})	References
DPS	136.00 ± 28.20	46.00 ± 3.10	This study
APS	55.00 ± 3.60	33.00 ± 1.70	This study
ARH	42.00 ± 7.40	37.00 ± 5.30	This study
C	11.00 ± 0.26	24.00 ± 0.02	This study
Bamboo	3.05 ± 0.20	11.51 ± 0.56	Azim et al. (2002b)
		39.89 ± 0.15	Khatoon et al. (2007)
Hizol	4.89 ± 0.26	8.30 ± 0.45	Azim et al. (2002b)
Kanchi	3.12 ± 0.20	8.83 ± 0.62	Azim et al. (2002b)
Bagasses	0.2–0.5	–	Keshavanath et al. (2001)
Jutestick	3.60 (max)	11.50 (max)	Keshavanath et al. (2001)
PVC	0.09–0.36		Keshavanath et al. (2001)
	–	13.81 ± 0.09	Khatoon et al. (2007)
Plastic sheet	–	21.79 ± 0.11	Khatoon et al. (2007)
Glass tubes	0.69 (max)	13.10 (max)	Moschini-Carlos et al. (2000)
Ceramic tile	–	15.42 ± 3.05	Khatoon et al. (2007)

increase the availability of carbon for the formation of periphyton (Singh et al., 2008), in turn resulting in rapid increases in periphyton biomass. The changes in the hydrophilicity of the DPS surface and biomass of the periphyton attached to DPS were similar, implying that the high periphyton biomass was caused by the high hydrophilicity.

Community-level physiological profiles (CLPPs) were assessed by Biolog EcoPlates using AWCD (Comte and Del Giorgio, 2009). AWCD is an important index of carbon source utilization by microorganisms (Li et al., 2012). To reflect the functional diversities (metabolic versatility) of periphyton, the Shannon index, McIntosh index, and Simpson index were calculated based on the AWCD data (Li et al., 2012). The results showed that modified carriers promoted functional diversity of microbial communities as described by Simpson index (23.27 for DPS, 20.47 for ARH, 21.41 for APS, 16.10 for C), Shannon index (3.20 for DPS, 2.58 for ARH, 2.88 for APS, 2.27 for C), and McIntosh index (4.84 for DPS, 4.56 for ARH, 4.84 for APS, 3.05 for C) (Fig. 9.4).

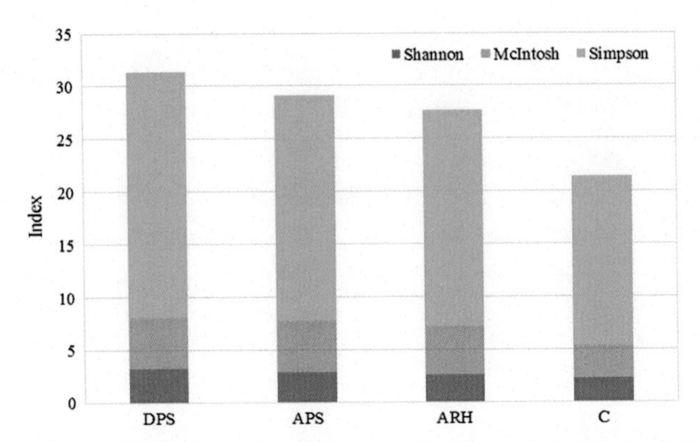

FIGURE 9.4 Functional diversity (metabolic versatility) of periphyton attached to different carriers.

Correlation analysis indicated that TN accumulated in periphyton and their carriers was positively related to the functional diversity index (TN $= 1.60 \times$ Shannon index $- 3.67$, $R^2 = 0.96$, $P < 0.05$; TN $= 0.59 \times$ McIntosh index $- 1.85$, $R^2 = 0.60$, $P < 0.05$; TN $= 0.20 \times$ Simpson index $- 3.34$, $R^2 = 0.85$, $P < 0.05$). There are significant correlations between functional diversity index and TP accumulated in periphyton and their carriers (TP $= 0.19 \times$ Shannon index $- 0.46$, $R^2 = 0.66$, $P < 0.05$; TP $= 0.05 \times$ McIntosh index $- 0.16$, $R^2 = 0.21$, $P < 0.05$; TP $= 0.02 \times$ Simpson index $- 0.37$, $R^2 = 0.48$, $P < 0.05$).

Recent research showed that the contaminant-removing ability of water/wastewater treatment systems based on microbial aggregates such as periphyton depends on the stability, activity, and functionality of periphyton (Ng et al., 2014). Over the past two decades, much research also has shown that surface water ecosystems with more species and higher diversity are more efficient in removing nutrients such as nitrate and ammonia from soil and water than ecosystems with fewer species (Azim and Asaeda, 2005; Biggs and Smith, 2002; Carey et al., 2007). There were positive relationships between nutrient removal and functional diversity in this study. The results imply that the amended agrowastes as carriers favor periphyton attachment, then maintain the periphyton on the amended carriers at a stable and sustainable level with high metabolic versatility (Larned, 2010).

9.3.5 Consideration of the Recycling of Used Periphyton and Its Carrier

Periphyton is a complex community of considerable diversity with a prevalence of microorganisms which have a net negative surface charge (Soni et al., 2008). Therefore, the hydrophilic surface will affect cell recognition of initial attachment to the carrier surface (Lower et al., 2001). The surface charges and roughness of carriers are limiting factors for periphyton attachment (Lakshmi et al., 2012). The artificial carrier (i.e., ceramsite) used in our study lacks surface charge, which is vital for periphyton development (Lakshmi et al., 2012). This study shows that using amended shells and husks as carriers, with surface charges, leads to increased periphytic biomasses and higher functional diversity than conventional carriers such as ceramsite. Moreover, the results in this study reveal that modified carriers with more hydrophilic surface "radicals" and nutrients promote the attachment of bacteria and will benefit the functional diversity of the periphyton. Using sulfuric acid to modify the agrowaste is not only for carbonization, which is rich in microporous structures (Wijeyekoon et al., 2004), but also retains the original inorganic nutrients such as nitrogen and phosphorus. The periphyton property results in this study showed that the modification was successful.

"Phosphorous removal" in this study was quantified as the accumulation (capture or immobilization) of P in the biomass. In contrast to N species, P species are typically removed from the system by removing excess biomass. This study has demonstrated that the periphyton growing on modified carriers have the ability to capture nutrients from the environment. The TN and TP contents of the periphyton on the amended agro-byproducts such as peanut shell and rice husk are 1.2–18 times higher than those in periphyton attached to ceramsite (Fig. 9.2). Therefore, the periphyton and their modified agrowaste carriers are a promising renewable and environmentally benign biofertilizer.

Essentially, fermentation is a biodegradation process used for decomposing a wide range of polluting substances (Oller et al., 2011), which is a good option when detoxifying contaminated agrowastes. In this study, the DPS was an environmentally benign carrier pretreated by fermentation for 1 year. Moreover, the periphyton attached to DPS might be beneficial in the detoxification of environmental contaminants such as microcystin-RR (Wu et al., 2010). Accordingly, periphyton and its carrier, especially DPS, and their captured nutrients are suggested as biofertilizer material.

In summary, this study showed that the use of the periphyton attached to modified agrowaste (DPS, APS, and ARH) increased the capacity of nutrient entrapment and removal from water due to the enhancement of periphyton biomass and metabolic activity. The biomass accumulated in amended agrowaste carriers with more hydrophilic groups is higher than in artificial carriers. After removing nutrients captured by periphyton out of water/wastewater treatment systems, the nutrient-enriched periphyton and its carrier, especially DPS and its periphyton, are suggested to be used as material in biofertilizer.

9.3.6 The Potential of Periphyton for Biofertilizer

9.3.6.1 The History From Microbial Fertilizer to Bio-organic Fertilizer Based on Microorganisms

The prototype of microbial fertilizer (MF) was the use of soil microorganisms to enhance agricultural yield due to the benefits of bacteria in the soil. In ancient times, the farmers in Rome found that rice production increased as long as legume plants had been planted in the rice-cropland, which was a result of the enrichment of bacteria (Zhuang, 2003). In the early 1900s, some US and German farmers put the soils that had been planted with legume or alfalfa plants into some newly reclaimed croplands to enhance agricultural production (Crews and Peoples, 2004). In 1838, J. B. Boussingault, a France agricultural chemist, found that N could be fixed by legume plants. He then built the first agricultural field experimental station in 1843 and analyzed relevant parameters (Manlay et al., 2007). Under the conditions of sand culturing legume plants between 1886 and 1888, H. Hellriegal, a German agriculturist, demonstrated that the N in the air could be fixed as long as *Rhizobium* formed (Fogarty, 1992). A Dutch researcher, M.W. Beijerinck successfully isolated *Rhizobium* in 1888, representing a breakthrough in terms of microbial fertilizers (Rodelas et al., 1999).

Many microbial fertilizer products have since been widely applied in agriculture globally. The first product of microbial fertilizer based on a type of soil bacteria was sold by a US company in 1898 (Arvanitoyannis, 2008). In the early 1900s, some organic fertilizers based on legume root *Rhizobium* were produced (Amarger, 2001). In the 1930s, many studies concerning self-fixed-nitrogen bacteria were conducted by scientists in different countries (Bowen and Rovira, 1999).

A combined fixed-nitrogen system in corn roots was discovered by Chinese and Brazilian researchers (Oelofse et al., 2010). This was followed by the discovery of the combined fixed-nitrogen systems in the rhizospheres of rice, cane, and some tropic pasturage crops (Ying et al., 1992). The nitrogen-fixed activities in these combined fixed-nitrogen systems existed widely were proven robust. In the middle of the 1980s, organic fertilizer based on nitrogen-fixed bacteria started to be applied in plot and field experiments in the US and Israel (Ladha et al., 2005). The development of fertilizers based on nitrogen-fixed bacteria led to fertilizers based on phosphorus/potassium-fixed bacteria in the 1960s (Bao et al., 2007; Epstein, 2003).

A type of fertilizer based on cyanobacteria (blue-green algae) was developed and has been applied extensively in rice cropland in Asia since the 1940s (Irisarri et al., 2001). This biofertilizer still plays an important role in improving rice yield and maintaining soil quality in the long term. Investigations into the role of blue-green algae in augmenting the fertility of rice soils and increasing rice production have been ongoing at the Central Rice Research Institute, Cuttack, since 1961. Some of the results have already been reported (Relwani, 1963, 1965; Relwani and Subrahmanyan, 1963; Subrahmanyan et al., 1964a,b). The nitrogen-fixing capacity, the extracellular N liberated and organic matter produced have also been investigated (Subrahmanyan et al., 1964a,b).

In recent years, much attention has focused on complex organic fertilizer based on different microorganisms. Heterocystous nitrogen-fixing blue-green algae consist of filaments containing two types of cells: the heterocysts, responsible for ammonia synthesis, and vegetative cells, which exhibit normal photosynthesis and reproductive growth. This unique biological system could be used for the conversion of solar energy into organic fertilizer, through cultivation of these algae in open ponds (Benemann, 1979). Complex organic fertilizer comprising of daily organic manure plus chemical fertilization provided an average 3 g carbon and 0.3 g N and approximately 0.3 g P m^{-2} per day. Twenty-four-hour net primary production added an average of 4.0 algal carbon m^{-2} (Schroeder et al., 1990).

9.3.6.2 The Definition of Bio-organic Fertilizer Based on Microorganisms

Bio-organic fertilizer based on microorganisms (BFM) is defined as a type of new organic fertilizer comprising of specific living microorganisms in microbial aggregates such as periphyton (Zhang et al., 2005). The microorganisms may be heterotrophic or photoautotrophic microorganisms or a complex mixture, including algae, bacteria, protozoa, metazoa, epiphytes, and detritus (Azim, 2005). BFM is formed by the mixture of microbial aggregates that are capable of entrapping nutrients or benefiting nutrient transformation. When BFM is applied in the soils of cropland or in ponds for fish cultivation, the living microorganisms in BFM will activate (or adjust or regulate) the environmental conditions such as increasing the soil porosity and affecting the nutrient (e.g., phosphorus) circle.

The application of BFM to cropland soils will promote the agricultural ecosystem in a more stable and sustainable way than conventional inorganic and organic fertilizers. The use of BFM not only increases the nutrient level but provides a range of necessary elements for crop growth. Moreover, BFM can enhance the activity of beneficial soil organisms, depleting the excessive nutrients in some contexts, avoiding many problems caused by inappropriate use of fertilizers. BFM can accelerate the transformation of soil organic matter into humus matter, increase the granular structure of soils, enhance the capacity of maintaining fertility and water, activate the nutrients fixed by soils, and then improve the efficiency of fertilizers.

In addition, the microorganisms in BFM might produce some beneficial secondary materials such as growth-stimulating hormones, indole acetic acid (IAA), gibberellic acid (GA), and many kinds of enzymes. These favorable materials might help to transform nutrients, decrease the occurrence of soil-borne disease, and restore polluted soils. Consequently, the development of BFM has great potential for use in sustainable agricultural processes, thereby leading to large increases in agricultural efficiency.

9.3.6.3 The Benefits to Agricultural Ecosystems

Many types of living microorganisms such as algae, bacteria, protozoa, metazoan, epiphytes, and detritus live in BFM as BFM is the substrate (or/and habitat) for these microorganisms. As a result, the application of BFM has great potential for the agricultural ecosystem, which is summarized as follows (Fig. 9.5).

Firstly, due to different microorganisms residing in BFM, the use of BFM will affect the richness of the soils, change the nutrient circle and optimize the structure (including composition and temporal-spatial characteristics) of soil ecosystems, thereby affecting soil health. Organic fertilizers, which mainly come from crop residues like rice bran, various oilseed cakes, and animal byproducts like meat bone meal, blood meal, fish meal, and crab meal, are sometimes distinguished from animal manure or compost based on animal waste (Lee, 2010). They contain specifically high levels of nutrients, e.g., N in oilseed cakes and blood meal and P in rice bran and meat bone meal, and are also high in organic matter content and a variety of micronutrients in general (Blatt, 1991; Cayuela et al., 2008), so that they have been widely used as alternative fertilizers for organically grown fields. Fertilization such as the application of BFM is one of the soil and crop management practices which exert a considerable influence on soil quality such as soil microbial properties, carbon contents, and enzyme activities (Li et al., 2008). Various organic materials have been recognized as soil amendments and disease controllers, including the control of brown spot disease and augmentation of bacterial numbers by rice bran (Osunlaja, 1989) and the increase in plant growth and reduction of nematode populations by oil cakes (Khan and Saxena, 1997).

There have, however, been problems like the accumulation of nitrate in vegetables and increased soil electronic conductivity in organic farming caused by excessive application of animal manure and organic fertilizer (Lee et al., 2004; Sohn et al., 1996). Secondly, it is likely that the application of BFM can prevent and/or carry some soilborne disease due to the reactions among the microorganisms and viruses causing soilborne diseases. In other words, BFM is a type of biocontrol agent (Zhao et al., 2014). It is well known that biological control was an alternative strategy for controlling *Fusarium* wilt disease (De Cal et al., 2000; Larkin and Fravel, 1998). Many antagonistic

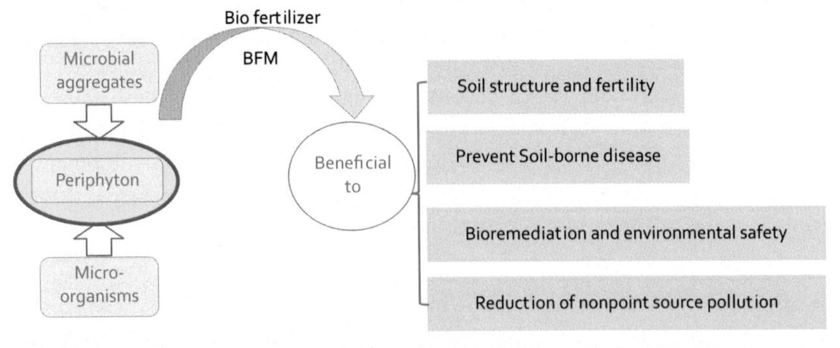

FIGURE 9.5 The potential benefits of the application of bio-organic fertilizer based on microorganisms (BFM) for the agricultural ecosystem.

strains of life in BFM have proven effective biocontrol agents in controlled laboratory or greenhouse conditions. *Rhizoctonia* (Muslim et al., 2003), *Bacillus* spp. (Gong et al., 2006), *Penicillium* spp. (Larena et al., 2003; Sabuquillo et al., 2006), *Aspergillus* spp. (Suárez-Estrella et al., 2007), and *Trichoderma* spp. (Rojo et al., 2007) are a few of these. There has been no biocontrol agent that is universally effective, because of the genetic diversity of responsible pathogens (Mishra et al., 2010).

In addition, antagonistic strains must be combined with a suitable substrate to improve their biocontrol efficacy because they can survive for a long time after being applied to soil (El-Hassan and Gowen, 2006). BFM not only plays an important role in providing a suitable substrate, but also serves as a growth-promoting medium (Raviv et al., 1998). The microorganisms in compost may produce antibiotics (Whipps, 1997), compete for nutrients and ecological niches (Hoitink and Boehm, 1999), and/or induce resistance in plants (Weller et al., 2002). Therefore, the development of a new bioorganic fertilizer with improved field consistency, such as BFM, is a continuous process and mandatory for several and individual ecological niches.

Thirdly, the application of BFM is environmentally benign. Chemical fertilizers tend to release many chemicals into the soil that contain nutrients helpful to soil but may also contain elements that are not easily biodegradable. These may go on to contaminate land and water. On the other hand, by definition, organic fertilizers such as BFM almost always have only biodegradable contents. When lawns and gardens are sprayed with chemical fertilizers, one has to be careful that family members, especially kids and pets who often play on lawns, do not ingest the harmful chemicals. However, there is no way of preventing local wildlife from being affected. Organic fertilizers raise no such concerns and can be used safely (http://edu.udym.com/five-advantages-of-organic-fertilizers/). In addition, the application of some organic fertilizers, such as complex microbial fertilizers, can be used against soilborne plant pathogenic fungi (Yan et al., 2004).

Last but not least, the application of BFM is inexpensive and cost-effective. BFM can be produced at home or on farms by using a mix of chicken, pig, cow, sheep, and horse manure along with wastes like leaves and dead plants. This is a great way of getting rid of wastes and certainly a cheaper alternative to purchasing chemical fertilizers (http://edu.udym.com/five-advantages-of-organic-fertilizers/).

9.3.6.4 Potential to Control Nonpoint Source Pollution

About 30–40% of fertilizer, such as nitrogen fertilizer, applied to cropland is absorbed in China's agricultural system, which is very low compared to the average fertilizer use efficiency (68%) in developed country agricultural systems (CAAS, 1994; Keeney, 1982). This leads to the loss of more than half of the fertilizer from cropland into downstream waters carried by runoff or leakage. Fertilizer loss is the main form of agricultural nonpoint source pollution. Using environmentally benign bio-organic fertilizer will prevent the movement of the excessive fertilizer.

The application of manure to agricultural fields also has the potential to contaminate surface waters through several different routes, including (1) runoff directly to drainage ditches and streams, and (2) percolation into groundwater that may later "daylight" as springs or contribute directly to surface waters through channels connecting groundwater and creeks, streams, and sloughs (Dowd et al., 2008). After decades of working to reduce emissions from point sources, pollution from nonpoint sources now constitutes the number one source of pollution in waterways, with agriculture being the single largest contributor (USEPA, 2000). How to best manage agricultural nonpoint source (NPS) pollution is still an issue of increasing importance to policy-makers in China. Therefore, the development of new fertilizer with low pollution loads is important for controlling nonpoint source agricultural pollution.

In China, many types of BFMs have been developed as a substitute for inorganic fertilizer and to decrease nutrient loss. As a result, the intensity and frequency of agricultural nonpoint source pollution discharge has decreased. Cyanobacteria are a popular biomaterial choice for producing organic fertilizers in China (Shen et al., 2005; Wu and Yang, 2012). Cyanobacteria are a phylum of bacteria that obtain their energy through photosynthesis (Stewart and Falconer, 2008). Cyanobacteria account for 20–30% of the earth's photosynthetic productivity and convert solar energy into biomass-stored chemical energy at the rate of ~450 TW (Oren, 2004). Cyanobacteria utilize the energy of sunlight to drive photosynthesis, a process where the energy of light is used to split water molecules into oxygen, protons, and electrons. Due to their ability to fix nitrogen in aerobic conditions, cyanobacteria are often found as symbionts with a number of other groups of organisms such as fungi (lichens), corals, pteridophytes (*Azolla*), angiosperms (*Gunnera*), and can even form a kind of biofilm (Spolaore et al., 2006).

Experiments have showed that cyanobacteria in Dianchi Lake, West China, are a good raw material for utilization (Tables 9.2 and 9.3) in fertilizer and feed (Shen et al., 2005).

TABLE 9.2 The Nutrient Content of Cyanobacteria (%)

Protein	17.8–26.4	Fat	1.49–4.67	Fiber	12.6–19.3
Ash	13.8–20.0	N	1.56–8.14	P_2O_5	0.71–1.95
Fe	0.05–0.1	Mn	0.05–0.1	Ca	0.58–2.5
Mg	0.18–0.46	K_2O	3.11–5.90	Na	0.12–0.93

TABLE 9.3 The Nutrient Content in Cyanobacteria and Other Feeds (Dry Weight)

	Protein (%)	LYS (%)	MET (%)	Vitamin E (mg kg^{-1})
Cyanobacteria	24.2	1.16	0.31	>50
Barley	10.8	0.37	0.13	38
Rice	8.5	0.29	0.14	13
Broomcorn	8.70	0.22	0.08	12
Chaff	12.1	0.86	0.25	
Bran	14.4	0.47	0.15	

Moreover, the concentration of heavy metals in cyanobacteria showed that the cyanobacteria could be safely used as fertilizer according to the national pollutants control standard in agriculture (Fig. 9.6) (Shen et al., 2005; Wu and Yang, 2012).

To remove the microcystins from cyanobacteria, the cyanobacteria were sequentially fermented under aerobic and anaerobic conditions (Wu and Yang, 2012). The residuals of the cyanobacteria fermentation were used as the raw materials for producing organic fertilizers. To balance the nutrients for crop growth, some inorganic fertilizers were added into the organic fertilizer based on cyanobacterial biomass (Fig. 9.7). This forms complex organic and inorganic fertilizer.

To test the effects of complex organic and inorganic fertilizer based on cyanobacterial biomass, a series of field experiments was conducted (Shen et al., 2005). The results showed that the organic and inorganic fertilizers from several cyanobacteria (*Microcystis*) produced higher yields of tobacco, celery, leek, and carnation compared with the efficiency of general compound

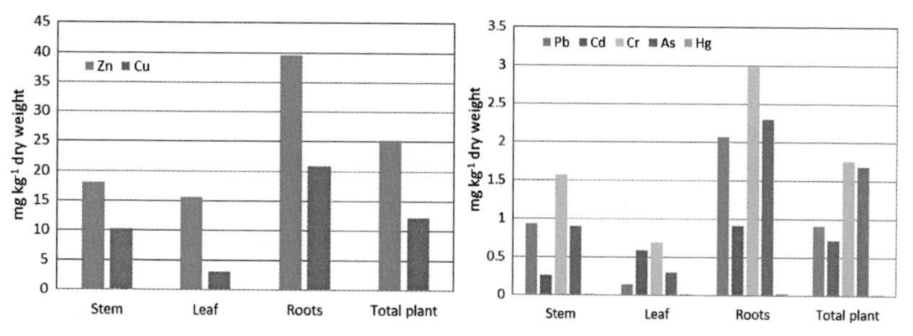

FIGURE 9.6 The content of heavy metals in cyanobacteria from Dianchi Lake, Western China (mg kg^{-1} dry weight) (Shen et al., 2005).

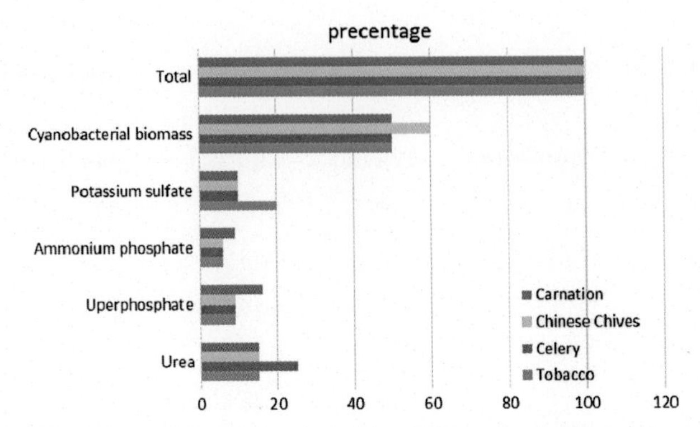

FIGURE 9.7 Prescription of complex organic and inorganic fertilizers for various crops (Shen et al., 2005).

fertilizers on the market. Doses of 900 kg hm^{-2} were best in tobacco experiments, with the tobacco yield, production value, and fine tobacco ratio increasing 7.28%, 5.04%, and 19.81%, respectively. Treated with three doses of 600 kg hm^{-2}, 900 kg hm^{-2}, and 1200 kg hm^{-2}, the celery output increased 9.5%, 17.1%, and 16.3%, respectively, more than in the control. Leek height, leaf breadth of Chinese chives increased significantly than control with treatment of 900 and 1200 kg hm^{-2}, and the yield increased 12.21% and 13.32%, respectively, of cyanobacterial fertilizer. The carnation output reached 4865 bunch/100 m^2, 11.6% higher than the control. No cyanobacterial toxin was determined in leeks and soil, suggesting that the cyanobacterial fertilizer was safe (Shen et al., 2005).

A study on the application of organic fertilizer based on cyanobacterial biomass on leek cropland examined the losses of N and P (Shen et al., 2005; Wu and Yang, 2012). The results showed that the application of the organic fertilizer could decrease nitrification, leading to a decline in nitrate movement from soil into groundwater. The nitrate concentration in the leakage in the organic fertilizer treatments was 32.9–60.2% lower than in the control. In addition, ammonia concentration in the runoff was reduced by 8.8–43.7% after the application of the organic fertilizer. The losses of TDP in the runoff declined from 12.3% to 53.6% after the organic fertilizer based on cyanobacteria was applied to leek croplands (Shen et al., 2005; Wu and Yang, 2012).

Overall, the application of BFM not only solves the difficult problem of reducing inputs of chemical fertilizer into cropland and the subsequent nonpoint source agricultural pollution into downstream waters, but also provides environmental, social, and economic benefits.

9.4 CONCLUDING REMARKS

Although the use of BFM to improve crop productivity and protect soil quality is commonplace, to date there has been no recent research focusing on the use of specific microorganisms or microbial aggregates (biofilms) to decrease nonpoint source pollution. Several proposals, including development of bio-organic fertilizer based on microorganism biomass (or substrates), including microalgae and combinations of several microorganisms, have the most potential for future commercial use. In terms of production of BFM, it is difficult to differentiate between the role of microalgae and other microorganisms, mostly bacteria.

Currently, there are major advantages to "greener" technologies. The public is constantly demanding green technologies for most aspects of daily life including food production and pollution control. The common decontamination technologies, wastewater included, that produce secondary pollution (like precipitation of P by metal salts in nonpoint source wastewater that are disposed of as toxic waste in landfill) are negatively perceived by the public. With public opinion on its side, it appears that the development and application of BFM is a prime candidate as a green technology.

From a scientific standpoint, BFM development has many advantages. (1) It controls and protects the dominant and useful microorganisms within the agricultural ecosystem itself. The microorganisms in the organism fertilizers can dominate the agricultural ecosystem despite competition with other microorganisms present. (2) It is possible to maintain a sustained agricultural ecosystem using different microorganisms in the BFM to simultaneously treat several contaminants in the soils and improve soil quality. This will be useful, especially for recalcitrant compounds that require specialized microalgae for degrading the pollutant. (3) The microorganisms in BFM have better plasmid stability within the agricultural ecosystem, allowing successful use of genetically modified special microorganisms designed for specific cleaning of polluted soils and avoiding the common failure of such genetically modified microorganisms in environmental systems. (4) From a practical view, the microorganisms in BFM systems use solar energy and need relatively small amounts of other inputs for operation. They are relatively easy to handle on a large scale and have been used by compound-producing industries for a very long time.

BFM applications in agricultural ecosystems without health hazards are environmentally friendly (promoting the green image of the public/company/government that uses them), produce no secondary pollution, and can be applied at large scales to decrease the output of nonpoint source pollution loads in intensive agricultural areas. In addition, the cost of BFM application is low, about one-third that of chemical fertilizers.

It is assumed that the application of BFM will be most useful when (1) several different contaminants in soils need to be treated simultaneously and the degraded soil quality needs to be improved on a large scale; (2) the discharge of some pollutants such as N and P outputs is very high, and the discharge mode of these pollutants is not fixed (including discharge intensity, frequency, and time); or (3) when complex degrading processes requiring specialized microorganisms are needed.

There are still a number of technical aspects of the application of BFM that could be developed, such as (1) improvement in the ability of BFM technology to create "real substrate conditions" for the living microorganisms in BFM, (2) development of more complex collectives of autotrophic and heterotrophic microorganisms to maintain a steady ecological state in the BFM, (3) optimizing selection of the proper microorganisms for specific applications such as removing some specific pollutants in soils, and (4) decreasing the volume of BFM to transport and operate easily. Solving these shortcomings will enhance the future potential of the application of BFM in commercial agriculture.

REFERENCES

Abdul Khalil, H.P.S., Bhat, A.H., Ireana Yusra, A.F., 2012. Green composites from sustainable cellulose nanofibrils: a review. Carbohydrate Polymers 87 (2), 963–979.

Amarger, N., 2001. Rhizobia in the field. Advances in Agronomy 73, 109–168.

Arvanitoyannis, I.S., 2008. Waste management in food packaging industries. In: Waste Management for the Food Industries, pp. 941–1045.

Azim, E., 2005. Periphyton: Ecology, Exploitation and Management. CABI Publishing.

Azim, M.E., 2009. Photosynthetic periphyton and surfaces. In: Likens, G.E. (Ed.), Encyclopedia of Inland Waters. Academic Press, Oxford, pp. 184–191.

Azim, M.E., Asaeda, T., 2005. Periphyton structure, diversity and colonization. In: Azim, M.E., Beveridge, M.C.M., van Dam, A.A. (Eds.), Periphyton: Ecology, Exploitation and Management. CABI Publishing, pp. 15–49.

Azim, M.E., Milstein, A., Wahab, M.A., Verdegam, M.C.J., 2003. Periphyton-water quality relationships in fertilized fishponds with artificial substrates. Aquaculture 228 (1–4), 169–187.

Azim, M.E., Verdegem, M.C.J., Khatoon, H., Wahab, M.A., van Dam, A.A., Beveridge, M.C.M., 2002a. A comparison of fertilization, feeding and three periphyton substrates for increasing fish production in freshwater pond aquaculture in Bangladesh. Aquaculture 212 (1–4), 227–243.

Azim, M.E., Wahab, M.A., Verdegem, M.C.J., van Dam, A.A., van Rooij, J.M., Beveridge, M.C.M., 2002b. The effects of artificial substrates on freshwater pond productivity and water quality and the implications for periphyton-based aquaculture. Aquatic Living Resources 15 (4), 231–241.

Bao, L.L., Li, D., Li, X.K., Huang, R.X., Zhang, J., LV, Y., Xia, G.Q., 2007. Phosphorus accumulation by bacteria isolated from a continuous-flow two-sludge system. Journal of Environmental Sciences 19, 391–395.

Benemann, J.R., 1979. Production of nitrogen fertilizer with nitrogen-fixing blue-green algae. Enzyme and Microbial Technology 1, 83–90.

Biggs, B.J.F., Smith, R.A., 2002. Taxonomic richness of stream benthic algae: effect of flood disturbance and nutrients. Limnology and Oceanography 47 (4), 1175–1186.

Blatt, C.R., 1991. Comparison of several organic amendments with a chemical fertilizer for vegetable production. Scientia Horticulture 47, 177–191.

Bowen, G.D., Rovira, A.D., 1999. The Rhizosphere and its management to improve plant growth. Advances in Agronomy 66, 1–102.

Briassoulis, D., Hiskakis, M., Babou, E., 2013. Technical specifications for mechanical recycling of agricultural plastic waste. Waste Management 33 (6), 1516–1530.

Burkholder, J.M., 1996. Interactions of benthic algae with their substrata. In: Lowe, R., Stevenson, R.J., Bothwell, M.L. (Eds.), Algal Ecology. Academic Press, San Diego, pp. 253–297.

CAAS, 1994. Chinese Academy of Agricultural Science, Edt. Chinese Fertilizer. Shanghai Sci-Tech Press, Shanghai, pp. 3–5.

Carey, R.O., Vellidis, G., Lowrance, R., Pringle, C.M., 2007. Do nutrients limit algal periphyton in small blackwater coastal plain streams? JAWRA Journal of the American Water Resources Association 43 (5), 1183–1193.

Cayuela, M.L., Sinicco, T., Mondini, C., 2008. Mineralization dynamics and biochemical properties during initial decomposition of plant and animal residues in soil. Applied Soil Ecology 41, 118–127.

Choi, K.-H., Dobbs, F.C., Wahab, M.A., 1999. Comparison of two kinds of Biolog microplates (GN and ECO) in their ability to distinguish among aquatic microbial communities. Journal of Microbiological Methods 36 (3), 203–213.

Comte, J., Del Giorgio, P.A., 2009. Links between resources, C metabolism and the major components of bacterioplankton community structure across a range of freshwater ecosystems. Environmental Microbiology 11 (7), 1704–1716.

Crews, T.E., Peoples, M.B., 2004. Legume versus fertilizer sources of nitrogen: ecological tradeoffs and human needs. Agriculture, Ecosystems & Environment 102, 279–297.

De Cal, A., García-Lepe, R., Melgarejo, P., 2000. Induced resistance by *Penicillium oxalicum* against of *Fusarium oxysporum* f. sp. *lycopersici*: histological studies of infected and induced tomato stems. Phytopathology 90, 260–268.

Dick, D.P., Santos, J.H.Z., Ferranti, E.M., 2003. Chemical characterization and infrared spectroscopy of soil organic matter from two southern Brazilian soils. Revista Brasileira De Ciencia Do Solo 27 (1), 29–39.

Dodds, W.K., 1989. Microscale vertical profiles of n(2) fixation, photosynthesis, o(2), chlorophyll a, and light in a cyanobacterial assemblage. Applied and Environmental Microbiology 55 (4), 882–886.

Dowd, B.M., Press, D., Huertos, M.L., 2008. Agricultural nonpoint source water pollution policy: the case of California's Central Coast. Agriculture, Ecosystems and Environment 128, 151–161.

El-Hassan, S.A., Gowen, S.R., 2006. Formulation and delivery of the bacterial antagonist *Bacillus subtilis* for management of lentil vascular wilt caused by *Fusarium oxysporum* f. sp. *lentis*. Phytopathology 154, 148–155.

Epstein, W., 2003. The roles and regulation of potassium in bacteria. Progress in Nucleic Acid Research and Molecular Biology 75, 293–320.

Fogarty, T.J., 1992. Organizational socialization in accounting firms: a theoretical framework and agenda for future research. Accounting, Organizations and Society 17, 129–149.

Ge, M., Li, B., Wang, L., Tao, Z., Mao, S., Wang, Y., Xie, G., Sun, G., 2014. Differentiation in MALDI-TOF MS and FTIR spectra between two pathovars of *Xanthomonas oryzae*. Spectrochimica Acta Part A: Molecular and Biomolecular Spectroscopy 133, 730–734.

Gong, M., Wang, J.D., Zhang, J., Yang, H., Lu, X.F., Pei, Y., Cheng, J.Q., 2006. Study of the antifungal ability of *Bacillus subtilis* strain PY-1 in vitro and identification of its antifungal substance (iturin A). Acta Biochimica et Biophysica Sinica 38, 233–240.

Graba, M., Sauvage, S., Moulin, F.Y., Urrea, G., Sabater, S., Sanchez-Pérez, J.M., 2013. Interaction between local hydrodynamics and algal community in epilithic biofilm. Water Research 47 (7), 2153–2163.

Graham, J.M., Arancibia-Avil, P., Graham, L.E., 1996. Effects of pH and selected metals on growth of the filamentous green alga *Mougeotia* under acidic conditions. Limnology & Oceanography 41 (2), 263–270.

Hoitink, H.A.J., Boehm, M.J., 1999. Biocontrol within the context of soil microbial communities: a substrate-dependent phenomenon. Annual Review of Phytopathology 37, 427–446.

Irisarri, P., Gonnet, S., Monza, J., 2001. Cyanobacteria in Uruguayan rice fields: diversity, nitrogen fixing ability and tolerance to herbicides and combined nitrogen. Journal of Biotechnology 91, 95–103.

Ivar do Sul, J.A., Costa, M.F., Wahab, M.A., 2014. The present and future of microplastic pollution in the marine environment. Environmental Pollution 185 (0), 352–364.

Keeney, D.R., 1982. Nitrogen management for maximum efficiency and minimum pollution. In: Severnson, F.J. (Ed.), Nitrogen in Agricultural Soils. American Society of Agronomy, Madison, Wisconsin, pp. 605–649.

Keshavanath, P., Ramesh, T.J., Gangadhar, B., Beveridge, M.C.M., Van Dam, A.A., Sandifer, P.A., 2001. On-farm evaluation of Indian major carp production with sugarcane bagasse as substrate for periphyton. Asian Fisheries Science 14 (4), 367–376.

Khan, R.A., Saxena, S.K., 1997. Integrated management of root knot nematode *Meloidogyne javanica* infecting tomato using organic materials and *Paecilomyces lilacininus*. Bioresource Technology 61, 247–250.

Khatoon, H., Yusoff, F., Banerjee, S., Shariff, M., Bujang, J.S., 2007. Formation of periphyton biofilm and subsequent biofouling on different substrates in nutrient enriched brackishwater shrimp ponds. Aquaculture 273 (4), 470–477.

Ladha, J.K., Pathak, H., Krupnik, T.J., Six, J., van Kessel, C., 2005. Efficiency of fertilizer nitrogen in cereal production: retrospects and prospects. Advances in Agronomy 87, 85–156.

Lakshmi, K., Muthukumar, T., Doble, M., Vedaprakash, L., Kruparathnam, Dineshram, R., Jayaraj, K., Venkatesan, R., 2012. Influence of surface characteristics on biofouling formed on polymers exposed to coastal sea waters of India. Colloids and Surfaces B: Biointerfaces 91 (0), 205–211.

Larena, I., Sabuquillo, P., Melgarejo, P., De Cal, A., 2003. Biocontrol of Fusarium and Verticillium wilt of tomato by *Penicillium oxalicum* under greenhouse and field conditions. Phytopathology 151, 507–512.

Larkin, R.P., Fravel, D.R., 1998. Efficacy of various fungal and bacterial biocontrol organisms for control of Fusarium wilt of tomato. Plant Disease 82, 1022–1028.

Larned, S.T., 2010. A Prospectus for periphyton: recent and future ecological research. Journal of the North American Benthological Society 29 (1), 182–206.

Lee, J., 2010. Effect of application methods of organic fertilizer on growth, soil chemical properties and microbial densities in organic bulb onion production. Scientia Horticulture 124, 299–305.

Lee, J.J., Park, R.D., Kim, Y.W., Shim, J.H., Chae, D.H., Rim, Y.S., Sohn, B.K., Kim, T.H., Kim, K.Y., 2004. Effect of food waste compost on microbial population, soil enzyme activity and lettuce growth. Bioresource Technology 93, 21–28.

Li, J., Zhao, B.Q., Li, X.Y., Jiang, R.B., So, H.B., 2008. Effects of long-term combined application of organic and mineral fertilizers on microbial biomass, soil enzyme activities and soil fertility. Agricultural Sciences in China 7, 336–343.

Li, T., Bo, L., Yang, F., Zhang, S., Wu, Y., Yang, L., 2012. Comparison of the removal of COD by a hybrid bioreactor at low and room temperature and the associated microbial characteristics. Bioresource Technology 108, 28–34.

Lowe, R.L., 1986. Periphyton response to nutrient manipulation in streams draining clearcut and forested watersheds. Journal of the North American Benthological Society 5 (3), 221–229.

Lower, S.K., Hochella Jr., M.F., Beveridge, T.J., 2001. Bacterial recognition of mineral surfaces: nanoscale interactions between Shewanella and alpha-FeOOH. Science 292 (5520), 1360–1363.

Lu, H., Wan, J., Li, J., Shao, H., Wu, Y., 2016. Periphytic biofilm: a buffer for phosphorus precipitation and release between sediments and water. Chemosphere 144, 2058–2064.

Lu, H., Yang, L., Shabbir, S., Wu, Y., 2014a. The adsorption process during inorganic phosphorus removal by cultured periphyton. Environmental Science and Pollution Research 21 (14), 8782–8791.

Lu, H., Yang, L., Zhang, S., Wu, Y., 2014b. The behavior of organic phosphorus under non-point source wastewater in the presence of phototrophic periphyton. PLoS One 9 (1), e85910.

Maher, W., Krikowa, F., Wruck, D., Louie, H., Nguyen, T., Huang, W.Y., 2002. Determination of total phosphorus and nitrogen in turbid waters by oxidation with alkaline potassium peroxodisulfate and low pressure microwave digestion, autoclave heating or the use of closed vessels in a hot water bath: comparison with Kjeldahl digestion. Analytica Chimica Acta 463 (2), 283–293.

Manlay, R.J., Feller, C., Swift, M.J., 2007. Historical evolution of soil organic matter concepts and their relationships with the fertility and sustainability of cropping systems. Agriculture, Ecosystems & Environment 119, 217–233.

Mazumder, S., Ghosh, S., Puri, I.K., 2011. Non-premixed flame synthesis of hydrophobic carbon nanostructured surfaces. Proceedings of the Combustion Institute 33 (2), 3351–3357.

Mishra, K., Kumar, A., Pandey, K., 2010. RAPD based genetic diversity among different isolates of *Fusarium oxysporum* f. sp. *lycopersici* and their comparative biocontrol. World Journal of Microbial Biotechnology 26, 1079–1085.

Moschini-Carlos, V., Henry, R., Pompêo, M.M., 2000. Seasonal variation of biomass and productivity of the periphytic community on artificial substrata in the Jurumirim Reservoir [2pt] (São Paulo, Brazil). Hydrobiologia 434 (1–3), 35–40.

Muslim, A.H., Horinouchi, R., Hyakumachi, M., 2003. Biological control of fusarium wilt of tomato with hypovirulent binucleate *Rhizoctonia* in greenhouse conditions. Mycoscience 44, 77–84.

Ng, K.K., Shi, X., Yao, Y., Ng, H.Y., 2014. Bio-entrapped membrane reactor and salt marsh sediment membrane bioreactor for the treatment of pharmaceutical wastewater: treatment performance and microbial communities. Bioresource Technology 171 (0), 265–273.

Oelofse, M., Høgh-Jensen, H., Abreu, L.S., Almeida, G.F., Qiao, Y.H., Sultan, T., de Neergaard, A., 2010. Certified organic agriculture in China and Brazil: market accessibility and outcomes following adoption. Ecological Economics 69, 1785–1793.

Oller, I., Malato, S., Sanchez-Perez, J.A., 2011. Combination of advanced oxidation processes and biological treatments for wastewater decontamination—a review. Science of the Total Environment 409 (20), 4141–4166.

Oren, A., 2004. A proposal for further integration of the cyanobacteria under the Bacteriological Code. International Journal of Systematic and Evolutionary Microbiology 54, 1895–1902.

Osunlaja, S.O., 1989. Effect of organic soil amendments on the incidence of brown spot disease in maize caused by *Physoderma maydis*. Journal of Basic Microbiology 29, 501–505.

Putz, R., 1997. Periphyton communities in Amazonian black- and whitewater habitats: community structure, biomass and productivity. Aquatic Sciences 59 (1), 74–93.

Raviv, M., Reuveni, R., Zaidman, B.Z., 1998. Improved medium for organic transplant. Biology and Agriculture Horticulture 16, 53–64.

Relwani, L.L., 1965. Response of paddy varieties to blue-green algae and methods of propagation. Biology and Agriculture Horticulture 5, 34.

Relwani, L.L., 1963. Role of blue-green algae on paddy yield. Current Science 32, 417–418.

Relwani, L.L., Subrahmanyan, R., 1963. Role of blue-green chemical nutrients and partial soil sterilization on paddy yield. Current Science 32, 441–443.

Renner, L.D., Weibel, D.B., 2011. Physicochemical regulation of biofilm formation. MRS Bulletin 36 (5), 347–355.

Rodelas, B., González-López, J., Pozo, C., Salmerón, V., Martínez-Toledo, M.V., 1999. Response of Faba bean (*Vicia faba* L.) to combined inoculation with *Azotobacter* and *Rhizobium leguminosarum* bv. *Viceae*. Applied Soil Ecology 12, 51–59.

Rojo, F.G., Reynoso, M.M., Ferez, M., Chulze, S.N., Torres, A.M., 2007. Biological control by *Trichoderma* species of *Fusarium solani* causing peanut crown root rot under field conditions. Crop Protection 26, 549–555.

Sabuquillo, P., De Cal, A., Melgarejo, P., 2006. Biocontrol of tomato wilt by *Penicillium oxalicum* formulations in different crop conditions. Biological Control 37, 256–265.

Schroeder, G.L., Wohlfarth, G., Alkon, A., Halevy, A., Krueger, H., 1990. The dominance of algal-based food webs in fish ponds receiving chemical fertilizers plus organic manures. Aquaculture 86, 219–229.

Shen, Y.W., Liu, Y.D., Wu, G.Q., Ao, H.Y., Qiu, C.Q., 2005. Efficiency test on organic and inorganic fertilizers with cyanobacteria (*Microcysits*) in several crops. Acta Hydrobiologica Sinica 29, 400–405.

Singh, C.K., Sahu, J.N., Mahalik, K.K., Mohanty, C.R., Mohan, B.R., Meikap, B.C., 2008. Studies on the removal of Pb(II) from wastewater by activated carbon developed from Tamarind wood activated with sulphuric acid. Journal of Hazardous Materials 153 (1–2), 221–228.

Sohn, S.M., Han, D.H., Kim, Y.H., 1996. Chemical characteristics of soils cultivated by the conventional farming, greenhouse cultivation and organic farming and accumulation of NO_3^- in Chinese cabbage and lettuce. Journal of Korean Organic Agriculture 5, 149–165.

Soni, K.A., Balasubramanian, A.K., Beskok, A., Pillai, S.D., 2008. Zeta potential of selected bacteria in drinking water when dead, starved, or exposed to minimal and rich culture media. Current Microbiology 56 (1), 93–97.

Spolaore, P., Joannis-Cassan, C., Duran, E., Isambert, A., 2006. Commercial applications of microalgae. Journal of Bioscience and Bioengineering 101, 87–96.

Stewart, I., Falconer, I.R., 2008. Cyanobacteria and cyanobacterial toxins. In: Walsh, P.J., Smith, S.L., Fleming, L.E. (Eds.), Oceans and Human Health: Risks and Remedies From the Seas. Academic Press.

Suárez-Estrella, F., Vargas-Garcia, C., Lopez, C.M.J., Capel, J.M., 2007. Antagonistic activity of bacteria and fungi from horticultural compost against *Fusarium oxysporum* f. sp. *melonis*. Crop Protection 26, 46–53.

Subrahmanyan, R., Relwani, L.L., Manna, G.B., 1964a. Observations on the role of blue-green algae on rice yield compared with that of conventional fertilizers. Crop Protection 33, 485–486.

Subrahmanyan, R., Relwani, L.L., Manna, G.B., 1964b. Role of blue-green algae and different methods of partial soil sterilization on rice yield. Process of Indian Academy for Science 60, 293–297.

USEPA, 2000. Environmental Protection Agency (EPA). Office of Transportation and Air Quality. National Water Quality Inventory. EPA, Washington, DC.

Vadeboncoeur, Y., 2009. Aquatic plants and attached algae. In: Likens, G.E. (Ed.), Encyclopedia of Inland Waters. Academic Press, Oxford, pp. 52–59.

Weller, D.M., Raaijmakers, J.M., Mc Spadden Gardener, B.B., Thomashow, L.S., 2002. Microbial populations responsible for specific soil suppressiveness to plant pathogens. Annual Review of Phytopathology 40, 309–348.

Whipps, J.M., 1997. Developments in the biological control of soilborne pathogens. Advance in Botany Research 26, 1–134.

Wijeyekoon, S., Mino, T., Satoh, H., Matsuo, T., 2004. Effects of substrate loading rate on biofilm structure. Water Research 38 (10), 2479–2488.

Wu, Y., He, J., Yang, L., 2010. Evaluating adsorption and biodegradation mechanisms during the removal of microcystin-RR by periphyton. Environmental Science & Technology 44 (16), 6319–6324.

Wu, Y., Xia, L., Liu, N., Gou, S., Nath, B., 2014b. Cleaning and regeneration of periphyton biofilm in surface water treatment systems. Water Science & Technology 69 (2), 235–243.

Wu, Y., Yang, L., 2012. A prospectus for bio-organic fertilizer based on microorganisms: recent and future research in agricultural ecosystem. In: Organic Fertilizers: Types, Production and Environmental Impact. Agriculture Issues and Policies. Nova Science Publisher, New York, ISBN 978-1-62081-422-2, 3rd Quarter (Chapter 8).

Wu, Y.H., Xia, L.Z., Yu, Z.Q., Shabbir, S., Kerr, P.G., 2014a. In situ bioremediation of surface waters by periphytons. Bioresource Technology 151, 367–372.

Yan, S.Z., Yang, Q.Y., Chen, Y.Y., 2004. Antagonism of complex microbial fertilizer and functional actinomycetes against soil-borne plant pathogenic fungi. Chinese Journal of Biological Control 20, 49–52.

Yang, X.-L., Jiang, Q., Song, H.-L., Gu, T.-T., Xia, M.-Q., 2015. Selection and application of agricultural wastes as solid carbon sources and biofilm carriers in MBR. Journal of Hazardous Materials 283 (0), 186–192.

Ying, J.F., Herridge, D.F., Peoples, M.B., Rerkasem, B., 1992. Effects of N fertilization on N_2 fixation and N balances of soybean grown after lowland rice. Plant Soil 147, 235–242.

Zhang, H., Li, G., Song, X., Yang, D., Li, Y., Qiao, J., Zhang, J., Zhao, S., 2013. Changes in soil microbial functional diversity under different vegetation restoration patterns for Hulunbeier Sandy Land. Acta Ecologica Sinica 33 (1), 38–44.

Zhang, Q., Qin, T., Zhang, H.Y., Luan, X.W., 2005. The progress of microbial fertilizer application. Xinjiang Agricultural Science (in Chinese) 42, 159–180.

Zhang, X., Mei, X., 2013. Periphyton response to nitrogen and phosphorus enrichment in a eutrophic shallow aquatic ecosystem. Chinese Journal of Oceanology and Limnology 31 (1), 59–64.

Zhao, J., Xiuwen, W., Hu, J., Liu, Q., Shen, D., Xiao, R., 2014. Thermal degradation of softwood lignin and hardwood lignin by TG-FTIR and Py-GC/MS. Polymer Degradation and Stability 108, 133–138.

Zhuang, S.D., 2003. The status, problem and strategy on the development and application of microbial fertilizer. Fujiang Agricultural Science & Technology 1, 34–45.

Chapter 10

The Removal of Heavy Metals by an Immobilized Periphyton Multilevel Bioreactor

10.1 INTRODUCTION

Growing attention is being given to the potential hazards posed by heavy metals to the environment (Vegliò and Beolchini, 1997). Heavy metal pollution (especially in wastewater and surface water) is of concern due to the toxic effects on flora and fauna (Quintelas et al., 2009b). Heavy metal pollution has negative effects on flora and fauna and is poisonous to humans (Miretzky et al., 2006). One of the characteristics of heavy metal pollution is the long-term exposure (bioaccumulation as well as biomagnification) to some flora and fauna (Gönen and Aksu, 2009). Therefore, removal of heavy metals from water and wastewater has now become an issue of major concern (Yang et al., 2009, 2016). The need for economical and effective methods to remove heavy metals from wastewater is necessary and urgent.

Cadmium (Cd), mercury (Hg), manganese (Mn), chrome (Cr), lead (Pb), and arsenic (As) are the most commonly observed heavy metals in both domestic and industrial wastewater as well as surface water. To date, many methods have been developed to remove Cd and Hg from water and wastewater, such as precipitation and electrochemistry (Pascal et al., 2007), ion exchange (Chen et al., 2008), electrodialysis, and solvent extraction (Wan Ngah and Hanafiah, 2008). Adsorption is a common process to entrap heavy metals (e.g., Cd and Hg) from water. Among the common adsorbents, activated carbon has been widely employed but is relatively expensive (Demirbas, 2008). Research efforts are now focused on generating low-cost alternatives to remove heavy metals (Wan Ngah and Hanafiah, 2008).

Biological technologies for the removal of heavy metals have recently been explored (Meylan et al., 2003; Yuncu et al., 2006). Bioaccumulation and biosorption are by far the most acceptable methods due to their environmental safety and low cost as well as easy deployment on an industrial scale (Davis et al., 2003). The removal efficiency of heavy metals by these two methods is dependent on the biological materials (sorbents), such as algae and bacteria

Periphyton. http://dx.doi.org/10.1016/B978-0-12-801077-8.00010-7

(Davis et al., 2003; Meylan et al., 2003). The bioreactor is a technology that is simultaneously based on the mechanisms of bioaccumulation and biosorption and has been widely used globally (Quintelas et al., 2009a). However, the presence of some heavy metals (e.g., As) in wastewater will inhibit the growth of organisms and then retard the removal of other heavy metals by biosorption (Hallberg et al., 1996). Moreover, although biosorption is a physicochemical process that is effective in removing heavy metals from aqueous solutions (Wan Ngah and Hanafiah, 2008), its adsorption of heavy metals is selective. As a result, it is difficult to simultaneously remove various heavy metals from heterogeneous wastewater with biosorption technology, thus limiting the practical application of the technology.

The composition of single bioreactors based on biosorption and bioaccumulation has been relatively simple, mostly dominated by specific microorganisms such as *Arthrobacter viscosus* (Quintelas et al., 2009a), *Escherichia coli* (Quintelas et al., 2009b), and *Trichoderma viride* (Morales-Barrera and Cristiani-Urbina, 2006) owing to the homogeneous conditions. Such simple periphytic biofilms require long maturation and acclimation times (Muñoz and Guieysse, 2006). The longer the maturation and acclimation periods, the greater the risk of death of these microorganisms from various contaminants (Zhang and Huck, 1996). Moreover, such simple periphytic biofilm composition may be susceptible to variable surface water conditions that make it difficult to form a stable and self-recycling micro-ecosystem in heterogeneous wastewater, thereby preventing the wastewater treatment system from reaching a self-sustaining state of continuous running. Therefore, it is important that a system carrying complex periphytic biofilms (i.e., heterotrophic and phototrophic microorganisms) be developed for practical wastewater treatment.

Heavy metal pollution often occurs in the areas of intensive industry. With the control of point source wastewater, nonpoint source wastewater from industrial parks, community areas, and storm water runoff has become the largest source of heavy metals (especially Cd and Hg) in the environment (Xiao and Ji, 2007). To date, it has been difficult to remove heavy metals such as Cd, Hg, Mn, Cr, Pb, and As from the wastewater and surface water due to the frequently changing hydraulic loadings and concentrations of the metals (Wu et al., 2010). Moreover, the cavities of the adsorbent or the pores of periphyton biofilm may be filled by suspended particles carried by the nonpoint source wastewater, thereby reducing the removal efficiency. In addition, Cd, Hg, Mn, Cr, Pb, and As are generally attached to the surfaces of road dust or other particulates (Brunner et al., 2008; Covelo et al., 2007). This also poses difficulties for the entrapment processes because entrapment efficiency is mainly dominated by the chemical nature of the metals, soil, and sediment particles, as well as the pH of the surrounding environment (Brunner et al., 2008; Covelo et al., 2007). Therefore, a pretreatment wastewater system should be designed with suitable remedial options to remove Cd, Hg, Mn, Cr, Pb, and As simultaneously from nonpoint source wastewater and surface water.

The processes of bioaccumulation and biosorption of heavy metals are affected by many factors such as pH, DO, nutrients, biomass, C/N ratio, and phosphorus species (Dynes et al., 2006; Vegliò and Beolchini, 1997; Yuncu et al., 2006). With the development of diversified industries, the composition of discharged wastewater is becoming increasingly complex, including such components as heavy metals, large quantities of nutrients, and organic pollutants (Gasperi et al., 2008; Wang et al., 2007). Therefore, the removal efficiency of heavy metals by bioaccumulation and biosorption will be inevitably affected by the physicochemical properties of the effluent such as the C/N ratio and the microbial species present (Baines and Fisher, 2008; Gaberell et al., 2003). It is well known that heavy metals are relatively difficult to remove due to their toxic and unbiodegradable properties (Quintelas et al., 2009a). Thus, it is necessary and of practical significance that an integrated technology is developed to simultaneously remove various toxic heavy metals from complex and heterogeneous wastewater and that the removal of heavy metals be prioritized in comparison with other pollutants.

We propose a multilevel treatment model based on biosorption and bioaccumulation to remove heavy metals from heterogeneous influent. In the proposed biointegrated solution, we believe that the most important issue is to bring the aquatic ecosystem of the multilevel bioreactor to a self-modulating and self-sustaining state. Three additional aspects should be taken into account: (1) this multilevel bioreactor should be promising for application on an industrial scale in terms of its operation, cost, and efficiency; (2) no potentially hazardous materials or chemicals will be brought into the environment; and (3) the microbial habitats should be benefited during the deployment of the multilevel bioreactor.

10.2 MATERIALS AND METHODS

10.2.1 Description of the Multilevel Bioreactor

The multilevel bioreactor consisted of five parts: a biological pond (biopond), an anaerobic biofilter, anoxic and aerobic fluidized beds, and an autotrophic system (Fig. 10.1). The influent flowed into the square biopond (30 m^3), where macrophytes (*Canna indica*, *Juncus minimus*, and *Cyperus alternifolius*) were planted every 0.5 m × 0.5 m, and then flowed into the anaerobic biofilter (96 m^3) that was filled with coarse gravel (diameter, 3–10 cm).

The wastewater in the overflow pool of the anaerobic filter was pumped into the settling tank (24 m^3), and then flowed sequentially into the anoxic fluidized bed (72 m^3), the aerobic fluidized bed (72 m^3), and the clarification tank (24 m^3). The periphytic biofilm substrates in the anoxic fluidized bed were Industrial Soft Carriers (Wuxi Guozhen Environmental Protection Co.,

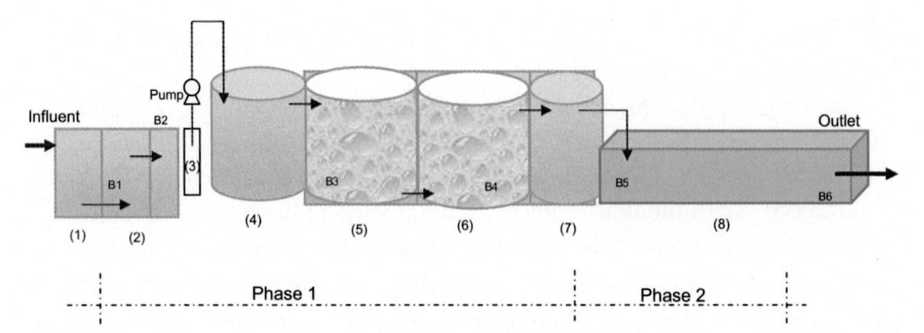

FIGURE 10.1 Schematic of the multilevel bioreactor. The process of the autotrophic biosystem consisted of (1) a biological pond (biopond), (2) an anaerobic biological filter (biofilter) filled with gravel, (3) an overflow pool, (4) a settling tank, (5) an anoxic fluidized bed, (6) an aerobic fluidized bed, (7) a clarification tank, and (8) an autotrophic system. The arrow refers to the direction of water flow. Periphytic biofilm sampling sites are from B1 to B6.

Ltd.), with a density of 0.3 m^3/m^3 water. The suspended periphytic biofilm substrates in the aerobic fluidized bed were Artificial Aquatic Mats (Wuhan Zhongke Environmental Engineering Co., Ltd.), also with a density of 0.3 m^3/m^3 water.

Finally, the water overflowed into the autotrophic system and then was discharged. The autotrophic system consisted of one ditch with a total length 230 m and an average width of 2.5 m (soil wall gradient 45°). A series of 0.04-m^3 nylon tanks, containing ceramsite adsorbent, was placed on the bottom of the ecological trunk channel at 2.0 m intervals for the adsorption of pollutants from the wastewater. Macrophytes (*Scirpus tabernaemontani*, *C. indica*, *Zizania latifolia*, *J. minimus*, *C. alternifolius*, *Zantedeschia aethiopica*, and *Acorus calamus*) were planted along the walls of the ecological ditches at 0.5 m intervals.

10.2.2 Characterization of Wastewater

The influent to the multilevel bioreactor was a combined effluent from the domestic wastewater of Liangjia Village, Kunming City, Western China, the washing wastewater from plastic bags that contained tobacco leaves, and the washing wastewater from the building materials of nine small companies. The quantity of domestic wastewater flowing to the multilevel bioreactor was around 120 $m^3\ d^{-1}$ and the quantity of the processing wastewater (two kind of washing wastewater) was about 80 $m^3\ d^{-1}$. The heterogeneous wastewater was hypereutrophic and polluted by heavy metals in high concentrations. The physicochemical parameters of this complex wastewater are given in Table 10.1.

10.2.3 Experimental Design

To obtain native microorganisms and facilitate large-scale industrial application, the periphytic biofilms were cultivated and incubated in the multilevel

TABLE 10.1 The Physicochemical Pre-experimental Parameters of the Complex Wastewater ($n = 23$)

	pH	DO (mg L^{-1})	TN (mg L^{-1})	NO_3–N (mg L^{-1})	NH_3–N (mg L^{-1})	TP (mg L^{-1})
Mean ± SD	7.6 ± 0.40	1.40 ± 0.30	27.20 ± 5.64	6.39 ± 2.72	18.16 ± 4.87	2.46 ± 0.80
	TDP (mg L^{-1})	**Mn (μg L^{-1})**	**Cr (VI) (μg L^{-1})**	**As (μg L^{-1})**	**Pb (μg L^{-1})**	**COD (mg L^{-1})**
Mean ± SD	0.66 ± 0.17	379.21 ± 16.38	105.32 ± 5.02	113.77 ± 2.52	107.76 ± 7.61	146.60 ± 45.49

bioreactor under natural conditions after the addition of 0.6 m^3 active sludge to the anoxic and aerobic fluidized beds. Two hundred cubic meters of complex wastewater were treated by the multilevel bioreactor every day. The running mode of liquid throughputs was in intervals of 12 h (12 h/12 h). The temperature ranged from 8 to 31°C during the entire experimental period.

To support the growth of native microbes in the autotrophic system, the sludge liquid in the bottom of the anoxic and the aerobic fluidized beds was pumped directly into the autotrophic system using a strong-pressure sludge pump during the nonloading intervals. To avoid the impact of the sludge on our data, the water samples were collected 10 days after the sludge had been discharged.

10.2.4 Samples and Analyses

Water samples were collected in triplicate. Total phosphorus (TP) and total dissolved phosphorus (TDP) were measured calorimetrically by the persulfate digestion–molybdophosphate reaction method. Total nitrogen (TN) was measured by persulfate digestion and by the oxidation-double wavelength (220 and 275 nm) method. Ammonia-N (NH_3–N) and nitrate-N (NO_3–N) in water were determined by ion-selective electrode potentiometry with preliminary distillation. We followed procedures that are described in detail elsewhere (APHA-AWWA-WEF, 1998). The dissolved oxygen (DO) and pH levels in water were measured in situ by a multimeter (YSI 52 dissolved oxygen and pH meters).

The cadmium (Cd) and lead (Pb) concentrations in water were measured using graphite furnace atomic absorption spectrometry (AA-7001, Beijing). The manganese (Mn) concentration in water was determined using flame atomic absorption spectrometry (AA-7001, Beijing). The mercury and arsenic (As) concentrations were determined with cold vapor atomic fluorescence spectrometry (QM201D, Jiangsu). Chromium (Cr(VI)) concentration was measured with 1,5-diphenyicarbazide spectrophotometry. These detailed procedures are described in the national standard methods of water and wastewater analyses, China (China EPA, 2002). Macrophyte samples (leaf and roots) from two sampling sites (B5 and B6) were collected quarterly in triplicate between June 2007 and May 2008. The samples were mixed together and dried at 103°C for 12 h before being digested with $HNO_3/HClO_4$ (5:1 v/v) at ~60–70°C. The digested samples were analyzed for Hg concentration using cold vapor atomic fluorescence spectrometry.

Fifteen periphytic biofilm samples (in triplicate) were collected at random locations from substrates in the anaerobic biofilter, the anoxic and aerobic fluidized beds, and the autotrophic system and then stored at 25–30°C until their moisture was about 85%. The periphytic biofilms were then weighed, and the total periphytic biofilm mass was estimated based on the periphytic biofilm weight sampled and the surface area of substrates.

Dice index (C_S) of similarity (LaPara et al., 2002) was used to evaluate the similarity of bacterial community structures based on ERIC-PCR fingerprints. Before the multilevel bioreactor was constructed, native periphytic biofilm samples were peeled off from the surfaces of stones in the entrance (A1), middle (A2), and end (A3) of the ditch (the locations of the biopond and autotrophic system). After the multilevel bioreactor was built, the periphytic biofilm samples were collected from the substrates along the sampling sites from B1 and B6 (Fig. 10.1). The Shannon diversity index (Eichner et al., 1999) was calculated to evaluate the bacterial community diversity in the biofilms. The use of Shannon diversity index to quantify bacterial diversity has been referred to previously (Miura et al., 2007).

Total DNA extraction and purification of periphytic biofilms was conducted for the ERIC-PCR analyses. The total DNA was isolated from the periphytic biofilm samples using a procedure modified from a previous report (Hill et al., 2002). One-milliliter periphytic biofilm sample aliquots were thawed in an ice-bath, and the cells were harvested by centrifugation at 9000 $\times$ g for 5 min. DNA was then purified by sequential extraction with Tris-equilibrated phenol, phenol, and chloroform-isoamyl alcohol (vol/vol/vol, 25:24:1), and chloroform isoamyl alcohol (vol/vol, 24:1) followed by precipitation with two volumes of ethanol. DNA was collected by centrifugation, air-dried, and dissolved in 50 μL sterile TE buffer. The detailed procedures have been previously described (Wei et al., 2004).

Community fingerprints of bacteria in the periphytic biofilm were obtained by using total bacterial DNA as templates for ERIC-PCR. The sequence of the ERIC primers was based on previous work (Li et al., 2006), E1 (ERIC-PCR): 5′-ATGTAAGCTCCTGGGGATTCAC-3′, E2 (ERIC-PCR): 5′-AAGTAAGT GACTGGGGTGAGCG-3′. The detailed procedures have been previously described (Li et al., 2006).

SPSS statistical software (version 12.0) was used for analyzing the data, and the level of statistical significance was set at $P < .05$. Statistically significant differences between the results were evaluated on the basis of standard deviation determinations and on the analysis of variance method (one-way ANOVA).

10.3 RESULTS AND DISCUSSION

10.3.1 Characterization of Periphytic Biofilms

During the period of stable running, the main composition of periphytic biofilms in different parts of the multilevel bioreactor was investigated with an optical microscope and a scanning electron microscope (SEM, Philips Xl30S-FEG) in November 2007. Table 10.2 shows that the dominant components of the periphytic biofilm communities in the multilevel bioreactor included bacteria (methanosarcina, diplobacillus, bacilli, brevibacterium and cocci),

TABLE 10.2 Dominant Components of the Periphytic Biofilms in Different Parts of the Multilevel Bioreactor

Parts of the Bioreactor	Main Component of the Periphytic Biofilm
Biopond	Methanosarcina, diplobacillus, and bacilli
Anaerobic biofilter	Methanosarcina, diplobacillus, and bacilli
Anoxic fluidized bed	Diplobacillus, bacilli, cocci, and brevibacterium
Aerobic fluidized bed	Bacteria (bacilli and cocci) and chladophora
Autotrophic system	Bacteria (bacilli and cocci), chladophora, diatoms (*Cyclostephanos dubius, Aulacoseira granulate, Stephanodiscus minutulus*), and cyanobacteria (*Microsystis aeruginosa* and *Aphanizomenon flos-aquae*)

chladophora, diatoms (*Cyclostephanos dubius*, *Aulacoseira granulate*, and *Stephanodiscus minutulus*) and cyanobacteria (*Microsystis aeruginosa* and *Aphanizomenon flos-aquae*). In addition, some microzooplankton, such as *Brachionus angular*, *B. calyciflorus*, *Polyarthra trigla*, *Keratella cochlearis*, and *K. valga*, were observed at the surface of the periphytic biofilms in the autotrophic system. These results indicate that the multilevel bioreactor is capable of simultaneously supporting varied microorganisms, including autotrophic and heterotrophic microbes.

10.3.2 The Removal of Heavy Metals

The DO in the influent to the multilevel bioreactor remained at 0–1.8 mg L^{-1} between April 2007 and May 2008 and the pH was maintained between 7.2 and 8.4. During the deployment of the multilevel bioreactor, the DO in the aerobic fluidized bed and autotrophic system was maintained at 7.2–9.8 mg L^{-1} and the pH was maintained at 6.8–8.2.

The multilevel bioreactor effectively and simultaneously removed heavy metals during the period from June 2007 to May 2008. Averaged across time points, removal rates were 79% for Mn, 76% for Cr(IV), 80% for Pb, and 75% for As. The removal efficiencies for the heavy metals were relatively steady, reflecting small fluctuations in removal rates ranging from 75–82% for Mn, 75–79% for Cr (VI), 77–82% for Pb, and 71–80% for As between June 2007 and May 2008 (Fig. 10.2).

The removal efficiencies of Cd and Hg were also investigated. Results showed the presence of high concentrations of heavy metals (especially Cd and Hg) in the wastewater with annual average of 420 and 2.0 μg L^{-1}, respectively (Fig. 10.2). The average removal efficiencies were 79% for Cd

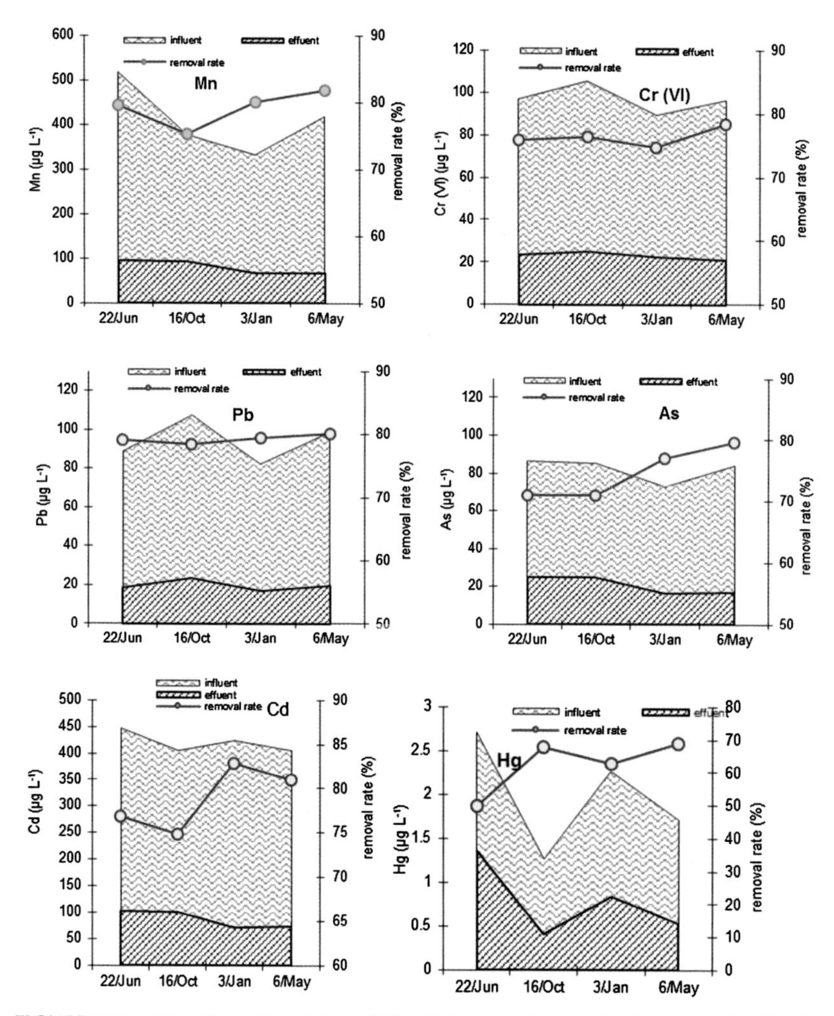

FIGURE 10.2 The efficiencies of the multilevel bioreactor in removing heavy metals (Mn, As, Cr(VI), Pb, Cd, and Hg) from June 2007 to May 2008.

and 62% for Hg during the experimental period (June 2007 to May 2008). However, the Cd removal efficiency decreased slightly during June to October 2007 (79% to 74%), while it increased in January 2008 and again decreased in May 2008. At the same time, however, the Hg removal efficiency behaved differently, increasing from 50% to 68% during June to October 2007, slightly decreasing in January followed by a slight increase in May 2008 (Fig. 10.2).

Biological measures, such as active sludge processes and bioreactors, are preferred measures when purifying wastewater with a high nutrient load because these biological measures are highly efficient, cost-effective, and environmentally friendly (Ahn et al., 2003; Wang et al., 2009). However, in many cases the applications of biological measures are limited due to certain toxic heavy metals (e.g., As) carried in the wastewater in "real-world" effluent (Hallberg et al., 1996). The performance of the multilevel bioreactor shows that the combination of a biopond, a biofilter, anoxic and aerobic fluidized beds, and an autotrophic system successfully resolved this issue. The development of the multilevel bioreactor has more potential for effectively treating practical wastewater than single bioreactors based on a specific periphytic biofilm community, such as *Arthrobacter viscosus* (Quintelas et al., 2009a), *Escherichia coli* (Quintelas et al., 2009b), or *Trichoderma viride* (Morales-Barrera and Cristiani-Urbina, 2006).

10.3.3 Bacterial Community Structure

The Shannon diversity indices increased from 1.3 to 1.8 (at sampling site B3), 1.5 to 2.1 (at sampling site B4), 1.7 to 2.7 (at sampling site B5), and 1.9 to 2.9 (at sampling site B6) during the experimental periods (Fig. 10.3). The Shannon diversity index at sampling sites B3, B4, B5, and B6 during May 2008 was significantly higher than June 2007 ($P < 0.05$). In addition, the Shannon diversity indices increased along the sampling sites (from B1 to B6) during both sampling periods, from 1.2 to 1.9 in June 2007 and from 1.1 to 2.9 in May 2008.

To evaluate the response of bacterial communities to water quality variations before and after the construction of the multilevel bioreactor, the Dice index of similarity of ERIC-PCR fingerprints for periphytic biofilm bacteria was investigated. The Dice indices of similarity in different sampling sites (A1, A2, and A3) did not vary greatly, from 98% to 99%, before the multilevel

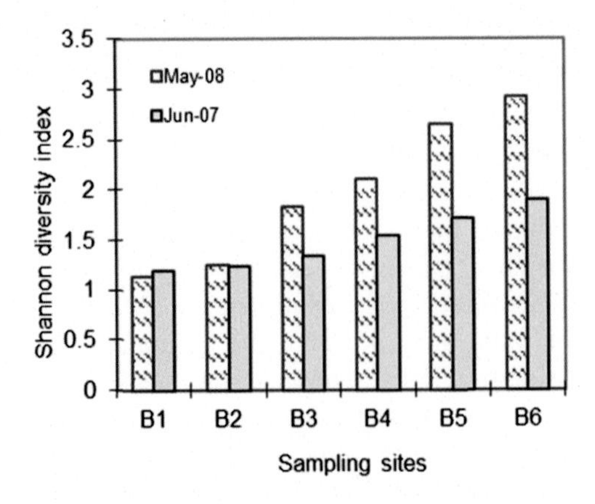

FIGURE 10.3 Changes in Shannon diversity indices of bacterial communities at different sampling sites (B1 to B6) between June 2007 and May 2008.

bioreactor was conducted, indicating that the bacterial community structures were similar before the experiment (Table 10.3).

There was a significant difference in the bacterial community structures between the anoxic/anaerobic treatment phase (biopond [B1], anaerobic biofilter [B2], and anoxic fluidized bed [B3]) and the aerobic treatment phase (aerobic fluidized bed [B4] and photoautotrophic system [B5 and B6]). In the anoxic and anaerobic phase, bacterial communities in different processes were very similar, 97.6–98.9% in June 2007 and 90.9–93.1% in May 2008. In the aerobic phase, the Dice indices of similarity of bacterial communities ranged from 88.7% to 91.6% in June 2007 and from 91.7% to 94.7% in May 2008 (Table 10.3). These results indicate that the application of the multilevel biological treatment had positive impacts on the bacterial community structures. This structure contributed to the conditions (i.e., DO, pH, heavy metals, and nutrients) in each part of the multilevel bioreactor that were modulated to beneficial states for the growth of diversified bacteria. The varied conditions such as the increase in DO level and the decrease of heavy metal concentrations caused the changes to the native bacterial community structure (Ciric et al., 2010; Wang et al., 2010a).

During the deployment of the multilevel bioreactor, the Dice indices along the sampling sites from B1 to B6 decreased from 100% to 64% in June 2007

TABLE 10.3 Dice Index of Similarity (%) of ERIC-PCR Fingerprints for Periphytic Biofilm Bacterial Communities (Wu et al., 2011)

Times	Sampling sites	B1 (A1)	B2 (A2)	B3 (A3)	B4	B5	B6
February-07	A1	100					
	A2	98.6	100				
	A3	98.3	99.1	100			
June-07	B1	100.0					
	B2	98.7	100.0				
	B3	97.6	98.9	100.0			
	B4	71.3	72.9	73.3	100.0		
	B5	70.1	70.4	71.2	90.1	100.0	
	B6	63.8	65.1	68.5	88.7	91.6	100.0
May-08	B1	100.0					
	B2	93.1	100.0				
	B3	91.8	90.9	100.0			
	B4	68.2	68.9	69.4	100.0		
	B5	61.7	67.2	68.3	93.3	100.0	
	B6	57.6	60.5	65.7	91.7	94.7	100.0

and from 100% to 58% in May 2008. The levels of bacterial community similarity in June 2007 were lower than those in May 2008 (Table 10.3). These results imply that with the running time and the distance of water flowing from B1 to B6, the habitat conditions for bacterial growth in the multilevel bioreactor increasingly improved. In turn, the robust bacteria could further enhance the pollution removal efficiency because bacterial activities are directly associated with removal efficiency (Lazarova and Manem, 1995; Quintelas et al., 2009b).

10.3.4 Benefits From the Disposal of Sludge

Highly concentrated sludge (as slurry liquid) was discharged into the photoautotrophic system during July 2007 and March 2008, with the volumes estimated based on the pump flux and time to be approximately 22 and 43 m^3, respectively. However, it was observed that the sludge disappeared within 1 week after being discharged into the photoautotrophic system (phase 2 of the multilevel bioreactor), thereby avoiding key environmental issues (such as toxic sludge pollution) associated with the current wastewater technology (i.e., mainly based on the A/O fluidized beds).

Many previous studies have reported that autotrophic systems, such as wetland systems, can treat and minimize sludge (Uggetti et al., 2010; Wang et al., 2010b). To remove the sludge, the sludge in the anoxic and aerobic fluidized beds was discharged directly into the autotrophic system. It was found that the new layer of high concentration sludge liquid (sediments) disappeared within one week after the high-concentration sludge liquid was discharged into the autotrophic system. This new layer supplied enough nutrients for the organisms in the autotrophic system and maintained the system in a self-sustaining state.

Moreover, the treatment of sludge by the autotrophic system might allow the wastes to be converted into a byproduct such as an organic fertilizer or soil conditioner suitable for native microbes and macrophytes (Uggetti et al., 2010). Most importantly, the introduction of autotrophic systems avoids the need to build specific facilities for the treatment of sludge and reduces the cost. According to a previous study, the costs of the sludge treatment systems range from 20% to 60% of the total operating cost of wastewater treatment plants (Uggetti et al., 2010).

10.3.5 Mechanisms Responsible for the Removal of Heavy Metals

To explore the relationships between periphyton biomass and heavy metal removal, the total periphyton biomass was estimated based on the specific samples (at 25–30°C, moisture content ~85%, estimated based on the weight and the specific surface area of the substrates). Results showed that the total biofilm masses generated in phase 1 of the multilevel bioreactor were 157 ± 20 kg (in June 2007), 153 ± 7.6 kg (in October 2007), 193 ± 29 kg (in January 2008), and 184 ± 18 kg (in May 2008). The changes in the generation of periphyton biomass and the Cd removal efficiency were very similar over time, and the correlations were very significant ($P < 0.05$). The linear relationship between periphyton mass and Cd removal rate is as follows: [Biomass] = 5.398 × [100 × Cd removal rate] − 254.1 ($n = 4$, $R^2 = 0.982$). This implies that the metal removal (i.e., Cd) was mainly due to the trapping of periphyton biomass in phase 1 of the multilevel bioreactor. In most cases, such metals were removed from the wastewater through adsorption onto biofilms (Quintelas et al., 2008, 2009b).

Macrophytes are capable of accumulating heavy metals via food chain transportation. Thus, the Hg content in macrophytes was determined. Results showed that the Hg concentrations in the macrophyte samples (collected from the photoautotrophic system) were 0.6 ± 0.03 $\mu g\ kg^{-1}$ and 1.3 ± 0.15 $\mu g\ kg^{-1}$ of the biomass for June 2007 and May 2008, respectively. This suggests that the removal of Hg from the wastewater was attributable to the acquisition of macrophytes in phase 2 of the multilevel bioreactor. Similarly, Hg concentrations at sampling site B6 were low compared with the sampling sites B1 and B4. The accumulation rate of Hg in the macrophytes was ~0.7 $\mu g\ kg^{-1}$ of the biomass $year^{-1}$. These results imply that the removal of Hg from the wastewater during the photoautotrophic phase 2 was due to the accumulation of Hg in the macrophytes.

To investigate which process of the multilevel bioreactor was primarily responsible for removing heavy metals, the multilevel bioreactor was divided into two phases. Phase 1 consisted of the biopond, anaerobic biofilter, and anoxic and aerobic fluidized beds. Phase 2 was the autotrophic system.

It is worthwhile noting that the Cr(VI), Mn, As, and Pb removed in phase 1 were 8%, 38%, 30%, and 36% more, respectively, than in phase 2 in June 2007. In May 2008, the proportions of the heavy metals removed in phase 1 were remarkably higher than those in phase 2, i.e., higher by 51% for Mn, by 54% for Cr(VI), by 52% for As, and by 70% for Pb (Fig. 10.3).

The heavy metal removal proportions in phase 1 increased markedly between June 2007 and May 2008 from 69% to 85% for Pb, from 65% to 76% for As, from 54% to 77% for Cr(VI), and from 68% to 76% for Mn. Accordingly, the proportions of the removed heavy metals in phase 2 decreased markedly from 31% to 15% for Pb, from 50% to 27% for As, from 46% to 23% for Cr (VI), and from 32% to 24% for Mn (Fig. 10.4).

Periphytic biofilms are an important point of entry for heavy metals that directly affect the food chain (Aouad et al., 2006; Hill and Larsen, 2005) and then easily bioconcentrate inorganic and organic contaminants from the surrounding water (Haack and Warren, 2003; Quintelas et al., 2009b). To investigate the contribution of periphytic biofilms to the removal of heavy

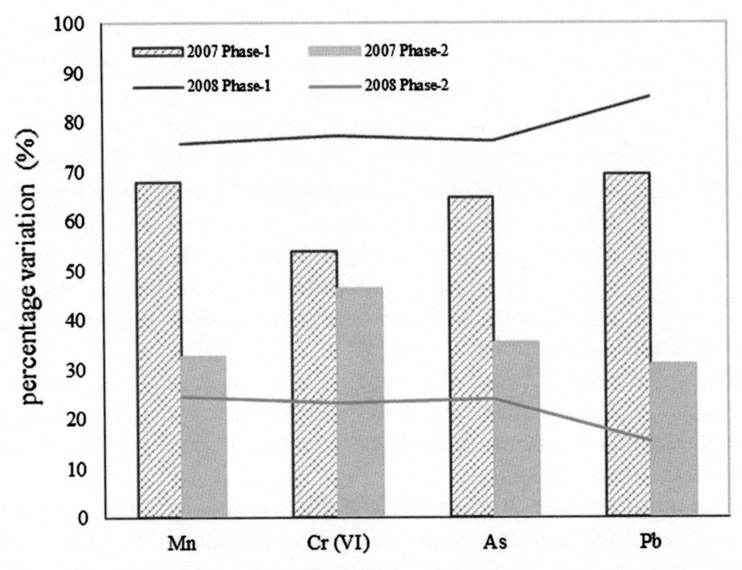

FIGURE 10.4 The proportions of heavy metals removal in phase 1 and phase 2 in the multilevel bioreactor in June 2007 and May 2008. Phase 1 comprised of the biopond, anaerobic biofilter, and anoxic and aerobic fluidized beds. Phase 2 was the autotrophic system.

metals, the periphytic biofilm masses were estimated four times. The periphytic biofilm biomasses in phase 1 were twice as higher as those in phase 2 between June 2007 and May 2008 (Fig. 10.5).

The majority of heavy metal removal (about 60–90%) occurred in phase 1, where the periphytic biofilm mass accounted for about 53–60% of the total

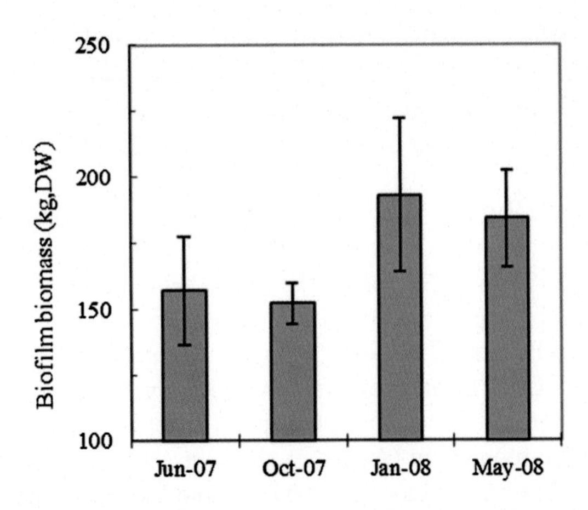

FIGURE 10.5 The periphytic biofilm masses (25–30°C, moisture ~85%) in phase 1 and phase 2 of the multilevel bioreactor (means ±SD, $n = 15$).

periphytic biofilm mass. Additionally, heavy metal removal in phase 2 decreased from June 2007 to May 2008. Therefore, the periphytic biofilm in the multilevel bioreactor was the main contributor to the removal of heavy metals from wastewater.

In addition, some macrophytes such as *Scirpus mariqueter* and *S. maritimus* have the potential to bioaccumulate heavy metals such as Pb and As (Madejón et al., 2006; Quan et al., 2007). Many studies have shown that As is known to be accumulated by iron plaque on macrophyte roots and easily taken up by macrophytes such as *Spirodela polyrhiza* L. (Rahman et al., 2008). During the running period of the multilevel bioreactor, macrophytes in the autotrophic system thrived. Therefore, it was concluded that the secondary mechanism for heavy metal (i.e., Cd, Hg, Mn, Cr(VI), Pb, and As) removal from wastewater is the bioaccumulation by macrophytes in the autotrophic system.

When the periphytic biofilm in the multilevel bioreactor ages and exfoliates, it carries the "immobilized" heavy metals into the sludge, which is then pumped into the autotrophic system. Thus, from a long-term view, the removal of heavy metals was accomplished by the extraction of macrophytes because heavy metals carried by the sludge could be bioaccumulated by the growing macrophytes in the autotrophic system (Madejón et al., 2006; Quan et al., 2007; Rahman et al., 2008) and then carried out of the multilevel bioreactor by the macrophyte harvest. Therefore, an additional sludge treatment system is not necessary for this integrated multilevel bioreactor.

10.4 CONCLUSION

The field-scale results indicate that the deployment of an environmentally benign, self-sustaining, easily deployed and cost-effective multilevel bioreactor can simultaneously remove heavy metals such as Cd, Hg, Mn, Cr (VI), As, and Pb via biosorption and bioaccumulation. The direct discharge of sludge, in turn, maintained the multilevel bioreactor in a self-modulating and self-sustaining state. The changes in Shannon and Dice indices of similarity of bacterial community structure indicate that the application of the multilevel bioreactor recovered the bacterial habitats responsible for the improvement of water quality. This study provides a promising, efficient and easily deployed biomeasure to simultaneously remove varied heavy metals from "real-world" wastewater on an industrial scale, such as treating the diffuse wastewater from industrial parks. This study also displays a valuable method for the concurrent culture of various microorganisms in wastewater treatment systems and the enhancement of habitat conditions. Improving upon the conventional wastewater treatment process (i.e., activated sludge method), the inclusion of an autotrophic system can help simultaneously remove varied heavy metals (i.e., Cd, Hg, Mn, Cr(VI), As, and Pb) and foster diverse microorganisms.

REFERENCES

Ahn, K.H., Song, K.G., Cho, E., Cho, J., Yun, H., Lee, S., Kim, J., 2003. Enhanced biological phosphorus and nitrogen removal using a sequencing anoxic/anaerobic membrane bioreactor (SAM) process. Desalination 157, 345–352.

Aouad, G., Crovisier, J.L., Geoffroy, V.A., Meyer, J.-M., Stille, P., 2006. Microbially-mediated glass dissolution and sorption of metals by *Pseudomonas aeruginosa* cells and biofilm. Journal of Hazardous Materials 136 (3), 889–895.

APHA-AWWA-WEF, 1998. Standard Methods for Examination of Water and Wastewater, twentieth ed. APHA, AWWA, and WEF, Washington, DC.

Baines, S.B., Fisher, N.S., 2008. Modeling the effect of temperature on bioaccumulation of metals by a marine bioindicator organism, *Mytilus edulis*. Environmental Science & Technology 42 (9), 3277–3282.

Brunner, I., Luster, J., Günthardt-Goerg, M.S., Frey, B., 2008. Heavy metal accumulation and phytostabilisation potential of tree fine roots in a contaminated soil. Environmental Pollution 152 (3), 559–568.

Chen, C.-Y., Lin, M.-S., Hsu, K.-R., 2008. Recovery of Cu(II) and Cd(II) by a chelating resin containing aspartate groups. Journal of Hazardous Materials 152 (3), 986–993.

ChinaEPA, 2002. The Standard Analyzed Methods of Water and Wastewater, fourth ed. Environmental Protection Agency (EPA) of China & Chinese Environmental Science Press, Beijing, China.

Ciric, L., Griffith, R.I., Philp, J.C., Whiteley, A., 2010. Field scale molecular analysis for the monitoring of bacterial community structures during on-site diesel bioremediation. Bioresource Technology 101, 5235–5241.

Covelo, E.F., Vega, F.A., Andrade, M.L., 2007. Simultaneous sorption and desorption of Cd, Cr, Cu, Ni, Pb, and Zn in acid soils: II. Soil ranking and influence of soil characteristics. Journal of Hazardous Materials 147 (3), 862–870.

Davis, T.A., Volesky, B., Mucci, A., 2003. A review of the biochemistry of heavy metal biosorption by brown algae. Water Research 37, 4311–4330.

Demirbas, A., 2008. Heavy metal adsorption onto agro-based waste materials: a review. Journal of Hazardous Materials 157 (2–3), 220–229.

Dynes, J.J., Tyliszczak, T., Araki, T., Lawrence, J., Swerhone, G.D.W., Leppard, G.G., Hitchcock, A.P., 2006. Speciation and quantitative mapping of metal species in microbial biofilms using scanning transmission X-ray microscopy. Environmental Science & Technology 40, 1556–1565.

Eichner, C.A., Erb, R.W., Timmis, K.N., Wagner-Döbler, I., 1999. Thermal gradient gel electrophoresis analysis of bioprotection from pollutant shock in the activated sludge microbial community. Applied and Environmental Microbiology 65, 102–109.

Gönen, F., Aksu, Z., 2009. Single and binary dye and heavy metal bioaccumulation properties of *Candida tropicalis*: use of response surface methodology (RSM) for the estimation of removal yields. Journal of Hazardous Materials 172 (2–3), 1512–1519.

Gaberell, M., Chin, Y.P., Hug, S.J., Sulzberger, B., 2003. Role of dissolved organic matter composition on the photoreduction of Cr(VI) to Cr(III) in the presence of iron. Environmental Science & Technology 37 (19), 4403–4409.

Gasperi, J., Garnaud, S., Rocher, V., Moilleron, R., 2008. Priority pollutants in wastewater and combined sewer overflow. Science of the Total Environment 407, 263–272.

Haack, E.A., Warren, L.A., 2003. Biofilm hydrous manganese oxyhydroxides and metal dynamics in acid rock drainage. Environmental Science & Technology 37, 4138–4147.

Hallberg, K.B., Dopson, M., Lindström, E.B., 1996. Arsenic toxicity is not due to a direct effect on the oxidation of reduced inorganic sulfur compounds by *Thiobacillus caldus*. FEMS Microbiology Letters 145, 409–414.

Hill, J.E., Seipp, R.P., Betts, M., Hawkins, L., van Kessel, A.G., Crosby, W.L., Hemmingsen, S.M., 2002. Extensive profiling of a complex microbial community by high-throughput sequencing. Applied and Environmental Microbiology 68, 3055–3066.

Hill, W., Larsen, I.L., 2005. Growth dilution of metals in microagal biofilms. Environmental Science & Technology 39, 1513–1518.

LaPara, T.M., Nakatsu, C.H., Pantea, L.M., Alleman, J.E., 2002. Stability of the bacterial communities supported by a seven-stage biological process treating pharmaceutical wastewater as revealed by PCR-DGGE. Water Research 36, 638–646.

Lazarova, V., Manem, J., 1995. Biofilm characterization and activity analysis in water and wastewater treatment. Water Research 29 (10), 2227–2245.

Li, H.Z., Li, X.Y., Zhao, Y.P., Huan, M.S., Yu, X.Z., Jin, C.X., Xu, Y.T., 2006. Analysis of structure changes of microbial community in medium biofilm by ERIC-PCR fingerprinting. Environmental Science 27, 2542–2546.

Madejón, P., Murillo, J.M., Marañón, T., Espinar, J.L., Cabrera, F., 2006. Accumulation of As, Cd and selected trace elements in tubers of *Scirpus maritimus* from Doñana marshes (South Spain). Chemosphere 64, 742–748.

Meylan, S., Behra, R., Sigg, L., 2003. Accumulation of copper and zinc in periphyton in response to dynamic variations of metal speciation in freshwater. Environmental Science & Technology 27, 5204–5212.

Miretzky, P., Saralegui, A., Cirelli, A.F., 2006. Simultaneous heavy metal removal mechanism by dead macrophytes. Chemosphere 62, 247–254.

Miura, Y., Hiraiwa, M.N., Ito, T., Itonaga, T., Watanabe, Y., Okabe, S., 2007. Bacterial community structures in MBRs treating municipal wastewater: relationship between community stability and reactor performance. Water Research 41, 627–637.

Morales-Barrera, L., Cristiani-Urbina, E., 2006. Removal of hexavalent chromium by *Trichoderma viride* in an airlift bioreactor. Enzyme and Microbial Technology 40 (1), 107–113.

Muñoz, R., Guieysse, B., 2006. Algal-bacterial processes for the treatment of hazardous contaminants: a review. Water Research 40 (15), 2799–2815.

Pascal, V., Laetitia, D., Joël, L., Marc, S., Serge, P., 2007. New concept to remove heavy metals from liquid waste based on electrochemical pH-switchable immobilized ligands. Applied Surface Science 253 (6), 3263–3269.

Quan, W.M., Han, J.D., Shen, A.L., Ping, X.Y., Qian, P.L., Li, C.J., Shi, L.Y., Chen, Y.Q., 2007. Uptake and distribution of N, P and heavy metals in three dominant salt marsh macrophytes from Yangtze River estuary, China. Marine Environmental Research 64, 21–37.

Quintelas, C., Fernandes, B., Castro, J., Figueiredo, H., Tavares, T., 2008. Biosorption of Cr (VI) by three different bacterial species supported on granular activated carbon-A comparative study. Journal of Hazardous Materials 153, 799–809.

Quintelas, C., Fonseca, B., Silva, B., Figueiredo, H., Tavares, T., 2009a. Treatment of chromium(VI) solutions in a pilot-scale bioreactor through a biofilm of *Arthrobacter viscosus* supported on GAC. Bioresource Technology 100, 220–226.

Quintelas, C., Rocha, Z., Silva, B., Fonseca, B., Figueiredo, H., Tavares, T., 2009b. Removal of Cd(II), Cr(VI), Fe(III) and Ni(II) from aqueous solutions by an *E. coli* biofilm supported on kaolin. Chemical Engineering Journal 149, 319–324.

Rahman, M.A., Hasegawa, H., Ueda, K., Maki, T., Rahman, M.M., 2008. Arsenic uptake by aquatic macrophyte *Spirodela polyrhiza* L.: interactions with phosphate and iron. Journal of Hazardous Materials 160, 356–361.

Uggetti, E., Ferrer, I., Llorens, E., García, J., 2010. Sludge treatment wetlands: a review on the state of the art. Bioresource Technology 101, 2905–2912.

Vegliò, F., Beolchini, F., 1997. Removal of metals by biosorption: a review. Hydrometallurgy 44, 301–316.

Wan Ngah, W.S., Hanafiah, M.A., 2008. Removal of heavy metal ions from wastewater by chemically modified plant wastes as adsorbents: a review. Bioresource Technology 99, 3935–3948.

Wang, F., Yao, J., Si, Y., Chen, H., Russel, M., Chen, K., Qian, Y., Zaray, G., Bramanti, E., 2010a. Short-time effect of heavy metals upon microbial community activity. Journal of Hazardous Materials 173 (1–3), 510–516.

Wang, R., Korboulewsky, N., Prudent, P., Domeizel, M., Rolando, C., Bonin, G., 2010b. Feasibility of using an organic substrate in a wetland system treating sewage sludge: impact of plant species. Bioresource Technology 101, 51–57.

Wang, Y., Li, X., Li, A., Wang, T., Zhang, Q., Wang, P., Fu, J., Jiang, G., 2007. Effect of municipal sewage treatment plant effluent on bioaccumulation of polychlorinated biphenyls and polybrominated diphenyl ethers in the recipient water. Environmental Science & Technology 41 (17), 6026–6032.

Wang, Y.L., Yu, S.L., Shi, W.X., Bao, R.L., Zhao, Q., Zuo, X.T., 2009. Comparative performance between intermittently cyclic activated sludge-membrane bioreactor and anoxic/aerobic-membrane bioreactor. Bioresource Technology 100, 3877–3881.

Wei, G., Pan, L., Du, H., Chen, J., Zhao, L., 2004. ERIC-PCR fingerprinting-based community DNA hybridization to pinpoint genome-specific fragments as molecular markers to identify and track populations common to healthy human guts. Journal of Microbiological Methods 59, 91–108.

Wu, Y., Hu, Z., Kerr, P.G., Yang, L., 2011. A multi-level bioreactor to remove organic matter and metals, together with its associated bacterial diversity. Bioresource Technology 102 (2), 736–741.

Wu, Y., Kerr, P.G., Hu, Z., Yang, L., 2010. Eco-restoration: simultaneous nutrient removal from soil and water in a complex residential-cropland area. Environmental Pollution 158 (7), 2472–2477.

Xiao, H., Ji, W., 2007. Relating landscape characteristics to non-point source pollution in mine waste-located watersheds using geospatial techniques. Journal of Environmental Management 82 (1), 111–119.

Yang, J., Tang, C., Wang, F., Wu, Y., 2016. Co-contamination of Cu and Cd in paddy fields: using periphyton to entrap heavy metals. Journal of Hazardous Materials 304, 150–158.

Yang, Z., Wang, Y., Shen, Z., Niu, J., Tang, Z., 2009. Distribution and speciation of heavy metals in sediments from the mainstream, tributaries, and lakes of the Yangtze River catchment of Wuhan, China. Journal of Hazardous Materials 166 (2–3), 1186–1194.

Yuncu, B., Sanin, F.D., Yetis, U., 2006. An investigation of heavy metal biosorption in relation to C/N ratio of activated sludge. Journal of Hazardous Materials 137, 990–997.

Zhang, S., Huck, P.M., 1996. Parameter estimation for biofilm processes in biological water treatment. Water Research 30 (2), 456–464.

Chapter 11

Photobioreactor–Wetland System Removes Organic Pollutants and Nutrients

11.1 INTRODUCTION

Human products such as chemical fertilizers, pesticides, herbicides, and plant hormones are often applied in intensive agricultural areas to maintain high yields (Haas et al., 2001). These chemicals usually constitute the majority of nonpoint-source pollutants to the downstream surface aquatic ecosystems when they are carried either by runoff or irrigation effluent (Wu et al., 2010a). These are some of the most predominant pollution sources causing eutrophication and harmful algal blooms (Pan et al., 2010). Furthermore, the migration of some natural organic matters, such as polycyclic aromatic hydrocarbons, into aquatic ecosystems poses a hazardous threat to some beneficial animals such as tadpoles (Guo et al., 2007; Novák et al., 2008) and destroys the food web and the balance of the aquatic ecosystem. Thus, the reduction of the inputs of organic contaminants and excessive nutrients into downstream aquatic ecosystems is of practical significance.

Natural organic matter is a complex assembly of organic compounds occurring in natural surface waters. The natural organic matter can directly affect the odor, color, and taste of water as well affect processes in drinking water and wastewater/water treatments (Chae et al., 2015; Cheng et al., 2016). The precursor of trihalomethanes (THMs) after chlorination of drinking water and wastewater/water treatments often occurs (Han et al., 2015; Peng et al., 2016; Sadrnourmohamadi and Gorczyca, 2015), which is often represented by UV_{254}nm-matter (ultraviolet absorbance at 254 nm) (Wu et al., 2011). Chlorination has been a popular means for disinfecting municipal drinking water and surface water in many countries, including China, for many decades. The addition of chlorine will continue to be the most common disinfection process. The added chlorine reacts with naturally occurring organic matter to form a wide range of undesired halogenated organic compounds, often referred to as disinfection byproducts. Among the most widely occurring byproducts are

Periphyton. http://dx.doi.org/10.1016/B978-0-12-801077-8.00011-9

THMs, haloacetic acids, haloacetonitriles, and haloketones (de la Rubia et al., 2008). Thus, this study investigated natural organic matter and its removal.

Several treatment processes or their combinations are capable of removing natural organic matter from water (de la Rubia et al., 2008). Therefore, some measures have been proposed to decrease the amount of organic contaminants and nutrients in aquatic ecosystems (Dorioz et al., 2006). These measures include (1) the reduction of the production of potential pollutants (e.g., reducing usage of agrochemicals) (Blanchoud et al., 2007), (2) the reduction of the migration of pollutants (e.g., improving irrigation management) (Kay et al., 2009) and (3) the acceleration of the sequestration and degradation of pollutants toward aquatic ecosystems (e.g., buffer zones and wetlands) (Schulz and Peall, 2001). Many specific technologies have been developed for the final treatment in the removal of aromatic compounds, and various measures have been introduced, such as the application of soybean peroxidase (Kinsley and Nicell, 2000), the utilization of ozone and photocatalytic processes (Fabbri et al., 2008), bioremediation via specific aromatic compound-degrading microorganisms (Häggblom, 1992), physical sequestration by clays (Liu et al., 2008), and powdered activated carbon (Dąbrowski et al., 2005). In some cases where microfiltration alone is inadequate, the natural organic matter was often pretreated by coagulation to meet water quality requirements (Jiang et al., 2016; Sillanpää and Matilainen, 2015). To meet specific demand, the application of advanced oxidation processes for the removal of organics from water is gaining importance in water treatment (Ganiyu et al., 2015; Zhang et al., 2016). For example, the utilization of active carbon to adsorb natural organic matter in the final stages of surface water treatment was evaluated (Zhang et al., 2016).

The aforementioned measures/technologies are useful and have great benefits for downstream environments. However, several new, complex problems have arisen due to the introduction of some "modern" farming techniques. For example, aromatic compounds have been brought into the soil with the application of some new pesticides, herbicides, and phytohormones, such as mecoprop [2-(2-methyl-4-chlorophenoxy) propionic acid] (Tett et al., 1994) and auxin (Kelley and Riechers, 2007). These aromatic compounds can move into the downstream aquatic ecosystems through runoff and irrigation. In addition, the long-term applications of chemical fertilizers have caused the degradation of soil quality and the decline of nutrient-holding capacities (Kong et al., 2008). This degradation might lead to excessive dissolved nutrients, such as dissolved nitrogen and phosphorus, flowing easily into downstream aquatic ecosystems, increasing the risk of eutrophication. To reduce the input of these organic contaminants and dissolved nutrients into downstream aquatic ecosystems, these pollutants should be removed by environmentally friendly biomeasures at the downstream catchments of intensive agricultural areas, thus improving the self-purifying capacities of aquatic systems.

We proposed a photobioreactor–wetland system and utilized the technology based on this system in the downstream catchments of an intensive agricultural area in the Kunming region of western China, where the self-purifying capacity had been weakened. To provide an amplified model of the photobioreactor–wetland system at an industrial scale, three additional considerations should be taken into account: (1) the measure should be environmentally friendly, requiring that it not introduce any hazardous materials or artificial chemicals into the environment; (2) the habitats of native flora and fauna should be recovered, and then the self-purifying capacity of the recovered ecosystem should be improved, and (3) the construction and operation of the photobioreactor–wetland system should be simple. The capital and operation costs should be affordable.

11.2 MATERIALS AND METHODS

11.2.1 Description of the Photobioreactor–Wetland System

To facilitate application at the industrial level, the wastewater treatment system was designed at a pilot scale and directly used for practical wastewater. The photobioreactor–wetland system was devised to remove organic matter and dissolved nutrients from upstream areas of surface waters and consisted of two parts (Fig. 11.1): a photobioreactor and a constructed wetland.

The photobioreactor covered an area of 1800 m^2 (180 m × 10 m) with a mean depth of ~1.7 m and a maximum depth of 2.6 m. Several aquatic plant species reported to have a high capacity for removing organic contaminants, including *Salix rosmarinifolia* L., *Myriophyllum verticillatum* L., *Pistia stratiotes* L., *Hydrilla verticillata* (L. f.) Royle, *Typha latifolia* L., and *Zizania latifolia*, were introduced into the photobioreactor from August to October 2006. During the experimental period, about 85–90% of the water

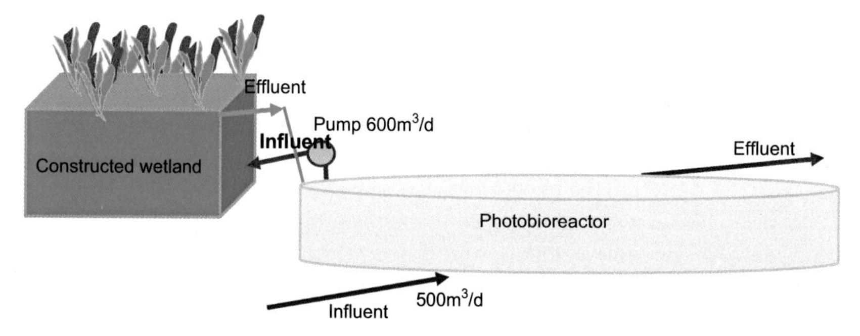

FIGURE 11.1 A schematic model of the photobioreactor–wetland system. The influent load of the photobioreactor–wetland system is 500 m^3 per day on nonrainy days (rainfall <10 mm). The load of the wetland is 600 m^3 per day. The influent of the wetland is from the photobioreactor, which in turn flows into the photobioreactor through the wetland outlet after purification by the wetland, thereby discharging the effluent of the photobioreactor–wetland system.

TABLE 11.1 The Species, Quantities, and Sources of Young Plants Used to Restore the Macrophyte Ecosystem

	Species	Quantities (Young Plants)	Sources of Young Plants
1	*Salix rosmarinifolia* L.	600	The aquatic willows (length 30–50 cm, diameter ≈1.5 cm) were collected from an area near the experimental photobioreactor
2	*Myriophyllum verticillatum* L.	18,000	The minor materials were collected from ditches and wetlands near the experimental area
3	*Pistia stratiotes* L.	40,000	40,000 were collected from Dianchi Lake, southeast China
4	*Hydrilla verticillata* (L. f.) Royle	12,000	Collected with silt from a cultural area near the east of Caohai Lake
5	*Typha latifolia* L.	5000	From a ditch near the experimental photobioreactor
6	*Zizania latifolia* (Griseb.) Stapf.	3000	From a ditch and wetland near the experimental photobioreactor
7	Other macrophytes[a]	20,000	From a ditch, photobioreactor, and wetland near the experimental area

[a]*Other macrophytes include Arundo donax L. var. "Versicolor," Cyperus alternifolius L., Iris pseudacorus L., Pontederia cordata L., Sagittaria pygmaea Miq., Scirpus validus cv. Zebrinus, Thalia dealbata Fraser ex Roscoe.*

area of the photobioreactor was covered by *M. verticillatum* L. and *P. stratiotes* L. A complete list of the introduced aquatic plants is shown in Table 11.1.

To purify the water in the photobioreactor, a constructed wetland area of 800 m^2 was built. The construction included subsurface flow and complete effluent percolation through artificial substrates. A wetland was built at a distance of 4.5 m from the photobioreactor that was 20 m wide, 40 m long, and 1.5 m deep, which was then attached to the photobioreactor. The 20-cm gravel (diameter: 15–25 mm) columns were filled at the bottom with a mixture of clay and sand layers of 60 cm (depth) for plant support. The six macrophytes, *Cyperus alternifolius*, *Scirpus tabernaemontani*, *Juncus effuses*, *Canna indica* Linn, *Pontederia cordata* and *Acorus gramineus Soland*, were planted at a density of nine rhizomes m^{-2}.

Filter-feeding fish (i.e., inhabiting different water depths) were added to the photobioreactor to keep the autotrophic photobioreactor in a self-modulated and self-sustained state. Accordingly, 200 kg of chub (*Squaliobarbus curriculus*) and 500 kg of bighead carp (*Hypophthalmichthys nobilis*) fingerlings with a length of 10–12 cm were introduced into the photobioreactor during the period from August to December 2006.

The wastewater (influent) directly flowed into the photobioreactor from a ditch in the intensive cropland. The hydraulic load of the experimental photobioreactor was 500 m^3 day^{-1} on nonrainy days (rainfall <10 mm) and the hydraulic retention time (HRT) was about 4.5 days. The inflow load of the wetland was 600 m^3 day^{-1}, and the HRT was 9.6 h. The wetland inflow was pumped from the photobioreactor while the outflow of the wetland was flowed back into the photobioreactor and sequentially discharged into the downstream ditch (Fig. 11.1).

11.2.2 Sampling and Analytical Methods

Water samples in triplicate (i.e., both influent and effluent) were collected from June 2006 to August 2008. The chemical oxygen demand (COD) in the water samples was measured by the potassium dichromate method. The nitrate nitrogen (NO_3-N), ammonia nitrogen (NH_4-N), and total dissolved phosphorus (TDP) in water were determined using the standard methods of the APHA (APHA et al., 1998). The samples for UV_{254} absorbance were filtered before being measured to eliminate the variations in UV absorption, which were caused by suspended particulate matter (Wu et al., 2005). In this study, the UV_{254}nm matter refers to all matters in the water having absorbance at 254 nm wavelength, including xenobiotic chemicals and natural compounds (e.g., natural humic substances). The dissolved oxygen (DO) and pH levels in water were measured in situ by a multimeter (YSI 52 dissolved oxygen and pH meters).

Sediment samples in triplicate, collected from the outlet of the photobioreactor (i.e., the inlet of the downstream ditch of the photobioreactor), were used to investigate the zoobenthos habitats. Each sample consisted of five grabs with surface areas of 50 cm^2 and depths of 10 cm. All fresh sediment samples were sieved through a 4.2-mm mesh to homogenize them. The homogenized samples were sieved through a 1.0-mm mesh to collect the zoobenthos. The material retained on the mesh was preserved in formalin (final concentration in sample: 8%). The intact zoobenthos were identified and counted. Identification procedures were similar to those of previous methods (Guo, 1995).

All water and sediment samples were collected 3 days after rainy days (rainfall ≥ 10 mm day^{-1}) to avoid the impact of precipitation on the data.

11.2.3 Data Analyses

Simpson's diversity index (D) (Simpson, 1949) was used to indicate the biodiversity of the photobioreactor habitat. The Simpson's diversity index was used as a community descriptor that represents the probability that two randomly selected individuals in the habitat belong to the same species (Simpson, 1949). When $D < 0.25$, the habitat is extremely polluted; when $0.25 \leq D < 0.50$, the habitat is heavily polluted; and when $0.50 \leq D < 0.75$, the habitat is moderately polluted. When D is between 0.75 and 1.00, the pollution level decreases from slightly polluted to clean (Wang et al., 2003).

The SPSS statistical software package (version 12.0) was used to analyze the data, and the level of statistical significance was set at $P = 0.05$. Statistically significant differences between the results were evaluated on the basis of standard deviation determinations and analysis of variance (ANOVA). The linear correlation between UV_{254} nm absorbance and COD was analyzed using the Pearson correlation.

11.3 RESULTS

11.3.1 Characteristics of Influent

Analyses of the influents showed that there was a good linear correlation between UV_{254} nm absorbance and COD concentration ($P < 0.05$). The linear equation used was Eq. (11.1) when the COD concentration was in the range of 11.3–180.4 mg L^{-1} (Fig. 11.2). Considering the measurement error of UV_{254} nm absorbance caused by suspended matter in water samples, the UV_{254} nm absorbance was calculated based on the COD concentration in this study as follows:

$$\text{COD} = 287.19\ \text{UV}_{254}\ \text{nm absorbance} - 17.485 \qquad (11.1)$$

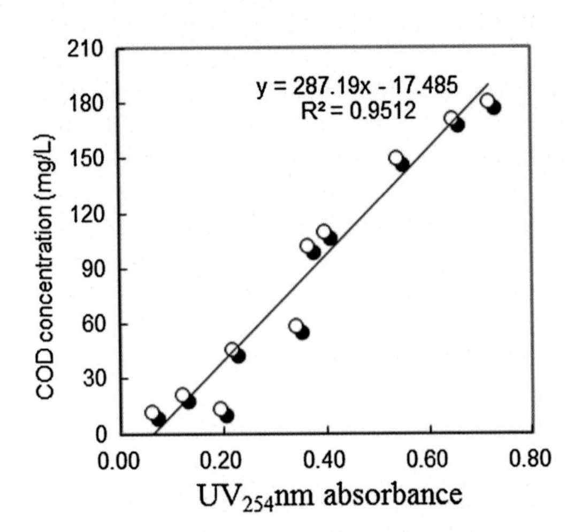

FIGURE 11.2 Correlation between COD concentration and UV_{254} nm absorbance in the influent ($n = 10$).

11.3.2 The Performance of the Photobioreactor–Wetland System

During the experiment, the DO in effluent was kept between 6.63 and 9.31 mg L^{-1}, and the pH ranged from 7.46 to 8.52. The average DO and pH values of the effluent were about 3.4 mg L^{-1} and 0.2 higher than those in the influent, respectively.

Before the photobioreactor–wetland system began running (from June to December 2006), the removal rate of UV_{254}nm matter was low, at 10–24%. However, during the running period of the photobioreactor–wetland system, the UV_{254}nm matter removal rates increased with time from 20% to 61% between January to November 2007 (adaptation period) and stabilized within a range of 60–81% between December 2007 and July 2008 (stabilized period) (Fig. 11.3). Furthermore, the overall average removal rate during this

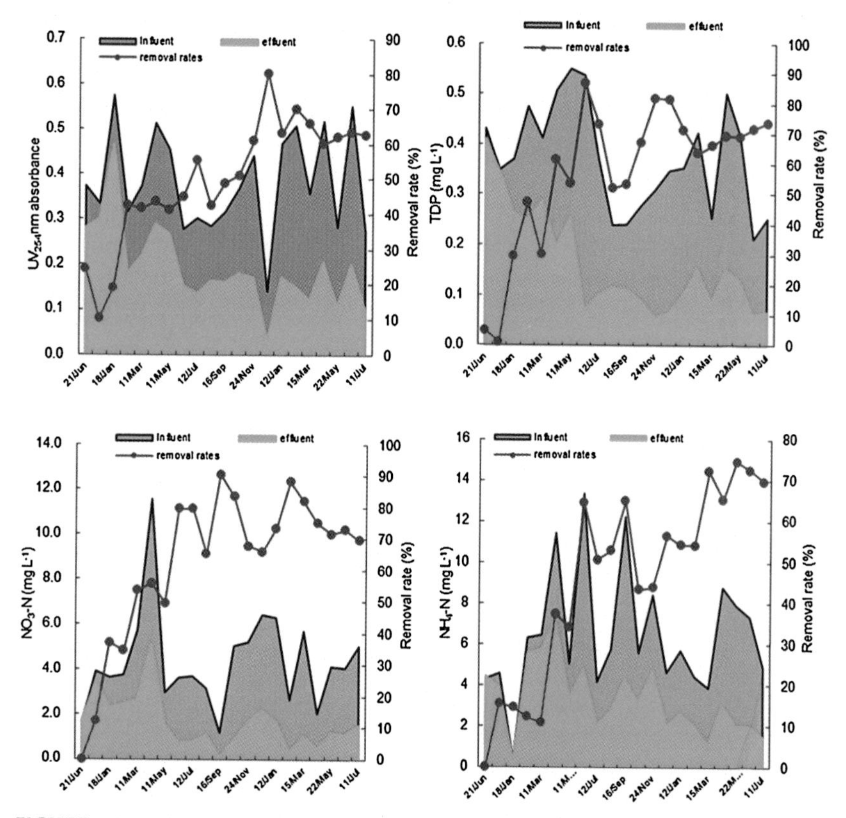

FIGURE 11.3 The removal efficiency of the photobioreactor–wetland system for treating UV_{254}nm matter, TDP, NO_3-N, and NH_4-N. The values from June to December 2006 were control values. The period from January to November 2007 was the adaptation period of the photobioreactor–wetland system. The photobioreactor–wetland system was operated under steady-state conditions after December 2007.

stabilized period was 66%, higher (by ~48%) than that at the beginning of the experiment to December 2006 (building period) (Fig. 11.3). The analysis showed that the average removal rate of UV_{254}nm absorbance in the stabilized period was significantly different from that during either the building period or the adaptation period ($P < 0.05$).

The overall average TDP removal rate was 3% before the photobioreactor–wetland system was run. The variation of the TDP removal rate in the adaptation period (i.e., from January to November 2007) was very large, ranging from 30% to 87% (Fig. 11.3). The overall average TDP removal rate in the stabilized period (i.e., from December 2007 to July 2008) was significantly (71%, $P < 0.05$) higher than those in the building (68%) and adaptation periods (by 13%), respectively.

The overall average NO_3-N removal rates ranged from 0% to 12% before the photobioreactor–wetland system began running (Fig. 11.3). The overall average NO_3-N increased from 37% to 91% between January and September 2007, followed by a rapid decrease from 91% to 68% between September and November 2007. During this adaptation period, the overall average NO_3-N removal rate was 63% (Fig. 11.3). It is worth noting that the removal rate increased remarkably from 65% to 88% between December 2007 and February 2008 and then rapidly decreased from 81% to 70% between March and July 2008. During this stabilized period, the overall average NO_3-N removal rate was 75% (Fig. 11.3). There were significant differences in the overall average NO_3-N removal rates among the building, adaptation, and stabilized periods of the photobioreactor-wetland system ($P < 0.05$).

The overall average NH_4-N removal rate was 8% during the period when the photobioreactor–wetland system was running, while the overall average NH_4-N removal rates were 39% and 65% during the adaptation and stabilized periods, respectively (Fig. 11.3). Analysis showed that the overall removal rate in the stabilized period of the photobioreactor–wetland system was significantly different from that in either the building or adaptation period ($P < 0.05$). The fluctuations of NH_4-N removal rates in the adaptation period were very large (ranging from 10% to 72%), while the changes of the NH_4-N in the stabilized period were relatively steady (ranging from 54% to 75%).

The capital cost of the photobioreactor–wetland system was estimated based on the local price level, which ranged from 100 to 120 US dollars per cubic meter of water, and the operation cost was about 0.015 US dollars per cubic water.

11.3.3 Response of Zoobenthos to Habitat Conditions

The species composition and biomass of the zoobenthos were examined in August 2006. Only 15 species of zoobenthos from 12 families were found at the inlet of the downstream ditch of the photobioreactor (Table 11.2). The average density and biomass of the zoobenthos were 960 individuals m^{-2} and

TABLE 11.2 The Species of Zoobenthos Identified in the Outlet of the Photobioreactor During August 2006 and 2008 (Wu et al., 2011)

Num.	Zoobenthos Species in August 2006	Zoobenthos Species in August 2008
1	*Glossiphonia complanata*	*Glossiphonia complanata*
2	*Lumbriculus variegatus*	*Lumbriculus variegatus*
3	*Cipangopaludina ventricosa*	*Cipangopaludina ventricosa*
4	*Ampullaria gigas spix*	*Ampullaria gigas spix*
5	*Margarya monody*	*Margarya monodi*
6	*Bellamya aeruginosa*	*Bellamya aeruginosa*
7	*Bellamya purificata*	*Bellamya purificata*
8	*Radix auricularia*	/
9	*Radix swinhoei*	*Radix swinhoei*
10	*Radix ovate*	*Radix ovata*
11	*Galba pervia*	*Galba pervia*
12	*Einfeldia* sp.	*Einfeldia* sp.
13	*Tokumagayuswika akamusi*	*Tokumagayuswika akamusi*
14	*Stratiomys* sp.	/
15	*Sphaerodema* sp.	/
16	/	*Procladius choreus*
17	/	*Parakiefferella sp.*
18	/	*Tanytarus sp.*
19	/	*Procladius sp.*
20	/	*Cipangopaludina dianchiensis*

245 g m^{-2}, respectively. The species composition and biomass of zoobenthos were also examined in August 2008. The average density and biomass of zoobenthos were 735 individuals m^{-2} and 214 g m^{-2}, respectively. *Radix auricularia*, *Stratiomys* sp., and *Sphaerodema* sp. were not found in August 2008. However, five new species, *Procladius choreus*, *Parakiefferella* sp., *Tanytarus* sp., *Procladius* sp., and *Cipangopaludina dianchiensis* appeared for the first time (Table 11.2).

To assess the effect of our technology on sediment conditions and water quality, Simpson's diversity indices of zoobenthos were calculated. The

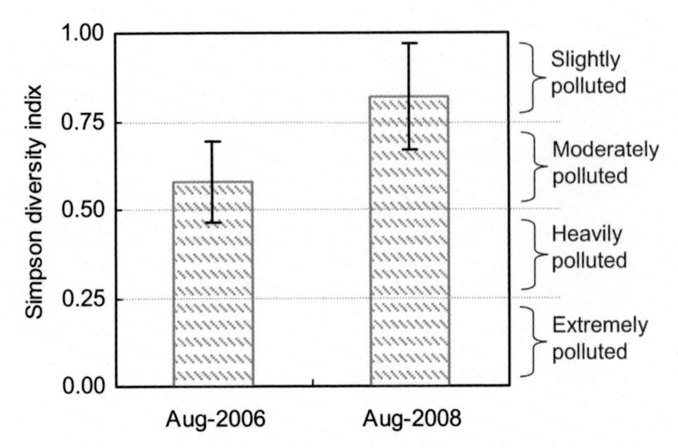

FIGURE 11.4 The assessment of the habitats of zoobenthos by the Simpson biodiversity index in the outlet of the photobioreactor (Wu et al., 2011).

pollution levels of zoobenthos habitats decreased between August 2006 and August 2008 (Fig. 11.4). In addition, the results of the Simpson diversity indices showed that the habitats of zoobenthos in the inlet of the downstream ditch of the photobioreactor improved from moderately polluted to slightly polluted.

11.4 DISCUSSION

UV_{254}nm absorbance has been used to monitor industrial wastewater effluents and organic pollution indicators such as lignin, tannin, humic substances, and carious aromatic compounds (Clesceri and Arnold, 2004). In this study, the influent was heterogeneous wastewater, which was combined with irrigation water, chemical fertilizers, pesticides, herbicides, and phytohormones. In some cases, when the UV_{254}nm represented the aromatic content of wastewater, the changes in the COD and aromatic content in the same wastewater samples were very similar (Balcıoğlu and Ötker, 2003; Ravikumar and Gurol, 1994). The significant correlation between the UV_{245}nm absorbance and COD concentration implies that the UV_{254}nm matter in this study might have originated primarily from aromatic compounds.

Aromatic compounds are a class of organic compounds that includes benzene ring molecular structures such as phenol, acetophenone, and isopropylbenzene (Alexieva et al., 2008). In addition to the natural origin of these compounds, anthropogenic inputs such as excessive use of various agrochemicals, including pesticides, herbicides, and phytohormones in agricultural fields, have contributed most to the introduction of aromatic compounds to environments. These aromatic compounds pose potential risks to humans and ecosystems (Yang et al., 2006). Thus, it is important to study the removal of these compounds in agricultural areas, especially in the upstream areas of surface waters.

The proposed photobioreactor–wetland system showed promising potential for the removal of aromatic compounds as seen by the dramatic reduction of UV_{254}nm absorbance in the stabilized period. In addition, the system exhibited some advantages in the removal of aromatic compounds. For instance, the removal of aromatic compounds in this system is a bioprocess that brings either a minimum amount or no potentially hazardous materials into the environment. Because it is inexpensive, effective, and easily deployed, the photobioreactor–wetland system can be used to remove aromatic compounds in undeveloped regions such as rural areas in developing countries.

The removal rate of TDP increased with the duration of the adaptation period of the photobioreactor–wetland system due to bioaccumulation (including uptake, assimilation, and biosorption). TDP is a preferential form of P in organisms. The depletion of TDP in water increased with the abundance of native organisms such as macrophytes and periphytic biofilms (Touchette and Burkholder, 2000). When the biomass of macrophytes and periphytic biofilms increased to a stable level, the P-nutrients demand of these organisms was kept at a specific level. For this reason, the TDP removal rates in the stabilized period of the photobioreactor–wetland system were kept at relatively steady.

The TDP removal rates rapidly decreased during the adaptation period between June and November 2007. This decrease might be attributed to the slowed growth of plants and the accelerating mineralization of dead plant tissues. The term "extractable biogenic phosphorus" has been proposed (Penn et al., 1995) because algae- and bacteria-produced inorganic P such as polyphosphate (Hupfer et al., 1995) are also included in this pool. Similarly, the lower the anoxic level, the larger the rate of decrease of the organic P content (Wu et al., 2005). This means that the anoxic conditions at the bottom of the constructed wetland have converted the available organic P into inorganic P. This conversion might have occurred because of the transformation of the pioneer bacterial community caused by the photobioreactor–wetland system (Wu et al., 2010b).

One of the reasons for the increased removal rates of NO_3-N during the adaptation period of the photobioreactor–wetland system is that the NO_3-N was extracted by the growing macrophytes. Multiple regression analysis indicated that NO_3-N and DO concentrations were significant predictors of denitrification rates; the increased DO concentration in water can accelerate the process of denitrification–nitrification (Piña-Ochoa and Álvarez-Cobelas, 2006). This suggests that the removal of NO_3-N was associated with the increased DO content in water during the operation of the photobioreactor–wetland system.

Previous studies showed that the habitation of aquatic plants (i.e., submerged macrophytes and phytoplankton) can have a positive impact on pH in water during periods of high productivity (Katsenovich et al., 2008). When pH levels become greater than 7, ammonium (NH_4^+) can be spontaneously converted into ammonia gas (NH_3), which allows it to diffuse from the vicinity of the air−water interface (El-Shafai et al., 2007). Considering that the NH_4-N removal rate increased with time, it is likely that most ammonia losses in the photobioreactor−wetland system are due to the uptake and assimilation by macrophytes and phytoplankton as well as the irreversible diffusion of gaseous ammonia from the water. Indeed, both macrophyte and phytoplankton uptake are integral to treatment within the photobioreactor (Koottatep and Polprasert, 1997). The decreases in NH_4-N removal rates during winter (3−18°C, from November 2007 to February 2008) revealed that the NH_4-N removal rate decreased with the decline in temperature. This finding is consistent with previous studies showing that the rate of ammonia removal decreased with decreases in temperature in biological treatment systems (Tseng and Wu, 2004).

When the photobioreactor−wetland system was in operation, the organic debris became fine (Wang et al., 2002) due to disturbance by fish and the cluster of macrophytes (Shrimpton et al., 2007), which lead to the debris not reaching the bottom as freely as it had previously. As a result, the water quality improved with the decrease of UV_{254}nm matter (aromatic compounds) and nutrients. The emergence of five new species in the inlet of the upstream ditch of the photobioreactor provided further evidence of water quality improvement because these species only thrive in clean water. Subsequently, several species of zoobenthos whose food consisted of macro-organic debris faced famine and died. In addition, the species of zoobenthos that had disappeared in the early part of the experiment reappeared when the water became clear (Wang et al., 2002, 2003). In addition, the fish (i.e., older filter-feeding fish) ate some of the zoobenthic animals (Ma et al., 2010) at the same time that the less-dominant species died out. The replanted macrophytes became the habitat for zoobenthos and zooplankton, which led the dominant species to recover and grow (Wu et al., 2010b).

11.5 CONCLUSION

This study showed that influent could be characterized using UV_{254}nm matter. There was a significant correlation between the UV_{254}nm absorbance and COD concentration in the influent, implying that the UV_{254}nm matter might originate mainly from aromatic compounds. The proposed photobioreactor−wetland system was implemented on a pilot scale for 2 years and proved to be highly effective in removing UV_{254}nm matter and dissolved nutrients such as TDP, NO_3-N, and NH_4-H. The increasing Simpson biodiversity indices of zoobenthos suggested that the photobioreactor−wetland system could provide a suitable habitat for zoobenthos due to the reduction of both organic contaminants and nutrients. This system is simple in terms of its construction, operation, and maintenance; therefore, it is suitable for use in rural areas to prevent the risk of pollution events such as eutrophication and harmful algal blooms in downstream surface waters.

REFERENCES

Alexieva, Z., Gerginova, M., Zlateva, P., Manasiev, J., Ivanova, D., Dimova, N., 2008. Monitoring of aromatic pollutants biodegradation. Biochemical Engineering Journal 40, 233–240.

APHA, AWWA, WPCF, 1998. Standard Methods for the Examination of Water and Wastewater, twentieth ed. American Public Health Association, Washington, DC.

Balcıoğlu, I., Ötker, M., 2003. Treatment of pharmaceutical wastewater containing antibiotics by O_3 and O_3/H_2O_2 processes. Chemosphere 50, 85–95.

Blanchoud, H., Moreau-Guigon, E., Farrugia, F., Chevreuil, M., Mouchel, J.M., 2007. Contribution by urban and agricultural pesticide uses to water contamination at the scale of the Marne watershed. Science of the Total Environment 375, 168–179.

Chae, S.-R., Noeiaghaei, T., Jang, H.-C., Sahebi, S., Jassby, D., Shon, H.-K., Park, P.-K., Kim, J.-O., Park, J.-S., 2015. Effects of natural organic matter on separation of the hydroxylated fullerene nanoparticles by cross-flow ultrafiltration membranes from water. Separation and Purification Technology 140, 61–68.

Cheng, X., Liang, H., Ding, A., Qu, F., Shao, S., Liu, B., Wang, H., Wu, D., Li, G., 2016. Effects of pre-ozonation on the ultrafiltration of different natural organic matter (NOM) fractions: membrane fouling mitigation, prediction and mechanism. Journal of Membrane Science 505, 15–25.

Clesceri, L.S., Arnold, E., 2004. Standard Methods for the Examination of Water and Wastewater, twentieth ed. American Public Health Association, Washington, DC.

Dąbrowski, A., Podkościelny, P., Hubicki, Z., Barczak, M., 2005. Adsorption of phenolic compounds by activated carbon-a critical review. Chemosphere 58, 1049–1070.

de la Rubia, Á., Rodríguez, M., León, V.M., Prats, D., 2008. Removal of natural organic matter and THM formation potential by ultra- and nanofiltration of surface water. Water Research 42 (3), 714–722.

Dorioz, J.M., Wang, D., Poulenerd, J., Trévisan, D., 2006. The effect of grass buffer strips on phosphorus dynamics – a review and synthesis as a basis for application in agricultural landscapes in France. Agriculture, Ecosystems and Environment 117, 4–21.

El-Shafai, S.A., El-Gohary, F.A., Nasr, F.A., Van der Steen, N.P., Gijzen, H.J., 2007. Nutrient recovery from domestic wastewater using a UASB-duckweed ponds system. Bioresource Technology 98, 798–807.

Fabbri, D., Prevot, A.B., Zelano, V., Ginepro, M., Pramauro, E., 2008. Removal and degradation of aromatic compounds from a highly polluted site by coupling soil washing with photocatalysis. Chemosphere 71, 59–65.

Ganiyu, S.O., van Hullebusch, E.D., Cretin, M., Esposito, G., Oturan, M.A., 2015. Coupling of membrane filtration and advanced oxidation processes for removal of pharmaceutical residues: a critical review. Separation and Purification Technology 156 (Part 3), 891–914.

Guo, W., He, M., Yang, Z., Lin, C., Quan, X., Wang, H., 2007. Distribution of polycyclic aromatic hydrocarbons in water, suspended particulate matter and sediment from Daliao River watershed, China. Chemosphere 68, 93–104.

Guo, X.W., 1995. Studies on Chironomid communities of Nanhu Lake (South Lake), Wuhan, China. Journal of Huazhong Agricultural University 14, 578–585.

Häggblom, M.M., 1992. Microbial breakdown of halogenated aromatic pesticides and related compounds. FEMS Microbiology Letters 103, 29–71.

Haas, G., Wetterich, F., Köpke, U., 2001. Comparing intensive, extensified and organic grassland farming in southern Germany by process life cycle assessment. Agriculture, Ecosystems and Environment 83, 43–53.

Han, Q., Yan, H., Zhang, F., Xue, N., Wang, Y., Chu, Y., Gao, B., 2015. Trihalomethanes (THMs) precursor fractions removal by coagulation and adsorption for bio-treated municipal wastewater: molecular weight, hydrophobicity/hydrophily and fluorescence. Journal of Hazardous Materials 297, 119–126.

Hupfer, M., Gachter, M.R., Giovanoli, R.R., 1995. Transformation of phosphorus species in settling seston and during early sediment diagenesis. Aquatic Science 57, 305–324.

Jiang, Y., Goodwill, J.E., Tobiason, J.E., Reckhow, D.A., 2016. Impacts of ferrate oxidation on natural organic matter and disinfection byproduct precursors. Water Research 96, 114–125.

Katsenovich, Y., Shapovalova, L., But, L., Ijitskaja, M., 2008. Evaluation of biological pond system modified with submerged planted dams. Ecological Engineering 33, 1–7.

Kay, P., Edwards, A.C., Foulger, M., 2009. A review of the efficacy of contemporary agricultural stewardship measures for ameliorating water pollution problem of key concern to the UK water industry. Agricultural Systems 99, 67–75.

Kelley, K.B., Riechers, D.E., 2007. Recent developments in auxin biology and new opportunities for anxinic herbicide research. Pesticide Biochemistry and Physiology 89, 1–11.

Kinsley, C., Nicell, J.A., 2000. Treatment of aqueous phenol with soybean peroxidase in the presence of polyethylene glycol. Bioresource Technology 73, 139–146.

Kong, W.D., Zhu, Y.G., Fu, B.J., Han, X.Z., Zhang, L., He, J.Z., 2008. Effect of long-term application of chemical fertilizers on microbial biomass and functional diversity of a black soil. Pedosphere 18, 801–808.

Koottatep, T., Polprasert, C., 1997. Role of plant uptake on nitrogen removal in constructed wetlands located in the tropics. Water Science and Technology 36, 1–8.

Liu, P., Zhu, D., Zhang, H., Shi, X., Sun, H., Dang, F., 2008. Sorption of polar and nonpolar aromatic compounds to four surface soils of eastern China. Environmental Pollution 156, 1053–1060.

Ma, H., Cui, F., Liu, Z., Fan, Z., He, W., Yin, P., 2010. Effect of filter-feeding fish silver carp on phytoplankton species and size distribution in surface water: a field study in water works. Journal of Environmental Sciences 22, 161–167.

Novák, J., Beníšek, M., Hilscherová, K., 2008. Disruption of retinoid transport, metabolism and signaling by environmental pollutants. Environment International 34, 898–913.

Pan, G., Li, L., Zhao, D., Chen, H., 2010. Immobilization of non-point phosphorus using stabilized magnetite nanoparticles with enhanced transportability and reactivity in soils. Environmental Pollution 158, 35–40.

Peng, D., Saravia, F., Abbt-Braun, G., Horn, H., 2016. Occurrence and simulation of trihalomethanes in swimming pool water: a simple prediction method based on DOC and mass balance. Water Research 88, 634–642.

Penn, M.R., Auer, M.T., Van Orman, E.L., Korienek, J.J., 1995. Phosphorus diagenesis in lake sediments: investigations using fractionation techniques. Marine Freshwater Research 46, 89–99.

Piña-Ochoa, E., Álvarez-Cobelas, M., 2006. Denitrification in aquatic environments: a cross system analysis. Biogeochemistry 81, 111–130.

Ravikumar, J.X., Gurol, M.D., 1994. Chemical oxidation of chlorinated organics by hydrogen peroxide in the presence of sand. Environmental Science and Technology 28, 394–400.

Sadrnourmohamadi, M., Gorczyca, B., 2015. Effects of ozone as a stand-alone and coagulation-aid treatment on the reduction of trihalomethanes precursors from high DOC and hardness water. Water Research 73, 171–180.

Schulz, R., Peall, S.K.C., 2001. Effectiveness of a constructed wetland for retention of nonpoint-source pesticide pollution in the Lourens River catchment, South Africa. Environmental Science and Technology 35, 422–426.

Shrimpton, L.M., Zydlewski, J.D., Heath, J.W., 2007. Effect of daily oscillation in temperature and increased suspended sediment on growth and smolting in juvenile Chinook salmon, *Oncorhynchus tshawytscha*. Aquaculture 273, 269–276.

Sillanpää, M., Matilainen, A., 2015. Chapter 3-NOM removal by coagulation. In: Natural Organic Matter in Water. Butterworth-Heinemann, pp. 55–80.

Simpson, E.H., 1949. Measurement of diversity. Nature 163, 688.

Tett, V.A., Willetts, A.J., Lappin-Scott, H.M., 1994. Enantioselective degradation of the herbicide mecoprop [2-(2-methyl-4-chlorophenoxy) propionic acid] by mixed and pure bacterial cultures. FEMS Microbiology Ecology 14, 191–199.

Touchette, B.W., Burkholder, J.M., 2000. Review of nitrogen and phosphorus metabolism in seagrasses. Journal of Experimental Marine Biology and Ecology 250, 133–167.

Tseng, K.F., Wu, K.L., 2004. The ammonia removal cycle for a submerged biofilter used in a recirculating eel culture system. Aquacultural Engineering 31, 17–30.

Wang, L.Z., Liu, Y.D., Chen, X.D., Xiao, B.D., 2003. Invertebrate community construction of zoobenthos and water quality assessment in Macun and Haidong Gulfs in Dianchi Lake. Reservoir Fisheries (in Chinese) 23, 47–49.

Wang, L.Z., Xu, X.Q., Zhou, W.B., 2002. A study on the zoobenthos in Macun and Haodongwan region of Dianchi Lake Yunnan. Journal of Yunnan University (in Chinese) 24, 134–139.

Wu, Y., Feng, M., Liu, J., Zhao, Y., 2005. Effects of polyaluminium chloride and copper sulfate on phosphorus and UV_{254} under different anoxic levels. Fresenius Environmental Bulletin 14, 406–412.

Wu, Y., He, J., Hu, Z., Yang, L., Zhang, N., 2011. Removal of UV 254 nm matter and nutrients from a photobioreactor-wetland system. Journal of Hazardous Materials 194, 1–6.

Wu, Y., Kerr, P.G., Hu, Z., Yang, L., 2010a. Eco-restoration: simultaneous nutrient removal from soil and water in a complex residential-cropland area. Environmental Pollution 158 (7), 2472–2477. Available online 27 April 2010.

Wu, Y., Kerr, P.G., Hu, Z., Yang, L., 2010b. Removal of cyanobacterial bloom from biopond-wetland system and the associated response of zoobenthic diversity. Bioresource Technology 101, 3903–3908.

Yang, H., Jiang, Z., Shi, S., 2006. Aromatic compounds biodegradation under anaerobic conditions and their QSBR models. Science of the Total Environment 358, 265–276.

Zhang, Y., Zhuang, Y., Geng, J., Ren, H., Xu, K., Ding, L., 2016. Reduction of antibiotic resistance genes in municipal wastewater effluent by advanced oxidation processes. Science of the Total Environment 550, 184–191.

Chapter 12

Hybrid Bioreactor Based on Periphyton: The Removal of Nutrients From Nonpoint Source Wastewater

12.1 INTRODUCTION

Nonpoint source pollution refers to water pollution from diffuse sources (Wilson et al., 1986). Nonpoint source water pollution negatively influences water bodies from sources such as polluted runoff from agricultural areas draining into a river, or windborne debris blowing out to sea. Nonpoint source pollution may derive from many different sources with no specific solution able to rectify the problem, making it difficult to regulate. Nonpoint source water pollution is difficult to control because it comes from the everyday activities of many different people, such as fertilizing a lawn, using a pesticide, or constructing a road or building (Zheng et al., 2014).

The input of nonpoint source pollution is becoming a main source of pollution events, such as black water agglomerates in rivers, eutrophication, and harmful algal blooms in downstream waters, in both developed and developing countries after the control of point source pollution (Wu et al., 2010b). The most important contributor of nonpoint source wastewater is nutrient losses from farmland, grassland, forest, and road surfaces (Ongley et al., 2010). Nutrients mainly refers to inorganic matter from runoff, landfills, livestock operations, and crop lands. The two primary nutrients of concern are phosphorus and nitrogen (Cha et al., 2016; Ouyang et al., 2014). Phosphorus is a main ingredient in many fertilizers used for agriculture as well as on residential and commercial properties, and may become a limiting nutrient in freshwater systems and some estuaries (Cho et al., 2016; Hao et al., 2012; Turner et al., 2013). Phosphorus is most often transported to water bodies via soil erosion because many forms of phosphorus tend to be adsorbed on to soil particles (Ferreira et al., 2016; Tomer et al., 2016; Zimmer et al., 2016). Nitrogen is the other key ingredient in fertilizers, and it generally becomes a pollutant in saltwater or brackish estuarine systems where nitrogen is a

Periphyton. http://dx.doi.org/10.1016/B978-0-12-801077-8.00012-0

limiting nutrient (Hosseini et al., 2016; Tong et al., 2016; Xing et al., 2016). Similar to phosphorus in freshwaters, excess amounts of bioavailable nitrogen lead to eutrophication and algal blooms (Tekile et al., 2015; Wu et al., 2015).

Nitrogen and phosphorus are the major pollutants in nonpoint source wastewater and are the leading causes of degeneration in water quality and the degradation of ecosystem function (Wu et al., 2010a). Therefore, it is very important that an integrated technology is developed to simultaneously remove nitrogen and phosphorus from nonpoint source wastewater to protect the water quality in downstream surface aquatic ecosystems.

To date, many ecological measures have been developed to reduce nonpoint source pollution (Wu et al., 2010a). They can be summarized into three classes: (1) ecological management based on the model of agricultural nonpoint source pollution at the watershed scale (Wang et al., 2008) such as the best management practices issued by the US Environmental Protection Agency (USEPA), (2) integrated ecological restoration to improve the self-purifying function of ecosystems, such as hierarchical eco-restoration (Wu et al., 2010a,b) and GIS-based ecological–economic modeling (Lant et al., 2005), and (3) the construction of environmental engineering such as vegetative strips between pollution sources and receiving water bodies (Duchemin and Hogue, 2009), riparian zones (Hefting et al., 2006), constructed wetlands (Kay et al., 2007), and agricultural drainage ditches (Moore et al., 2010).

The application of the aforementioned measures helps to filter nonpoint source wastewater, promote sedimentation of the suspended particles and the pollutants bound to them, and recover ecological system function (Duchemin and Hogue, 2009; Wu et al., 2010a,b). These mitigation measures are suited to the current socioeconomic context in watershed (Duchemin and Hogue, 2009; Gunes, 2008), in which the adoption of simple and inexpensive agri-environmental practices is advocated with a view to protecting water quality and managing farm fertilizer use.

Nonpoint source wastewater originating in urban/suburban areas is caused by both natural factors (i.e., rainfall) and human activities (i.e., the diffuse wastewater discharges of industrial parks and irrigation) (Chen et al., 2008; Martínez et al., 2000). These human activities make the composition of nonpoint source wastewater more complex and heterogeneous. For example, some heavy metals might be introduced by industrial activities, which lead to changes in the properties of nonpoint source wastewater. Moreover, the concentrations of some components of the nonpoint source wastewater might be increased due to the combination of high loading diffuse wastewater from industrial production in the urban/suburban areas. In addition, the problem of controlling nonpoint source wastewater in suburban/urban areas is complicated by the vast territory and the multiple types of land use (Duchemin and Hogue, 2009). Thus, it is necessary to explore new technology or integrate current technologies to control and manage nonpoint source wastewater in urban/suburban areas.

The activated sludge process, such as the A^2/O process, has been widely applied to treat high loading wastewater and can simultaneously remove organic matter, nutrients (e.g., nitrogen and phosphorus) and metals (e.g., chromium and arsenic) because it is highly efficient and environmentally friendly (Samaras et al., 2009; Ying et al., 2010). Photoautotrophic systems such as wetlands and ecological ditches are often used to purify low loading wastewater because it is inexpensive and environmentally benign. However, the treatment of sludge in wetlands enhances the cost and limits its application in many developing or underdeveloped areas. The ecological ditches typically require a large area of land for large-scale engineering, which conflicts with the local residents' need for cropland and industrial land use (Wu et al., 2010a). Thus, it is practical to combine the advantages of these two technologies and overcome the shortages of both in treating "real-world" wastewater.

Therefore, we propose an integrated technology-hybrid bioreactor combined A^2/O and ecological ditch to simultaneously remove high-loading nitrogen and phosphorus from heterogeneous nonpoint source wastewater and recover the microbial habitats. In the proposed biointegrated solution, we believe the most important issue is to bring the aquatic ecosystem in the hybrid bioreactor to a self-modulating and self-sustaining state. To facilitate industrial-scale application, three additional considerations should be taken into account: (1) the hybrid bioreactor should be easily employed, inexpensive, and highly efficient; (2) the technology should be environmentally benign; and (3) the ecosystem should be self-cycling and the microbial habitats should be recovered.

12.2 MATERIALS AND METHODS

12.2.1 Hybrid Bioreactor

The hybrid bioreactor consisted of A^2/O processes and an ecological ditch (eco-ditch) (Fig. 12.1). There are eight different compartments within the hybrid bioreactor: (1) the depositional tank ($\sim 30\ m^3$) was planted with macrophytes (*Canna indica*, *Juncus minimus*, and *Cyperus alternifolius* species) with a planting density of $0.5\ m \times 0.5\ m$; (2) the anaerobic tank ($\sim 96\ m^3$) was filled with coarse gravels (diameter 3–10 cm); (3) the overflow pool ($\sim 4\ m^3$) was to reduce suspended materials; (4) the settling tank ($\sim 24\ m^3$) was to further reduce suspended materials; (5) the anoxic fluidized bed ($\sim 72\ m^3$) contained biofilm substrates (Industrial Soft Carriers, Wuxi Guozhen Environmental Protection Co. Ltd.) with a density of $0.3\ m^3$; (6) the aerobic fluidized bed ($\sim 72\ m^3$) contained suspended biofilm substrates "Artificial Aquatic Mats" (Wuhan Zhongke Environmental Engineering Co. Ltd.), also having density of $0.3\ m^3$; (7) the clarification tank ($\sim 24\ m^3$) was to reduce the suspended materials; and (8) The eco-ditch had a total length of 230 m and average width of 2.5 m (soil wall gradient 45°). A series of $0.04\text{-}m^3$

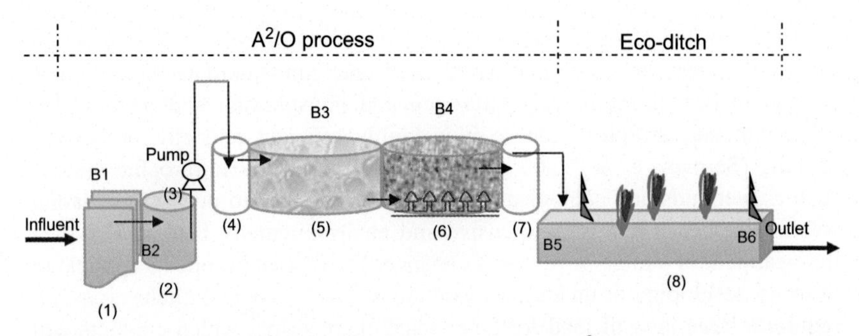

FIGURE 12.1 The schematic of the hybrid bioreactor combined A^2/O process and eco-ditch. The A^2/O process consisted of (1) deposition pool, (2) anaerobic tank filled with gravel, (3) overflow pool, (4) settling tank, (5) anoxic fluidized bed, (6) aerobic fluidized bed, and (7) clarification tank. The (8) eco-ditch is a photoautotrophic system. The *arrow* refers to direction of water flow. Water and biofilm sampling sites are from B1 to B6.

nylon tanks, containing ceramsite adsorbent (Kunming Yuxi Materials Co. Ltd.), was placed on the bottom of the ecological trunk channel at 2.0 m intervals for the adsorption of pollutants from the wastewater. Macrophytes, including *Scirpus tabernaemontani*, *Canna indica*, *Zizania latifolia*, *J. minimus*, *C. alternifolius*, *Zantedeschia aethiopica*, and *Acorus calamus*, were planted along the walls of the eco-ditches at 0.5-m intervals.

The flow of the wastewater (influent) in the hybrid bioreactor was as follows: (1) the influent flowed into the deposition pool and then into the anaerobic tank; (2) the wastewater was pumped into the settling tank, and then flowed into the anoxic fluidized bed, aerobic fluidized bed, and clarification tank; (3) the water overflowed into the eco-ditch, and then was discharged.

12.2.2 Experimental Design

The nonpoint source wastewater to the hybrid bioreactor was combined as follows: effluent of the diffuse wastewater from communities, processing wastewater from an industrial park, and wastewater from road surface runoff, Liangjia Village, Kunming, western China. The loading of nonpoint source wastewater was approximately 200 m^3 day^{-1} on nonrainy days. Therefore, the quantity of the processing wastewater was around 80 m^3 daily.

To obtain native microorganisms and facilitate large-scale industrial application, the microorganism aggregates (biofilms) were inoculated in the hybrid bioreactor. A total of 0.6 m^3 active sludge from a domestic wastewater treatment plant was put into the anoxic and aerobic fluidized beds. The biofilms were cultivated and incubated in the hybrid bioreactor under natural conditions for 15 days before the experiment. The 200 m^3 nonpoint source wastewater was treated by the hybrid bioreactor every day. The running mode of liquid throughputs was in 12-h intervals (12 h/12 h). The natural air temperature ranged from 8 to 31°C during the entire experimental period.

To support the growth of native microorganisms in the eco-ditch, the high concentration sludge (sediments) liquid at the bottom of the anoxic and aerobic fluidized beds was pumped directly into the eco-ditch using a strong-pressure sludge-pump in interval time. To avoid the effects of sludge (sediments) impact on data collection, the water was sampled 10 days after the sludge was discharged.

12.2.3 Samples and Analyses

To avoid the impact of rain on the data, the samples were collected 3 days after rainy days (rainfall $>$10 mm). Water samples were collected in triplicate from sampling sites B1 to B6 (Fig. 12.1). Total phosphorus (TP) and total dissolved phosphorus (TDP) were measured calorimetrically by the persulfate digestion—molybdophosphate reaction method. Total nitrogen (TN) was measured by the persulfate digestion and oxidation-double wavelength (220 and 275 nm) method. Ammonia-N (NH_4-N) and nitrate-N (NO_3-N) were determined by ion-selective electrode potentiometry with preliminary distillation. Detailed procedures are described according to the standard method (APHA-AWWA-WEF, 1998). The DO and pH levels in water were measured in situ by a multimeter (YSI 52 m).

The arsenic (As) concentration in water was determined with cold vapor atomic fluorescence spectrometry (QM201D, Jiangsu). Chromium (Cr(VI)) concentration was measured with 1,5-diphenyicarbazide spectrophotometry. These detailed procedures are described in the national standard methods of water and wastewater analyses for China (ChinaEPA, 2002b).

Biofilm samples were collected in triplicate at random locations from substrates in the anaerobic tank, anoxic and aerobic fluidized beds, as well as the eco-ditch and kept at 25—30°C until their moisture levels were reduced to ~85%. The biofilm was then weighed and the total biofilm biomass in the anaerobic tank and anoxic and fluidized beds was estimated based on the biofilm weight and specific surface area of the substrates.

Dice index (Cs) of similarity (LaPara et al., 2002) was used to evaluate the similarity of bacterial community structures based on ERIC-PCR fingerprints. Before the hybrid bioreactor was started, native biofilm samples were peeled off the surfaces of stones in the entrance (A1), middle (A2), and end (A3) of the ditch. After the hybrid bioreactor had started, the biofilm samples were collected from the substrates along the sampling sites from B1 and B6 (Fig. 12.1).

The quantified methods of Dice index have been reported previously (Miura et al., 2007). Total DNA extraction and purification of biofilms were conducted for the ERIC-PCR analyses. Total DNA was isolated from the biofilm samples following a procedure modified from a previous study (Hill et al., 2002). One-mL biofilm sample aliquots were thawed in an ice-bath, and the cells were harvested by centrifugation at 9000 $\times$ g for 5 min. DNA was then purified by sequential extraction with Tris-equilibrated phenol, phenol, and chloroform-isoamyl alcohol (vol/vol/vol, 25:24:1), and chloroform isoamyl alcohol (vol/vol, 24:1) followed by precipitation with two volumes of ethanol. DNA was collected by centrifugation, air-dried and dissolved in 50-μL sterile TE buffer. The detailed procedures are described in Wei et al. (2004).

Community fingerprints were obtained for bacteria in the biofilm using total bacterial DNA as templates for ERIC-PCR. The sequence of the ERIC primers was based on previous work (Li et al., 2006), E1 (ERIC-PCR): 5′-ATGTAAGCTCCTGGGGATTCA C-3′; E2 (ERIC-PCR): 5′-AAGTAAGT-GACTGGGGTGAGCG-3′. The detailed procedures have been described previously (Li et al., 2006).

SPSS statistical software (version 12.0) was used for analyzing the data, and the level of statistical significance was set at $P < 0.05$. Statistically significant differences between the results were evaluated on the basis of standard deviation determinations and on the analysis of variance method (one-way ANOVA).

12.3 RESULTS AND DISCUSSION

12.3.1 Characteristics of the Nonpoint Source Wastewater

The physicochemical parameters of the nonpoint source wastewater are shown in Table 12.1. The nonpoint source wastewater had a high nutrient concentration including TN (27.20 mg L^{-1}) and TP (2.46 mg L^{-1}). In addition, the nonpoint source wastewater had high As and Cr(VI) concentrations. The concentration of As (113.77 μg L^{-1}) and Cr (VI) (105.32 μg L^{-1}) were higher than the baseline values (100 μg L^{-1} for As and 100 μg L^{-1} Cr(VI)) of the worst level of National Water Quality Standard of Water Environments, China (China EPA, 2002a), implying that the nonpoint source wastewater was heavily polluted by As and Cr(VI).

The pH in the influent of the hybrid bioreactor (B1 sampling site in deposition pool) ranged from 7.2 to 8.4 between April 2007 and May 2008. The DO was from 0 to 1.8 mg L^{-1} between April 2007 and May 2008. When the hybrid bioreactor was in use, the pH in aerobic fluidized bed and eco-ditch ranged from 6.8 to 8.2 and the DO ranged from 7.2–9.8 mg L^{-1}.

12.3.2 Microorganisms in the Hybrid Bioreactor

The microscopic studies showed the presence of bacteria (e.g., methanosarcina, diplobacillus, bacilli, brevibacterium, and cocci), chladophora, diatoms (e.g., *Cyclostephanos dubius*, *Aulacoseira granulate*, and *Stephanodiscus minutulus*) and cyanobacteria (e.g., *Microsystis aeruginosa* and *Aphanizomenon flos-aquae*) within the hybrid bioreactor. These observations indicate that the hybrid bioreactor could simultaneously culture heterotrophic and autotrophic microorganisms.

TABLE 12.1 The Physicochemical Pre-experimental Parameters of the Complex Wastewater ($n = 23$) (Wu et al., 2011)

	pH	DO ($mg\ L^{-1}$)	TN ($mg\ L^{-1}$)	NO_3-N ($mg\ L^{-1}$)	NH_4-N ($mg\ L^{-1}$)
Mean ± SD	7.6 ± 0.40	1.40 ± 0.30	27.20 ± 5.64	6.39 ± 2.72	18.16 ± 4.87
	TDP ($mg\ L^{-1}$)	TP ($mg\ L^{-1}$)	COD ($mg\ L^{-1}$)	Cr (VI) ($\mu g\ L^{-1}$)	As ($\mu g\ L^{-1}$)
Mean ± SD	0.66 ± 0.17	2.46 ± 0.80	146.60 ± 45.49	105.32 ± 5.02	113.77 ± 2.52

12.3.3 The Removal of Phosphorus and Nitrogen

The average removal rates in the A^2/O process were 60% and 54% for TP and TDP, respectively. The variations in TP and TDP removal rates were from 50% to 81% and from 44% to 70%, respectively (Fig. 12.2A and B). The average removal rates in the eco-ditch were 56% for TD and 48% for TDP during the experimental period (Fig. 12.2C and D). The overall average TP and TDP removal rates of the hybrid bioreactor were 81% and 74%, respectively.

The TP and TDP concentrations in the effluent of the hybrid bioreactor were from 0.16 to 1.01 mg L^{-1} and from 0.09 to 0.42 mg L^{-1} between April 2006 and May 2008, respectively (Fig. 12.2E and F). The variations in TP and

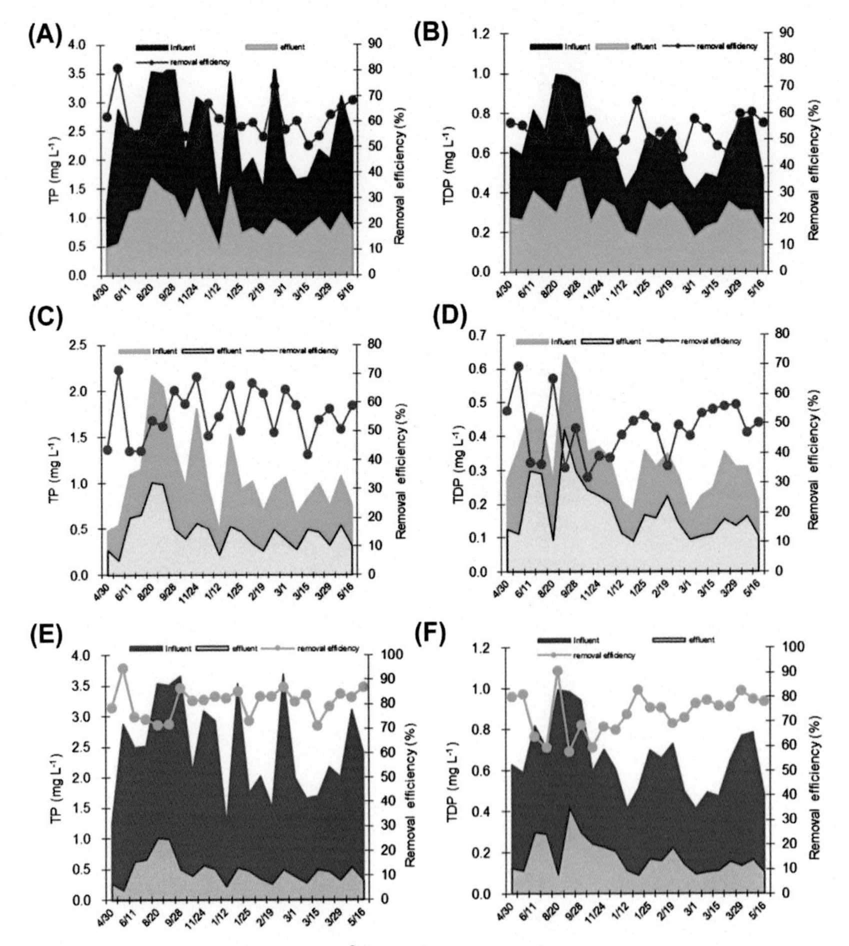

FIGURE 12.2 The efficiency of the A^2/O process (A, B), eco-ditch (C, D), and the overall hybrid bioreactor (E, F) in removing total phosphorus (TP) and total dissolved phosphorus (TDP).

TDP removal rates in the hybrid bioreactor were relatively steady, from 71% to 95% and from 58% to 90%, respectively (Fig. 12.2E and F), implying that the hybrid bioreactor is capable of removing TD and TDP from nonpoint source wastewater.

The TN content in the influent was dominated by NH_4-N. The proportion of NH_4-N to TN ranged from 51% to 85%. The average TN, NO_3-N, and NH_4-N removal rates in the A^2/O process were 63%, 56%, and 69%, respectively, whilst their removal rates in the eco-ditch were 52%, 53%, and 58%, respectively (Figs. 12.3 and 12.4).

The overall average TN, NO_3-N, and NH_4-N removal rates in the hybrid bioreactor were 82%, 79%, and 86%, respectively (Fig. 12.3). The TN, NO_3-N, and NH_4-N removal rates of the hybrid bioreactor were steady during the experimental period, resulting from the small variations in TN and NO_3-N removal rates ranging from 72% to 89%, 71% to 86%, and 81% to 94%, respectively (Fig. 12.3), implying that the hybrid bioreactor could efficiently remove TN, NO_3-N, and NH_4-N from nonpoint source wastewater.

Biological measures such as A^2/O, active sludge processes, and membrane bioreactors are preferred measures in purifying wastewater of high nutrient

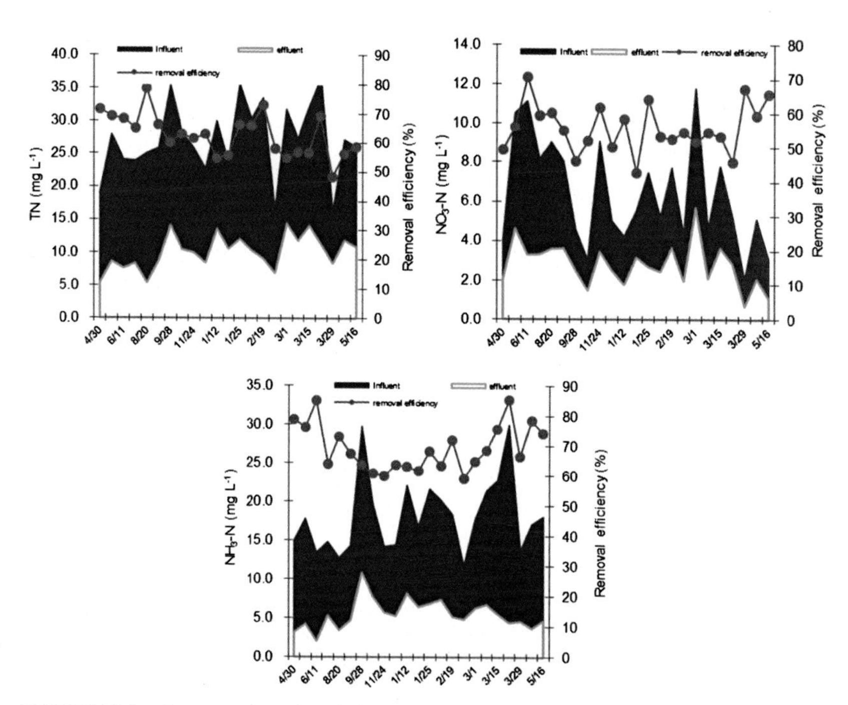

FIGURE 12.3 Concentration of total nitrogen (TN), nitrate (NO_3-N), and ammonia (NH_4-N) in influent and effluent and their removal rates in A^2/O process of the hybrid bioreactor from April 2007 to May 2008.

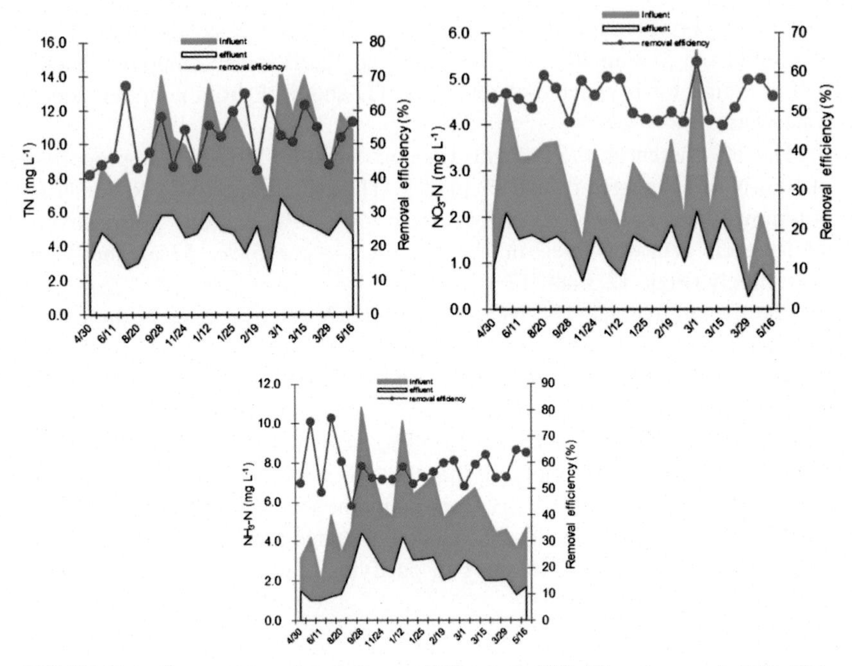

FIGURE 12.4 Concentration of total nitrogen (TN), nitrate (NO_3-N), and ammonia (NH_4-N) in influent and effluent and their removal rates in the eco-ditch of the hybrid bioreactor from April 2007 to May 2008.

loads because these biological measures are highly efficient, cost-effective, and environmentally friendly (Ahn et al., 2003; Wang et al., 2009). However, the common problem in many cases is when the wastewater contains some toxic metals (e.g., As, Cr (VI)), nutrient removal efficiency by biological measures is negatively affected (Lin et al., 2010; Schaller et al., 2010). In this study, the combination of the A^2/O process and eco-ditch successfully resolved this problem. This suggests that the highly effective hybrid bioreactor has vast potential for simultaneously removing nitrogen and phosphorus from practical nonpoint source wastewater.

The average proportion of the TP and TDP amounts removed in the A^2/O process were very high, about 70% and 68% of the total, respectively. The average proportions of the TN, NO_3-N and NH_4-N amounts removed in A^2/O process were 68–89%, 60–82%, and 63–92%, respectively. These results implied that the nutrient amounts (i.e., TP, TDP, TN, NO_3-N and NH_4-N) removed in the A^2/O process were more than those in the eco-ditch.

The removed amounts of P-nutrients (including TP and TDP) and N-nutrients (including TN, NO_3-N, and NH_4-N) were higher in the A^2/O process than in the eco-ditch. Two reasons could explain this. One is that the feed concentrations of P- and N-nutrients in the influent in the A^2/O process were higher than in the eco-ditch. Generally, the higher the contamination concentration of the influent, the higher the removal rate. The other is that the combination of the deposition pool, anaerobic tank, and anoxic and aerobic fluidized beds is more capable of removing nutrients than the eco-ditch. Many studies have shown that the removal efficiencies of A^2/O processes based on biofilm (such as the combination of anaerobic tanks and anoxic and aerobic fluidized beds) are greater than the eco-ditch under similar hydraulic and nutrient loadings (Ding et al., 2006; Martínez et al., 2000; Nootong and Shieh, 2008).

Between April 2007 and May 2008, the overall average proportions of NH_4-N to TN in the influent and effluent significantly decreased from 68% to 49% ($P < 0.05$), whilst the overall average proportions of NO_3-N to TN in the influent and effluent increased from 24% to 30% (Fig. 12.5).

The oxygen level in the influent increased from anaerobic (0 mg L^{-1}) to aerobic (7.2–9.8 mg L^{-1}) levels along the water pathway. This change in DO levels accelerates the nitrification process. The robust nitrification translated more NH_4-N into nitrogen gas. This process led to a significant decrease in the proportions of NH_4-N to TN (from 68% to 50%) along the sampling sites from B1 to B6.

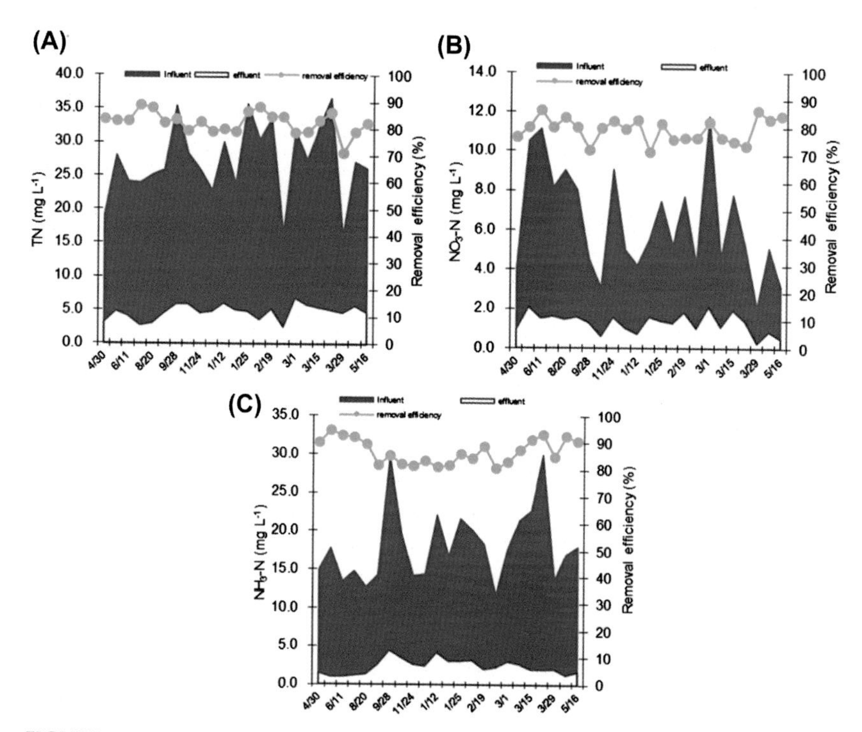

FIGURE 12.5 Removal efficiencies of the hybrid bioreactor in removing (A) total nitrogen (TN), (B) nitrate (NO_3-N), and (C) ammonia (NH_4-N) from April 2007 to May 2008.

12.3.4 The Contribution of Biofilm to Nutrient Removal

After the sludge from the A^2/O process was pumped into the eco-ditch, a new layer of high-concentration sludge liquid (sediments) formed immediately, but this new layer of sludge disappeared within 1 week. The high-concentration sludge liquid discharged into the eco-ditch was estimated twice during the experimental period, about 22 and 43 m^3 (estimated based on the pump flux and time) in July 2007 and March 2008, respectively.

Many studies have reported that eco-ditches, such as wetland systems, could treat and minimize sludge (Uggetti et al., 2010; Wang et al., 2010). The direct discharge of high-concentration sludge liquid into the eco-ditch of the hybrid bioreactor supplied the nutrients and habitats for native microbes and macrophytes. This kept the eco-ditch in a self-cycling state. Most importantly, the input of sludge from A^2/O process to the eco-ditch avoids the need to build specific facilities for sludge treatment. This process saves on capital and operational costs.

Biofilm is a basic element in bio-treatment technology such as the A^2/O process and its biomass is directly associated with the removal efficiency of pollutants. Estimates of the biofilm biomasses in the hybrid reactor showed that the biofilm biomasses in the A^2/O process ranged from 152 to 157 kg in 2007 and from 184 to 193 kg in 2008. The biofilm biomass in the eco-ditch ranged from 60 to 65 kg in 2007 and from 85 to 91 kg in 2008 (Fig. 12.6).

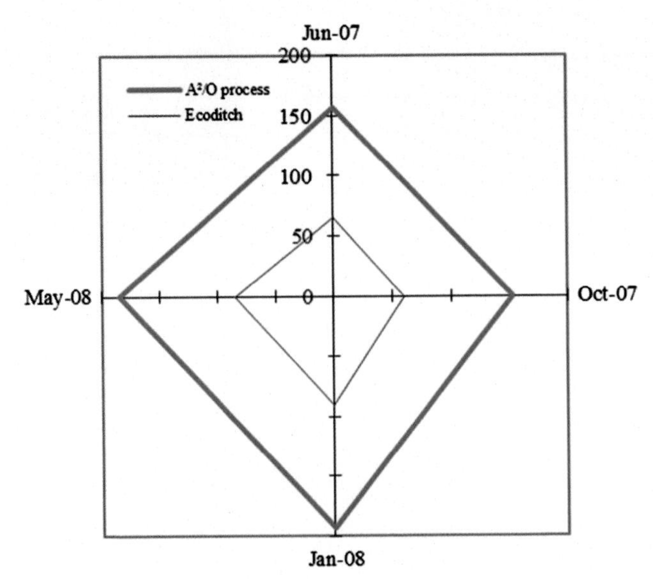

FIGURE 12.6 The biofilm biomasses (25–30°C, moisture ~85%) in the A^2/O process and eco-ditch of the hybrid bioreactor (means ±SD, kg) ($n = 15$).

The biofilm biomass in the A^2/O process was about 68–72% higher than in the eco-ditch in 2007 and 2008. Similarly, the majority of nutrients (about 60–92%) was removed in the A^2/O process of the hybrid bioreactor. These similarities in the distributions of biofilm biomass and nutrient removal in the A^2/O process might be associated. For example, the high biofilm mass tends to bioconcentrate (including biosorption, digestion, precipitation, and bio-accumulation) more nitrogen and phosphorus. Moreover, the average proportion of nutrients removed in the eco-ditch decreased with time from June 2007 to May 2008 despite the abundance of the macrophytes. Therefore, it was implied that the biofilms in the hybrid bioreactor were the main contributor to nutrient removal.

The dynamics of pollutants adsorbed to or desorbed from the active sites on the biofilm surface can occur concomitantly, such as that of nutrients freely transporting into biofilms (Scinto and Reddy, 2003), due to the special porous structure of biofilms (Wimpenny and Colasanti, 1997). Indeed, biofilm is an important point of entry for pollutants (such as excessive nitrogen and phosphorus) that directly affect food webs (Aouad et al., 2006; Hill and Larsen, 2005), and then easily bioconcentrate inorganic and organic nutrients from the surrounding water (Haack and Warren, 2003; Quintelas et al., 2009). Therefore, it was concluded that the nutrient removal from nonpoint source wastewater is due to the adsorption and assimilation of the biofilms.

12.3.5 The Response of Bacteria to Water Quality Improvement

To evaluate the response of bacterial communities to their habitat conditions, the Dice index of similarity of ERIC-PCR fingerprints for biofilm bacteria was calculated before and after the hybrid bioreactor was used. Before use, the nonpoint source influent directly entered into a ditch (the locations from the deposition pool to eco-ditch). The Dice indices in A1, A2, and A3 were from 98% to 99%, implying that the bacterial community structures in these three sampling sites were similar. This result indicated that the nonpoint source wastewater was homogeneous before the hybrid bioreactor was used.

During the running of the hybrid bioreactor, obvious differences in the bacterial community structures between anaerobic and anoxic treatment phases (deposition pool, anaerobic tank, and anoxic fluidized bed) and aerobic treatment phases (aerobic fluidized bed and eco-ditch) occurred. The Dice indices of similarity of bacterial communities between B1–B3 and B4–B6 sampling sites were only 58–73%, suggesting that the hybrid bioreactor had a large impact on bacterial community structures.

The Dice indices of similarity of bacterial communities along the sampling sites from B1 to B6 decreased from 100% to 64% in June 2007 and from 100% to 58% in May 2008 (Fig. 12.7). In addition, the decease rates in June 2007 did not differ in May 2008 (Fig. 12.7). These changes implied that the bacterial habitats improved along the sampling sites.

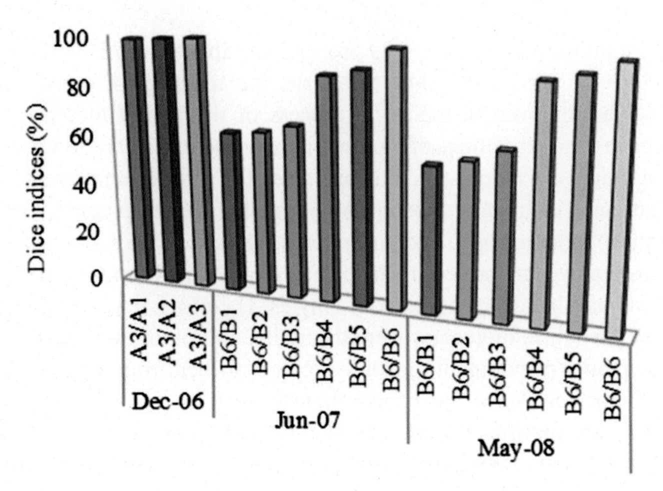

FIGURE 12.7 Dice index of similarity (%) of ERIC-PCR fingerprints for biofilm bacterial communities.

This optimization of bacterial community structure was due to the improvement of water quality, with the sampling site distance- and time-dependent. Indeed, previous studies showed that the bacterial community structure changed with changes in environmental factors, such as nutrients, DO, pH, and heavy metals (Ciric et al., 2010; Nyyssonen et al., 2009). In this study, the oxygen content significantly increased in the aerobic fluidized beds and eco-ditch, which probably led to the different microbial communities in these processes. Moreover, the nutrients deceased with the sampling site distance, which could also lead to the optimization of bacterial community structure. In turn, robust bacteria enhance the removal efficiencies of contaminants such as excessive nitrogen and phosphorus. In many cases, the contaminant removal efficiency is directly associated with bacterial structure and activities (Rajakumar et al., 2008).

12.4 CONCLUSION

The field-scale results indicated that the employment of the environmentally benign and self-sustaining hybrid bioreactor could simultaneously culture heterotrophic and autotrophic microorganisms and efficiently remove TN, NO_3-N, NH_4-N, TP, and TDP. In addition, this hybrid bioreactor avoided the need to build a sludge treatment system and lowered capital and operational costs. The sludge in the A^2/O process of the hybrid bioreactor was directly discharged into the eco-ditch, which in turn maintained the eco-ditch in self-modulating and self-sustaining states. The changes in the Dice indices of similarity indicated that the application of the hybrid bioreactor improved the bacterial habitat conditions, in turn demonstrating improvement of water quality. This study provides a promising, highly effective and easily deployed biomeasure to remove high-loading nitrogen and phosphorus from nonpoint source wastewater at an industrial scale and enhance biological habitats such as bacterial habitats.

REFERENCES

Ahn, K.H., Song, K.G., Cho, E., Cho, J., Yun, H., Lee, S., Kim, J., 2003. Enhanced biological phosphorus and nitrogen removal using a sequencing anoxic/anaerobic membrane bioreactor (SAM) process. Desalination 157, 345–352.

Aouad, G., Crovisier, J.L., Geoffroy, V.A., Meyer, J.-M., Stille, P., 2006. Microbially-mediated glass dissolution and sorption of metals by *Pseudomonas aeruginosa* cells and biofilm. Journal of Hazardous Materials 136 (3), 889–895.

APHA-AWWA-WEF, 1998. Standard Methods for Examination of Water and Wastewater, twentieth ed. APHA, AWWA, and WEF, Washington, DC.

Cha, S.M., Lee, S.W., Cho, K.H., Lee, S.H., Kim, J.H., 2016. Determination of best management timing of nonpoint source pollutants using particle bins and dimensionless time in a single stormwater runoff event. Ecological Engineering 90, 251–260.

Chen, T., Liu, X., Zhu, M., Zhao, K., Wu, J., Xu, J., Huang, P., 2008. Identification of trace element sources and associated risk assessment in vegetable soils of the urban-rural transitional area of Hangzhou, China. Environmental Pollution 151 (1), 67–78.

ChinaEPA, 2002a. National Water Quality Standard of Water Environments, P. R. China, Vol. GB3838-2002. China Environmental Protection Agency, Beijing.

ChinaEPA, 2002b. The Standard Analyzed Methods of Water and Wastewater, fourth ed. Environmental Protection Agency (EPA) of China & Chinese Environmental Science Press, Beijing, China.

Cho, D.-H., Ramanan, R., Heo, J., Shin, D.-S., Oh, H.-M., Kim, H.-S., 2016. Influence of limiting factors on biomass and lipid productivities of axenic *Chlorella vulgaris* in photobioreactor under chemostat cultivation. Bioresource Technology 211, 367–373.

Ciric, L., Griffith, R.I., Philp, J.C., Whiteley, A., 2010. Field scale molecular analysis for the monitoring of bacterial community structures during on-site diesel bioremediation. Bioresource Technology 101, 5235–5241.

Ding, Y.W., Wang, L., Wang, B.Z., Wang, Z., 2006. Removal of nitrogen and phosphorus in a combined A^2/O-BAF system with a short aerobic SRT. Journal of Environmental Sciences 18, 1082–1087.

Duchemin, M., Hogue, R., 2009. Reduction in agricultural non-point source pollution in the first year following establishment of an integrated grass/tree filter strip system in southern Quebec (Canada). Agriculture, Ecosystems and Environment 131 (1–2), 85–97.

Ferreira, R.V., Serpa, D., Cerqueira, M.A., Keizer, J.J., 2016. Short-time phosphorus losses by overland flow in burnt pine and eucalypt plantations in north-central Portugal: a study at micro-plot scale. Science of the Total Environment 551–552, 631–639.

Gunes, K., 2008. Point and nonpoint sources of nutrients to lakes – ecotechnological measures and mitigation methodologies – case study. Ecological Engineering 34 (2), 116–126.

Haack, E.A., Warren, L.A., 2003. Biofilm hydrous manganese oxyhydroxides and metal dynamics in acid rock drainage. Environmental Science and Technology 37, 4138–4147.

Hao, Z., Li, Y., Cai, W., Wu, P., Liu, Y., Wang, G., 2012. Possible nutrient limiting factor in long term operation of closed aquatic ecosystem. Advances in Space Research 49 (5), 841–849.

Hefting, M., Beltman, B., Karssenberg, D., Rebel, K., van Riessen, M., Spijker, M., 2006. Water quality dynamics and hydrology in nitrate loaded riparian zones in the Netherlands. Environmental Pollution 139 (1), 143–156.

Hill, J.E., Seipp, R.P., Betts, M., Hawkins, L., van Kessel, A.G., Crosby, W.L., Hemmingsen, S.M., 2002. Extensive profiling of a complex microbial community by high-throughput sequencing. Applied and Environmental Microbiology 68, 3055–3066.

Hill, W., Larsen, I.L., 2005. Growth dilution of metals in microagal biofilms. Environmental Science and Technology 39, 1513–1518.

Hosseini, V.A., Wessman, S., Hurtig, K., Karlsson, L., 2016. Nitrogen loss and effects on microstructure in multipass TIG welding of a super duplex stainless steel. Materials & Design 98, 88–97.

Kay, D., Aitken, M., Crowther, J., Dickson, I., Edwards, A.C., Francis, C., Hopkins, M., Jeffrey, W., Kay, C., McDonald, A.T., McDonald, D., Stapleton, C.M., Watkins, J., Wilkinson, J., Wyer, M.D., 2007. Reducing fluxes of faecal indicator compliance parameters to bathing waters from diffuse agricultural sources: the Brighouse Bay study, Scotland. Environmental Pollution 147 (1), 138–149.

Lant, C.L., Kraft, S.E., Beaulieu, J., Bennett, D., Loftus, T., Nicklow, J., 2005. Using GIS-based ecological-economic modeling to evaluate policies affecting agricultural watersheds. Ecological Economics 55 (4), 467–484.

LaPara, T.M., Nakatsu, C.H., Pantea, L.M., Alleman, J.E., 2002. Stability of the bacterial communities supported by a seven-stage biological process treating pharmaceutical wastewater as revealed by PCR-DGGE. Water Research 36, 638–646.

Li, H.Z., Li, X.Y., Zhao, Y.P., Huan, M.S., Yu, X.Z., Jin, C.X., Xu, Y.T., 2006. Analysis of structure changes of microbial community in medium biofilm by ERIC-PCR fingerprinting. Environmental Science 27, 2542–2546.

Lin, Y.-P., Cheng, B.-Y., Shyu, G.-S., Chang, T.-K., 2010. Combining a finite mixture distribution model with indicator kriging to delineate and map the spatial patterns of soil heavy metal pollution in Chunghua County, central Taiwan. Environmental Pollution 158 (1), 235–244.

Martínez, M.E., Sánchez, S., Jiménez, J.M., Yousfi, F.E.I., Muñoz, L., 2000. Nitrogen and phosphorus removal from urban wastewater by the microalga *Scenedesmus obliquus*. Bioresource Technology 73, 263–272.

Miura, Y., Hiraiwa, M.N., Ito, T., Itonaga, T., Watanabe, Y., Okabe, S., 2007. Bacterial community structures in MBRs treating municipal wastewater: relationship between community stability and reactor performance. Water Research 41, 627–637.

Moore, M.T., Kröger, R., Locke, M.A., Cullum, R.F., Steinriede Jr., R.W., Testa Iii, S., Lizotte Jr., R.E., Bryant, C.T., Cooper, C.M., 2010. Nutrient mitigation capacity in Mississippi Delta, USA drainage ditches. Environmental Pollution 158 (1), 175–184.

Nootong, K., Shieh, W.K., 2008. Analysis of an upflow bioreactor system for nitrogen removal via autotrophic nitrification and denitrification. Bioresource Technology 99, 6292–6298.

Nyyssonen, M., Kapanen, A., Piskonen, R., Lukkari, T., Itavaara, M., 2009. Functional genes reveal the intrinsic PAH biodegradation potential in creosote-contaminated groundwater following in situ biostimulation. Applied and Microbiological Biotechnology 84, 169–182.

Ongley, E.D., Xiaolan, Z., Tao, Y., 2010. Current status of agricultural and rural non-point source pollution assessment in China. Environmental Pollution 158 (5), 1159–1168.

Ouyang, W., Song, K., Wang, X., Hao, F., 2014. Non-point source pollution dynamics under long-term agricultural development and relationship with landscape dynamics. Ecological Indicators 45, 579–589.

Quintelas, C., Rocha, Z., Silva, B., Fonseca, B., Figueiredo, H., Tavares, T., 2009. Removal of Cd(II), Cr(VI), Fe(III) and Ni(II) from aqueous solutions by an *E. coli* biofilm supported on kaolin. Chemical Engineering Journal 149, 319–324.

Rajakumar, S., Ayyasamy, P.M., Shanthi, K., Thavamani, P., Velmurugan, P., Song, Y.C., Lakshmanaperumalsamy, P., 2008. Nitrate removal efficiency of bacterial consortium (*Pseudomonas* sp. KW1 and *Bacillus* sp. YW4) in synthetic nitrate-rich water. Journal of Hazardous Materials 157, 553–563.

Samaras, P., Papadimitriou, C.A., Vavoulidou, D., Yiangou, M., Sakellaropoulos, G.P., 2009. Effect of hexavalent chromium on the activated sludge process and on the sludge protozoan community. Bioresource Technology 100 (1), 38–43.

Schaller, J., Mkandawire, M., Gert Dudel, E., 2010. Heavy metals and arsenic fixation into freshwater organic matter under *Gammarus pulex* L. influence. Environmental Pollution 158 (7), 2454–2458.

Scinto, L.J., Reddy, K.R., 2003. Biotic and abiotic uptake of phosphorus by periphyton in a subtropical freshwater wetland. Aquatic Botany 77 (3), 203–222.

Tekile, A., Kim, I., Kim, J., 2015. Mini-review on river eutrophication and bottom improvement techniques, with special emphasis on the Nakdong River. Journal of Environmental Sciences 30, 113–121.

Tomer, M.D., Moorman, T.B., Kovar, J.L., Cole, K.J., Nichols, D.J., 2016. Eleven years of runoff and phosphorus losses from two fields with and without manure application, Iowa, USA. Agricultural Water Management 168, 104–111.

Tong, Y., Chen, L., Chi, J., Zhen, G., Zhang, Q., Wang, R., Yao, R., Zhang, W., Wang, X., 2016. Riverine nitrogen loss in the Tibetan Plateau and potential impacts of climate change. Science of the Total Environment 553, 276–284.

Turner, R.D.R., Will, G.D., Dawes, L.A., Gardner, E.A., Lyons, D.J., 2013. Phosphorus as a limiting factor on sustainable greywater irrigation. Science of the Total Environment 456–457, 287–298.

Uggetti, E., Ferrer, I., Llorens, E., García, J., 2010. Sludge treatment wetlands: a review on the state of the art. Bioresource Technology 101, 2905–2912.

Wang, J., Da, L., Song, K., Li, B.-L., 2008. Temporal variations of surface water quality in urban, suburban and rural areas during rapid urbanization in Shanghai, China. Environmental Pollution 152 (2), 387–393.

Wang, R., Korboulewsky, N., Prudent, P., Domeizel, M., Rolando, C., Bonin, G., 2010. Feasibility of using an organic substrate in a wetland system treating sewage sludge: impact of plant species. Bioresource Technology 101, 51–57.

Wang, Y.L., Yu, S.L., Shi, W.X., Bao, R.L., Zhao, Q., Zuo, X.T., 2009. Comparative performance between intermittently cyclic activated sludge-membrane bioreactor and anoxic/aerobic-membrane bioreactor. Bioresource Technology 100, 3877–3881.

Wei, G., Pan, L., Du, H., Chen, J., Zhao, L., 2004. ERIC-PCR fingerprinting-based community DNA hybridization to pinpoint genome-specific fragments as molecular markers to identify and track populations common to healthy human guts. Journal of Microbiological Methods 59, 91–108.

Wilson, B.N., Barfield, B.J., Warner, R.C., 1986. Simple models to evaluate non-point pollution sources and controls. In: Aldo, G., Franco, Z. (Eds.), Developments in Environmental Modelling, vol. 10. Elsevier, pp. 231–263.

Wimpenny, J.W.T., Colasanti, R., 1997. A unifying hypothesis for the structure of microbial biofilms based on cellular automaton models. FEMS Microbiological Ecology 22 (1), 1–16.

Wu, H., Huo, Y., Hu, M., Wei, Z., He, P., 2015. Eutrophication assessment and bioremediation strategy using seaweeds co-cultured with aquatic animals in an enclosed bay in China. Marine Pollution Bulletin 95 (1), 342–349.

Wu, Y., Hu, Z., Yang, L., 2010a. Hierarchical eco-restoration: a systematical approach to removal of COD and dissolved nutrients from an intensive agricultural area. Environmental Pollution 158 (10), 3123–3129.

Wu, Y., Kerr, P.G., Hu, Z., Yang, L., 2010b. Eco-restoration: simultaneous nutrient removal from soil and water in a complex residential-cropland area. Environmental Pollution 158 (7), 2472–2477.

Wu, Y.H., Hu, Z.Y., Yang, L.Z., Graham, B., Kerr, P.G., 2011. The removal of nutrients from non-point source wastewater by a hybrid bioreactor. Bioresource Technology 102 (3), 2419–2426.

Xing, W., Yang, P., Ren, S., Ao, C., Li, X., Gao, W., 2016. Slope length effects on processes of total nitrogen loss under simulated rainfall. CATENA 139, 73–81.

Ying, C., Umetsu, K., Ihara, I., Sakai, Y., Yamashiro, T., 2010. Simultaneous removal of organic matter and nitrogen from milking parlor wastewater by a magnetic activated sludge (MAS) process. Bioresource Technology 101 (12), 4349–4353.

Zheng, Y., Han, F., Tian, Y., Wu, B., Lin, Z., 2014. Chapter 5-Addressing the uncertainty in modeling watershed nonpoint source pollution. In: Sven Erik Jørgensen, N.-B.C., Fu-Liu, X. (Eds.), Developments in Environmental Modelling, vol. 26. Elsevier, pp. 113–159.

Zimmer, D., Kahle, P., Baum, C., 2016. Loss of soil phosphorus by tile drains during storm events. Agricultural Water Management 167, 21–28.

Chapter 13

Investigation of Adsorption and Absorption Mechanisms During Copper (II) Removal by Periphyton

13.1 INTRODUCTION

Industrialization and urbanization have generated huge amounts of wastewater containing high concentrations of heavy metals (Cheng, 2003; Grossl et al., 1994). Heavy metal contamination is a serious ecological problem and poses a heavy risk to flora and fauna as well as humans. These heavy metals can accumulate in the tissues of living organisms resulting in long-term adverse effects on biological systems (Barakat, 2011; Shamshad et al., 2015). Copper (Cu) is a heavy metal that is an essential nutrient for plant and animal growth, but can be toxic at high concentrations (Cheng, 2003). With the rapid human population growth and development, Cu (II) pollution in natural water bodies is becoming increasingly severe (Cheng, 2003).

As with other heavy metals, Cu^{2+} contamination mainly originates from industrial and agricultural emissions (Barakat, 2011; Grossl et al., 1994). Cu^{2+} has been found in municipal wastes as a byproduct of the metal mining and processing industries and from agricultural sources such as fertilizers, fungicidal sprays, and animal waste (Gong and Donahoe, 1997). Strict environmental protection legislation and public environmental concerns have led to the invention of techniques for heavy metal wastewater treatment. Many techniques have been developed to remove heavy metals from wastewater including chemical precipitation, ion exchange, adsorption, membrane filtration, coagulation and flocculation, flotation and biological assimilation (Babel and Kurniawan, 2003; Barakat, 2011; Deng et al., 2007; Fu and Wang, 2011; Wu et al., 2005, 2010b). Among these techniques, adsorption and membrane filtration are the most effective methods, especially at low heavy metal concentrations (Barakat, 2011; Fu and Wang, 2011). The use of membrane filtration at large scales however, is limited due to its high cost and low permeate flux. Conventional adsorbents such as activated carbon can be less

Periphyton. http://dx.doi.org/10.1016/B978-0-12-801077-8.00013-2

effective for the removal of metals (Babel and Kurniawan, 2003; Fu and Wang, 2011). Recently, algae and periphyton-based biomeasures are becoming widely accepted in heavy metal removal because of their high heavy metal assimilation and adsorption capacities, high temporal efficiency, and low cost (Deng et al., 2007; Little et al., 1991; Wu et al., 2014).

Many microorganisms, including bacteria (Fein et al., 2001), yeasts (Wang and Chen, 2006) and algae (Romera et al., 2007), can be used as biomaterials for heavy metal adsorption from wastewater. Periphyton, a new type of biomaterial, can be used to remove heavy metals from contaminated wastewater and has emerged as a potential alternative to conventional techniques. Periphyton is a complex microbial assemblage of algae, bacteria, protozoa, metazoa, epiphytes, and detritus (Wu et al., 2012, 2014). Periphyton commonly attaches to submerged surfaces in aquatic ecosystems and plays a vital role in natural aquatic ecosystems through its influence on primary production, nutrient recycling, and self-purification of surface water ecosystems (Larned, 2010; Lu et al., 2014). Periphyton is capable of assimilating various contaminants from surface water bodies, such as phosphorus, microcystins, hormones, and heavy metals (Dodds, 2003; Li et al., 2012; Wicke et al., 2008; Writer et al., 2011; Wu et al., 2014).

Moreover, periphyton composed of photoautotrophic microorganisms (e.g., cyanobacteria and diatoms) can secrete extracellular polymeric substances (EPS), which can greatly strengthen the adsorption of contaminants onto periphyton (Sheng et al., 2010). Cell walls of periphyton microorganisms consist mainly of proteins, polysaccharides, and lipids, offer many functional groups (such as carboxylate, hydroxyl, thiol, sulfonate, amino and imidazole groups) for binding metal ions (Gong et al., 2005). The EPS of periphyton enriches hydroxyl groups, carboxylic groups, acetylated amino acids, and some noncarbohydrates, e.g., phosphate and sulfate (De Philippis et al., 2011). These EPS groups can effectively bind with heavy metal ions through ion exchange or complexation (Fang et al., 2011). A small number of studies found that EPS exhibited great ability to complex heavy metals, by mechanisms including proton exchange, global electric field, or microprecipitation of metals (Comte et al., 2008; Fang et al., 2010; Guibaud et al., 2009). In addition, the large surface area and high binding affinity of EPS are effective in removing heavy metals from wastewater.

The high porosity and numerous void spaces in periphyton with photoautotrophic microorganisms can facilitate the transport of nutrients and heavy metals into the internal periphyton region to be adsorbed or assimilated (Donar et al., 2004; Lu et al., 2014; Wu et al., 2010a). Dissolved metal (Cu^{2+}) removal bioprocesses generally are bioaccumulation (active) by living cells and biosorption (passive) by nonliving biomass. Living biomass systems (active uptake) often require the addition of nutrients. The biosorption capacities are generally similar between live and dead biomass of a specific type.

Biosorption utilizing the ability of periphyton to accumulate Cu^{2+} from wastewater is considered a competitive, effective, and economically attractive method. In our previous research, adsorption was the main mechanism of microcystin and inorganic phosphorus removal by cultured periphyton (Lu et al., 2014; Wu et al., 2010a). We therefore propose that Cu^{2+} removal by periphyton will also be dominated by adsorption.

In this study, Cu^{2+} removal mechanisms and adsorption characteristics onto periphyton including kinetic, isothermal, and thermodynamic qualities were investigated. The main objectives were: (1) determine the Cu^{2+} removal capability of periphyton; (2) identify whether the Cu^{2+} is removed by adsorption or absorption; and (3) describe the Cu^{2+} adsorption process with kinetic, isotherm, and mathematical models. This study evaluates the potential of this environmentally benign biomaterial in removing Cu^{2+} from wastewater and provides an insight into Cu^{2+} adsorption mechanisms by periphyton or similar microbial aggregates.

13.2 MATERIALS AND METHODS

13.2.1 Periphyton Culture

Fiber carriers (FCs, 0.2×0.1 m) were fixed at the bottom of glass tanks ($0.5 \times 0.35 \times 0.24$ m) with a density of 0.3 m^2 per cubic meter of water. They were sterilized with 95% alcohol then rinsed three times with fresh double-distilled water. The glass tanks were then filled with artificial wastewater which was composed of WC medium (Beutler et al., 2002) and water rich in periphyton from Xuanwu Lake, a hypereutrophic lake in Nanjing, China. The property of the artificial wastewater was as follows: pH 7.92 ± 0.03, TN 1.26 ± 0.02 mg L^{-1}, NO_3-N 0.73 ± 0.02 mg L^{-1}, NH_4+-N 0.53 ± 0.01 mg L^{-1}, TP 0.12 ± 0.01 mg L^{-1}, TDP 0.035 ± 0.001 mg L^{-1}. The glass tanks were kept in a greenhouse at 25–30°C for 45 days, until brown and microporous periphyton formed on the FCs with a thickness of 5–8 mm. The morphology of the periphyton was characterized at 400× magnifications under an optical microscope (H600L, Nikon, Japan) and a scanning electron microscope (Quanta200, FEI, USA). The periphyton and its carrier were then used in the following experiments.

13.2.2 Cu Removal Experiment

To prevent the interference of other substances from real wastewater and to simultaneously provide essential nutrients for periphyton growth, a simulated wastewater with various Cu^{2+} concentrations was used ($C_6H_{12}O_6$ 169 mg L^{-1}, NaCl 63 mg L^{-1}, $(NH_4)_2SO_4$ 63 mg L^{-1}, KH_2PO_4 44 mg L^{-1}, $NaHCO_3$ 94 mg L^{-1}, $MgSO_4 \cdot 7H_2O$ 94 mg L^{-1}, $CaCl_2 \cdot 2H_2O$ 31 mg L^{-1}, $FeSO_4 \cdot 7H_2O$ 3.25 mg L^{-1}, $CuSO_4 \cdot 5H_2O$ 7.86–78.6 mg L^{-1}). To investigate the adsorption process of the cultured periphyton, 16.0 g of fresh periphyton (at 25–30°C, moisture 85 ± 5%) with its carrier (FCs) was fixed at the bottom of six 500-mL beakers filled with 250 mL of simulated wastewater with Cu^{2+} concentration of 5.0 mg L^{-1}. To investigate whether adsorption or assimilation was the main mechanism of Cu^{2+} removal by periphyton, 0.5 g L^{-1} of NaN_3 was added to three of the beakers to inhibit microbial activity of periphyton. The other three beakers were left as controls. All simulated microcosms (beakers) were incubated at 30°C with a light intensity of 2500 lux under a 14/10 h light/dark cycle in an incubator. The Cu^{2+} concentrations of the wastewater were determined after 1, 2, 4, 6, 8, 12, 24, 36, and 48 h.

13.2.3 Cu Adsorption Experiment

To investigate adsorption kinetics and isotherms, adsorption thermodynamic experiments were carried out at 303.15K in triplicate. A fixed amount, 8 g, of wet periphyton (moisture 85 ± 5%) was added to 250-mL Erlenmeyer flasks with 100 mL wastewater of Cu^{2+} concentrations 2, 5, 10, and 20 mg L^{-1} (see Table 13.2). The microbial activity of periphyton was inhibited by the addition of 0.5 g L^{-1} NaN_3. An equilibration time of 48 h was used. Cu^{2+} concentration was determined after 2, 6, 12, 24, 36, and 48 h. This experiment was conducted in an incubator with a light intensity of 2500 lux under a 14/10 h light/dark cycle.

13.2.4 Sample Analysis

For Cu^{2+} concentration measurements, 5 mL of wastewater was collected by syringe from each flask, then filtered through a 0.25-μm micromembrane. Cu^{2+} concentration was determined following an inductively coupled plasma atomic emission spectrometry method (Optima 8000, Perkin Elmer, USA). To estimate the periphyton dry weight (DW), moisture was removed by oven-drying wet samples at 80°C for 72 h.

The microbial activity and metabolic diversity of the periphyton with and without the addition of NaN_3 for 36 h were investigated using the Biolog ECO Microplate (Hayward, CA, USA). The ECO Microplate is comprised of three replicate wells of 31 types of carbon sources. The wells were filled with a redox-sensitive tetrazolium dye which turns purple as a result of respiratory electron transport in metabolically active cells (Choi and Dobbs, 1999). Therefore, the plate color is directly proportional to respiratory activity (Li et al., 2012). For all samples, 50-mL aliquots were used for each Biolog and 150-μL aliquots were added into each well of every ECO Microplate, and analyzed according to Choi and Dobbs (1999). The plates were incubated at 25°C and average well color development (AWCD) was evaluated using a Biolog Microplate Reader at 590 nm every 24 h for 7 days.

13.2.5 Data Analysis

Each experiment was conducted in triplicate and the results were expressed as mean ±1 SD. A one-way ANOVA was used to compare the microbial activity of periphyton with and without NaN_3 treatment (STATISTICA 7.0) and a significance level of $P < 0.05$ was applied throughout.

The Shannon index (H) (Lu et al., 2014) evaluated the species diversity of the periphyton community with and without the addition of NaN_3 using Eq. (13.1).

$$H = -\sum p_i \ln p_i \tag{13.1}$$

For adsorption kinetics and isotherms, the amount of Cu^{2+} adsorbed onto periphyton at time t and the equilibrium time were obtained by means of Eqs. (13.2) and (13.3) (Deng et al., 2007).

$$q_t = \frac{(C_0 - C_t)V}{m} \tag{13.2}$$

$$q_e = \frac{(C_0 - C_e)V}{m} \tag{13.3}$$

The kinetics of adsorption is one of the most important characteristics in defining the efficiency of adsorption. Various kinetic models have been proposed by different research groups where the adsorption has been treated as a pseudo-first-order (Cheng et al., 2016), a pseudo-second-order (Largitte and Pasquier, 2016) and intraparticle diffusion (Largitte and Pasquier, 2016). The pseudo-first-order adsorption, Eq. (13.4), pseudo-second-order adsorption, Eq. (13.5) (Lu et al., 2014), and the intraparticle diffusion model, Eq. (13.6), were used to identify the adsorption mechanism (Özcan et al., 2005).

$$\log(q_e - q_t) = \log q_e - \frac{k_1}{2.303} t \tag{13.4}$$

$$\frac{t}{q_t} = \frac{1}{k_2 q_e^2} + \frac{t}{q_e} \tag{13.5}$$

$$q_t = k_{id} t^{0.5} + C \tag{13.6}$$

Boyd's model (Tang et al., 2012) was used to investigate the rate-limiting step of the adsorption process, Eq. (13.7). When the calculated B_t values are plotted against time t (h), the linearity of the plots can be used to identify whether film diffusion or intraparticle diffusion controls the adsorption process. A nonlinear plot indicates that the adsorption process is mainly governed by film diffusion (Ngah and Hanafiah, 2008).

$$B_t = -0.4977 - \ln\left(1 - \frac{q_t}{q_e}\right) \tag{13.7}$$

The Elovich equation (Özacar and Sengil, 2005) is used to describe the chemisorption Eq. (13.8):

$$q_t = a_1 + b_1 \ln t \tag{13.8}$$

The Langmuir adsorption isotherm assumes that adsorption takes place at specific homogeneous sites within the adsorbent and has been successfully applied to adsorption processes of monolayer adsorption (Langmuir, 1918). The Freundlich isotherm is an empirical equation employed to describe heterogeneous systems (Freundlich, 1906). Langmuir and Freundlich equations were applied to fit the isotherm data as in Eqs. (13.9) and (13.10), respectively.

$$q_e = \frac{k_{\mathrm{L}} q_{\mathrm{m}} C_e}{1 + K_{\mathrm{L}} C_e} \tag{13.9}$$

$$q_e = K_{\mathrm{F}} C_e^{1/n} \tag{13.10}$$

The effect of isotherm shape has been discussed (Carvalho et al., 2011) with a view to predicting whether an adsorption system is favorable or unfavorable. Based on further analysis of the Langmuir equation, the dimensionless parameter of the equilibrium or adsorption intensity (R_{L}) can be expressed by Eq. (13.11).

$$R_{\mathrm{L}} = \frac{1}{1 + K_{\mathrm{L}} C_o} \tag{13.11}$$

where K_{L} is the Langmuir constant ($\mathrm{dm^3\ mol^{-1}}$) and C_{o} is the highest initial Cu (II) concentration ($\mathrm{mg\ L^{-1}}$). The values of R_{L} calculated from above equation are incorporated in Fig. 13.4. R_{L} is one of the most reliable indicators of the adsorption intensity of a surface. There are four possibilities of the R_{L} value: (1) irreversible adsorption, $R_{\mathrm{L}} = 0$; (2) favorable adsorption, $0 < R_{\mathrm{L}} < 1$; (3) linear adsorption, $R_{\mathrm{L}} = 1$; (4) unfavorable adsorption, $R_{\mathrm{L}} > 1$ (Ho et al., 2002).

The Dubinin–Radushkevich (D–R) isotherm is more general than the Langmuir isotherm because it does not assume a homogeneous surface or constant adsorption potential (Dubinin and Radushkevich, 1947). The D–R isotherm model was used to determine the adsorption type (physical or chemical). The linear form of the D–R model is expressed by Eq. (13.12) (Wu et al., 2010a).

$$\ln q_e = \ln q_{\mathrm{m}} - \beta \varepsilon^2 \tag{13.12}$$

where β is a constant related to the mean free energy of adsorption per mole of the adsorbate ($\mathrm{mol^2\ J^{-2}}$), q_{m} the theoretical saturation capacity and ε is the Polanyi potential, which is equal to $RT \ln\left(1 + \frac{1}{C_e}\right)$, where R ($\mathrm{J\ mol^{-1}\ K^{-1}}$) is the gas constant and T(K) is the absolute temperature. Hence, by plotting $\ln q_e$

versus ε^2, it is possible to obtain the value of q_m (mol g^{-1}) from the intercept and the value of β from the slope. Table 13.2 indicates the D–R isotherm for Cu adsorption onto periphyton.

Finally, the constant β gives an idea about the mean free energy E (kJ mol^{-1}) of adsorption per molecule of the adsorbate when it is transferred to the surface of the solid from infinity in the solution. Thermodynamic parameters including adsorption energy (E) and Gibbs free energy (ΔG°) were evaluated to study the feasibility of the process, and are expressed by the following equations.

$$E = \frac{1}{\sqrt{2\beta}} \tag{13.13}$$

$$\Delta G^\circ = -RT\ln k_d \tag{13.14}$$

$$k_d = \frac{C_o - C_e}{C_e} \tag{13.15}$$

If the magnitude of E is between 8 and 16 kJ mol^{-1}, the adsorption process follows ion exchange (Helfferich, 1962). For values of $E < 8$ kJ mol^{-1}, the adsorption process is of a physical nature (Onyango et al., 2004). The numerical value of adsorption of the mean free energies between 8.975 and 8.177 kJ mol^{-1} corresponds to a boundary of physisorption and the predominance of van der Waals forces.

13.3 RESULTS AND DISCUSSION

13.3.1 Periphyton Characteristics

After 45 days cultivation, many microorganisms appeared on the FCs and formed a dense and microporous periphyton (Fig. 13.1A). This periphyton was then used in the experiments. Under the optical microscope, it was observed that the periphyton community diversity was high with filamentous green

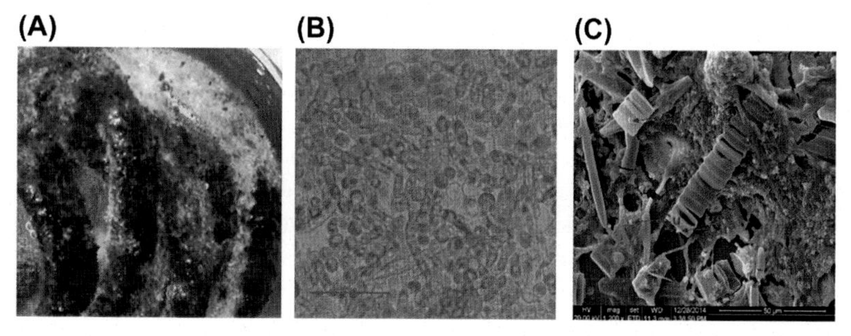

FIGURE 13.1 Morphological characteristics of periphyton observed in water (A) and under optical microscope (B) and SEM (C).

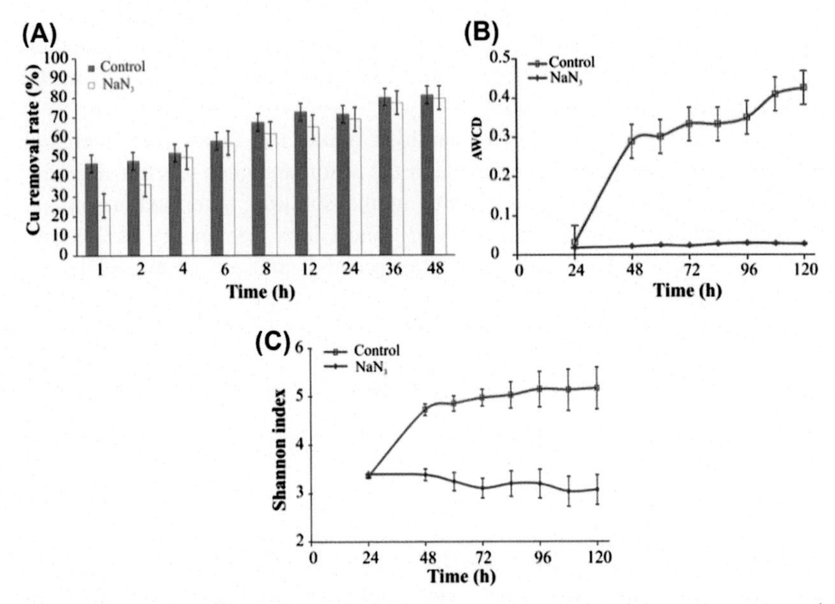

FIGURE 13.2 (A) Cu^{2+} removal rate by periphyton at an initial concentration of 5 mg L^{-1}, (B) microbial activity (represented by AWCD), and (C) microbial community functional diversity (Shannon index) of periphyton with and without addition of 0.5 g L^{-1} NaN_3 for 36 h over incubation time at 25°C.

algae and cyanobacteria as the representative microbial community (Fig. 13.1B). Under the scanning electron microscope (SEM), diatoms, bacteria, and protozoa were observed on the periphyton (Fig. 13.1C). The Shannon indexes of periphyton diversity based on Biolog analyses showed that periphyton was comprised of several types of microorganisms (Fig. 13.2B and C).

13.3.2 Cu Removal Process Dominated by Adsorption

The rate of Cu (II) removed by periphyton ranged from 25.6% to 79.6%, while the control ranged from 46.8% to 81.1% after 1–48 h. There is no significant difference between the experimental results. Cu (II) concentration showed a sharp decrease in the first hour (Fig. 13.2A) in the control group with a removal rate of 50.2% at an initial Cu (II) concentration of 5.0 mg L^{-1}. Thereafter, the Cu (II) removal efficiency increased gradually to 81.1% after 48 h (Fig. 13.2A), which demonstrated that periphyton was efficient in Cu (II) removal. This result indicated that periphyton had a high binding affinity for Cu (II) and could effectively remove Cu (II) from wastewater.

To determine whether adsorption dominated the Cu^{2+} removal process, NaN_3 was used to impede microbial activity by restraining microbial respiration and inhibiting assimilation (Saisho et al., 2001). AWCD values (representing microbial activities) of the NaN_3-treated group were significantly lower ($P < 0.001$) than the control group without NaN_3 addition (Fig. 13.2B). Specifically, the AWCD values in the NaN_3 treatment were close to zero over the incubation time, while that of the control group increased to almost 0.5. This indicates that microbial respiration in periphyton was completely inhibited by NaN_3 and thus the assimilation of Cu (II) by periphyton was trivial. Moreover, the Shannon index of periphyton diversity based on the ECO Microplate analyses showed that the NaN_3-treated periphyton was significantly less diverse than the control group ($P < 0.001$, Fig. 13.2C). In contrast to the high Cu (II) removal efficiency (50.2%) of the control group during the first hour, the NaN_3-treated group only had a Cu (II) removal efficiency of 29.1%. This further suggested that Cu (II) adsorption by periphyton contributed to 58.0% of Cu (II) removal in the first hour and can be considered the dominant mechanism.

In addition to micropores, the periphyton also had many heavy metal binding sites on its surface (Fig. 13.1C). Thus, Cu^{2+} could combine with the heavy metal binding sites. Moreover, the microorganisms composing the periphyton may also have marked effects on its architectural characteristics (Donar et al., 2004). For example, the microspaces constructed among complex cells such as algae, bacteria, and protozoa are relatively larger than those constructed by single species (James et al., 1995). This difference may provide more microspaces or heavy adsorption sites for capturing Cu^{2+}. It also provides strong evidence of the function of periphyton in self-purification of aquatic systems (Sabater et al., 2002).

13.3.3 Adsorption Characteristics and Kinetics

The influence of the initial Cu^{2+} concentration in the solution on the rate of adsorption onto periphyton was investigated. To describe the adsorption process, the widely used pseudo-first-order kinetic, pseudo-second-order kinetic, and Elovich models were used to fit the data collected from the different initial Cu^{2+} concentrations. It can be seen that the calculated adsorption capacity of phenol (q_e) and the correlation coefficients (R^2) could be used to determine the applicability of the adsorption kinetics model. The regression coefficients (R^2) of the pseudo-second-order kinetic model were high ($R^2 > 0.99$, Table 13.1) for all Cu^{2+} concentrations, and were higher than those of the other models ($R^2 = 0.88-0.99$). Accordingly, kinetic parameters q_e showed the same trend that increased 2.213 and 1.750 to 20.620 and 16.835, respectively, with Cu^{2+} concentration rise. While the Cu^{2+} concentration increased, the k did not increase. Therefore, it can be concluded that the adsorption of Cu^{2+} onto periphyton strongly followed the pseudo-second-order kinetic model. In general, the pseudo-second-order model is appropriate to represent the adsorption kinetics.

TABLE 13.1 Kinetic Model Parameters for Cu^{2+} Adsorption by Periphyton Under Different Initial Cu^{2+} Concentrations

	Pseudo-First-Order Kinetic Model			Pseudo-Second-Order Kinetic Model			Elovich Equation Model		
Cu (mg L^{-1})	k_1 (h^{-1})	q_1 (mg g^{-1})	R^2	k_2 (mg g^{-1} h^{-1})	q_2 (mg g^{-1})	R^2	a_1 (mg g^{-1} h^{-1})	b_1 (mg g^{-1})	R^2
2	0.051	2.213	0.895	0.080	1.750	0.999	0.428	0.672	0.957
5	0.078	5.445	0.979	0.059	4.325	0.999	2.550	1.407	0.940
10	0.064	8.933	0.892	0.031	8.177	0.997	6.862	2.014	0.957
20	0.046	20.620	0.885	0.008	16.835	0.996	5.834	5.834	0.928
Cu (mg L^{-1})	Intra-Particle Diffusion Model			Boyd Model					
	k_{id} (mg g^{-1} $h^{-0.5}$)	C (mg g^{-1})	R^2	k_{id} (mg g^{-1} $h^{-0.5}$)	C	R^2			
2	0.276	0.965	0.775	0.057	−0.243	0.903			
5	0.570	3.707	0.743	0.084	−0.222	0.969			
10	0.873	8.254	0.864	0.072	−0.126	0.895			
20	2.493	10.012	0.812	0.051	−0.220	0.888			

This result is similar to previous studies on the adsorption of Cu^{2+} and other heavy metal ions (El-Sikaily et al., 2007; Huang et al., 2014; Ngah and Hanafiah, 2008)˙ Moreover, it was obvious that the q_2 values (Table 13.1) were very close to the experimental q_e values (Fig. 13.3B), which increased with increasing adsorbent. During the experiment q_t showed the same increasing trend from 1 to 48 h at different initial Cu^{2+} concentrations, and it reached as high as 3.267, 8.300, 15.778, and 31.520 mg L^{-1}, with initial Cu^{2+} of 2, 5, 10, and 20 mg L^{-1}, respectively. In addition, models fit relatively better at low initial Cu^{2+} concentrations of 2–5 mg L^{-1} than at high Cu^{2+} concentrations. This was probably because the adsorption was saturated at high Cu^{2+} concentrations. For instance, the Cu^{2+} adsorption rate decreased from 1 h (1.5 mg L^{-1} h^{-1}) to 48 h (0.008 mg L^{-1} h^{-1}) with an initial Cu^{2+} concentration of 5 mg L^{-1} as shown in Fig. 13.2A. Additionally, high Cu^{2+} concentrations can be toxic to periphyton, thereby terminating the adsorption process (Serra et al., 2009).

The adsorption process occurs in several steps, including external diffusion, intraparticle diffusion, and the actual adsorption on the surface (Kim et al., 2011). The intraparticle diffusion model is used to identify the rate-controlling step in the phenol adsorption process. In a solid–liquid adsorption system, the solute molecule transfer process is usually characterized as intraparticle diffusion, boundary layer diffusion, or both (Dawood and Sen, 2012). An intraparticle diffusion model and Boyd plots were used to investigate the Cu^{2+} adsorption process onto periphyton. The determination coefficients (R^2) of the intraparticle diffusion model were 0.743–0.864 for all Cu^{2+} concentrations (Table 13.1, Fig. 13.3A), indicating that intraparticle diffusion was not the only adsorption mechanism (Tang et al., 2012).

The intraparticle diffusion plots were multilinear, containing at least three linear segments (Fig. 13.3B). According to the intraparticle diffusion model, if the plot of q_t versus $t^{0.5}$ presents a multilinearity correlation then at least three steps occurred during the adsorption process (Sun and Yang, 2003). The first step is the transport of molecules from the bulk solution to the adsorbent external surface by diffusion through the boundary layer (film diffusion). The second step is the diffusion of the molecules from the external surface into the pores of the adsorbent. The third step is the final equilibrium stage, during which the molecules are adsorbed on active sites on the internal surface of the pores and the intraparticle diffusion starts to slow down following the reduction of solute concentration (Hameed and El-Khaiary, 2008; Kiran et al., 2006; Sun and Yang, 2003; Tang et al., 2012). Therefore, the Cu^{2+} adsorption process by periphyton in this study could be divided into three steps: external mass transfer at the beginning (0–12 h), intraparticle diffusion in the middle period (12–24 h), and equilibrium during the last period (24–48 h). The linear portions of curves did not pass through the origin, suggesting that pore diffusion was not the step controlling the overall rate of mass transfer at the

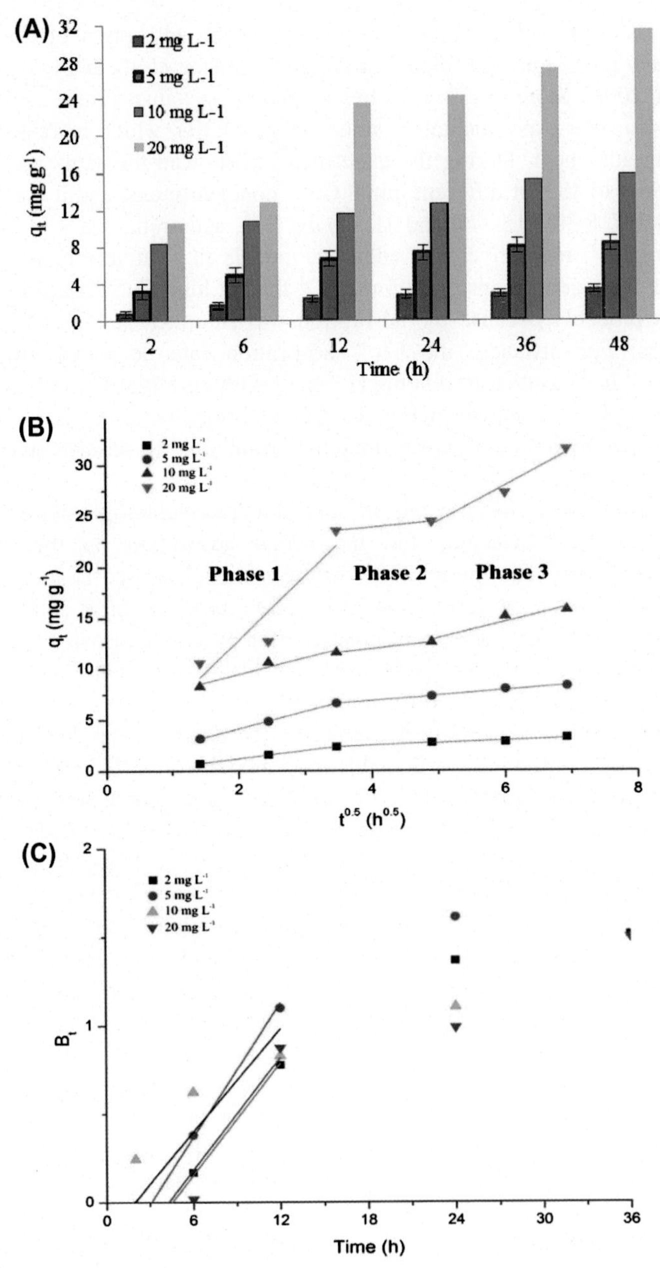

FIGURE 13.3 (A) The amount of Cu^{2+} absorbed onto periphyton over time, (B) intraparticle diffusion, and (C) Boyd plots for Cu^{2+} adsorption at initial Cu^{2+} concentrations of 2–20 mg L^{-1} under an illumination of 2500 lux at 30°C.

beginning of adsorption (Tang et al., 2012). The boundary layer effect may control the rate of mass transfer in the time period of the first linear segment.

The linearity of the plot of B_t against time in the Boyd model can be used to distinguish intraparticle diffusion and boundary layer effect (film diffusion) rates of adsorption (Tang et al., 2012). If a plot of B_t versus t is a straight line passing through the origin, then adsorption will fit the boundary layer effect. In this study, the plots were linear ($R^2 = 0.89-0.97$) in the beginning, but did not pass through the origin (Fig. 13.3C). This indicates external mass transfer was the rate-limiting process at the beginning of the adsorption process and intraparticle diffusion was the rate-limiting process in the latter stages.

Furthermore, the intercept (C) of the intraparticle diffusion model represents the boundary layer effect (film diffusion). A high C value indicates a great contribution of surface sorption in the rate-controlling step (Özcan et al., 2005a). The C values increased from 0.97 to 10.01 when the Cu^{2+} concentration changed from 2 to 20 mg L^{-1} (Table 13.1, Fig. 13.3B), reflecting the larger boundary layer effect at a relative higher Cu^{2+} concentration.

13.3.4 Adsorption Isotherms

The equilibrium adsorption isotherm is one of the most important types of data needed to understand the mechanism of the adsorption system. Several isotherm equations are available and three important isotherms were selected in this study, namely Freundlich, Langmuir, and Dubinin–Radushkevich (D–R) isotherms. The Langmuir, Freundlich, and D–R parameters for the adsorption of Cu are listed in Table 13.2. Briefly, all of the isotherm models fit very well when the R^2 values are compared (Table 13.2). The correlation coefficients (R^2) of the Freundlich, Langmuir, and D–R isotherm models were 0.993, 0.656, and 0.821, respectively.

The capacity of the adsorption isotherm is fundamental, and plays a crucial role in determining maximum adsorption capacity. It also provides a panorama of the course taken by the system under study in a concise form, indicating how Cu^{2+} can be adsorbed. As shown in Fig. 13.4A, the Cu^{2+} adsorption capacities (q_e) at initial concentrations of 2, 5, 10, and 20 mg L^{-1} were 1.1, 3.0 mg g^{-1}, 5.7 mg g^{-1}, and 11.9 mg g^{-1}, respectively.

TABLE 13.2 Parameters of the Langmuir, Freundlich, and Dubinin–Radushkevich (D–R) Isotherms at 30°C

Langmuir			Freundlich			D–R		
K_L (g^{-1})	q_m (mg g^{-1})	R^2	K_F (g^{-1})	n	R^2	β	q_m (mg g^{-1})	R^2
70.4	13.33	0.656	8.55	1.11	0.993	$1*10^{-6}$	19.83	0.821

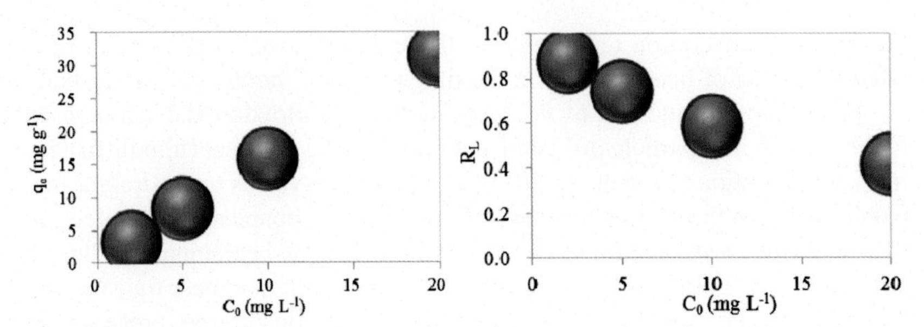

FIGURE 13.4 Variation in adsorption intensity with initial Cu^{2+} concentrations of 2–20 mg L^{-1} under an illumination of 2500 lux at 30°C.

The data from the adsorption experiment with initial Cu^{2+} concentrations of 2–20 mg L^{-1} were fit to the Freundlich, Langmuir, and D–R isotherm models (Table 13.2). A satisfactory empirical isotherm can be used for nonideal adsorption because of the high coefficients of determination (R^2). One of the Freundlich constants, K_F, indicates the adsorption capacity of the adsorbent. The other Freundlich constant, n, is a measure of the deviation from linearity of the adsorption. If a value for n is equal to unity, the adsorption is linear. If a value for n is below unity, this implies that the adsorption process is chemical, if a value for n is above unity, adsorption is favorable, a physical process. It is generally accepted that n values between 1 and 10 represent good adsorption potential of the adsorbent (García-Calzón and Díaz-García, 2007). In this study, the Freundlich constant (n) of 1.11 indicates that Cu^{2+} adsorption onto periphyton was favorable. The equilibrium data of Cu adsorption onto periphyton were fit to the linear Freundlich equation. The linear plots of ln q_e versus ln C_e were examined to determine K_F and n (Table 13.2). The high coefficients ($R^2 = 0.993$) of the plots indicated that the Cu^{2+} adsorption isotherm can be well described by the linear Freundlich equation.

The adsorption intensity (R_L) decreased from 0.877 to 0.415 when the initial Cu^{2+} concentration increased from 2 to 20 mg L^{-1} (Fig. 13.4B). An R_L between 0 and 1 indicated that the periphyton was an effective adsorbent of Cu^{2+} from an aqueous solution (Ho et al., 2002). The equilibrium data can also be described by the linear D–R isotherm model to distinguish physical adsorption from chemical adsorption ($R^2 = 0.821$, Table 13.2). The β value was 1×10^{-6} mol^2 kJ^{-2} at 0.25 g periphyton biomass.

In many cases, the Freundlich and D–R equations have been regarded as empirical, rather than mechanistic (e.g., surface models) adsorption isotherms models. In this study, data collected from the adsorption process fit both of the two empirical equations well. This indicates that the Cu^{2+} adsorption process of periphyton has mechanistic relevance (or surface adsorption properties) and that the model parameters included in these empirical equations have specific thermodynamic significance.

13.3.5 Thermodynamic Parameters of Adsorption

Because K_L is an equilibrium constant, its dependence on temperature can be used to estimate thermodynamic parameters, such as changes in the standard free energy (ΔG°) associated with the adsorption process. Generally, the change of free energy for physisorption is between -20 and 0 kJ mol^{-1}, but chemisorption is a range of -80 to -400 kJ mol^{-1} (Özcan et al., 2005).

The mean sorption energy (E) can be used to distinguish chemical and physical adsorptions. An E between 8 and 16 kJ mol^{-1} indicates that the adsorption process follows chemical ion exchange, and an E less than 8 kJ mol^{-1} indicates that the adsorption is physical in nature (Wu et al., 2010a). In this study, the E value of 0.7 kJ mol^{-1} demonstrates that the adsorption process of Cu^{2+} by periphyton was physical in nature. The thermodynamic results based on Eqs. 13.13 and 13.14 showed that ΔG° varied between -3.309 and -3.996 kJ mol^{-1} with Cu^{2+} concentration changing from 2 to 20 mg L^{-1}. As all the ΔG° values were less than zero, this indicates that Cu^{2+} adsorption onto periphyton was a spontaneous process, thereby demonstrating that the process is stable energetically (Tang et al., 2012).

13.4 CONCLUSIONS

This study implies that periphyton removal of Cu^{2+} from surface water is a promising method, especially for the removal of copper after "emergency" accidents, by taking out copper with periphyton based on adsorption mechanisms. The Freundlich, Langmuir, and Dubinin–Radushkevich (D–R) models provided evidence that the adsorption of Cu^{2+} by periphyton was the primary removal mechanism. Cu^{2+} adsorption by periphyton occurring in the latent adaptation period had mechanistic relevance and was a physical and spontaneous process. Compared to physical and chemical methods, periphyton is an environmentally friendly natural material for removing Cu^{2+} from wastewater. Therefore, Cu^{2+} recovery and reuse technologies based on periphyton may have wide application potential. It should be considered, however, that the adsorption might be reversible (Serra et al., 2009) and Cu^{2+} desorption or release from periphyton to the water column needs further investigation. Furthermore, periphyton might detach under high flow conditions (Lu et al., 2014) during its long-term application.

The presence of heavy metals in the periphyton biomass after adsorption is problematic in the further utilization and disposal of the biomass, such as biofertilizer and animal feed (Ahluwalia and Goyal, 2007; Renuka et al., 2015). Thus, the retrieval of heavy metals from biomass and their recovery into usable forms are critical (Ahluwalia and Goyal, 2007). Mining of the heavy metal-enriched periphyton biomass is a commonly used method and the regenerated biomass could be used as biofertilizer or biogas through biorefinery (Renuka et al., 2015; Zhu, 2015). Further investigation of the best methods to dispose of heavy metals in periphyton are needed.

REFERENCES

Ahluwalia, S.S., Goyal, D., 2007. Microbial and plant derived biomass for removal of heavy metals from wastewater. Bioresource Technology 98 (12), 2243–2257.

Babel, S., Kurniawan, T.A., 2003. Low-cost adsorbents for heavy metals uptake from contaminated water: a review. Journal of Hazardous Materials 97 (1), 219–243.

Barakat, M.A., 2011. New trends in removing heavy metals from industrial wastewater. Arabian Journal of Chemistry 4 (4), 361–377.

Beutler, M., Wiltshire, K.H., Meyer, B., Moldaenke, C., Lüring, C., Meyerhöfer, M., Hansen, U.P., Dau, H., 2002. A fluorometric method for the differentiation of algal populations in vivo and in situ. Photosynthesis Research 72 (1), 39–53.

Carvalho, J., Araujo, J., Castro, F., 2011. Alternative low-cost adsorbent for water and wastewater decontamination derived from eggshell waste: an overview. Waste and Biomass Valorization 2 (2), 157–167.

Cheng, S., 2003. Heavy metal pollution in China: origin, pattern and control. Environmental Science and Pollution Research 10 (3), 192–198.

Cheng, W.P., Gao, W., Cui, X., Ma, J.H., Li, R.F., 2016. Phenol adsorption equilibrium and kinetics on zeolite X/activated carbon composite. Journal of the Taiwan Institute of Chemical Engineers 62, 192–198.

Choi, K.-H., Dobbs, F.C., 1999. Comparison of two kinds of Biolog microplates (GN and ECO) in their ability to distinguish among aquatic microbial communities. Journal of Microbiological Methods 36 (3), 203–213.

Comte, S., Guibaud, G., Baudu, M., 2008. Biosorption properties of extracellular polymeric substances (EPS) towards Cd, Cu and Pb for different pH values. Journal of Hazardous Materials 151 (1), 185–193.

Dawood, S., Sen, T.K., 2012. Removal of anionic dye Congo red from aqueous solution by raw pine and acid-treated pine cone powder as adsorbent: equilibrium, thermodynamic, kinetics, mechanism and process design. Water Research 46 (6), 1933–1946.

De Philippis, R., Colica, G., Micheletti, E., 2011. Exopolysaccharide-producing cyanobacteria in heavy metal removal from water: molecular basis and practical applicability of the biosorption process. Applied Microbiology and Biotechnology 92 (4), 697–708.

Deng, L., Su, Y., Su, H., Wang, X., Zhu, X., 2007. Sorption and desorption of lead (II) from wastewater by green algae *Cladophora fascicularis*. Journal of Hazardous Materials 143 (1–2), 220–225.

Dodds, W.K., 2003. The role of periphyton in phosphorus retention in shallow freshwater aquatic systems. Journal of Phycology 39 (5), 840–849.

Donar, C., Condon, K., Gantar, M., Gaiser, E., 2004. A new technique for examining the physical structure of everglades floating periphyton mat. Nova Hedwigia 78 (1–2), 107–119.

Dubinin, M., Radushkevich, L., 1947. Equation of the characteristic curve of activated charcoal. Zhurnal Nevropatologii I Psikhiatrii Imeni S.S. Korsakova 1 (1), 875.

El-Sikaily, A., Nemr, A.E., Khaled, A., Abdelwehab, O., 2007. Removal of toxic chromium from wastewater using green alga *Ulva lactuca* and its activated carbon. Journal of Hazardous Materials 148 (1–2), 216–228.

Fang, L., Huang, Q., Wei, X., Liang, W., Rong, X., Chen, W., Cai, P., 2010. Microcalorimetric and potentiometric titration studies on the adsorption of copper by extracellular polymeric substances (EPS), minerals and their composites. Bioresource Technology 101 (15), 5774–5779.

Fang, L., Wei, X., Cai, P., Huang, Q., Chen, H., Liang, W., Rong, X., 2011. Role of extracellular polymeric substances in Cu (II) adsorption on *Bacillus subtilis* and *Pseudomonas putida*. Bioresource Technology 102 (2), 1137–1141.

Fein, J.B., Martin, A.M., Wightman, P.G., 2001. Metal adsorption onto bacterial surfaces: development of a predictive approach. Geochimica et Cosmochimica Acta 65 (23), 4267–4273.

Freundlich, H., 1906. Over the adsorption in solution. The Journal of Physical Chemistry 57 (385), e470.

Fu, F., Wang, Q., 2011. Removal of heavy metal ions from wastewaters: a review. Journal of Environmental Management 92 (3), 407–418.

García-Calzón, J.A., Díaz-García, M.E., 2007. Characterization of binding sites in molecularly imprinted polymers. Sensors and Actuators B: Chemical 123 (2), 1180–1194.

Gong, C., Donahoe, R.J., 1997. An experimental study of heavy metal attenuation and mobility in sandy loam soils. Applied Geochemistry 12 (3), 243–254.

Gong, R., Ding, Y., Liu, H., Chen, Q., Liu, Z., 2005. Lead biosorption and desorption by intact and pretreated *Spirulina maxima* biomass. Chemosphere 58 (1), 125–130.

Grossl, P.R., Sparks, D.L., Ainsworth, C.C., 1994. Rapid kinetics of Cu (II) adsorption/desorption on goethite. Environmental Science and Technology 28 (8), 1422–1429.

Guibaud, G., van Hullebusch, E., Bordas, F., d'Abzac, P., Joussein, E., 2009. Sorption of Cd (II) and Pb (II) by exopolymeric substances (EPS) extracted from activated sludges and pure bacterial strains: modeling of the metal/ligand ratio effect and role of the mineral fraction. Bioresource Technology 100 (12), 2959–2968.

Hameed, B.H., El-Khaiary, M.I., 2008. Kinetics and equilibrium studies of malachite green adsorption on rice straw-derived char. Journal of Hazardous Materials 153 (1–2), 701–708.

Helfferich, F.G., 1962. Ion Exchange. Courier Corporation.

Ho, Y.S., Huang, C.T., Huang, H.W., 2002. Equilibrium sorption isotherm for metal ions on tree fern. Process Biochemistry 37 (12), 1421–1430.

Huang, B., Li, Z., Huang, J., Guo, L., Nie, X., Wang, Y., Zhang, Y., Zeng, G., 2014. Adsorption characteristics of Cu and Zn onto various size fractions of aggregates from red paddy soil. Journal of Hazardous Materials 264, 176–183.

James, G., Beaudette, L., Costerton, J., 1995. Interspecies bacterial interactions in biofilms. Journal of Industrial Microbiology 15 (4), 257–262.

Kim, T.Y., Cho, S.Y., Kim, S.J., 2011. Adsorption equilibrium and kinetics of copper ions and phenol onto modified adsorbents. Adsorption 17 (1), 135–143.

Kiran, I., Akar, T., Ozcan, A.S., Ozcan, A., Tunali, S., 2006. Biosorption kinetics and isotherm studies of Acid Red 57 by dried *Cephalosporium aphidicola* cells from aqueous solutions. Biochemical Engineering Journal 31 (3), 197–203.

Langmuir, I., 1918. The adsorption of gases on plane surfaces of glass, mica and platinum. Journal of the American Chemical Society 40 (9), 1361–1403.

Largitte, L., Pasquier, R., 2016. A review of the kinetics adsorption models and their application to the adsorption of lead by an activated carbon. Chemical Engineering Research and Design 109, 495–504.

Larned, S.T., 2010. A prospectus for periphyton: recent and future ecological research. Journal of the North American Benthological Society 29 (1), 182–206.

Li, T., Bo, L., Yang, F., Zhang, S., Wu, Y., Yang, L., 2012. Comparison of the removal of COD by a hybrid bioreactor at low and room temperature and the associated microbial characteristics. Bioresource Technology 108, 28–34.

Little, B., Wagner, P., Ray, R., Pope, R., Scheetz, R., 1991. Biofilms: an ESEM evaluation of artifacts introduced during SEM preparation. Journal of Industrial Microbiology 8 (4), 213–221.

Lu, H., Yang, L., Shabbir, S., Wu, Y., 2014. The adsorption process during inorganic phosphorus removal by cultured periphyton. Environmental Science and Pollution Research 21 (14), 8782–8791.

Ngah, W.W., Hanafiah, M., 2008. Biosorption of copper ions from dilute aqueous solutions on base treatedrubber (*Hevea brasiliensis*) leaves powder: kinetics, isotherm, and biosorption mechanisms. Journal of Environmental Sciences 20 (10), 1168–1176.

Onyango, M.S., Kojima, Y., Aoyi, O., Bernardo, E.C., Matsuda, H., 2004. Adsorption equilibrium modeling and solution chemistry dependence of fluoride removal from water by trivalent-cation-exchanged zeolite F-9. Journal of Colloid and Interface Science 279 (2), 341–350.

Özacar, M., Sengil, İ.A., 2005. A kinetic study of metal complex dye sorption onto pine sawdust. Process Biochemistry 40 (2), 565–572.

Özcan, A.S., Erdem, B., Özcan, A., 2005. Adsorption of Acid Blue 193 from aqueous solutions onto BTMA-bentonite. Colloids and Surfaces A: Physicochemical and Engineering Aspects 266 (1–3), 73–81.

Renuka, N., Sood, A., Prasanna, R., Ahluwalia, A.S., 2015. Phycoremediation of wastewaters: a synergistic approach using microalgae for bioremediation and biomass generation. International Journal of Environmental Science and Technology 12 (4), 1443–1460.

Romera, E., González, F., Ballester, A., Blázquez, M., Munoz, J., 2007. Comparative study of biosorption of heavy metals using different types of algae. Bioresource Technology 98 (17), 3344–3353.

Sabater, S., Guasch, H., Romaní, A., Muñoz, I., 2002. The effect of biological factors on the efficiency of river biofilms in improving water quality. Hydrobiologia 469 (1–3), 149–156.

Saisho, D., Nakazono, M., Tsutsumi, N., Hirai, A., 2001. ATP synthesis inhibitors as well as respiratory inhibitors increase steady-state level of alternative oxidase mRNA in *Arabidopsis thaliana*. Journal of Plant Physiology 158 (2), 241–245.

Serra, A., Corcoll, N., Guasch, H., 2009. Copper accumulation and toxicity in fluvial periphyton: the influence of exposure history. Chemosphere 74 (5), 633–641.

Shamshad, I., Khan, S., Waqas, M., Ahad, N., Khan, K., 2015. Removal and bioaccumulation of heavy metals from aqueous solutions using freshwater algae. Water Science & Technology 71 (1), 38–44.

Sheng, G.-P., Yu, H.-Q., Li, X.-Y., 2010. Extracellular polymeric substances (EPS) of microbial aggregates in biological wastewater treatment systems: a review. Biotechnology Advances 28 (6), 882–894.

Sun, Q., Yang, L., 2003. The adsorption of basic dyes from aqueous solution on modified peat–resin particle. Water Research 37 (7), 1535–1544.

Tang, H., Zhou, W., Zhang, L., 2012. Adsorption isotherms and kinetics studies of malachite green on chitin hydrogels. Journal of Hazardous Materials 209, 218–225.

Wang, J., Chen, C., 2006. Biosorption of heavy metals by *Saccharomyces cerevisiae*: a review. Biotechnology Advances 24 (5), 427–451.

Wicke, D., Böckelmann, U., Reemtsma, T., 2008. Environmental influences on the partitioning and diffusion of hydrophobic organic contaminants in microbial biofilms. Environmental Science and Technology 42 (6), 1990–1996.

Writer, J.H., Ryan, J.N., Barber, L.B., 2011. Role of biofilms in sorptive removal of steroidal hormones and 4-nonylphenol compounds from streams. Environmental Science and Technology 45 (17), 7275–7283.

Wu, Y., Feng, M., Liu, J., Zhao, Y., 2005. Effects of polyaluminum chloride and copper sulfate on phosphorus and UV254 under different anoxic levels. Fresenius Environmental Bulletin 14 (5), 406–412.

Wu, Y., He, J., Yang, L., 2010a. Evaluating adsorption and biodegradation mechanisms during the removal of microcystin-RR by periphyton. Environmental Science and Technology 44 (16), 6319–6324.

Wu, Y., Zhang, S., Zhao, H., Yang, L., 2010b. Environmentally benign periphyton bioreactors for controlling cyanobacterial growth. Bioresource Technology 101 (24), 9681–9687.

Wu, Y., Li, T., Yang, L., 2012. Mechanisms of removing pollutants from aqueous solutions by microorganisms and their aggregates: a review. Bioresource Technology 107 (0), 10–18.

Wu, Y., Xia, L., Yu, Z., Shabbir, S., Kerr, P.G., 2014. In situ bioremediation of surface waters by periphytons. Bioresource Technology 151, 367–372.

Zhu, L., 2015. Biorefinery as a promising approach to promote microalgae industry: an innovative framework. Renewable and Sustainable Energy Reviews 41 (0), 1376–1384.

Chapter 14

Simultaneous Removal of Cu and Cd From Soil and Water in Paddy Fields by Native Periphyton

14.1 INTRODUCTION

There has been increasing pollution of cropland by heavy metals around the world, especially in China. In general, the combined pollution of heavy metals (e.g., Zn, Pb, Cd, Cu, and As) occurs in intensive agricultural areas such as paddy fields (Arini et al., 2012; Li et al., 2014). Heavy metals contamination in croplands can result in dietary exposure through soil–plant–food chain transfer, causing elevated levels of toxic metals in human organs. Compared with single heavy metals, the combined toxicity of multiple heavy metals (Ippolito et al., 2015) might pose higher potential risks to organisms and humans as well as ecosystem health (Bian et al., 2014; Huang et al., 2014; Zhao et al., 2010). For instance, the coexistence of Cu and Cd poses higher toxicity to animals such as *Hemigrapsus crenulatus* (Lee et al., 2010) as well as to humans (Hashemi-Manesh, 2010) than does each metal individually.

Many approaches have been proposed to remove Cu and Cd from the environment (Covelo et al., 2007b; Wang et al., 2005; Wu et al., 2015), which can be divided into three groups. The first group is to immobilize (inactivate) heavy metals and thus reduce their bioavailabilities by using specific materials such as biochar and rice straw ash (Hsu et al., 2015; Koptsik, 2014). The second is to "eat" the heavy metals with the use of hyperaccumulation plants, followed by removal of the plants from the land (Wang et al., 2015; Wei et al., 2005). The third types involve physical and chemical methods to regenerate the exhausted washing solution, such as evaporation and reversal osmosis (Di Palma et al., 2003). Other approaches have also been conducted to remove the Cu and Cd from the wash waste solution, such as electrochemical treatment (Gómez et al., 2010), photocatalysis (Fabbri et al., 2009; Parra et al., 2002), advanced oxidation processes (Bandala et al., 2008), and selective adsorption by activated carbons (Ahn et al., 2008) or biochar (Regmi et al.,

Periphyton. http://dx.doi.org/10.1016/B978-0-12-801077-8.00014-4

2012). All of these methods are costly in maintenance and operational terms because they need special apparatus and materials.

Compared with these physical and chemical methods, the biological method may be a cost-effective choice for the regeneration of the wash waste solutions. Periphyton, a ubiquitous microbial aggregate that is widely present in paddy fields, wetlands, and rivers (Wu et al., 2014; Žižek et al., 2011), may be better than other single communities, such as algae (Chen et al., 2015; Lavoie et al., 2012), plants (Saeed and Iqbal, 2003), and bacteria (Luo et al., 2011), in tolerating the toxicity of Cu and Cd (Machado et al., 2015; Ouyang et al., 2016; Wang et al., 2015). The periphyton communities are enclosed in a matrix of extracellular polymeric substances (EPS) (Cui et al., 2012), which are composed of protein, polysaccharides, and other organic and inorganic substances (Wang et al., 2014). These substances might chelate metals such as Cu and Cd, so it would be possible to use native periphyton in the effective removal of Cu and Cd at a low cost.

Some native specific functional microbial communities such as algae (Davis et al., 2003), fungi (Xie et al., 2014), and bacteria (Johansson et al., 2014) can immobilize heavy metals in the environment. Although this group had proved to be promising for in situ remediation of heavy metals contamination (Koptsik, 2014), the removal process was dominated by a single process—adsorption (Bere et al., 2012; Bradac et al., 2010; Soldo and Behra, 2000). The adsorbed heavy metals would be released into the environment again under certain conditions. Moreover, the single dominant species in these microbial communities is generally vulnerable to shifts in environmental conditions such as pH, temperature, nutrients, and multiple contamination (Yan et al., 2011). Because Cd and Cu often coexist in the soil and water in paddy fields (Pokrovsky et al., 2008), developing a multifunctional in situ remediation technology for the removal of combined heavy metals pollution in paddy fields is important and necessary.

Compared with single microbial communities, microbial aggregates (e.g., periphyton), contain many types of organisms and more complex structure debris (Wu et al., 2012). There are many diverse microporous structures with varying sizes in the periphyton (Higgins et al., 2003; Rectenwald and Drenner, 1999), which allow the transformation of nutrient substrates such as phosphorus and nitrogen and other contaminants, such as heavy metals, in the special cavities (Cui et al., 2012; Khatoon et al., 2007; Obolewski et al., 2011). Different adsorption and desorption processes for nutrients and contaminants (microcystin-RR) might occur because of dissimilar active sites on the surface of periphyton (Scinto and Reddy, 2003; Wu et al., 2010).

Previous studies have investigated the response of microbial communities to heavy metal pollution. For example, the photosynthesis of *Penicillium chrysogenum* and *Bacillus licheniformis* ceased due to the pollution of heavy metals such as Co, Cd, and Pb (Paraneeiswaran et al., 2015; Usman et al., 2013). However, the periphyton with more-complex components contain many microbial species such as microalgae, cyanobacteria, bacteria, fungi, actinomycetes, and protozoa (Quinlan et al., 2011; Wu et al., 2014), forming a relatively stable microecosystem of hierarchical structure. Periphyton is sensitive to habitat quality and can respond quickly to any changes in their habitats. The feedback of periphyton to the contamination of heavy metals might be different from the responses of a single species microbial community.

The interactions between heavy metals (e.g., Cu and Cd) could strengthen or exacerbate toxic impacts on specific organisms (Henriques et al., 2015; Pokrovsky et al., 2008), via altering the absorption kinetics and dynamics of metals (Henriques et al., 2015). As periphyton is very complex with different organisms, and combined pollution is popular in most areas with paddy fields in China, it is very difficult to distinguish the specific role of individual organisms. Hence, in this study, the periphyton will be considered as an ensemble when examining the combined effects of Cu and Cd.

The objectives of this study were to investigate (1) the efficiency of soil-washing solutions to extract Cd and Cu, (2) the capacity of periphyton to entrap Cu and Cd from soil-washing solutions, (3) the relationship between heavy metals (Cu and Cd) uptake and the carbon metabolization of periphyton, (4) the potential to remediate heavy metals–polluted soil and soil-washing solutions, (5) the responses of periphyton to co-contamination of Cu and Cd in paddy fields, and (6) the concentrations of Cu and Cd in rice and the germination of rice in the presence of periphyton. The findings will contribute to an effective option for simultaneously eliminating Cu and Cd from paddy fields, in turn helping to restore unbalanced wetland systems.

14.2 MATERIALS AND METHODS

14.2.1 Experiment 1

14.2.1.1 Preparation of Heavy Metals–Polluted Soils

The surface soil samples (depth 0–30 cm) were collected from a paddy field in Jiangsu Province (Table 14.1). The soils were air-dried (5% DW) and ground to pass through a 1-mm sieve, then homogenized for use. To simulate the heavily polluted soil in typical croplands in China, the collected soil was cultured with two common heavy metals, Cu and Cd, for 1 month (Chaiyaraksa and Sriwiriyanuphap, 2004). Briefly, 1000 mL of $CuSO_4 \cdot 5H_2O$ and $Cd(NO_3)_2 \cdot 8H_2O$ solutions at concentrations of 20 mg L^{-1} Cu/Cd were added to 1000 g dry soil. The samples were mixed for homogeneity by shaking (180 rpm) at room temperature (25°C) for 1 month. Sequentially, the samples were dried at 103–105°C for 8 h and passed through a 1-mm sieve again. At this stage, the total Cu and Cd concentrations in the soils were 815.5 and 138.95 μg kg^{-1}, respectively.

TABLE 14.1 Characteristics of Soil Samples Collected From a Paddy Field in Jiangsu Province

Parameters	Values
Land use	Rice paddy
Soil texture	Loam
pH	6.1 ± 0.3
Organic carbon (%)	2.5 ± 0.2
Sand (%)	25.5 ± 0.4
Silt (%)	20.1 ± 2.4
Clay (%)	55.4 ± 5.2
Cation exchange capacity (CEC) [cmol (+) kg^{-1}]	9.7

14.2.1.2 Periphyton Collection and Cultivation

Periphyton were collected from six paddy fields in Changshu City, southeast Jiangsu Province (120°33′–121°03′E, 31°33′–31°50′N). To reduce bias and error, all the selected sampling sites had similar physical and chemical conditions (such as light, shade, substrata, water depth, and flow velocity). Periphyton was peeled off from soil surfaces using a brush sterilized by 0.1 M HCl solution and mixed together for indoor culture. The periphyton were cultured in an incubator with light power of 240 W, light/dark regimen of 12 h/12 h, and air temperature of 25 ± 1°C. The chemical composition and preparation of the WC medium for periphyton culture were according to our previous study (Yang et al., 2015). When the periphyton matured (i.e., the color turned to deep green), they were used for experiments.

14.2.1.3 Soil Washing

The soil-washing experiments were carried out in a specific system composed of three parts: filtering tank (inner diameter: 20 mm, height: 300 mm), sedimentary tank (inner diameter: 30 mm, height: 90 mm), and periphyton tank (length × width × height = 20 cm × 10 cm × 10 cm) (Fig. 14.1). In the filtering tank, the soils (thickness: 200 mm) were covered by sand (size: 1.0–3.5 mm, sand layer thickness: 50 mm). To reduce the discharge of suspended materials, a layer of cotton (thickness: 50 mm) was set at the bottom of the sand. In the periphyton tank, 8.5 g periphyton with its carriers (Jineng Environmental Protection Company of Yixing, China) were cultured. The volume of periphyton and its carrier was about one-third of the total volume of the tank.

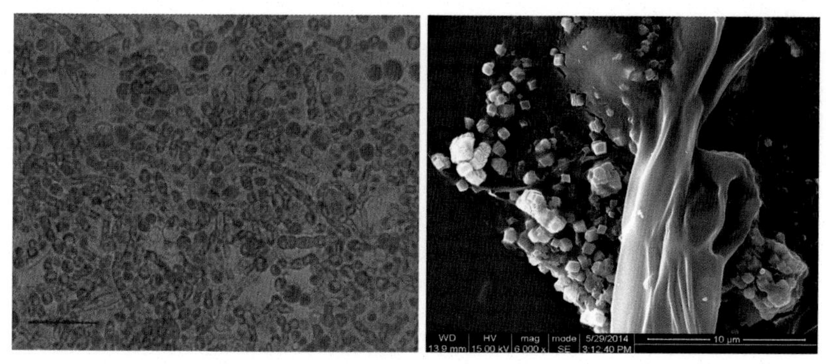

FIGURE 14.1 Compositions of the periphyton observed by optical microscope and SEM (Yang et al., 2016).

Three widely used washing agents were selected for washing Cu- and Cd-polluted soil: water, disodium salt of Na_2EDTA (0.2 M), and mixed solution of Mehlick 3 (0.2 M HOAc, 0.25 M NH_4NO_3, 0.015 M NH_4F, 0.013 M HNO_3, 0.001 M Na_2EDTA, pH 2.5 ± 0.1). The agents were individually added into the filtering tank at regular intervals (100 mL once per day, velocity 1.2 mL min^{-1}). The soil leachate flowed from the sedimentary tank to periphyton bioreactor, then discharged. The system was run at room temperature (20–25°C). The hydraulic retention time was 6 h for the first stage and 2 days for the second stage. After the soil leachate purification system was running at a steady state, the data in the second stage were collected once every 8 days for 96 days.

14.2.2 Experiment 2

14.2.2.1 Periphyton Sample Collection and Culture

Periphyton was collected from the same six paddy fields as in Experiment 1 using the same methodology. The mixed periphyton were moved into a spiral glass pipe for expanding culture in the laboratory. The length of the spiral glass pipe was 5.0 cm. The external and internal diameters of the spiral glass pipe were 4.0 mm and 2.0 mm, respectively. The total length of the spiral glass pipe was 76.4 cm.

Prior to experiments, the periphyton was colonized in an open system in glass tanks with tap water in ambient temperature (10–25°C) for 3 weeks. The periphyton were then transported to six spiral glass pipes (length: 76.4 cm, internal diameter: 2.0 mm) with WC media at a velocity of 0.01 mL s^{-1}. The periphyton in the spiral pipes were cultured in an incubator with light power of 240 W, light/dark regimen of 12 h/12 h, and air temperature of 25 ± 1°C. When the periphyton matured (i.e., the color was deep green), it was used in the following experiments.

14.2.2.2 Accumulation Experiment

To investigate the accumulation of Cu and Cd by mature periphyton, a series of experiments were conducted in spiral glass pipes as mentioned earlier under different conditions. The mixture solution containing 10 mg L^{-1} Cu and 10 mg L^{-1} Cd was prepared by dissolving $CuSO_4 \cdot 5H_2O$ and $Cd(NO_3)_2 \cdot 4H_2O$ into deionized water. The target mixture solution was then bumped into the spiral glass pipe at a velocity of 0.01 mL s^{-1}. When the periphyton was mature, the influent and effluent samples were collected at 0, 2, 18, 30, 42, 54, 72, 102, 144, 168, and 240 h for Cu and Cd determination in aqueous phase. The spiral glass assembly with periphyton was placed in the incubator with light power of 240 W, light/dark regimen of 12 h/12 h, and air temperature of 25 ± 1°C.

14.2.2.3 Rice Seed Germination Experiment

The rice germination experiment investigated the effect of periphyton on rice submerged in the mixture solution of Cu and Cd. The bosomy with similarly sized rice seeds were selected for the experiment. The rice seeds were immersed in 1% NaClO for 20 min followed by washing with deionized water to remove the residual NaClO. Then, the seeds were placed on filter paper in a petri dish for culture under different conditions. There were 20 grain rice seeds in each dish. There were three experimental groups: one treatment group and two control groups. The treatment group was cultured in the petri dish containing 5 g periphyton and 15 mL of the mixture solution of Cu and Cd. The first control group was cultured with 15 mL deionized water alone, and the second control group was cultured in the presence of 15 mL of the mixture solution of Cu and Cd but with no periphyton. Each trial was conducted in triplicate. The petri dishes were transferred to an artificial incubator with temperature of 25 ± 1°C and light/dark regimen of 12 h/12 h. To avoid drying in the incubator, 15 mL of the corresponding solution (water or Cu + Cd mixture) was added to each dish daily. The number of the germinating rice seeds was counted daily. The concentrations of Cu and Cd in rice and the liquid phase were measured by inductively coupled–plasma mass spectrometry (ICP-MS) after 7 days' culture.

14.2.2.4 Sampling and Analysis

The concentrations of Cu and Cd in the influent and effluent of the spiral glass pipes were measured by ICP-MS after filtration. The speciation of Cu and Cd in the samples was obtained using the visual MINTEQ 3.1 software (Zhang et al., 2008). The accumulated Cu and Cd in periphyton were calculated using Eq. (14.1).

$$Q_t = V(C_0 - C_t)/B_t \qquad (14.1)$$

where Q_t is the entrapped content of heavy metal (Cu/Cd) at the time t; V is the total volume (L) of heavy metal solution bumped into the spiral glass tube until time t; C_0 is the original concentration of the Cu/Cd in influent (mg L^{-1}); C_t is the concentration of Cu/Cd in the effluent (mg L^{-1}); and B is the biomass (g) of periphyton at time t.To evaluate the amount of heavy metal (Cu and Cd) accumulation in periphytic cells, 4 M ethylenediaminetetraacetate (EDTA) was added to desorb Cu and Cd adsorbed on the surface of the periphyton. Then, the accumulated metals (Cu or Cd) in the periphytic cells were calculated using Eq. (14.2).

$$Q_{t,c} = V(C_0 - C_{t,c})/B_t \tag{14.2}$$

where $Q_{t,c}$ is the accumulated content (mg kg^{-1}) of the heavy metal (Cu or Cd) at t time; and $C_{t,c}$ is the concentration (mg L^{-1}) of Cu or Cd in the effluent when EDTA was added to the influent.

To avoid destroying the community structure of the periphyton on the inner side of the spiral glass tube, the periphyton together with the glass pipe were placed in an oven at 60°C for 30 min after air drying at 20–25°C for 6 h. The net mass of periphyton biomass was calculated from the difference of the total weight of glass pipe containing periphyton and the glass pipe alone.

The periphyton samples were collected at the beginning and the end of the experiment. The community structure of the periphyton was observed by optical microscope and scanning electron microscope (SEM). The maximum electron transfer rate (ETR_{max}), total chlorophyll, algal species, and chlorophyll in the periphyton were determined using a Phyto-PAM fluorescence analyzer (Waltz, Germany). The biomass of each species of algae was calculated by the concentration of its algal chlorophyll multiplied by the volume of the periphyton in the spiral glass tube. The biomass of the heterotrophs was calculated by the difference between total biomass of the periphyton and the biomass of all algae.

The carbon source utilization activity of periphyton was assayed by the commercially available Biolog ECO Microplates (Hayward, CA, USA). The detailed analysis process is described in our previous study (Li et al., 2012). The functional diversities of the periphyton were obtained based on the activities of different carbon source use.

A new modified procedure, which combined the BCR (Nemati et al., 2011) and the five-step sequential extraction method (Orecchio and Polizzotto, 2013), was used to assess Cu and Cd speciation distribution in the clean, contaminated, and treated soils after each washing step. Five operationally defined fractions were determined for each metal: acetic acid soluble (Fraction 1), reducible (Fraction 2), organic affinity (Fraction 3), sulfide (Fraction 4), and residual (Fraction 5). The five-step sequential extraction procedures are summarized in Table 14.2. All data reported are averaged values of three replicates ($n = 3$).

TABLE 14.2 Sequential Extraction Scheme for the Cu and Cd Speciation of Soil

Metallic Fractions	Reagents and Conditions
Fraction 1 Soluble metal that is easily extracted by acetic acid	40 mL of 0.11 M HAc agitation during 16 h at room temperature
Fraction 2 Reductive metal that is easily interchangeable with water by sorption–desorption processes	40 mL of 0.04 M $NH_2OH \cdot HCl$ continuous stir 16 h at room temperature
Fraction 3 Metal bound to organic matter, that is, found adsorbed to living organisms, detritus, coatings on proteins, fats, mineral particles, etc.	40 mL of 0.01 M NaOH agitation during 12 h + 40 mL of 0.1 M $NH_2OH \cdot HCl$ continuous stir 16 h at room temperature
Fraction 4 Metal bounded to sulfide that is easily soluble under oxidizing conditions.	10 mL of 30% H_2O_2 60 min at 85°C; 10 mL 30% H_2O_2 60 min at 85°C + 3 mL 1 M HAc continuous agitation 30 min at room temperature
Fraction 5 Residual metal found in elementary form that should be extracted under hard acid conditions	10 mL of conc. HF + 1 mL of conc. $HClO_3$ 90 min at 95°C (triplicate), agitating occasionally

14.2.2.5 Metal Stability and Mobility in Soils

Metal stability was calculated using the I_R index (Gusiatin and Klimiuk, 2012) to describe the relative binding intensity of metals based on sequential extraction and is defined as:

$$I_R = \frac{\sum_{i=1}^{k} i^2 F_i}{k^2}$$

where i is the index number of the extraction step, progressing from 1 (for the weakest) to the most aggressive extractant ($k = 5$), and F_i is relative content of the considered metal in fraction i. The range of I_R values is 0–1: higher value indicates the metal in soils as residual fractions; low values represent distributions with a high proportion of soluble forms; and intermediate values represent patterns involving metal partitioning among all solid-phase components.

To assess the relative mobility and bioavailable form of metals in soil, the M_F factor was used, a ratio of metal concentration in the mobile fraction to the sum of all fractions:

$$M_F = \frac{F_1 + F_2}{F_1 + F_2 + F_3 + F_4 + F_5} \times 100\%$$

High M_F values indicate relatively high mobility and biological availability of heavy metals in soils.

Statistical Package for Social Science (SPSS 17) was used for analyses of the Biolog data. The linear relationships between Cu and Cd in periphyton (cells) and between average well-color development (AWCD) and the content of Cu and Cd were examined using regression analyses in Excel (2010). Differences between treatments were evaluated by using one-way ANOVA. p was set at .05 for all analyses.

14.3 RESULTS AND DISCUSSION

14.3.1 Evolution of Periphyton Community

Algae, bacteria (such as *Antinomies* and *Nitrobacteria*), fungi (such as *Trichobacteria* and *Candida*), and protozoa were the dominant species of the periphytic communities observed by optical microscope and SEM at the beginning of the experiment (Fig. 14.1).

14.3.2 Cu and Cd Extraction From Soil by Different Washing Solutions

The original concentrations of the Cu and Cd in the soil were 815 mg kg^{-1} and 138.95 mg kg^{-1}, respectively. The removal rate of the Cu and Cd in the soil was 22% and 40% after water washing, 38.5% and 87.9% after Na_2EDTA solution washing, and 52.1% and 91.6% after Mehlick 3 solution washing, respectively (Fig. 14.2). Clearly, the Na_2EDTA solution and the Mehlick 3 solution were better than the distilled water for extracting heavy metals from the soil.

Heavy metals mobility and bioavailability are increasingly used instead of total metal content to determine the success of soil remediation and to meet soil standards (Lee et al., 2015; Zhang et al., 2014). The I_R and M_F indices were used to assess metal binding intensity and mobility (Nowack et al., 2010). The three washing solutions significantly influenced the metal binding intensity in soils (Fig. 14.2). For Cu, the binding intensity (I_R) in the soil after water washing ($I_R = 0.52$) and the Mehlick 3 solution washing ($I_R = 0.71$) was higher than in the original soil ($I_R = 0.33$), which indicated that Mehlick 3 solution and distilled water could take the free Cu from the soil, thus decreasing the toxicity of soil to plants. For Cd, the I_R in the soil after Mehlick 3 ($I_R = 0.55$) and Na_2EDTA solution ($I_R = 0.38$) washing were higher than the I_R in the original soil ($I_R = 0.14$). Thus, the results showed that three washing solutions could remove Cu and Cd from the soil, but Mehlick 3 solution was efficient for both metals.

I_R was strongly correlated with M_F. Generally, the lower the relative binding intensity, the higher is the metal mobility. M_F reflects the variation of labile fractions, and so to some extent the M_F is more sensitive than I_R. In the

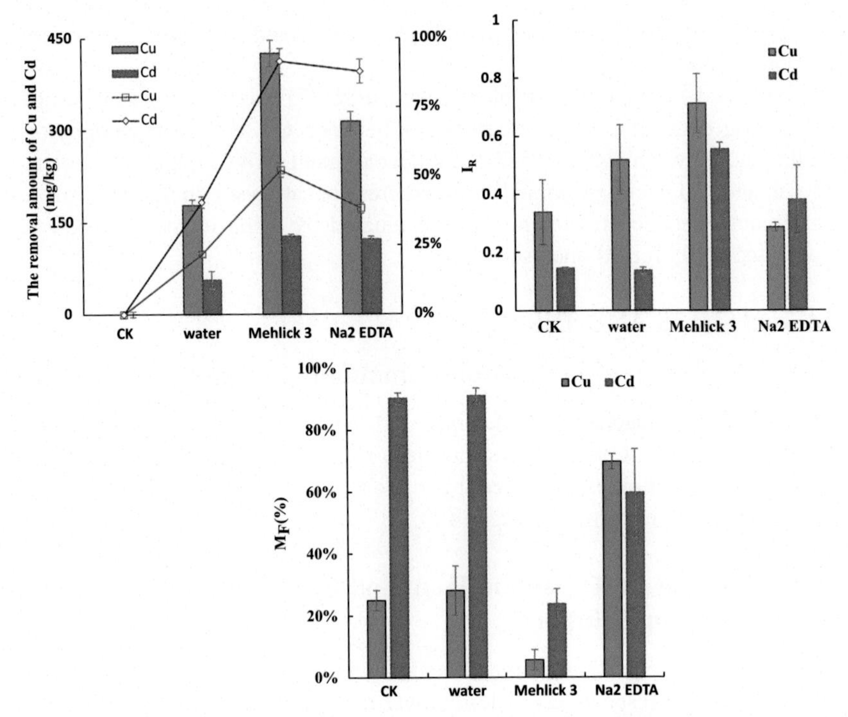

FIGURE 14.2 The changes in the reduced partition index (I_R) and mobility factor (M_F) for Cu and Cd in soils after solution washing.

original soils, the value of mobility was Cd > Cu. The original M_F of Cd and Cu was 90% and 25%, respectively (Fig. 14.2). The final M_F of Cd and Cu in the soil after the Mehlick 3 solution washing decreased to 23.71% and 5.58%, respectively, which indicated that the Mehlick 3 solution was better at removing the Cu and Cd from the soil and decreasing the metal mobility than the other washing solutions.

14.3.3 Heavy Metal Removal and Their Speciation Distribution

Cu was more easily accumulated by the periphyton than was Cd. When the initial concentrations of Cu and Cd were 10 mg L^{-1} each, the removal rates of Cu and Cd from the aqueous phase fluctuated from 39.2% to 68.8% and 22.5% to 49.9%, respectively (Fig. 14.3A).

During the experiment, different Cu- and Cd-hydrates such as $CuOH^+$, $Cu_2(OH)_2^{2+}$, $CdOH^+$, and $Cu_3(OH)_4^{2+}$ were analyzed. The results of the heavy metals speciation analysis shows that $CuOH^+$ and $Cu_2(OH)_2^{2+}$ were the dominant form of the Cu-hydrate after 2 h, accounting for 19% of the total Cu(II) (Fig. 14.3B). They sharply decreased to 4.5% at 18 h and then remained stable. They increased to 2.9% again at 102 h.

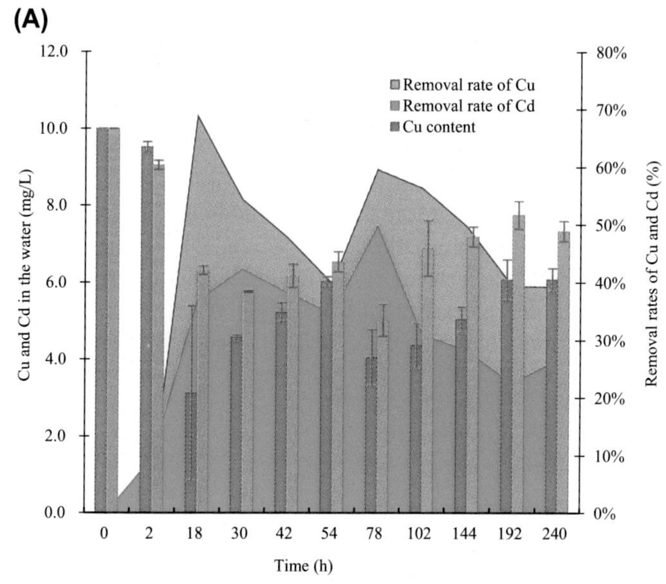

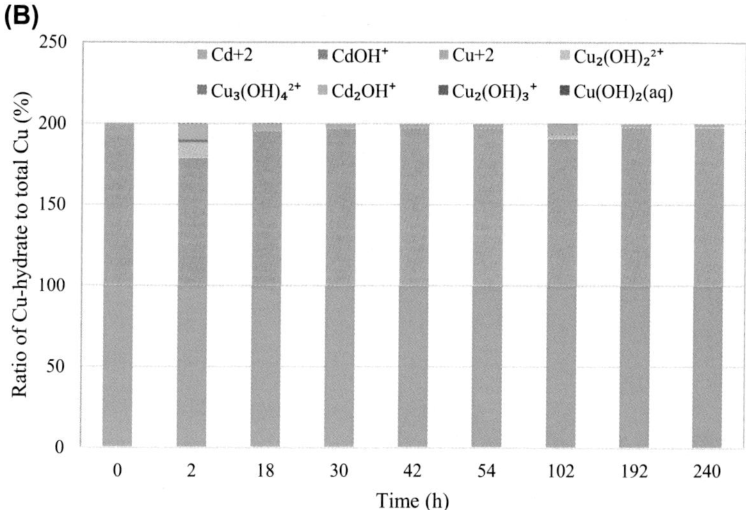

FIGURE 14.3 Concentration profiles of Cu/Cd and their corresponding removal rates (A); and the proportions of different species of Cu- and Cd-hydrates to the total Cu(II) and Cd(II) in the aqueous phase during the removal of Cu and Cd by periphyton (B).

$Cu_3(OH)_4^{2+}$ and Cu_2OH^{3+} were detected only at 2 and 102 h. $CdOH^+$ was the only species of Cd-hydrate observed, taking up 0.03% of the total Cd^{2+} amount (Fig. 14.3B). The contents of $Cu(OH)_2$ (aq) were very low during the whole experiment, only near 0.08% at 2 h and 0.01% at 18 h of the total Cu(II) (Fig. 14.3B).

14.3.4 Cu and Cd Accumulation in Periphytic Surface and Cells

The contents of heavy metals entrapped by the periphyton fluctuated from 145.20 to 342.42 mg kg^{-1} for Cu and from 101.75 to 236.29 mg kg^{-1} for Cd after 2 h. The concentrations of Cd accumulated in the periphytic cells (the contents of heavy metal in periphyton extracted by 4 M EDTA) varied from 42.93 to 174 mg kg^{-1} after 2 h (Fig. 14.4). More Cu was observed on the periphyton surface than Cd. More Cd was found in the periphytic cells. No Cu was detected in the periphytic cells at 54, 192, and 240 h. These results showed that the periphytic cells were more inclined to accumulate Cd than Cu.

To investigate the interaction of Cu and Cd, the relationships between Cu and Cd on the periphyton surface and in the cells were studied during periphyton culture in the mixture solution of Cu and Cd. A significant

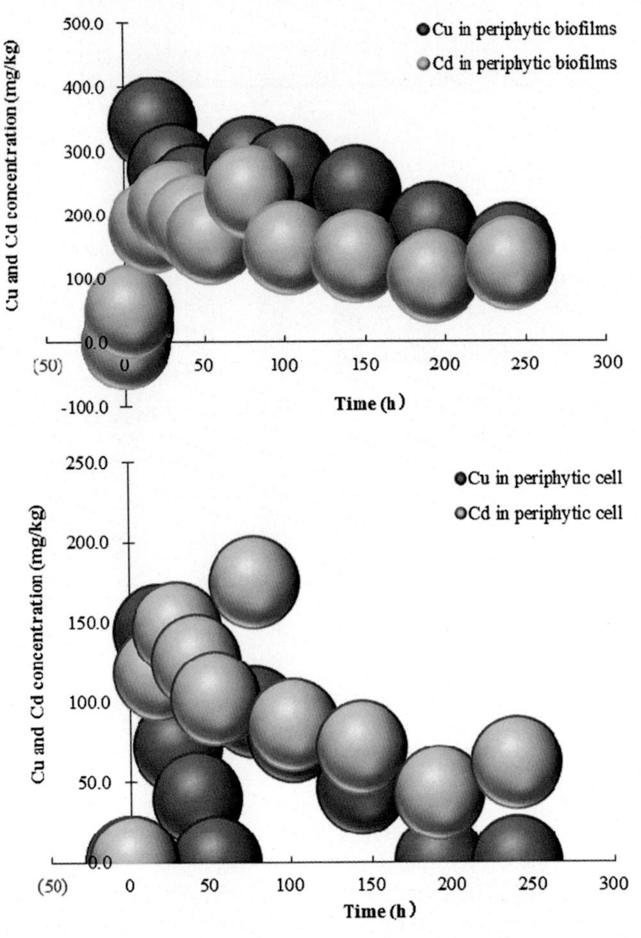

FIGURE 14.4 The contents of Cu and Cd on the periphyton surface and in their cells.

relationship between Cu and Cd contents on the periphyton surface ($p < .05$) was observed, with the equation of Cd (mg kg^{-1}) = 0.5884 × Cu (mg kg^{-1}) + 23.561 ($R^2 = 0.80$). The relationship between Cu and Cd contents in the periphytic cells could be described as: Cd (mg g^{-1}) = 0.7809 × Cu (mg g^{-1}) + 50.775 ($R^2 = 0.47$, $p < 0.05$). It was noted that the correlation coefficient of Cu-Cd in the cell was lower than that on the periphyton surface. The results showed that the Cu and Cd could be entrapped by periphyton simultaneously, but the presence of Cd had negative effects on the Cu accumulation by periphytic cells.

14.3.5 Effect of Periphyton on Physical Factors

The electrical conductivity (EC), dissolved oxygen (DO), and pH of the washing solution were determined to assess the variation in each environmental parameter when the periphyton removed the Cu and Cd from the washing solution. In all the washing solutions, the value of the EC increased slowly. It increased from 348.85 to 559.93 μs cm^{-1} in water, from 4469.2 to 5487.2 μs cm^{-1} in the Na_2EDTA solution, and from 22,759 to 27,416 μs cm^{-1} in the Mehlick 3 solution. The pH of the three washing solutions remained stable (pH = 7.5 in water, pH = 4.7 in Na_2EDTA solution, pH = 2.93 in Mehlick 3 solution). The DO decreased from 3.7 to 1.3 μmol L^{-1} in the Na_2EDTA solution and floated between 6.3 and 4.44 μmol L^{-1} in the water solution (Fig. 14.5). In the Mehlick 3 solution, DO decreased from 0.97 to 0.07 μmol L^{-1} and then increased to 1.07 μmol L^{-1}. The results showed that the periphyton in different solutions maintained activity and steady states through changing the chemical factors in the ambient environment.

To further probe the role of periphyton, the protein and chlorophyll contents were determined. Protein content increased, which indicated that the periphyton secreted more protein to entrap the Cu and Cd in the soil-washing solution (Fig. 14.5). The chlorophyll content also increased, reflecting the dynamic photosynthesis, which indicated periphyton self-regulation under the stimulation of Cu and Cd in the soil-washing solution.

To understand the tolerance of the periphyton to the toxicity of Cu and Cd, we conducted Pearson analyses among the EC, pH, DO, AWCD, Cu, and Cd (Table 14.2). In the three solutions, there was a positive linear relationship between the content of Cu and Cd ($r > 0.8$, $p < .05$). In the periphyton washing solution, there was a positive relationship between EC, pH, Cu, and Cd (Table 14.3).

AWCD was only correlated to the concentration of Cu and Cd ($r > 0.53$), which indicated that the periphyton were attracted by the toxicity of the Cu and Cd (Table 14.3). In the periphyton−Na_2EDTA solution, the EC had a good relationship with pH ($r > 0.8$) and the AWCD was only correlated to pH, which indicated that the pH is the limiting factor for periphyton growth. In the periphyton−Mehlick 3 solution, the AWCD was correlated with the

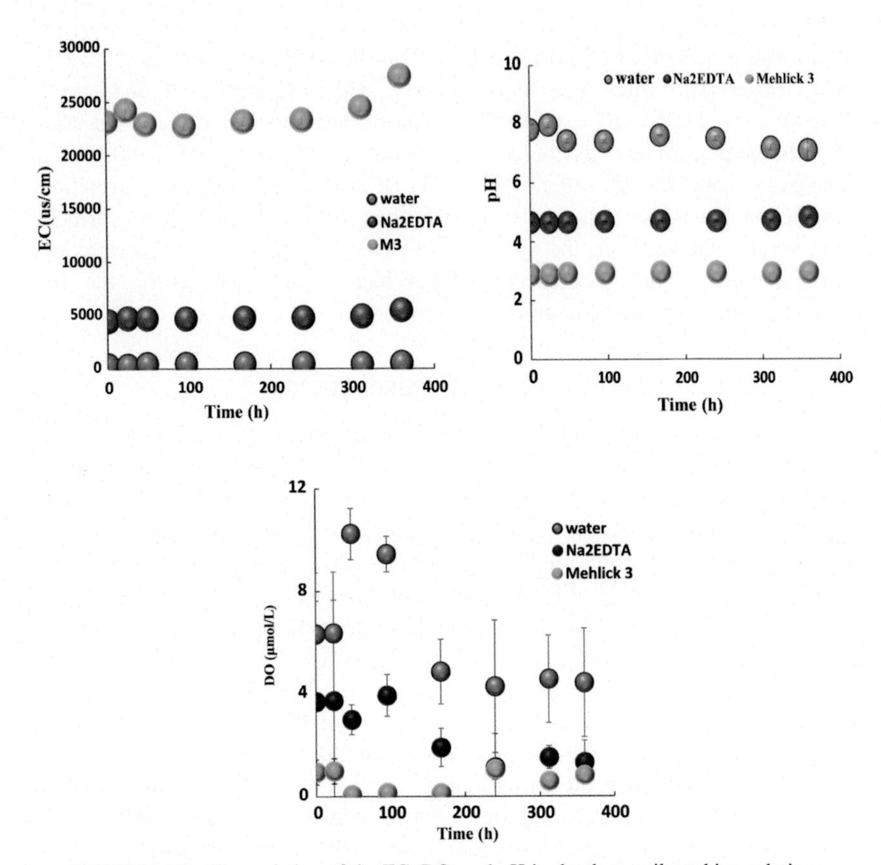

FIGURE 14.5 The variation of the EC, DO, and pH in the three soil-washing solutions.

concentration of Cu and DO, which indicated that the toxicity of the Cu blocked the periphyton growth. In every solution, however, the periphyton had the ability to self-adjust depending on the different conditions and act as a buffer to slow down the dispersal of Cu and Cd to the soil or water.

14.3.6 Effect of Heavy Metals on Periphyton Growth

Total chlorophyll of periphyton was determined in the presence or absence of heavy metals to examine the effect of heavy metals contamination on periphyton growth (Fig. 14.6). During the first 2 h after exposure to heavy metals pollution, little variation in the total chlorophyll was observed. After 2 h, it decreased sharply, from 2.08 to 0.13 mg L^{-1} at the end of the experiment. The total chlorophyll in the control treatment increased from 2.77 to 5.61 mg L^{-1} during the experimental period. It indicated that the presence of Cu and Cd inhibited the growth of periphyton. A slight increase in chlorophyll was observed in the middle stages (from 18 to 54 h) and the late stages (from

TABLE 14.3 Pearson Analysis of EC, pH, DO, AWCD, Cu, and Cd in the Three Soil-Washing Solutions

Solution/P (* <0.05, ** <0.01)		EC	pH	DO	AWCD	Cu	Cd
Water	EC	1	−0.885**	−0.248	−0.031	−0.842**	−0.704
	pH	−0.885**	1	0.110	−0.231	0.856**	0.745*
	DO	−0.248	0.110	1	−0.344	0.486	0.343
	AWCD	−0.031	−0.231	−0.344	1	−0.538	−0.589
	Cu	−0.842**	0.856**	0.486	−0.538	1	0.937**
	Cd	−0.704	0.745*	0.343	−0.589	0.937**	1
Na_2EDTA	EC	1	0.955**	−0.642	−0.189	−0.519	−0.374
	pH	0.955**	1	−0.697	−0.635	−0.329	−0.316
	DO	−0.642	−0.697	1	0.482	−0.136	−0.067
	AWCD	−0.189	−0.635	0.482	1	−0.256	0.239
	Cu	−0.519	−0.329	−0.136	−0.256	1	0.803*
	Cd	−0.374	−0.316	−0.067	0.239	0.803*	1
Mehlick 3 solution	EC	1	0.224	0.391	−0.208	−0.466	−0.411
	pH	0.224	1	−0.278	−0.194	0.161	−0.078
	DO	0.391	−0.278	1	−0.569	0.027	0.151
	AWCD	−0.208	−0.194	−0.569	1	−0.450	−0.006
	Cu	−0.466	0.161	0.027	−0.450	1	0.725*
	Cd	−0.411	−0.078	0.151	−0.006	0.725*	1

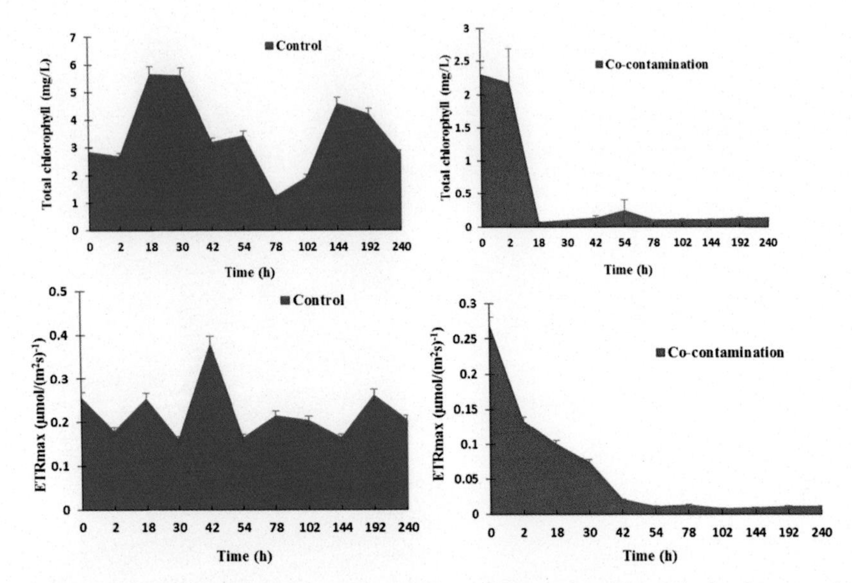

FIGURE 14.6 Changes in the total chlorophyll and ETR_{max} of periphyton in the absence and presence of heavy metals contamination.

78 to 240 h) in the presence of heavy metals, implying that the periphyton had the ability to self-regulate.

Maximum electron transport rate (ETR_{max}) is a critical parameter for characterizing the rate of photosynthesis of microbial aggregates such as multiple algal communities and the state of algal growth (Wu et al., 2011). The results showed that the ETR_{max} of periphyton in the control fluctuated from 0.205 to 0.379 μmol ($m^{-2}s^{-1}$) during the experiment. After exposure to heavy metals pollution, ETR_{max} sharply decreased from ~0.268 at the beginning to 0.008 μmol ($m^{-2}s^{-1}$) after 102 h (Fig. 14.6). Interestingly, a slight increase in ETR_{max} after 102 h was observed, from 0.008 to 0.015 μmol ($m^{-2}s^{-1}$) at 240 h, possibly due to the adaption of periphyton to the contaminated condition.

Blue algae (cyanobacteria), green algae, and brown algae were the dominant algae identified by chlorophyll content (Fig. 14.7). The biomass of these three algae rapidly declined with time, and the brown algae was the main component of the periphyton in the late stage (from 102 to 240 h) in the presence of heavy metals contamination. In the control treatment, green and brown algae were the dominant species after 18 h. These results indicate that only the brown algae could bear the combined pollution of Cu and Cd, though it was also inhibited to some degree. The biomass of heterotrophs was calculated by the difference between total biomass (represented by total chlorophyll) and autotrophic organisms (represented by green, blue, and brown algal chlorophyll). The results showed that the biomass of heterotrophs in the control fluctuated periodically due to the aging–regeneration shifts of

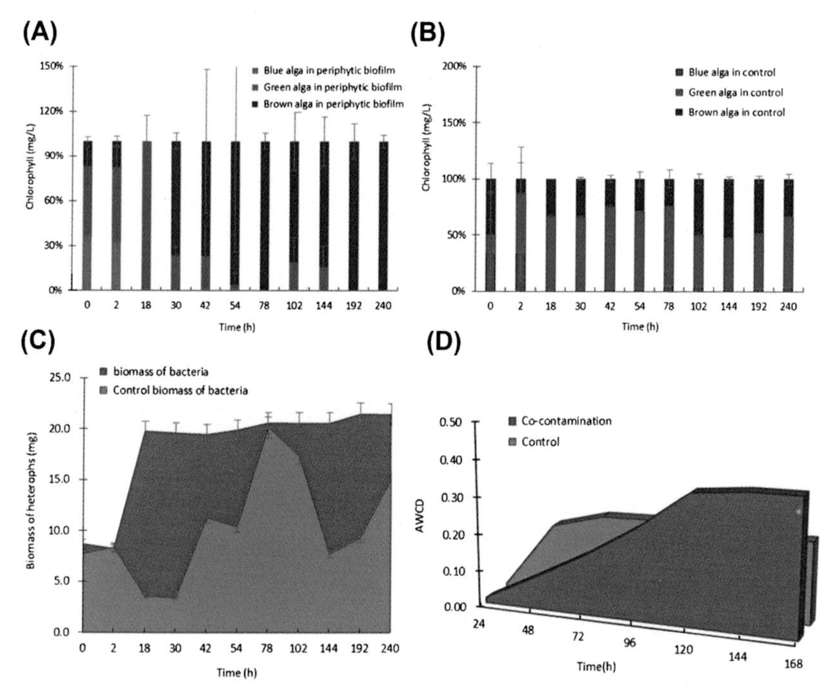

FIGURE 14.7 Changes in the properties of the periphytic community, including (A, B) biomasses represented by chlorophyll contents in different algae in the control and the treatment with Cu and Cd solution, (C) the biomass of heterotrophs in the periphyton during culture in Cu and Cd solution, and (D) the utilization ability of different carbon sources by microorganisms in the periphyton at the end of the experiment (represented by AWCD).

the periphyton. In the treatment with Cu and Cd, the biomass of heterotrophs sharply increased from 8.23 to 21.43 mg after 2 h (Fig. 14.7). These results showed that the dominant periphyton species changed from autotrophic organisms (i.e., algae) to heterotrophs to adapt to the contaminated conditions. On the other hand, compared with the control, the utilization ability of amine and amino acids by microorganisms increased in the periphyton (Fig. 14.7), suggesting that the proportion of heterotrophs in the periphyton increased gradually (Li et al., 2012).

14.3.7 Mechanisms Involved in Heavy Metals Removal by Periphyton

Microbial metabolism activity based on the utilization of carbon sources is usually characterized by the AWCD determined by using Biolog EcoPlates (Li et al., 2012). The AWCD values of periphyton were detected in the control and the contaminated treatment groups (Fig. 14.8). The AWCD significantly decreased from the beginning to the middle of the experiment ($p < 0.05$) and

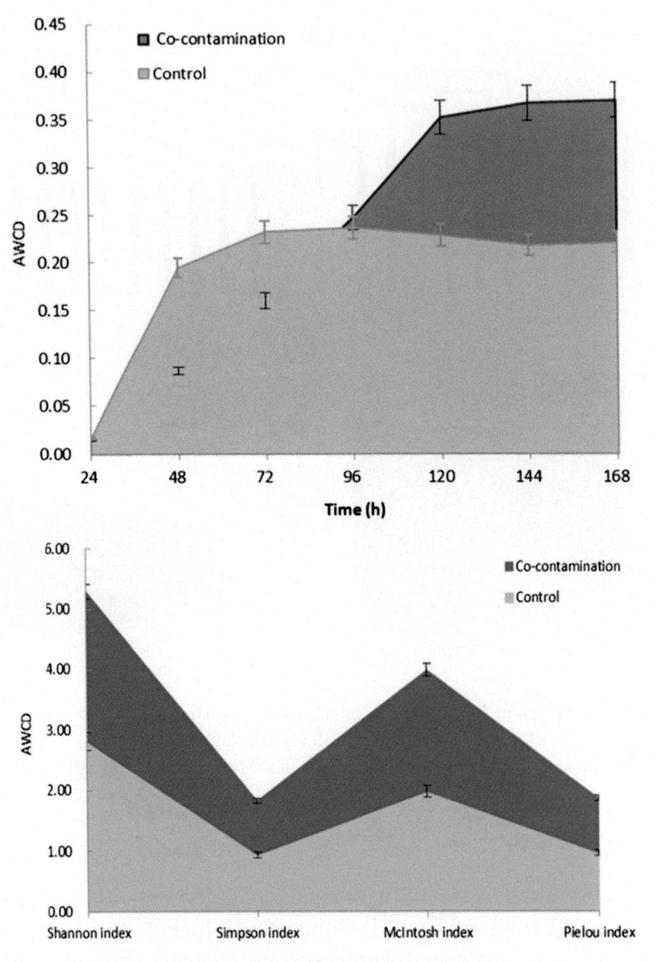

FIGURE 14.8 Changes in the carbon metabolic potential and functional diversity of periphytic biofilm (including Shannon, Simpson, McIntosh, and Pielou indices).

then increased with incubation time in each period (the middle stage and the end of the experiment). There was a positive linear relationship between AWCD and the Cu and Cd contents in the periphyton, AWCD = 0.0219 × (Cu and Cd, mg kg^{-1}) + 0.0384 ($R^2 = 0.70$, $p < 0.05$). These results showed that although the presence of Cu and Cd decreased carbon utilization by the periphyton, the periphyton exhibited a vast potential to adapt to the heavy metals contamination conditions.

Diversity indices (Simpson, Shannon, McIntosh, and Pielou indices) calculated based on AWCD values were used to characterize the function of periphyton (Li et al., 2012; Rusznyák et al., 2008). Although the Shannon, Simpson, and Pielou indices decreased with the increasing incubation time in the presence of Cu and Cd, the average McIntosh index increased from 1.98 at the beginning to 2.01 at the middle stage of the experiment (Fig. 14.8). These results implied that the periphyton had experienced a self-adjustment period to maintain the evenness of the different periphytic communities (Li et al., 2012), which is similar to the aforementioned results that the dominant species changed from photoautotrophs to heterotrophs.

EPS are the main chemical components in the surface of the periphyton (Wu et al., 2014), which has an affinity with heavy metal ions. The high molecular organic matter, such as protein and polysaccharide, in the EPS might change the adsorption–desorption processes of Cu and Cd. In addition, the negatively charged groups of the protein containing amino, amide, and carboxyl groups in the EPS might lead to competition adsorption of the coexisting metal ions such as Cu(II) and Cd^{2+} (Covelo et al., 2007a,b; Vega et al., 2006). As a result, the entrapped/accumulated contents of Cu and Cd in the periphyton and cells differed in this study.

14.3.8 Accumulation of Heavy Metals in Rice

The rice germination experiments were conducted to investigate the effects of periphyton on heavy metals accumulation in rice and rice germination. The results showed that when the water (without heavy metals and periphyton) was used to culture rice, the concentration of Cu and Cd in rice was very low, 0.49 mg L^{-1} for Cd and 10.08 mg L^{-1} for Cu, respectively; the number of rice seeds germinating increased with time, implying that the rice seeds used were healthy.

The concentration of Cu and Cd accumulated in the rice in the treatment group (with heavy metals and periphyton) was lower than those in the control group (without periphyton) (Fig. 14.9). Cu was more easily accumulated by rice than was Cd. Although the number of the rice seeds germinating increased with incubation time, the number of germinating rice seeds treated by heavy metals alone was lower than in treatment with periphyton. The results suggest that periphyton reduces the toxicity of Cu and Cd to the rice.

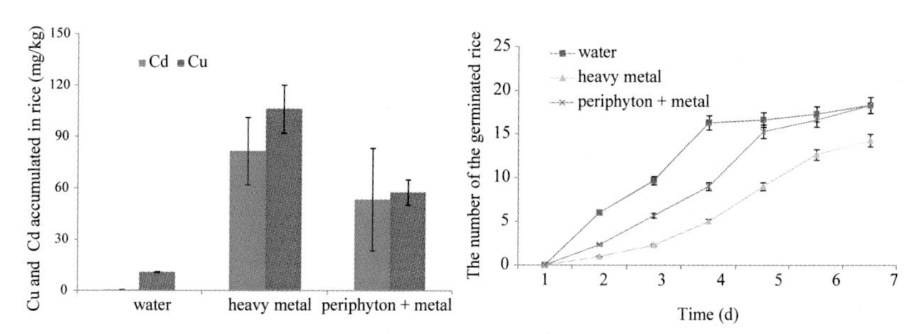

FIGURE 14.9 The variation in heavy metal accumulation in rice and the number of rice seeds germinating under different conditions (Yang et al., 2016).

14.3.9 Implications for In Situ Remediation of Heavy Metals Contamination in Paddy Fields

Contrary to many organic pollutants, metals cannot be destroyed and are nondegradable. For this reason, metal-polluted soils must always be treated and remediated to decrease the risks of pollution. Soil washing was proposed to remediate the heavy metals pollution in soil and was based on both physical separation (e.g., size separation or magnetic fractionation) and chemical leaching. However, after washing, the waste flush solution contains many potential contaminants, such as the bacteria, chelated metals, and metal ions, because of the complex soil properties. Common technologies also focus on "point-to-point" soil pollution, such as phytoremediation (Cassina et al., 2012; Sun et al., 2011) and biochar (Puga et al., 2015; Shen et al., 2016). Soil is a habitat for many creatures like bacterial fungi and other small organisms, which compose an autarkic circle for ecosystems. For this reason, the communities change the ambient conditions, such as pH and oxidation–reduction potential in the soil, which affect the efficiency of remediation methods (Sullivan et al., 2013). Microbial aggregates may be an alternative resource for remediating heavy metals pollution in the soil. First, the microbes exist everywhere in soils; second, the microbes would change the speciation of the heavy metals (Xu et al., 2016); and third, the various microbial communities, such as natural metal accumulators, could adsorb different heavy metals such as Fe (Cai et al., 2015), Cu,Cd, and As.

The periphyton could remove the Cu and Cd from the soil wash waste solution, but in different solutions the effects of the periphyton were different. The Cu and Cd in the soil were washed out by the washing solutions, but the chemical interactions in the soil were different due to the different solutions. The possible chemical reactions would be as follows:

$$\mathrm{EDTA} - 2\mathrm{Na} + \frac{\mathrm{Cu}^{2+}}{\mathrm{Cd}^{2+} \; =} \mathrm{EDTA} - \mathrm{Cd(or\ Cu)} + \mathrm{Na_2SO_4} \tag{14.3}$$

$$\mathrm{EDTA} - 2\mathrm{Na} + \frac{\mathrm{Cu}^{2+}}{\mathrm{Cd}^{2+} \xrightarrow{\mathrm{NH_4^+}}} (\mathrm{NH_4^+})_2[\mathrm{EDTA} - \mathrm{Cd\ (or\ Cu)}] + \mathrm{Na_2SO_4} \tag{14.4}$$

Eq. (14.3) occurred while washing the soil with Na_2EDTA solution, and Eq. (14.4) occurred while washing the soil with Mehlick 3 solution. The distilled water could only take away the Cu and Cd that were free in the soil. The EDTA-Cu/EDTA-Cd may be the primary content in the Na_2EDTA solution, and $(NH^{4+})_2$[EDTA-Cu]/$(NH_4^+)_2$ [EDTA-Cd] may be the primary content in the Mehlick 3 solution. For the Na_2EDTA solution, the chelating agent of Na_2EDTA (Ren et al., 2014) depends on the species of Cu and Cd, so the EDTA-Cu/EDTA-Cd may be the primary species in the periphyton. However, the pH (Sharma et al., 2014; Wu et al., 2014) also restricts factors in the

Mehlick 3 solution, which directly affects the activities of the periphyton rather than the chelating agent of EDTA.

The periphyton first interacts with ions dissolved in aquatic phase through surface complexation, and functional groups present on periphyton cell surfaces (contained EPS) can become negatively charged to buffer the surrounding environment by donating protons (Warren and Haack, 2001). The functional groups were found in different biomacromolecules such as peptidoglycans, teichoic acids, proteins, and lipoproteins. The varying amounts of protein and chlorophyll showed that the periphyton absorb the Cu and Cd ions through the production of protein and increasing photosynthesis. The different carbon utilizations also show that the periphyton structures were shifted in response to Cu and Cd. The periphyton structure also affected the entrapped Cu and Cd in the soil-washing solution (Bere and Tundisi, 2012; Komulainent and Morozov, 2010). This study has shown that periphyton accumulated Cu and Cd from the soil-washing solution. To some extent, the relationship between the growth of the periphyton and the Cu and Cd toxicity becomes stable. It was clear that the periphyton resisted the toxicity of Cu and Cd. When taking pH, EC, and DO into consideration, the periphyton tended to react differently to different soil-washing solutions and survived via biochemical and biological reactions with Cu and Cd.

Due to growing mine activities and the abuse of the use of chemical fertilizers on paddy fields, increasing attention has been paid to the remediation of heavy metals-contaminated paddy fields. The reduction in heavy metals bioavailability and the accumulation of heavy metals by hyperaccumulators are two effective measures (Koptsik, 2014). The immobilization and accumulation of Cu and Cd by environmentally friendly materials such as rice straw ash and biochar (Hsu et al., 2015; Huang et al., 2011) or hyperaccumulators such as *Solatium nigrum* L. (Wei et al., 2005) and *Commelina communis* (Wang et al., 2004) have been widely employed.

Extra-supplied materials, such as rice straw ash and biochar, might adsorb and/or desorb other contaminants such as Cd, Cu, Pb, and Zn (Hsu et al., 2015; Huang et al., 2011; Lu et al., 2014) or nutrients such as phosphorus (Yao et al., 2011), leading to shifts in the balance of the element biogeochemical cycles in paddy fields. Additionally, the in situ remediation of contaminated paddy fields using hyperaccumulators might be impractical.

In this study, the removal process of Cu and Cd by periphyton from water was a periodic cycling of adsorption–desorption. From a long-term perspective, the use of periphyton is a stable measure for the removal of heavy metals such as Cu and Cd in paddy fields. The capacity of the periphyton to accumulating heavy metals is higher than some previously used methods. For instance, the maximum capacity of entrapping Cd by the Cd-hyperaccumulator *Solatium nigrum* L. was 124.6 mg kg^{-1} (Wei et al., 2005) which is lower than in our study (up to 174 mg kg^{-1}). Moreover, the employment of periphyton is capable of simultaneously removing Cd and Cu, while specific heavy metal hyperaccumulators often have the ability to accumulate single heavy metal species (Koptsik, 2014; Wei et al., 2005).

The toxicity of Cu and Cd in the paddy fields causes ecological risks to rice safety and human health (Liu et al., 2006). The multiple contaminations also caused heavy damage to the periphyton in the early stages. Although the periphyton was stressed by Cu and Cd, the activity and functionality of the periphyton recovered to some degree after adaptation. The periphyton exhibited clear adaption to the combined pollution of Cu and Cd, as shown by the changes in photosynthesis, microbial composition, utilization of carbon sources, and functional diversity.

In the study, the pH decreased from 6.63 to 5.6 during the introduction of periphyton, which is beneficial in sustaining the heavy metal ions in hydrate form in the early stages of the experiment according to the hydrolysis reaction $[M^{+n}(aq) + nH_2O \leftrightarrow M(OH)_n(s) + nH^+]$. With the culture of the periphyton in the Cu and Cd solutions, the periphyton contents of Cu- and Cd-hydrates significantly declined ($p < 0.05$). It is well known that ionic states of Cu and Cd can be easily accumulated by organisms. This indicates that the periphyton in paddy fields has substantial potential to reduce the risk of Cu and Cd accumulation in rice.

From an ecological point of view, the microorganisms in periphyton have the ability to self-acclimate and thus can adapt to various contamination levels (Wu et al., 2010, 2012, 2014), leading to a relatively steady entrapment/accumulation of heavy metals in periphyton. In the paddy fields environment, the sediments might provide a habitat for the native microorganisms, which, in turn, supply microbial sources for the aggregation of periphyton on the surface of the sediments, and thus facilitate the removal of heavy metal pollution.

14.4 CONCLUSIONS

In this study, native periphyton was proposed to purify the soil-washing solutions through accumulating Cu and Cd. The periphyton entrapped the heavy metals from the soil-washing solution with a highest removal of 80%. Moreover, the periphyton could buffer the environmental factors such as pH, EC, and DO to avoid fluctuating heavily. Although Cu and Cd had a negative effect on periphyton growth, the periphyton exhibited significant potential to recover its activity and functionality by adjusting its composition and environmental factors. The periphyton in paddy fields formed Cu- and Cd-hydrates in the early stages of the experiment and then reduced the proportion of hydrates in the later stages of the experiment. The periphyton could enrich Cu and Cd from water with a high capacity. As a result, the Cu and Cd contents in rice were low and the number of rice seeds germinating was high, compared to the control (without periphyton). Cd was more inclined to be accumulated by the periphytic cells than Cu. The findings in this study provide a promising and environmentally benign biomeasure to simultaneously eliminate Cu and Cd in paddy fields and offer some valuable insights into the adaption of periphyton to ecosystems contaminated by heavy metals.

REFERENCES

Ahn, C.K., Kim, Y.M., Woo, S.H., Park, J.M., 2008. Soil washing using various nonionic surfactants and their recovery by selective adsorption with activated carbon. Journal of Hazardous Materials 154 (1–3), 153–160.

Arini, A., Feurtet-Mazel, A., Morin, S., Maury-Brachet, R., Coste, M., Delmas, F., 2012. Remediation of a watershed contaminated by heavy metals: a 2-year field biomonitoring of periphytic biofilms. Science of the Total Environment 425, 242–253.

Bandala, E.R., Velasco, Y., Torres, L.G., 2008. Decontamination of soil washing wastewater using solar driven advanced oxidation processes. Journal of Hazardous Materials 160 (2–3), 402–407.

Bere, T., Chia, M.A., Tundisi, J.G., 2012. Effects of Cr III and Pb on the bioaccumulation and toxicity of Cd in tropical periphyton communities: implications of pulsed metal exposures. Environmental Pollution 163, 184–191.

Bere, T., Tundisi, J.G., 2012. Effects of cadmium stress and sorption kinetics on tropical freshwater periphytic communities in indoor mesocosm experiments. Science of the Total Environment 432 (16), 103–112.

Bian, R., Joseph, S., Cui, L., Pan, G., Li, L., Liu, X., Zhang, A., Rutlidge, H., Wong, S., Chia, C., Marjo, C., Gong, B., Munroe, P., Donne, S., 2014. A three-year experiment confirms continuous immobilization of cadmium and lead in contaminated paddy field with biochar amendment. Journal of Hazardous Materials 272, 121–128.

Bradac, P., Wagner, B., Kistler, D., Traber, J., Behra, R., Sigg, L., 2010. Cadmium speciation and accumulation in periphyton in a small stream with dynamic concentration variations. Environmental Pollution 158 (3), 641–648.

Cai, Y.A., Li, D., Liang, Y., Luo, Y., Zeng, H., Zhang, J., 2015. Effective start-up biofiltration method for Fe, Mn, and ammonia removal and bacterial community analysis. Bioresource Technology 176, 149–155.

Cassina, L., Tassi, E., Pedron, F., Petruzzelli, G., Ambrosini, P., Barbafieri, M., 2012. Using a plant hormone and a thioligand to improve phytoremediation of Hg-contaminated soil from a petrochemical plant. Journal of Hazardous Materials 231–232, 36–42.

Chaiyaraksa, C., Sriwiriyanuphap, N., 2004. Batch washing of cadmium from soil and sludge by a mixture of Na2S2O5 and Na2EDTA. Chemosphere 56 (11), 1129–1135.

Chen, B., Li, F., Liu, N., Ge, F., Xiao, H., Yang, Y., 2015. Role of extracellular polymeric substances from Chlorella vulgaris in the removal of ammonium and orthophosphate under the stress of cadmium. Bioresource Technology 190, 299–306.

Covelo, E.F., Vega, F.A., Andrade, M.L., 2007a. Competitive sorption and desorption of heavy metals by individual soil components. Journal of Hazardous Materials 140 (1–2), 308–315.

Covelo, E.F., Vega, F.A., Andrade, M.L., 2007b. Simultaneous sorption and desorption of Cd, Cr, Cu, Ni, Pb, and Zn in acid soils: II. Soil ranking and influence of soil characteristics. Journal of Hazardous Materials 147 (3), 862–870.

Cui, J., Shan, B., Tang, W., 2012. Effect of periphyton community structure on heavy metal accumulation in mystery snail (*Cipangopaludina chinensis*): a case study of the Bai River, China. Journal of Environmental Sciences 24 (10), 1723–1730.

Davis, T.A., Volesky, B., Mucci, A., 2003. A review of the biochemistry of heavy metal biosorption by brown algae. Water Research 37 (18), 4311–4330.

Di Palma, L., Ferrantelli, P., Merli, C., Biancifiori, F., 2003. Recovery of EDTA and metal precipitation from soil flushing solutions. Journal of Hazardous Materials 103 (1–2), 153–168.

Fabbri, D., Crime, A., Davezza, M., Medana, C., Baiocchi, C., Prevot, A.B., Pramauro, E., 2009. Surfactant-assisted removal of swep residues from soil and photocatalytic treatment of the washing wastes. Applied Catalysis B: Environmental 92 (3–4), 318–325.

Gómez, J., Alcántara, M.T., Pazos, M., Sanromán, M.Á., 2010. Soil washing using cyclodextrins and their recovery by application of electrochemical technology. Chemical Engineering Journal 159 (1–3), 53–57.

Gusiatin, Z.M., Klimiuk, E., 2012. Metal (Cu, Cd and Zn) removal and stabilization during multiple soil washing by saponin. Chemosphere 86 (4), 383–391.

Hashemi-Manesh, M., 2010. The toxic effect of copper on human low density lipoprotein oxidation. Toxicology Letters 196, S298–S299.

Henriques, I., Araújo, S., Pereira, A., Menezes-Oliveira, V.B., Correia, A., Soares, A.M.V.M., Scott-Fordsmand, J.J., Amorim, M.J.B., 2015. Combined effect of temperature and copper pollution on soil bacterial community: climate change and regional variation aspects. Ecotoxicology and Environmental Safety 111, 153–159.

Higgins, S.N., Kling, H.J., Hecky, R.E., Taylor, W.D., Bootsma, H.A., 2003. The community composition, distribution, and nutrient status of epilithic periphyton at five rocky littoral zone sites in Lake Malawi, Africa. Journal of Great Lakes Research 29 (Suppl. 2), 181–189.

Hsu, S.-H., Wang, S.-L., Huang, J.-H., Huang, S.-T., Wang, M., 2015. Effects of rice straw ash amendment on Cd solubility and distribution in a contaminated paddy soil under submergence. Paddy and Water Environment 13 (1), 135–143.

Huang, B., Li, Z., Huang, J., Guo, L., Nie, X., Wang, Y., Zhang, Y., Zeng, G., 2014. Adsorption characteristics of Cu and Zn onto various size fractions of aggregates from red paddy soil. Journal of Hazardous Materials 264, 176–183.

Huang, J.-H., Hsu, S.-H., Wang, S.-L., 2011. Effects of rice straw ash amendment on Cu solubility and distribution in flooded rice paddy soils. Journal of Hazardous Materials 186 (2–3), 1801–1807.

Ippolito, A., Kattwinkel, M., Rasmussen, J.J., Schäfer, R.B., Fornaroli, R., Liess, M., 2015. Modeling global distribution of agricultural insecticides in surface waters. Environmental Pollution 198, 54–60.

Johansson, C.H., Janmar, L., Backhaus, T., 2014. Toxicity of ciprofloxacin and sulfamethoxazole to marine periphytic algae and bacteria. Aquatic Toxicology 156, 248–258.

Khatoon, H., Yusoff, F., Banerjee, S., Shariff, M., Bujang, J.S., 2007. Formation of periphyton biofilm and subsequent biofouling on different substrates in nutrient enriched brackishwater shrimp ponds. Aquaculture 273 (4), 470–477.

Komulainent, S.F., Morozov, A.K., 2010. Heavy metal dynamics in the periphyton in small rivers of Kola Peninsula. Water Resources 37 (6), 874–878.

Koptsik, G.N., 2014. Problems and prospects concerning the phytoremediation of heavy metal polluted soils: a review. Eurasian Soil Science 47 (9), 923–939.

Lavoie, I., Lavoie, M., Fortin, C., 2012. A mine of information: benthic algal communities as biomonitors of metal contamination from abandoned tailings. Science of the Total Environment 425, 231–241.

Lee, J.A., Marsden, I.D., Glover, C.N., 2010. The influence of salinity on copper accumulation and its toxic effects in estuarine animals with differing osmoregulatory strategies. Aquatic Toxicology 99 (1), 65–72.

Lee, P.-K., Choi, B.-Y., Kang, M.-J., 2015. Assessment of mobility and bio-availability of heavy metals in dry depositions of Asian dust and implications for environmental risk. Chemosphere 119, 1411−1421.

Li, T., Bo, L., Yang, F., Zhang, S., Wu, Y., Yang, L., 2012. Comparison of the removal of COD by a hybrid bioreactor at low and room temperature and the associated microbial characteristics. Bioresource Technology 108, 28−34.

Li, Z., Ma, Z., van der Kuijp, T.J., Yuan, Z., Huang, L., 2014. A review of soil heavy metal pollution from mines in China: pollution and health risk assessment. Science of the Total Environment 468−469, 843−853.

Liu, X., Wu, J., Xu, J., 2006. Characterizing the risk assessment of heavy metals and sampling uncertainty analysis in paddy field by geostatistics and GIS. Environmental Pollution 141 (2), 257−264.

Lu, K., Yang, X., Shen, J., Robinson, B., Huang, H., Liu, D., Bolan, N., Pei, J., Wang, H., 2014. Effect of bamboo and rice straw biochars on the bioavailability of Cd, Cu, Pb and Zn to Sedum plumbizincicola. Agriculture, Ecosystems & Environment 191, 124−132.

Luo, S., Xiao, X., Xi, Q., Wan, Y., Chen, L., Zeng, G., Liu, C., Guo, H., Chen, J., 2011. Enhancement of cadmium bioremediation by endophytic bacterium Bacillus sp. L14 using industrially used metabolic inhibitors (DCC or DNP). Journal of Hazardous Materials 190 (1−3), 1079−1082.

Machado, M.D., Lopes, A.R., Soares, E.V., 2015. Responses of the alga *Pseudokirchneriella subcapitata* to long-term exposure to metal stress. Journal of Hazardous Materials 296, 82−92.

Nemati, K., Bakar, N.K.A., Abas, M.R., Sobhanzadeh, E., 2011. Speciation of heavy metals by modified BCR sequential extraction procedure in different depths of sediments from Sungai Buloh, Selangor, Malaysia. Journal of Hazardous Materials 192 (1), 402−410.

Nowack, B., Schulin, R., Luster, J., 2010. Metal fractionation in a contaminated soil after reforestation: temporal changes versus spatial variability. Environmental Pollution 158 (10), 3272−3278.

Obolewski, K., SkorbiŁowicz, E., SkorbiŁowicz, M., Glińska-Lewczuk, K., Maria Astel, A., Strzelczak, A., 2011. The effect of metals accumulated in reed (*Phragmites australis*) on the structure of periphyton. Ecotoxicology and Environmental Safety 74 (4), 558−568.

Orecchio, S., Polizzotto, G., 2013. Fractionation of mercury in sediments during draining of Augusta (Italy) coastal area by modified Tessier method. Microchemical Journal 110, 452−457.

Ouyang, F., Zhai, H., Ji, M., Zhang, H., Dong, Z., 2016. Physiological and transcriptional responses of nitrifying bacteria exposed to copper in activated sludge. Journal of Hazardous Materials 301, 172−178.

Paraneeiswaran, A., Shukla, S.K., Prashanth, K., Rao, T.S., 2015. Microbial reduction of [Co(III) −EDTA]− by *Bacillus licheniformis* SPB-2 strain isolated from a solar salt pan. Journal of Hazardous Materials 283, 582−590.

Parra, S., Malato, S., Pulgarin, C., 2002. New integrated photocatalytic-biological flow system using supported TiO_2 and fixed bacteria for the mineralization of isoproturon. Applied Catalysis B: Environmental 36 (2), 131−144.

Pokrovsky, O.S., Pokrovski, G.S., Feurtet-Mazel, A., 2008. A structural study of cadmium interaction with aquatic microorganisms. Environmental Science & Technology 42 (15), 5527−5533.

Puga, A.P., Abreu, C.A., Melo, L.C.A., Beesley, L., 2015. Biochar application to a contaminated soil reduces the availability and plant uptake of zinc, lead and cadmium. Journal of Environmental Management 159, 86−93.

Quinlan, E.L., Nietch, C.T., Blocksom, K., Lazorchak, J.M., Batt, A.L., Griffiths, R., Klemm, D.J., 2011. Temporal dynamics of periphyton exposed to tetracycline in stream mesocosms. Environmental Science & Technology 45 (24), 10684–10690.

Rectenwald, L.L., Drenner, R.W., 1999. Nutrient removal from wastewater effluent using an ecological water treatment system. Environmental Science & Technology 34 (3), 522–526.

Regmi, P., Garcia Moscoso, J.L., Kumar, S., Cao, X., Mao, J., Schafran, G., 2012. Removal of copper and cadmium from aqueous solution using switchgrass biochar produced via hydrothermal carbonization process. Journal of Environmental Management 109, 61–69.

Ren, H.-Y., Liu, B.-F., Kong, F., Zhao, L., Xie, G.-J., Ren, N.-Q., 2014. Enhanced lipid accumulation of green microalga Scenedesmus sp. by metal ions and EDTA addition. Bioresource Technology 169, 763–767.

Rusznyák, A., Vladár, P., Molnár, P., Reskóné, M.N., Kiss, G., Márialigeti, K., Borsodi, A.K., 2008. Cultivable bacterial composition and BIOLOG catabolic diversity of biofilm communities developed on *Phragmites australis*. Aquatic Botany 88 (3), 211–218.

Saeed, A., Iqbal, M., 2003. Bioremoval of cadmium from aqueous solution by black gram husk (*Cicer arientinum*). Water Research 37 (14), 3472–3480.

Scinto, L.J., Reddy, K.R., 2003. Biotic and abiotic uptake of phosphorus by periphyton in a subtropical freshwater wetland. Aquatic Botany 77 (3), 203–222.

Sharma, K., Derlon, N., Hu, S., Yuan, Z., 2014. Modeling the pH effect on sulfidogenesis in anaerobic sewer biofilm. Water Research 49, 175–185.

Shen, Z., Som, A.M., Wang, F., Jin, F., McMillan, O., Al-Tabbaa, A., 2016. Long-term impact of biochar on the immobilisation of nickel (II) and zinc (II) and the revegetation of a contaminated site. Science of the Total Environment 542 (Part A), 771–776.

Soldo, D., Behra, R., 2000. Long-term effects of copper on the structure of freshwater periphyton communities and their tolerance to copper, zinc, nickel and silver. Aquatic Toxicology 47 (3), 181–189.

Sullivan, T.S., McBride, M.B., Thies, J.E., 2013. Soil bacterial and archaeal community composition reflects high spatial heterogeneity of pH, bioavailable Zn, and Cu in a metalliferous peat soil. Soil Biology and Biochemistry 66, 102–109.

Sun, Y., Zhou, Q., Xu, Y., Wang, L., Liang, X., 2011. Phytoremediation for co-contaminated soils of benzo[a]pyrene (B[a]P) and heavy metals using ornamental plant *Tagetes patula*. Journal of Hazardous Materials 186 (2–3), 2075–2082.

Usman, A.R.A., Almaroai, Y.A., Ahmad, M., Vithanage, M., Ok, Y.S., 2013. Toxicity of synthetic chelators and metal availability in poultry manure amended Cd, Pb and As contaminated agricultural soil. Journal of Hazardous Materials 262, 1022–1030.

Vega, F.A., Covelo, E.F., Andrade, M.L., 2006. Competitive sorption and desorption of heavy metals in mine soils: influence of mine soil characteristics. Journal of Colloid and Interface Science 298 (2), 582–592.

Wang, C., Hu, X., Chen, M.-L., Wu, Y.-H., 2005. Total concentrations and fractions of Cd, Cr, Pb, Cu, Ni and Zn in sewage sludge from municipal and industrial wastewater treatment plants. Journal of Hazardous Materials 119 (1–3), 245–249.

Wang, H., Shan, X-q., Wen, B., Zhang, S., Wang, Z-j, 2004. Responses of antioxidative enzymes to accumulation of copper in a copper hyperaccumulator of *Commelina communis*. Archives of Environmental Contamination and Toxicology 47 (2), 185–192.

Wang, M., Kuo-Dahab, W.C., Dolan, S., Park, C., 2014. Kinetics of nutrient removal and expression of extracellular polymeric substances of the microalgae, *Chlorella* sp. and *Micractinium* sp., in wastewater treatment. Bioresource Technology 154, 131–137.

Wang, X., Wang, X., Zhang, J., Bu, Y., Yan, X., Chen, J., Huang, J., Zhao, J., 2015. Direct toxicity assessment of copper (II) ions to activated sludge process using a p-benzoquinone-mediated amperometric biosensor. Sensors and Actuators B: Chemical 208, 554–558.

Warren, L.A., Haack, E.A., 2001. Biogeochemical controls on metal behaviour in freshwater environments. Earth-Science Reviews 54 (4), 261–320.

Wei, S., Zhou, Q., Wang, X., Zhang, K., Guo, G., Ma, L., 2005. A newly-discovered Cd-hyperaccumulator *Solatium nigrum* L. Chinese Science Bulletin 50 (1), 33–38.

Wu, Q., Leung, J.Y.S., Geng, X., Chen, S., Huang, X., Li, H., Huang, Z., Zhu, L., Chen, J., Lu, Y., 2015. Heavy metal contamination of soil and water in the vicinity of an abandoned e-waste recycling site: implications for dissemination of heavy metals. Science of the Total Environment 506–507, 217–225.

Wu, Y., He, J., Yang, L., 2010. Evaluating adsorption and biodegradation mechanisms during the removal of microcystin-RR by periphyton. Environmental Science & Technology 44 (16), 6319–6324.

Wu, Y., Li, T., Yang, L., 2012. Mechanisms of removing pollutants from aqueous solutions by microorganisms and their aggregates: a review. Bioresource Technology 107, 10–18.

Wu, Y., Liu, J., Yang, L., Chen, H., Zhang, S., Zhao, H., Zhang, N., 2011. Allelopathic control of cyanobacterial Blooms by periphyton biofilms. Environmental Microbiology 13 (3), 604–615.

Wu, Y., Xia, L., Yu, Z., Shabbir, S., Kerr, P.G., 2014. In situ bioremediation of surface waters by periphytons. Bioresource Technology 151, 367–372.

Xie, Y., Luo, H., Du, Z., Hu, L., Fu, J., 2014. Identification of cadmium-resistant fungi related to Cd transportation in bermudagrass [*Cynodon dactylon* (L.) Pers.]. Chemosphere 117, 786–792.

Xu, L., Wu, X., Wang, S., Yuan, Z., Xiao, F., Yang, M., Jia, Y., 2016. Speciation change and redistribution of arsenic in soil under anaerobic microbial activities. Journal of Hazardous Materials 301, 538–546.

Yan, R., Yang, F., Wu, Y., Hu, Z., Nath, B., Yang, L., Fang, Y., 2011. Cadmium and mercury removal from non-point source wastewater by a hybrid bioreactor. Bioresource Technology 102 (21), 9927–9932.

Yang, J., Tang, C., Wang, F., Wu, Y., 2016. Co-contamination of Cu and Cd in paddy fields: using periphyton to entrap heavy metals. Journal of Hazardous Materials 304, 150–158.

Yang, J., Tang, C., Wang, F., Wu, Y.H., 2015. Co-contamination of Cu and Cd in paddy fields: using periphyton to entrap heavy metals. Journal of Hazardous Materials 304.

Yao, Y., Gao, B., Inyang, M., Zimmerman, A.R., Cao, X., Pullammanappallil, P., Yang, L., 2011. Removal of phosphate from aqueous solution by biochar derived from anaerobically digested sugar beet tailings. Journal of Hazardous Materials 190 (1–3), 501–507.

Zhang, C., Yu, Z.-G., Zeng, G.-M., Jiang, M., Yang, Z.-Z., Cui, F., Zhu, M.-Y., Shen, L.-Q., Hu, L., 2014. Effects of sediment geochemical properties on heavy metal bioavailability. Environment International 73, 270–281.

Zhang, Y., Jiang, J., Chen, M., 2008. MINTEQ modeling for evaluating the leaching behavior of heavy metals in MSWI fly ash. Journal of Environmental Sciences 20 (11), 1398–1402.

Zhao, K., Liu, X., Xu, J., Selim, H.M., 2010. Heavy metal contaminations in a soil-rice system: identification of spatial dependence in relation to soil properties of paddy fields. Journal of Hazardous Materials 181 (1–3), 778–787.

Žižek, S., Milačič, R., Kovač, N., Jaćimović, R., Toman, M.J., Horvat, M., 2011. Periphyton as a bioindicator of mercury pollution in a temperate torrential river ecosystem. Chemosphere 85 (5), 883–891.

Chapter 15

Removal of COD by a Spiral Periphyton Bioreactor and Its Associated Microbial Community

15.1 INTRODUCTION

At present, human society is facing two severe water environmental problems: the limitation of water resources and water pollution. The latter is increasingly serious in many countries due to improper economic development patterns (Azizullah et al., 2011; Dimitrova et al., 1998; Agrawal, 1999; Zhao et al., 2010). The exponential human population increase and rapid industrialization have produced tremendous volumes of wastewater containing various organic pollutants since the middle of the 20th century (Demirel et al., 2005; Liu et al., 2016a; Qian et al., 2012). Consequently, surface and ground water ecosystems have been greatly impacted or threatened by organic pollution (Ali et al., 2012; Pitarch et al., 2016).

Biological measures based on microorganisms or microbial assemblages treating wastewater and surface water represent one of the most widely used environmentally friendly methods in wastewater and water treatments. These biological measures can usually be divided into aerobic and anaerobic treatment processes according to the dissolved oxygen (DO) levels in the wastewater and water (Chan et al., 2009). These measures also can be divided into activated sludge and biofilm treatment processes based on microorganism attachment states in suspension or fixation (Fadi, 1999). Among the conventional biological wastewater treatments, activated sludge plants and anaerobic treatments are commonly employed in organic pollutant removal, but the large sludge production and high energy consumption are significant drawbacks (Contreras et al., 2002; Demirel et al., 2005; Di Trapani et al., 2011). Therefore, innovative biomeasures with low energy requirements and environmentally benign technologies are urgently needed to treat organic pollution.

Periphyton. http://dx.doi.org/10.1016/B978-0-12-801077-8.00015-6

The long-term chemical oxygen demand (COD) removal by biological wastewater treatments in low-temperature conditions is often limited due to seasonal changes in air temperature. This leads to the majority of these studies being conducted under ~15°C conditions (Elmitwalli et al., 2002; Mahmoud et al., 2004). Wastewater, however, is still discharged in winter and the temperature is often lower than ~15°C. In addition, more research focuses on low-temperature anaerobic biological wastewater treatments than aerobic treatments, which implies that studies of aerobic biotechnology to remove COD at lower than ~15°C are needed.

Periphyton, mainly composed of benthic microorganisms including photoautotrophic microalgae, bacteria, fungi, protozoa, and small multicellular animals, is an important ecological component of surface water and plays critical roles in primary productivity, nutrient transformation, and food source biomass production in aquatic ecosystems (Bere and Tundisi, 2012; Chetelat et al., 1999; Larned, 2010; Wu et al., 2014b). In recent years, periphyton has been increasingly considered for use in wastewater treatment and the bioremediation of polluted natural water ecosystems as it can be easily contrived and/or incorporated in bioreactors, such as algal turf scrubbers (Adey et al., 2013; Liu et al., 2016a) and hybrid floating treatment beds (Liu et al., 2016c). Periphyton-based bioreactors have been investigated for the removal of various contaminants including heavy metals and nutrients and the degradation of organic pollutants (Adey et al., 2013; Bere and Tundisi, 2012; Hamelin et al., 2015; Wu et al., 2014b). The periphyton-based bioreactors are easy to construct and resistant to external environmental changes without producing any secondary pollution and/or toxic secretions (Craggs et al., 1996; Larned, 2010; Wu et al., 2010). Moreover, the autotrophs (e.g., microalgae) of periphyton biofilm can produce oxygen through photosynthesis and restore the DO levels of the wastewater to facilitate the aerobic degradation of organic matter by heterotrophic bacteria (Craggs et al., 1996; Markou and Georgakakis, 2011; Wu et al., 2014a). Consequently, no or little aeration is needed for COD removal.

Despite these advantages, periphyton-based systems have limitations. For instance, periphyton community compositions are usually not stable and nutrient and COD removal efficiency can be negatively influenced by low temperatures and solar irradiance (Craggs et al., 1996; Larras et al., 2013; Wu et al., 2014b). Consequently, it is not practical to use these systems during cold seasons or in cold regions. In addition, the pollutant removal efficiency increases slowly during the start-up phase of a periphyton bioreactor and sloughing of the periphyton biofilm has negative effects on effluent water quality (Adey et al., 2013; Liu et al., 2016a; Sandefur et al., 2011). Therefore, new types of periphyton bioreactors able to operate under broad temperature ranges and with less periphyton shedding are needed to quickly adapt to different types of wastewater, to remove multiple organic pollutants with less biofilm detachment, and to conserve energy (Larned, 2010; Wu et al., 2014b).

Moreover, most previous research on pollutant removal from wastewater using periphyton-based systems focused on their performance at the community level (Adey et al., 2013; Guzzon et al., 2008; Hamelin et al., 2015). Little attention has been paid to the taxonomic components of the periphyton community in detail and their specific roles in the metabolic activities of the associated microbial community (Bere and Tundisi, 2012). Therefore, a more thorough investigation of the dominant species in a periphyton community and identification of their functions in carbon metabolic processes can improve the COD removal capability and sustainability of a periphyton bioreactor via the regulation of periphyton community compositions (Liu et al., 2016a; Williams et al., 2010; Wu et al., 2014b).

Accordingly, the primary objectives in this study were to (1) design a new type of periphyton bioreactor that can be backwashed and efficiently remove COD from wastewater over a broad temperature range; (2) domesticate the periphyton community to remove high-concentration COD at low temperatures; and (3) investigate microorganism characteristics (functional diversity, species composition, and their roles in organic carbon degradation) of the periphyton community during domestication to high COD loading and low temperature. This study will provide a new approach to periphyton bioreactors to quickly remove high-concentration organic matter at different temperatures and provide and better understanding of the relationship between microbial diversity and the functional microorganisms during the domestication process.

15.2 MATERIALS AND METHODS

15.2.1 Spiral Periphyton Bioreactor

The core components of the spiral periphyton bioreactor (SPR) in this study were one spiral poly ethylene (PE) pipe with periphyton attached (total length: 49.4 m; diameter: 2 mm), one pump, one clarification tank, and one influent tank (Fig. 15.1). During the operation, the wastewater or growth medium is

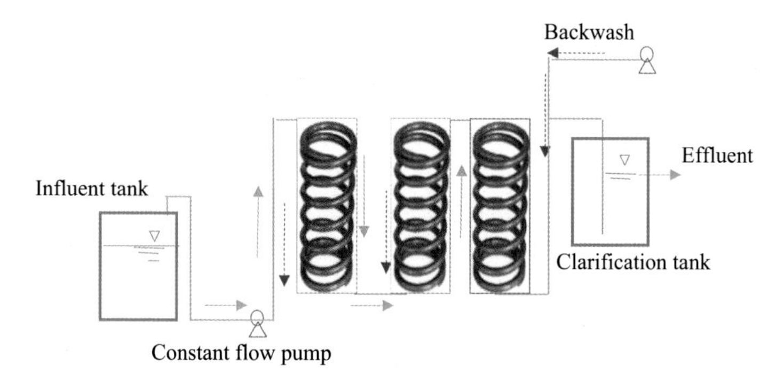

FIGURE 15.1 A schematic drawing of the SPR. The *blue arrows* represent the wastewater direction, and the *red arrows* represent the backwash direction.

pumped from the influent tank into the spiral PE pipe and clarification tank and then discharged. To prevent pipe blockage and periphyton shedding, backwash components (a pump and connecting tubes) were added to the SPR for regular cleaning and periphyton regeneration.

15.2.2 Experimental Design

This study was carried out in two stages: periphyton formation on the PE pipe (October to November 2014) and the trial operation stage (November 2014 to January 2015).

During the periphyton formation stage, lake water rich in microalgae and bacteria collected from the eutrophic Xuanwu Lake, Nanjing, China, was added to the PE pipe. Seven days later, the lake water was refreshed. Seven days after that, a green periphyton community was observed by the naked eye in the PE pipe and the SPR was started with Woods Hole culture medium (WC medium) at a speed of 309 $\mu L\ min^{-1}$ for 14 days. The spiral PE pipe was set up in an artificial climatic box (QHX-300BS-III) at different temperatures (from 4°C to 30°C) with a light intensity of 2500–5000 lux and light/dark cycle of 12 h:12 h.

For the trial operation stage, the SPR was set up under the same conditions as the formation stage and lasted 44 days. To investigate the periphyton shedding of SPR, distilled water was used in the first 8 days, constituting phase I (days −8 to 0). Then, to verify the ability of the SPR to remove organic matter from wastewater of various concentrations, the influent COD concentration of the synthetic wastewater was gradually increased from 107.3 $mg\ L^{-1}$ on day 0 to 375.2 $mg\ L^{-1}$ (day 8), 562.6 $mg\ L^{-1}$ (day 18), and 738.2 $mg\ L^{-1}$ (day 28) constituting phases II, III, IV, and V, respectively. The synthetic wastewater in this study was mainly composed of CH_3COONa (0–945.75 $mg\ L^{-1}$), NH_4Cl (0–77.39 $mg\ L^{-1}$), and KH_2PO_4 (0–39.55 $mg\ L^{-1}$). Trace elements of $MgSO_4 \cdot 7H_2O$ (22.6 $mg\ L^{-1}$), $CaCl_2 \cdot 2H_2O$ (2.8 $mg\ L^{-1}$), and $FeCl_3 \cdot 6H_2O$ (0.2 $mg\ L^{-1}$) as described by Qian et al. (2012) were added to the synthetic wastewater. $NaHCO_3$ was used to regulate the pH of the synthetic wastewater from 7.2 to 7.6. The synthetic wastewater was pumped through the spiral PE pipe at a constant rate of 463.5 $\mu L\ min^{-1}$ (hydraulic retention time: 5.6 h) and then out to the clarification tank (Fig. 15.1). From phase II on, the influent wastewater was replaced every 2 days by the effluent accumulated in the clarification tank during the previous 2 days until the end of each phase. Accordingly, 2 days was considered as one cycle. At the end of stage II, the temperature of the influent tank was gradually decreased from 30°C on day −8 to 4°C on day 36 by using a water bath or refrigerator. To minimize periphyton shedding and stimulate periphyton growth, the SBR was backwashed with distilled water for 30–60 min in the opposite direction to the influent and cleaned every 7 days. The speed of the backwash water was controlled between 1.63 and 2.03 $\mu L\ min^{-1}$, with the intensity of the backwash dependent on the amount of periphyton shed.

15.2.3 Samples and Analyses

COD of the influent and effluent was determined every 2 days by collecting 10-mL water samples according to the standard potassium dichromate digestion method (GB11914-89) of the Ministry of Environmental Protection of China (Wei, 2002). To determine the periphyton biomass, 5 mm of the PE pipe was cut every 4–8 days and the periphyton was washed into an Erlenmeyer flask and diluted to 10 mL using distilled water. The biomass COD was determined following the method of Bullock et al. (1996), and the periphyton biomass on the PE pipe was expressed as mg BioCOD mm^{-1}.

To determine the functional diversity of the microbial community during the operation process, periphyton samples were collected by cutting 10 mm of the PE pipe and washing with 10 mL of distilled water on days −8 and 22 of the operational stage, constituting the nondomesticated and 30-day domesticated microbial community, respectively. The community-level substrate utilization was assayed by using a Biolog ECO Microplate (Hayward, CA, USA). The ECO microplate contains three replicate wells of 31 carbon substrates (Choi and Dobbs, 1999). The wells contain a redox-sensitive tetrazolium dye that turns purple following respiratory electron transport in metabolically active cells. Therefore, the plate color is proportional to microbial respiratory activity and can reflect the characteristics of the microbial community (Choi and Dobbs, 1999; Qian et al., 2012). A 150-μL aliquot was added to each well of the ECO microplate. The ECO microplate was then incubated at 25°C and the average well color development (AWCD) was determined using a Biolog Microplate Reader at 590 nm every 24 h for 7 days.

Periphyton samples collected by cutting and washing the 10 mm of PE pipe on days −8, 7, and 22 of the operation stage were used for microbial diversity level analysis via MiSeq sequencing. First, DNA was extracted from the water with washed periphyton biomass using E.Z.N.A. Water DNA Kit (D5525-02), and the A260/280 value of the extracted DNA was maintained at ~2.0. Then, primers 515 forward (5′-GTGCCAGCMGCCGCGG-3′) from Turner et al. (1999) and 907 reverse (5′-CCGTCAATTCMTTTRAGTTT-3′) from Morales and Holben (2009) were used for polymerase chain reaction (PCR) amplification of 16S rRNA in ABI GeneAmp 9700 (ABI, USA). Each 20-μL reaction mixture for PCR consisted of 4 μL of 5× FastPfu Buffer, 2 μL of dNTPs (2.5 mM), 0.4 μL of forward primer (5 μM), 0.4 μL of reverse primer (5 μM), 0.4 μL of FastPfu Polymerase, 20 ng of template DNA, and 12.8 μL of ddH_2O. The PCR product was then checked with agarose gel electrophoresis and extracted with use of an AxyPrep DNA gel extraction kit (AxyGen, China). The PCR product was subsequently quantified on QuantiFluor-ST (Promega, USA) and sequenced using the MisSepence. These were done by Shanghai Majorbio (http://www.majorbio.com/).

After sequencing, the raw sequence data were processed and the PE reads were separated and overlapped to assemble the final sequences with a minimum overlap length of 10 bp for each two PE reads. The PE reads with more than two mismatches within the 10-bp region were removed during the overlap step (Wang et al., 2012). The sequence trim was processed using Trimmomatic. Operational taxonomic unit (OTU) grouping was then determined at a similarity of 97% using Usearch (version 7.1 http://drive5.com/uparse/). The taxonomy assignment of tags was performed on OTUs using an RDP classifier (http://sourceforge.net/projects/rdp-classifier/), and the taxonomic structure compositions of the periphyton community were calculated and expressed as relative abundances. Based on OTUs, the rank–abundance distribution curve and Shannon–Wiener Index curve were analyzed using Mothur to evaluate the community diversity levels (Wang et al., 2012). A Venn diagram was analyzed using the R statistical program to investigate the diversity similarity of the three periphyton communities (Shade and Handelsman, 2012).

15.3 RESULTS AND DISCUSSION

15.3.1 Periphyton Shedding and Chemical Oxygen Demand Removal of SPR

The periphyton biomass in the PE pipe increased gradually from 3.63 to 10.03 mg BioCOD mm^{-1} from phase I to V (Fig. 15.2A). After the influent COD loading increased from 107.28 to 738.15 mg L^{-1}, the maximal COD removal rate of each phase varied between 33.79 and 335.4 mg L^{-1}. The highest COD removal quantity of 335.4 mg L^{-1} was achieved during phase IV (day 18) with a COD loading of 562.6 mg L^{-1} and a temperature of 15°C.

During phase I, the effluent had a much higher COD concentration of 86.12–100.08 mg L^{-1} than the influent, which was only 2–4 mg L^{-1}. Most likely, this was due to the shedding of periphyton biomass from the PE pipe. The shedding of biofilm, including periphyton, during growth is a common phenomenon (Horn et al., 2003; Sandefur et al., 2011) that will result in the overestimation of the effluent COD concentration and underestimation of the COD removal efficiency by the periphyton bioreactor. In the study by Horn et al. (2003), the detached biofilm biomass was positively correlated with biofilm density. In this study, the periphyton biomass of the PE pipe showed a gradual increase during the operational stage (Fig. 15.2A), so the minimum periphyton shedding must appear during phase I, which was 86.12–100.08 mg BioCOD L^{-1}. Therefore, the shed periphyton biomass should be subtracted from the effluent COD when calculating the COD removal efficiency and a value of 86.12 mg BioCOD L^{-1} was used in this study.

From phase II on, the periphyton biomass gradually increased and the maximal growth was observed during phase III from 5.35 to 8.42 mg BioCOD mm^{-1}. Consequently, the maximal COD removal rate increased from 130 mg L^{-1} in phase III to 335 mg L^{-1} in phase IV. During phases IV and V, although the

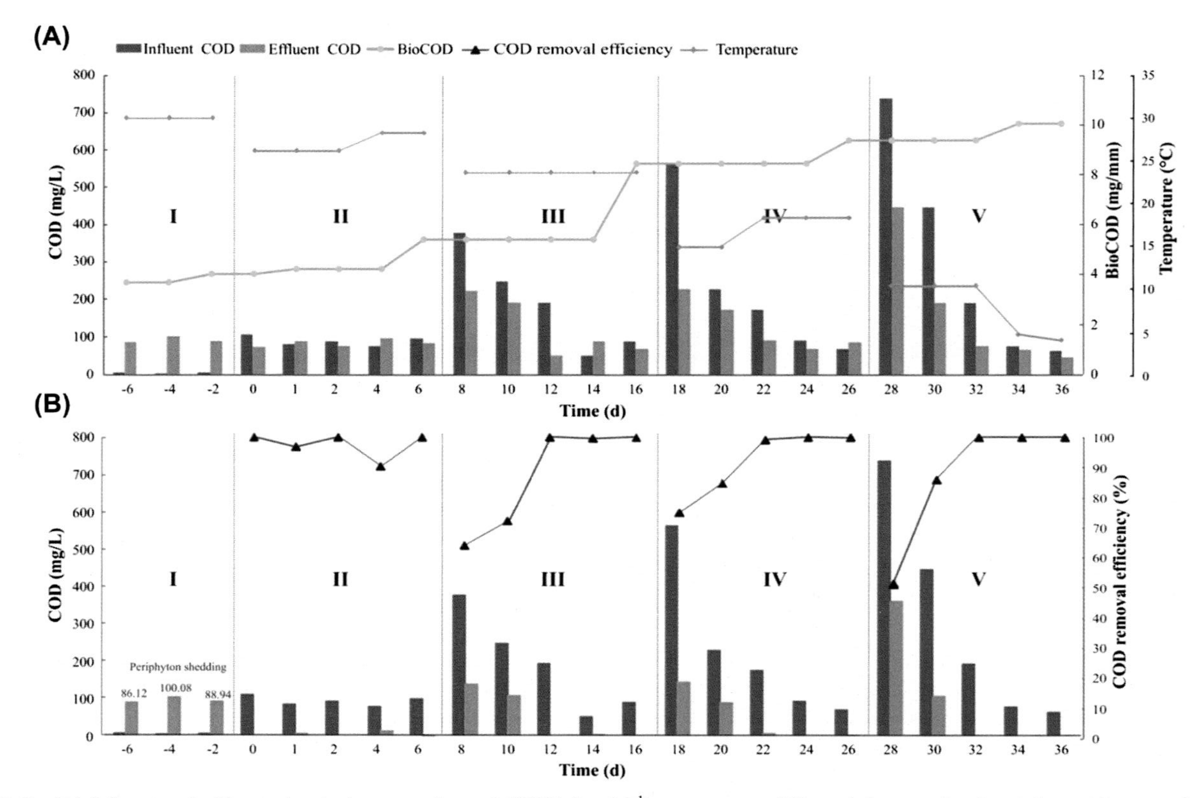

FIGURE 15.2 (A) Influent and effluent chemical oxygen demand (COD) (mg L^{-1}), temperature (°C), and the associated periphyton biomass of the PE pipe (mg BioCOD mm^{-1}) over time. (B) Influent COD (mg L^{-1}), effluent COD (mg L^{-1}), and COD removal efficiency (%) after subtracting a periphyton shedding of 86.12 mg BioCOD L^{-1} over time (light intensity: 2500–5000 lux; light/dark cycle: 12 h:12 h) (Shangguan et al., 2015).

temperature decreased from 18°C to 10°C and then to 4°C, the periphyton maintained a high COD removal rate of 114–293 mg L^{-1} at a hydraulic retention time of 5.6 h. Moreover, in contrast to the high COD concentration of the effluent in phases I and II, the effluent COD of phase V was as low as 43.9 mg L^{-1}. This was due to the periodical backwash of the SPR. The backwash could remove the mature periphyton biomass from the PE pipe and stimulate further growth and regeneration of periphyton. Thus, the backwash of this SPR reduced the influence of shedding periphyton biomass on effluent COD and prompted a continuous and sustainable COD removal efficiency even at a low temperature of 4–10°C.

After subtracting the periphyton shedding biomass of 86.12 mg BioCOD L^{-1}, the COD removal efficiency at a hydraulic retention time of 5.6 h was 64.1–100% from phase II to phase IV (Fig. 15.2B). During phase V, although the COD removal efficiency of the first cycle slightly decreased (51.33%) compared with phase IV (74.92%), the COD removal efficiency reached as high as 85.9% after two cycles (equal to an HRT of 11.2 h) and almost 100% after three cycles (equal to an HRT of 16.8 h). These results imply that the periphyton has a quick adaptation capability to low temperatures of 4°C and simultaneously maintains a high COD removal efficiency. Although previous studies (Feng et al., 2009; Qian et al., 2012) reported that microorganisms cultured at low temperature could have a high COD removal capacity, these high COD removal efficiencies only appeared under low-temperature conditions (such as 4°C). In this study, the COD removal by SPR was highly effective under continuous temperature shifts from 30°C to 4°C. Therefore, this SPR could efficiently remove COD of 375–738 mg L^{-1} from wastewater at a broad temperature range of 4–30°C with a hydraulic retention time of 5.6–16.8 h after a short domestication and with periodical cleaning through backwash.

15.3.2 Functional Diversity of the Periphyton Community

15.3.2.1 Comparison of AWCD

The AWCD of the periphyton collected on days −8 and 22 showed no visible changes in the first 24 h (Fig. 15.3A). This was probably due to the low initial microbial density and minimal carbon utilization. After 24 h, the AWCD increased quickly to 96 h and thereafter with a relatively slow rate. Furthermore, the periphyton from day 22 had a higher AWCD than that of day −8 and thus a higher utilization capability of carbon sources. In the first 120 h, the periphyton from day 22 had an S-shape growth curve, while the periphyton from day −8 did not have a distinct limiting point. This was probably because the periphyton community diversity was highly dependent on the wastewater composition (Liu et al., 2016b; Wu et al., 2011). Specifically, the periphyton collected on day −8 was not adapted to the various carbon sources, while after 30 days' domestication, the periphyton was well adapted to various organic carbon sources and pollutants. In addition, the AWCD experiment was carried

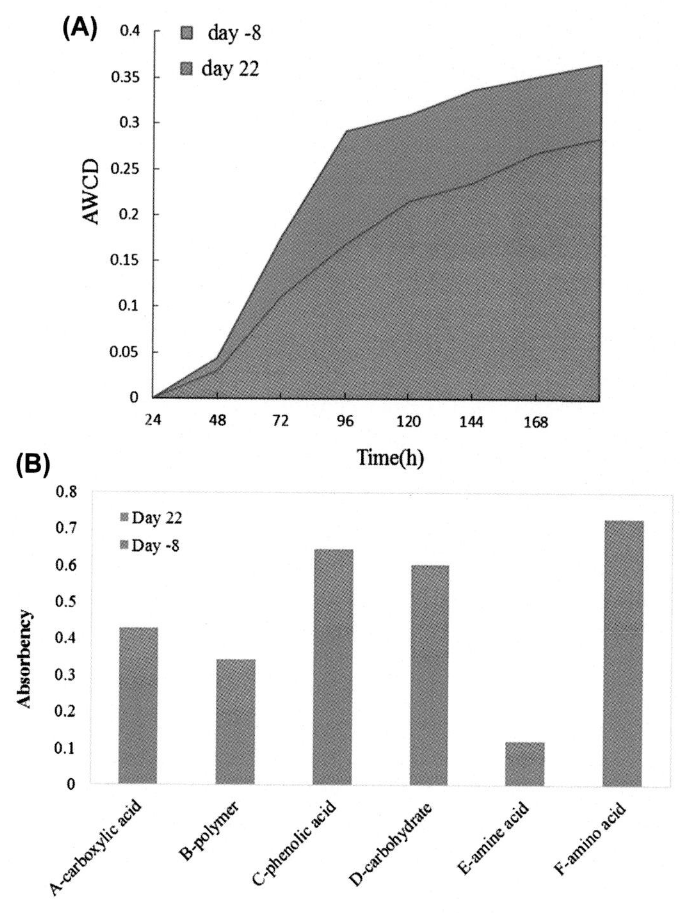

FIGURE 15.3 (A) Changes in AWCD of periphyton community with (day 22) and without (day −8) domestication over incubation time at 25°C on Biolog ECO microplate. (B) The metabolic capability of six main types of carbon sources (amine, amino acid, carbohydrate, carboxylic acid, phenolic acid, and polymer) by the periphyton communities from days −8 and 22 after 96-h incubation at 25°C on a Biolog ECO microplate.

out at a lower incubation temperature (25°C) than that on day −8 of the operation stage (30°C). The periphyton from day 22 had already been domesticated at a broad temperature range and could quickly adapt to temperature changes. The periphyton collected on day −8 needed time to adapt to the sudden temperature decrease and therefore grew at a slower rate than the periphyton from day 22. In addition, microorganisms grown at a low temperature usually have higher metabolic activity and adaptability than those growing at high temperature (Qian et al., 2012). Therefore, after 30 days' domestication to increasing COD loading and decreasing temperature, the periphyton community had a higher functional diversity and could adapt to the single carbon source environment more quickly than nondomesticated periphyton. This was in accordance with the contrasting COD removal performance of the SPR between phases I and II and phases IV and V.

15.3.2.2 Metabolic Capability of Different Types of Carbon Sources

Many studies have demonstrated that the morphology, staining, and metabolic activity of microorganisms are very typical in the later stages of the exponential phase (Qian et al., 2012; Subbalakshmi and Sitaram, 1998). In this study, the periphyton community from day 22 entered its stationary phase around 96 h (Fig. 15.3A), and consequently, the 96 h Biolog data were selected for detailed metabolic capability analysis of carbon sources. As described by Choi and Dobbs (1999), the 31 carbon substrates of the ECO microplate can be classified into six main types: amine, amino acid, carbohydrate acid, carboxylic acid, phenolic acid, and polymer. Accordingly, the metabolic capability of the carbon sources was evaluated on the level of these six main types.

The periphyton from day 22 had greatly higher utilization efficiency of all six types of carbon sources than did periphyton from day −8 (Fig. 15.3B). The microorganisms from the 0 and 30 days' domesticated periphyton communities showed a little difference in their carbon sources utilization efficiency. Generally, the periphyton community collected on day 22 had higher utilization efficiency of all six carbon sources than that from day −8. Specifically, the periphyton from day 22 had the highest utilization efficiency of phenolic acid and amino acid and then carbohydrate, carboxylic acid, polymer, and amine. The periphyton collected on day −8 used more amino acid than other carbon sources; carbohydrates were the second most used carbon source, followed by phenolic acid, polymer, carboxylic acid, and amine. Again, results implied that the 30-day domesticated periphyton had higher metabolic capability in using carbons than the periphyton without domestication (Qian et al., 2012).

15.3.3 Microbial Diversity Levels of Periphyton Community

15.3.3.1 Taxonomic Compositions of Periphyton

In this study, the MiSeq sequencing technology was used to investigate the different diversity levels of the microbial communities in detail based on 44,701,497 bases of 112,992 sequences of three periphyton community samples from days −8, 7, and 22. Cyanobacteria were the largest components of the periphyton community, and their relative abundance increased from 50% to 80.4% from days −8 to 7 and then decreased to 12.7% on day 22 (Fig. 15.4A). Cyanobacteria are therefore the main contributor to the high effluent COD during phases I and II (Fig. 15.2). This also indicated that the periphyton had sufficient oxygen produced by cyanobacteria (Berman-Frank et al., 2003; Markou and Georgakakis, 2011) and thus promoted the growth

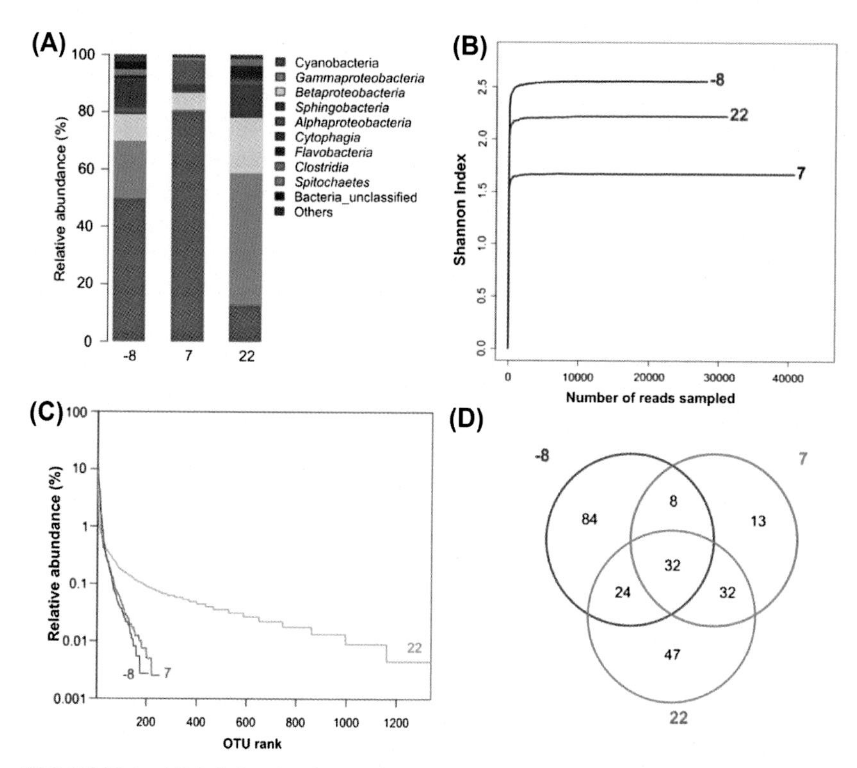

FIGURE 15.4 (A) Relative abundance of the taxonomic components (%). (B) Shannon−Wiener Index. (C) Rank−abundance curve. (D) Microbial community similarity represented by Venn diagrams of the periphyton community from days −8, 7, and 22 based on microbial diversity level analysis via MiSeq sequencing (Shangguan et al., 2015).

of the aerobic microbial community and their metabolic activities in degrading organic matter (Subashchandrabose et al., 2011). Consequently, the periphyton bioreactor is suitable for the aerobic treatment of wastewater and could also reduce the energy consumption needed for aeration in organic matter degradation.

The growth of other microorganisms were relatively suppressed during days −8 to 7 due to the large quantity of cyanobacteria that assimilated most of the nutrients from wastewater through photosynthesis (Dodds, 2003; Markou and Georgakakis, 2011). *Proteobacteria* including *Alphaproteobacteria*, *Betaproteobacteria*, and *Gammaproteobacteria* were the second largest components after Cyanobacteria in the periphyton community with a relative abundance of 14.9−66.9% (Fig. 15.4A). Specifically, *Alphaproteobacteria* had its highest relative abundance of 8.6% in the periphyton community on day 7. *Betaproteobacteria* was the only group with a relative abundance higher than 5% (5.9−19.4%) during the whole process. *Gammaproteobacteria* exhibited great changes in its relative abundance with an initial value of 20% on day −8 to 0.46% on day 7 and then an increase to 46.19% by day 30. *Sphingobacteria* showed a gradual increase in its relative abundance of 0.8% from day 8 to 11.5% by day 22. In summary, following the increase in COD loading and decrease in temperature, aerobic bacteria were more relatively abundant on day 22 than on days −8 and 7, which was in accordance with the COD removal performance in phases IV and V (Fig. 15.2). This indicates that the 30 days' domestication and operation of the periphyton selected the best functional bacterial species to grow and thus to degrade the organic matter. The detailed microbial community compositions also explained the differences in AWCD and the metabolic utilization of the six types of carbon sources after 0 and 30 days' domestication (Fig. 15.3).

Proteobacteria has been widely reported as the dominant bacteria in municipal wastewater and activated sludge (Amorim et al., 2014; Baudišová et al., 2012), and *Betaproteobacteria* play a critical role in degradation of organic matter in sewage treatment plants (Liu et al., 2007). Some members of *Gammaproteobacteria* (e.g., *Thiomicrospira crunogena*) have a wide temperature tolerance (Scott et al., 2006; Williams et al., 2010), and some other members such as *Acinetobacter* and *Pseudomonas* are efficient in COD removal (Olukanni et al., 2006; Pronk et al., 2015). In the work of Sinkko et al. (2013), *Sphingobacteria*, *Alphaproteobacteria*, and *Gammaproteobacteria* showed strong positive correlations with the organic carbons in coastal surface sediments, and *Sphingobacteria* was important in the degradation of biopolymer sedimentary organic matter. Therefore, in this study, *Proteobacteria*, including *Alphaproteobacteria*, *Betaproteobacteria*, and *Gammaproteobacteria*, and *Sphingobacteria* must be the key functional microorganisms in COD removal by the SPR and could be further regulated to optimize the COD removal efficiency by this SPR.

15.3.3.2 Microbial Community Diversity Indices

Microbial community diversity concerns both taxon richness and evenness (Wang et al., 2012), so the Shannon−Wiener Index and rank−abundance distribution curve were used to investigate species richness and evenness of the periphyton community during the 30 days' domestication (Bates et al., 2013). The Shannon−Wiener Index (Fig. 15.4B) showed that species richness decreased from days −8 to 7, probably due to the increasing COD loading and decreasing temperature. After 15 days' acclimation to the wastewater, the periphyton community started to restore its species diversity and higher species richness was observed on day 22 than on day 7. This indicates that the periphyton bacterial community may need 15−30 days to adapt to wastewater and that the periphyton has a strong ability to adapt to the environment and restore its diversity and functions. Furthermore, results suggest that the species richness of periphyton is not strongly influenced by decreasing temperature, even to 4°C, and this is in accordance with the reports by Qian et al. (2012).

The species evenness of the periphyton community increased greatly following the domestication process (Fig. 15.4C). Although the species evenness on day 7 was slightly higher than on day −8, it was much higher on day 22 than days −8 and 7. Furthermore, the species richness of the periphyton community on day 22 was lower than on day −8 (Fig. 15.4B), but the species evenness on day 22 was much higher than that on day −8 (Fig. 15.4C). This is probably caused by the periodical increase in COD loading of the influent and/or the gradual decrease of temperature.

The Venn diagram is a measure of the similarities between different communities (Fouts et al., 2012; Shade and Handelsman, 2012). In this study, the similarity analysis of microbial communities was done on the OTUs. The numbers of OTUs of the periphyton communities from days −8, 7, and 22 were 148, 85, and 135, respectively (Fig. 15.4D). This was in accordance with the species richness determined by the Shannon−Wiener Index (Fig. 15.4B). The periphyton communities of days −8 and 7 had 40 OTUs in common, while the periphyton of days 7 and 22 shared 64 OTUs. This indicates that the periphyton community had greater changes from days −8 to 7 than from days 7 to 22. This was probably because the periphyton community gradually adapted to the wastewater in the first 15 days and subsequently restored the microbial functional diversity and vitality.

15.4 CONCLUSIONS

This innovative SPR could be quickly started up after periphyton formation on the PE pipe and a short domestication (less than 15 days) and easily backwashed to stimulate the regeneration of periphyton for efficient and continuous COD removal at a broad temperature range of 4−30°C. After 30 days' domestication to increasing COD loading and decreasing temperature, the periphyton community had higher microbial diversity and metabolic activities and, thus, COD removal capabilities than before. Cyanobacteria were the main phototrophic components of the periphyton community and produced an aerobic environment to prompt the biodegradation of organic matters, while *Proteobacteria* including *Alphaproteobacteria, Betaproteobacteria*, and *Gammaproteobacteria* and *Sphingobacteria* were the pivotal functioning microorganisms in degrading organic matter. More thorough investigations of microbial diversity and key functioning microorganisms of periphyton communities can significantly contribute to the improvement of COD removal by periphyton based bioreactors including this SPR.

REFERENCES

Adey, W.H., Laughinghouse, H.D., Miller, J.B., Hayek, L.A.C., Thompson, J.G., Bertman, S., Hampel, K., Puvanendran, S., 2013. Algal turf scrubber (ATS) floways on the Great Wicomico River, Chesapeake Bay: productivity, algal community structure, substrate and chemistry. Journal of Phycology 49 (3), 489−501.

Agrawal, G.D., 1999. Diffuse agricultural water pollution in India. Water Science and Technology 39 (3), 33–47.

Ali, I., Asim, M., Khan, T.A., 2012. Low cost adsorbents for the removal of organic pollutants from wastewater. Journal of Environmental Management 113 (0), 170–183.

Amorim, C.L., Maia, A.S., Mesquita, R.B.R., Rangel, A.O.S.S., van Loosdrecht, M.C.M., Tiritan, M.E., Castro, P.M.L., 2014. Performance of aerobic granular sludge in a sequencing batch bioreactor exposed to ofloxacin, norfloxacin and ciprofloxacin. Water Research 50, 101–113.

Azizullah, A., Khattak, M.N.K., Richter, P., Häder, D.-P., 2011. Water pollution in Pakistan and its impact on public health — A review. Environment International 37 (2), 479–497.

Bates, S.T., Clemente, J.C., Flores, G.E., Walters, W.A., Parfrey, L.W., Knight, R., Fierer, N., 2013. Global biogeography of highly diverse protistan communities in soil. The ISME Journal 7 (3), 652–659.

Baudišová, D., Benáková, A., Wanner, F., 2012. Changes in prokaryotic community composition in the small wastewater treatment plant of Zbytiny during treatment processes. Silva Gabreta 18 (2), 79–90.

Bere, T., Tundisi, J.G., 2012. Cadmium and lead toxicity on tropical freshwater periphyton communities under laboratory-based mesocosm experiments. Hydrobiologia 680 (1), 187–197.

Berman-Frank, I., Lundgren, P., Falkowski, P., 2003. Nitrogen fixation and photosynthetic oxygen evolution in cyanobacteria. Research in Microbiology 154 (3), 157–164.

Bullock, C.M., Bicho, P.A., Zhang, Y., Saddler, J.N., 1996. A solid chemical oxygen demand (COD) method for determining biomass in waste waters. Water Research 30 (5), 1280–1284.

Chan, Y.J., Chong, M.F., Law, C.L., Hassell, D.G., 2009. A review on anaerobic–aerobic treatment of industrial and municipal wastewater. Chemical Engineering Journal 155 (1–2), 1–18.

Chetelat, J., Pick, F., Morin, A., Hamilton, P., 1999. Periphyton biomass and community composition in rivers of different nutrient status. Canadian Journal of Fisheries and Aquatic Sciences 56 (4), 560–569.

Choi, K.-H., Dobbs, F.C., 1999. Comparison of two kinds of biolog microplates (GN and ECO) in their ability to distinguish among aquatic microbial communities. Journal of Microbiological Methods 36 (3), 203–213.

Contreras, E.M., Bertola, N.C., Giannuzzi, L., Zaritzky, N.E., 2002. A modified method to determine biomass concentration as COD in pure cultures and in activated sludge systems. Water SA 28 (4), 463–468.

Craggs, R.J., Adey, W.H., Jessup, B.K., Oswald, W.J., 1996. A controlled stream mesocosm for tertiary treatment of sewage. Ecological Engineering 6 (1–3), 149–169.

Demirel, B., Yenigun, O., Onay, T.T., 2005. Anaerobic treatment of dairy wastewaters: a review. Process Biochemistry 40 (8), 2583–2595.

Di Trapani, D., Christensso, M., Odegaard, H., 2011. Hybrid activated sludge/biofilm process for the treatment of municipal wastewater in a cold climate region: a case study. Water Science and Technology 63 (6), 1121–1129.

Dimitrova, I., Kosturkov, J., Vatralova, A., 1998. Industrial surface water pollution in the region of Devnya, Bulgaria. Water Science and Technology 37 (8), 45–53.

Dodds, W.K., 2003. The role of periphyton in phosphorus retention in shallow freshwater aquatic systems. Journal of Phycology 39 (5), 840–849.

Elmitwalli, T.A., Sklyar, V., Zeeman, G., Lettinga, G., 2002. Low temperature pre-treatment of domestic sewage in an anaerobic hybrid or an anaerobic filter reactor. Bioresource Technology 82 (3), 233–239.

Fadi, G., 1999. Activated sludge biofilm wastewater treatment system. Water Research 33 (1), 230–238.

Feng, H., Hu, L., Mahmood, Q., Fang, C., Qiu, C., Shen, D., 2009. Effects of temperature and feed strength on a carrier anaerobic baffled reactor treating dilute wastewater. Desalination 239 (1), 111–121.

Fouts, D.E., Szpakowski, S., Purushe, J., Torralba, M., Waterman, R.C., MacNeil, M.D., Alexander, L.J., Nelson, K.E., 2012. Next generation sequencing to define prokaryotic and fungal diversity in the bovine rumen. PloS One 7 (11), e48289.

Guzzon, A., Bohn, A., Diociaiuti, M., Albertano, P., 2008. Cultured phototrophic biofilms for phosphorus removal in wastewater treatment. Water Research 42 (16), 4357–4367.

Hamelin, S., Planas, D., Amyot, M., 2015. Mercury methylation and demethylation by periphyton biofilms and their host in a fluvial wetland of the St. Lawrence River (QC, Canada). Science of the Total Environment 512, 464–471.

Horn, H., Reiff, H., Morgenroth, E., 2003. Simulation of growth and detachment in biofilm systems under defined hydrodynamic conditions. Biotechnology and Bioengineering 81 (5), 607–617.

Larned, S.T., 2010. A prospectus for periphyton: recent and future ecological research. Journal of the North American Benthological Society 29 (1), 182–206.

Larras, F., Lambert, A.-S., Pesce, S., Rimet, F., Bouchez, A., Montuelle, B., 2013. The effect of temperature and a herbicide mixture on freshwater periphytic algae. Ecotoxicology and Environmental Safety 98, 162–170.

Liu, J., Danneels, B., Vanormelingen, P., Vyverman, W., 2016a. Nutrient removal from horticultural wastewater by benthic filamentous algae *Klebsormidium* sp., *Stigeoclonium* spp. and their communities: from laboratory flask to outdoor Algal Turf Scrubber (ATS). Water Research 92, 61–68.

Liu, J., Liu, W., Wang, F., Kerr, P., Wu, Y., 2016b. Redox zones stratification and the microbial community characteristics in a periphyton bioreactor. Bioresource Technology 204, 114–121.

Liu, J., Wang, F., Liu, W., Tang, C., Wu, C., Wu, Y., 2016c. Nutrient removal by up-scaling a hybrid floating treatment bed (HFTB) using plant and periphyton: from laboratory tank to polluted river. Bioresource Technology 207, 142–149.

Liu, X., Zhang, Y., Min, Y., Wang, Z., Lv, W., 2007. Analysis of bacterial community structures in two sewage treatment plants with different sludge properties and treatment performance by nested PCR-DGGE method. Journal of Environmental Sciences 19 (1), 60–66.

Mahmoud, N., Zeeman, G., Gijzen, H., Lettinga, G., 2004. Anaerobic sewage treatment in a one-stage UASB reactor and a combined UASB-Digester system. Water Research 38 (9), 2348–2358.

Markou, G., Georgakakis, D., 2011. Cultivation of filamentous cyanobacteria (blue-green algae) in agro-industrial wastes and wastewaters: a review. Applied Energy 88 (10), 3389–3401.

Morales, S.E., Holben, W.E., 2009. Empirical testing of 16S rRNA gene PCR primer pairs reveals variance in target specificity and efficacy not suggested by in silico analysis. Applied and Environmental Microbiology 75 (9), 2677–2683.

Olukanni, O., Osuntoki, A., Gbenle, G., 2006. Textile effluent biodegradation potentials of textile effluent-adapted and non-adapted bacteria. African Journal of Biotechnology 5 (20), 1980–1984.

Pitarch, E., Cervera, M.I., Portolés, T., Ibáñez, M., Barreda, M., Renau-Pruñonosa, A., Morell, I., López, F., Albarrán, F., Hernández, F., 2016. Comprehensive monitoring of organic micro-pollutants in surface and groundwater in the surrounding of a solid-waste treatment plant of Castellón, Spain. Science of The Total Environment 548, 211–220.

Pronk, M., Abbas, B., Kleerebezem, R., Van Loosdrecht, M., 2015. Effect of sludge age on methanogenic and glycogen accumulating organisms in an aerobic granular sludge process fed with methanol and acetate. Microbial Biotechnology 8 (5), 853–864.

Qian, H., Li, J., Pan, X., Sun, Z., Ye, C., Jin, G., Fu, Z., 2012. Effects of streptomycin on growth of algae *Chlorella vulgaris* and *Microcystis aeruginosa*. Environmental Toxicology 27 (4), 229–237.

Sandefur, H.N., Matlock, M.D., Costello, T.A., 2011. Seasonal productivity of a periphytic algal community for biofuel feedstock generation and nutrient treatment. Ecological Engineering 37 (10), 1476–1480.

Scott, K.M., Sievert, S.M., Abril, F.N., Ball, L.A., Barrett, C.J., Blake, R.A., Boller, A.J., Patrick, S., Clark, J.A., Davis, C.R., 2006. The genome of deep-sea vent chemolithoautotroph *Thiomicrospira crunogena* XCL-2. PLoS Biology 4 (12), e383.

Shade, A., Handelsman, J., 2012. Beyond the Venn diagram: the hunt for a core microbiome. Environmental Microbiology 14 (1), 4–12.

Shangguan, H., Liu, J., Zhu, Y., Tong, Z., Wu, Y., 2015. Start-up of a spiral periphyton bioreactor (SPR) for removal of COD and the characteristics of the associated microbial community. Bioresource Technology 193, 456–462.

Sinkko, H., Lukkari, K., Sihvonen, L.M., Sivonen, K., Leivuori, M., Rantanen, M., Paulin, L., Lyra, C., 2013. Bacteria contribute to sediment nutrient release and reflect progressed eutrophication-driven hypoxia in an organic-rich continental sea. PLoS One 8 (6), e67061.

Subashchandrabose, S.R., Ramakrishnan, B., Megharaj, M., Venkateswarlu, K., Naidu, R., 2011. Consortia of cyanobacteria/microalgae and bacteria: Biotechnological potential. Biotechnology Advances 29 (6), 896–907.

Subbalakshmi, C., Sitaram, N., 1998. Mechanism of antimicrobial action of indolicidin. FEMS Microbiology Letters 160 (1), 91–96.

Turner, S., Pryer, K.M., Miao, V.P.W., Palmer, J.D., 1999. Investigating deep phylogenetic relationships among cyanobacteria and plastids by small subunit rRNA sequence analysis. Journal of Eukaryotic Microbiology 46 (4), 327–338.

Wang, Y., Sheng, H.-F., He, Y., Wu, J.-Y., Jiang, Y.-X., Tam, N.F.-Y., Zhou, H.-W., 2012. Comparison of the levels of bacterial diversity in freshwater, intertidal wetland, and marine sediments by using millions of illumina tags. Applied and Environmental Microbiology 78 (23), 8264–8271.

Wei, F., 2002. Monitoring and Analysis Methods of Water and Wastewater. China Environmental Science Press, Beijing.

Williams, K.P., Gillespie, J.J., Sobral, B.W., Nordberg, E.K., Snyder, E.E., Shallom, J.M., Dickerman, A.W., 2010. Phylogeny of gammaproteobacteria. Journal of Bacteriology 192 (9), 2305–2314.

Wu, Y., He, J., Yang, L., 2010. Evaluating adsorption and biodegradation mechanisms during the removal of microcystin-RR by periphyton. Environmental Science & Technology 44 (16), 6319–6324.

Wu, Y., Hu, Z., Kerr, P.G., Yang, L., 2011. A multi-level bioreactor to remove organic matter and metals, together with its associated bacterial diversity. Bioresource Technology 102 (2), 736–741.

Wu, Y., Xia, L., Liu, N., Gou, S., Nath, B., 2014a. Cleaning and regeneration of periphyton biofilm in surface water treatment systems. Water Science and Technology 69 (2), 235–243.

Wu, Y., Xia, L., Yu, Z., Shabbir, S., Kerr, P.G., 2014b. In situ bioremediation of surface waters by periphytons. Bioresource Technology 151, 367–372.

Zhao, Y., Yang, Z., Li, Y., 2010. Investigation of water pollution in Baiyangdian Lake, China. Procedia Environmental Sciences 2 (0), 737–748.

Chapter 16

The Removal of Methyl Orange by Periphytic Biofilms: Equilibrium and Kinetic Modeling

16.1 INTRODUCTION

Industrial effluent, such as dyes, surfactants, minerals, and certain metals, from industrial processing poses a serious threat to aquatic biota and ecosystems. Depending on the amount and composition of effluents, textile waste is the most damaging contaminant among all the industrial wastes (Vandevivere et al., 1998). The textile industry produces about 10,000 commercially available dyes, and the production rate of these dyes is greater than 7×10^5 tons per annum (Pearce et al., 2003; Robinson et al., 2001), with about 10−15% of these dyes discharged as effluents during manufacturing and processing procedures (Feng et al., 2012; Gong et al., 2007). Most of these dyes, particularly azo dyes containing one or more nitrogen double bonds (−N=N−), are recalcitrant and xenobiotic in nature, thus not only affecting the aesthetic value of water but also negatively impacting aquatic biota by reducing light penetration for photosynthetic bacteria and plants. They result in the addition of aromatics, metals, and chlorides, along with color, thus exacerbating the toxicity levels in aquatic ecosystem (Fu and Viraraghavan, 2001; Robinson et al., 2001). Due to their miscellaneous molecular structures, they are resistant to degradation by physical, chemical, and microbial methods (Ramalho et al., 2002), and the metabolites produced after degradation processes are carcinogenic and mutagenic (García-Montaño et al., 2008; Hameed et al., 2008). Further, effluents from textile industries are accompanied by higher chemical oxygen demand (COD) and biochemical oxygen demand (BOD) (Wong and Yu, 1999). Due to their toxicity and the visibility of even a small amount of dye in industrial effluents, the management of these dyes is a considerable problem that has gained significant attention from scientists in the past few decades.

Periphyton. http://dx.doi.org/10.1016/B978-0-12-801077-8.00016-8

Different treatment technologies are used for the treatment of dyes, including chemical oxidation, flocculation, ozonation, photolysis, ion exchange, irradiation, precipitation, electrochemical treatment, and adsorption. Unfortunately, these conventional physicochemical techniques are not efficacious in eliminating dye wastes from effluents and have certain technoeconomical limitations such as excessive consumption of chemicals, production of concentrated sludge, higher operational costs, lower proficiency in dye removal, generation of wide variety of toxic metabolites, dye specificity, the need for secondary processing, and being sensitive to variable dye concentrations (Jadhav et al., 2011; Robinson et al., 2001; Tuttolomondo et al., 2014). Government authorities in developed countries are rigorous about the alarming escalation of dye concentration in industrial effluents, forcing the innovation of a novel technology that is not only feasible and cost-efficient but also eco-friendly.

Microbiological degradation of dyes is a cost-effective and eco-friendly technology compared with different physicochemical approaches and primarily involves dye metabolism in the presence of different oxidative and reductive enzymes (Pointing, 2001; Zimmermann et al., 1982) and reduced electron carriers (Gingell and Walker, 1971). Microbial decolorization is composed of three phases: biosorption, bioaccumulation, and biodegradation (Arunarani et al., 2013). Several studies have already reported on the biodegradation and/or adsorption of dyes by microbiological approaches such as fungi (El-Rahim et al., 2009; Patel and Suresh, 2008), algae (Daneshvar et al., 2007; Khataee et al., 2011), and bacteria (Phugare et al., 2011; Shah, 2013), where bacteria were found to be the most effectual dye degraders (Eichlerová et al., 2006; Stolz, 2001). Most of these microbial approaches are dye specific and cannot be scaled up with ease for industrial-scale wastewater treatment (Dafale et al., 2008b). Using a biofilm instead of a pure culture is beneficial in the adsorption and degradation of dyes and can be upscaled more easily. Further, a combination of aerobic and anaerobic treatment might be more useful to concurrently improve water quality via degradation of dyes and decreasing COD (Dafale et al., 2008a).

Periphyton are a multifaceted group of cyanobacteria, algae, protozoa, and organic debris, mainly dominated by phototrophic microorganisms. The adhesion of this multilayered structure is facilitated by extracellular polymeric substances (EPS), secreted by microbial communities of periphytic biofilm (Lu et al., 2014a,b). Periphyton are found attached to different substrates in aquatic ecosystems and have a profound effect on aquatic ecosystems by influencing primary production, food webs, organic matter, and nutrient recycling (Battin et al., 2003). These biofilms are well known for their quick, large-scale responses to even trivial changes in the environment (Gaiser et al., 2005; Ghaedi et al., 2014). Previous studies have shown high efficacy of these biofilms in the treatment of organic and inorganic phosphorus (Lu et al., 2014a,b), microcystin-RR (Wu et al., 2010), hormones (Writer et al., 2011), and toxic metals (Dong et al., 2003), but the role of these biofilms has never been discussed in the degradation and adsorption of dyes. In this study, we used three different kinds of periphytic biofilms to investigate the degradation optimality and kinetic parameters of bioadsorption and degradation of an azo dye, methyl orange (MO).

The main objectives of this investigation were (1) to study the feasibility of different types of periphytic biofilms, (epiphyton, metaphyton and epipelon) for the degradation of MO, (2) analyze experimental data using different kinetic equations, and (3) determine whether the mechanism of decolorization is adsorption and/or biodegradation. The findings of this study will (1) evaluate a novel eco-friendly technology for the removal of dyes from wastewater, (2) provide insight into the dye removal efficiency of different types of periphytic biofilms, and (3) assist in understanding biosorption and biodegradation kinetic of dye removal by these biological assemblages.

16.2 MATERIAL AND METHODS

16.2.1 Dyestuff and Chemicals

Analytical-grade MO and other chemicals were purchased from Sinophorm Chemical Reagent, Shanghai. Stock solution ($500\ mg\ L^{-1}$) of MO was prepared in 1 L of deionized water. This stock solution was further diluted to make the different concentrations used in the experiments.

16.2.2 Periphyton Biofilms and Culture Conditions

The source of periphytic biofilms for this study was Xuanwu Lake, East China. Three different types of periphyton were collected in situ from rocks (metaphyton), plants (epiphyton), and sediments (epipelon). These periphytic biofilms were cultured in three indoor glass tanks (length $\times$ width $\times$ height; 100 cm $\times$ 100 cm $\times$ 60 cm) and allowed to grow on periphyton substrate—Industrial Soft Carriers (length $\times$ width $\times$ diameter; 9 cm $\times$ 1 cm $\times$ 1 cm) (Jineng Environmental Protection Company of Yixing, China). The tanks were first thoroughly washed with 95% alcohol and rinsed with water. The three kinds of periphytic biofilms were inoculated into three different tanks with substrate in artificially simulated water (WC medium). WC medium was composed of different macronutrients and micronutrients (macronutrients; $NaNO_3$, $CaCl_2 \cdot 2H_2O$, $MgSO_4 \cdot 7H_2O$, $NaHCO_3$, $Na_2SiO_3 \cdot 9H_2O$, K_2HPO_4, H_3BO_3, micronutrients; vitamin B12, thiamine, biotin, some WC trace elements, Na_2EDTA, $FeCl_3 \cdot 6H_2O$, $CuSO_4 \cdot 5H_2O$, $ZnSO_4 \cdot 7H_2O$, $CoCl_2 \cdot 6H_2O$, $MnCl_2 \cdot 4H_2O$, $Na_2MoO_4 \cdot 2H_2O$). The periphytic biofilms were allowed to grow for almost 50 days (until the surface of the substrate was evenly covered with biofilm) at 25–30°C in a greenhouse to reduce the effect of environmental variations.

16.2.3 Characterization of Periphyton

The biofilms were morphologically studied by the use of scanning electron microscope (SEM) (Quanta 200, FEI, Netherlands) and phase contrast microscope (PCM). Biofilm microbial diversity was analyzed by the BiologTM ECO Microplates method (Balser and Wixon, 2009). The valves were inoculated with 150-mL aliquots of scrapped periphyton in sterilized conditions, and color development was observed at 590 nm every 24 h for 7 days. The microbial activity of biofilm was determined by average well-color development (AWCD) and was calculated by using Eq. (16.1).

$$\mathrm{AWCD} = \frac{\sum(C-R)}{n} \tag{16.1}$$

where C indicates the rate of color production in each well, R is the absorbance value in the control well without a carbon source, and n is the number of substrates ($n = 31$).

Shannon diversity index, Simpson index, McIntosh index, and Pielou index were calculated to determine the functional diversity of the biofilms.

$$\text{Shannon diversity index } = H' = -\sum p_i \ln p_i \tag{16.2}$$

$$\text{Simpson index} = D = \frac{1}{\sum (p_i)^2} \tag{16.3}$$

$$\text{McIntosh index} = U = \sqrt{\sum n_i{}^2} \tag{16.4}$$

$$\text{Pielou index} = J = \frac{H'}{\ln S} \tag{16.5}$$

where p_i is the proportion of relative absorbance of each plate, n_i was obtained by subtracting blank value from absorbance of each substrate, and S is the total number of species.

Substrate-related diversity indices were determined statistically with ANOVA by using SPSS. The ratio of dry to wet weight of biofilms was found to be 5%, which was taken as standard; therefore, the periphytic biomass provided has already been converted by this factor.

16.2.4 Batch Adsorption Experiment

The adsorption experiments were carried out using different concentrations of MO in a 250-mL conical flask with 100 mL of dye and 0.4 mg of periphyton in each flask at 30°g at pH 7 for 7 days. The dye concentrations were 0, 0.1, 0.2, 0.4, 0.8, and 1.2 g L^{-1}. The degradation capability of periphytic biofilms was also evaluated at different pH (3–11) values. Then, 5-mL aliquots of dye, degraded by different biofilms, were collected at different time intervals and centrifuged at 4500 rpm for 10 min. The supernatants were collected, filtered through 0.45-μm membrane filters, and analyzed by UV–visible spectrophotometer

(Shimadzu, UV2450, Japan). All experiments were performed in triplicate, and the mean results are shown with standard deviations (±SD).

The amount of dye adsorbed at equilibrium time (q_e) was calculated by Eq. (16.6).

$$q_e = \frac{(C_0 - C_e)V}{m} \tag{16.6}$$

where C_0 and C_e (mg L^{-1}) are the initial concentration and concentration at time t, respectively, m (g) is the corrected mass of periphyton, and V (L) is the volume of solution.

To determine the optimum time for specific decolorization, MO and periphyton biomass were collected for different time intervals (24–168 h). The amount of MO adsorbed at different intervals was measured and the amount of MO adsorbed at time t was determined by Eq. (16.7).

$$q_t = \frac{(C_0 - C_t)V}{m} \tag{16.7}$$

where C_t is the concentration of MO at time t.

16.2.4.1 Biosorption Kinetics

Pseudo-first-order and pseudo-second-order models have been extensively used to interpret adsorption studies and can be represented by Eqs. (16.8) and (16.9) (Lu et al., 2014a,b).

$$\log(q_e - q_t) = \log q_e - \frac{k_1}{2.303}t \tag{16.8}$$

$$\frac{t}{q_t} = \frac{1}{k_2 {q_e}^2} + \frac{t}{q_e} \tag{16.9}$$

where q_t is the amount of dye (mg g^{-1}) adsorbed at time t, and k_1 and k_2 (min^{-1}) are rate constants of the pseudo-first-order and pseudo-second-order model, respectively. The k_1 value can be determined by the linear plot between log ($q_e - q_t$) and time t at different concentrations while k_2 can be determined by a plot of t/q_t versus t.

The intraparticle diffusion model and Elovich equation were also used to analyze the data. The intraparticle diffusion equation is seen in Eq. (16.10).

$$q_t = k_{id}t^{0.5} + C \tag{16.10}$$

where k_{id} (mg g^{-1} $min^{0.5}$) is the diffusion rate constant, which is obtained by plotting a linear graph between $t^{0.5}$ and q_t.

The Elovich equation is generally expressed as follows:

$$\frac{dq_t}{d_t} = \alpha \exp(-\beta q_t) \tag{16.11}$$

where α (mg g^{-1} min^{-1}) is the rate of adsorption and β (mg g^{-1}) is the rate of desorption.

The equation was integrated to the following form:

$$q_t = \beta(\ln\alpha\beta) + \ln(t) \tag{16.12}$$

The constants were obtained from the slope and intercept of a linear graph of q_t versus ln (t).

16.2.4.2 Adsorption Isotherm

To evaluate the applicability of an adsorption process as a unit process, analysis of its equilibrium data is essential. The two most common equations used for this purpose are the Langmuir and Freundlich equations, which were applied to describe the adsorption equilibrium between the amount of MO adsorbed on periphytic biofilms and the amount of MO in solution at constant temperature. The equations are generally expressed as:

$$\text{Langmuir model: } \frac{C_e}{q_e} = \frac{1}{q_m} C_e + \frac{1}{K_L q_m} \tag{16.13}$$

$$\text{Freundlich model: } \ln q_e = \ln K_F + \frac{1}{n} \ln C_e \tag{16.14}$$

where K_L is the Langmuir constant and K_F is the Freundlich constant which indicate sorption capacity and intensity, respectively. The Langmuir equation can also be expressed in terms of the dimensionless parameter of the equilibrium or adsorption intensity (R_L) by the following equation:

$$R_L = \frac{1}{1 + K_L C_0} \tag{16.15}$$

The values of R_L are a reliable source to determine the efficacy of adsorption reactions. The reaction isotherms would be either linear ($R_L = 1$), favorable ($0 < R_L < 1$), unfavorable ($R_L > 1$), or reversible ($R_L = 0$).

16.2.5 Degradation Experiment

For the biodegradation experiments, 50 mg L^{-1} of dye was exposed to epiphyton, metaphyton, and epilithon in 250-mL conical flasks. After complete decolorization of dyes, the sample was collected and centrifuged at 10,000 rpm to isolate cells from the degraded product. After filtering through 0.45-μm membrane filters, the metabolites were extracted using equal amounts of diethyl ether. These extracts were dried on a rotary evaporator and the degradation of dye was confirmed by preparing pellets with KBr (in a ratio of 1:10) and Fourier transform infrared (FTIR) spectra were recorded (Nicolet 360, Thermo Electron Co., USA) using a spectral range of 4000–400 cm^{-1}.

16.3 RESULTS AND DISCUSSION

16.3.1 Periphyton Characterization

The biofilms were composed of green algae and cyanobacteria along with some bacteria, protozoa, and unicellular diatomic species and were dominated by autotrophic algal species and cyanobacteria. Epiphytic biofilm was more abundant in cyanobacteria, including some members of *Cladophora glomerata*, *Synechococcus, Oedogonium* sp., *Fragilaria* sp., and *Spirogyra* sp. and unicellular diatoms compared with epilithon and metaphyton. The order of abundance of cyanobacteria and autotrophic algal species was epiphyton > metaphyton > epilithon (Fig. 16.1).

AWCD followed similar patterns for different samples with the passage of time but also varied between the different periphyton biofilms (Fig. 16.2). Higher AWCD values were observed for epiphyton, indicating that epiphyton has a higher metabolic activity than metaphyton and epilithon. The diversity indices estimated the microbial community richness, dominance, homogeneity, and species evenness (Table 16.1). There was no significant difference in diversity indices between the three types of biofilm, demonstrating that all

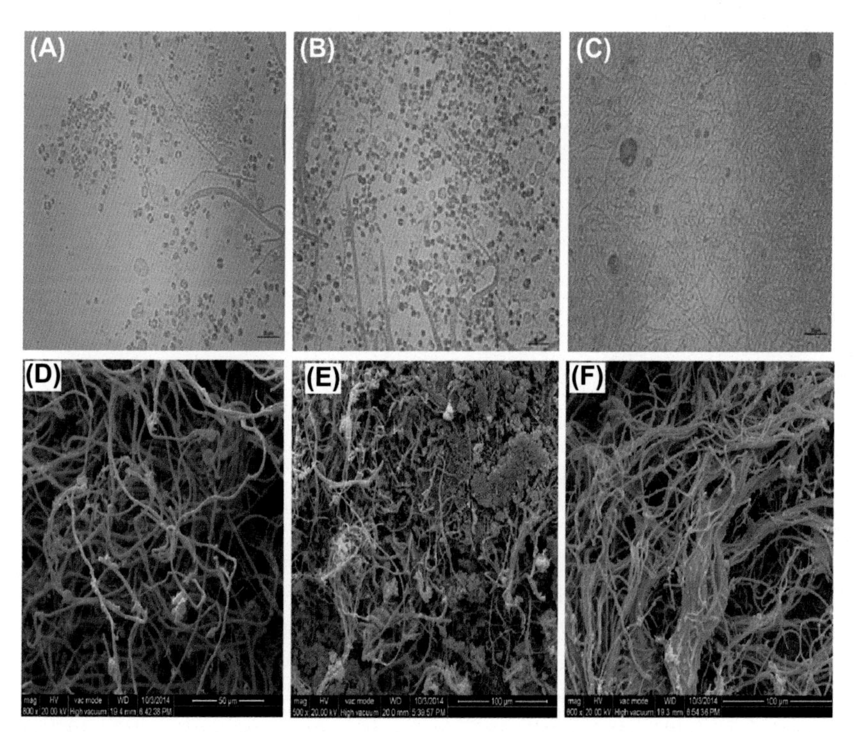

FIGURE 16.1 Periphytic biofilms observed under phase-contrast microscope (A) metaphyton, (B) epilithon, (C) epiphyton and SEM (D) metaphyton, (E) epilithon, and (F) epiphyton.

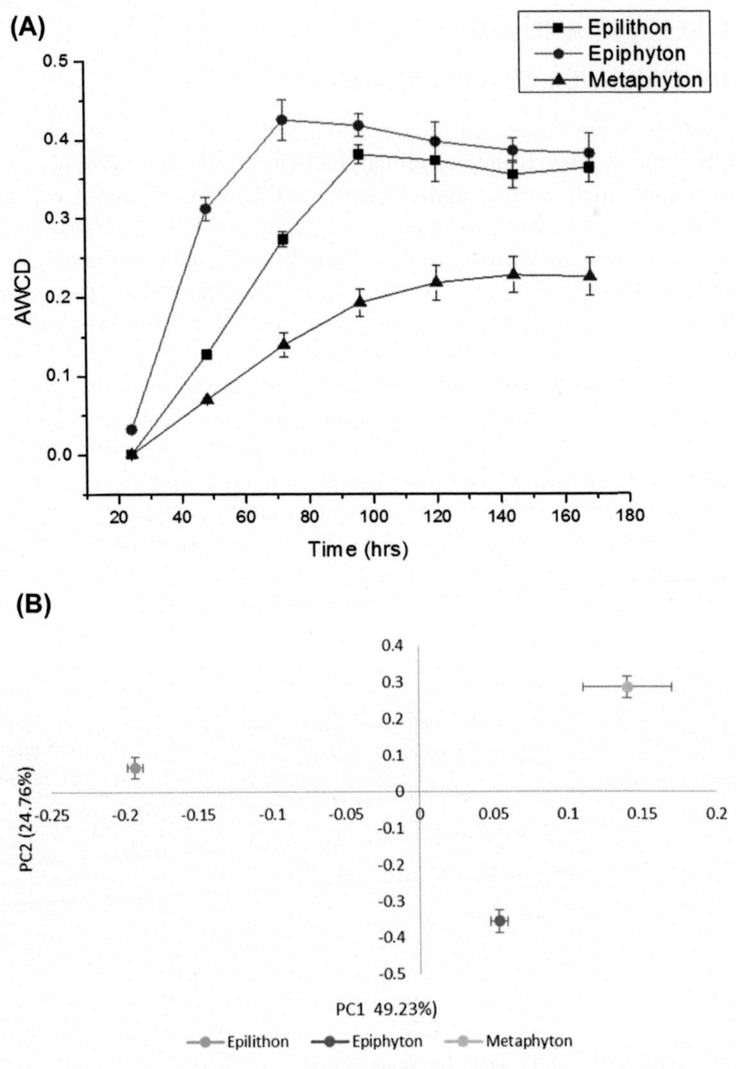

FIGURE 16.2 (A) The AWCD of the Biolog EcoPlates at 590 nm for three different kinds of periphytic biofilm, epiphyton, metaphyton, and epilithon, and (B) PCA on the basis of utilization of carbon sources inoculated with metaphyton, epilithon, and epiphyton after 96-h incubation.

biofilms had high microbial diversity and richness. The Simpson index was slightly higher in metaphyton than the other two types (Table 16.1). Principal component analysis (PCA) was used to further investigate the consumption of carbon sources by biofilms. There was a distinctive difference in all three types of biofilm after 96 h of incubation (Fig. 16.2). The PCA explained a total variance of 73.99% with the first principal component (PC1) explaining

TABLE 16.1 Diversity Indices for Periphytic Biofilms (Epilithon, Epiphyton, and Metaphyton)

Biofilm	Shannon Index	Simpson Index	McIntosh Index	Pielou Index
Epiphyton	2.664 ± 0.230	12.639 ± 2.106	3.288 ± 0.775	0.893 ± 0.038
Epilithon	2.578 ± 0.663	13.525 ± 7.736	3.165 ± 2.629	0.908 ± 0.063
Metaphyton	2.856 ± 0.141	15.883 ± 2.280	3.619 ± 1.345	0.931 ± 0.038

49.23% of the variation compared with 24.76% for the second principal component (PC2) in the Biolog EcoPlates data set. The rate of substrate utilization by these microbial biofilms was significantly different from each other. Previously, periphytic biofilms were reported to be composed of internal voids, micropores and tunnels that facilitate back and forth transfer of nutrients and oxygen (Lu et al., 2014a,b). These micropores and tunnels might be the primary sites for adsorption and further biodegradation of dyes by certain enzymes produced by periphyton microbial communities. The adsorption and biodegradation rates would be affected by variable microbial diversity and the arrangement of these channels in different periphytic biofilms.

16.3.2 Removal of Methyl Orange by Periphyton

16.3.2.1 Effect of Biomass

Different concentrations of biofilm (0 [control], 0.1, 0.2, 0.4, 0.6, and 1.2 g L^{-1}) were used to treat 25 mg L^{-1} dye concentration and investigate the effect of different biomass concentration on dye decolorization (Fig. 16.3). The rates of decolorization of dye ranged from 25% to 100%, 28% to 100%, and 11% to 100% for epilithon, epiphyton, and metaphyton, respectively. These outcomes proved that periphyton have an affinity for MO and are effectual in its removal. Although all the biofilms completely decolorized dye, the removal rate increased more obviously in epiphyton than metaphyton and epilithon. The order of adsorption of MO on different biofilms was as follows: epiphyton > metaphyton > epilithon. The rate of removal was amplified with increasing biomass concentration. Greater amounts of biomass provide greater surface area leading to an increase in adequate MO binding and adsorption sites on the biofilm surface of biofilm. The decolorization rate was slower at lower concentrations due to insufficient binding and adsorption sites and required more time to reach an equilibrium stage. The optimum concentration of biomass for dye removal was 0.4 g L^{-1}.

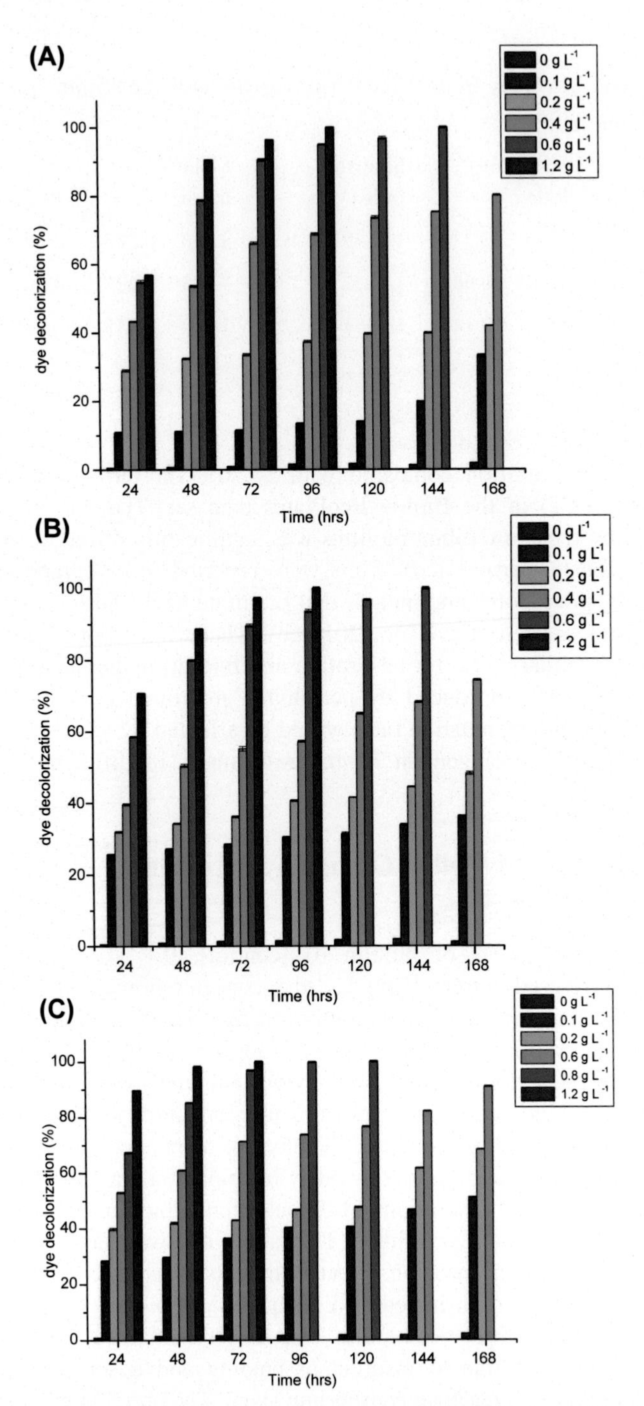

FIGURE 16.3 Effect of different biomass concentrations on the decolorization of MO at different time intervals, (A) metaphyton, (B) epilithon, and (C) epiphyton (experimental conditions: pH = 7, temperature = 30℃, initial concentration = 25 mg L^{-1}).

16.3.2.2 Effect of pH

pH is one of the most important factors to be considered in bioadsorption, and changes in pH have significant effects on the adsorption of dye. The effect of pH on dye removal was evaluated by changing the pH of the MO solution (25 mg L^{-1}) from 3 to 11. Metaphyton, epiphyton, and epipelon demonstrated a wide range of decolorization abilities at different pH values. The order of MO removal by periphytic biofilms was epiphyton > epipelon > metaphyton. These biofilms showed a wide range of dye degradability but the highest adsorption/degradation was observed in pH range 5–9: 62–99% for epiphyton, 50–100% for epiphyton, and 19–83% for metaphyton (Fig. 16.4). The results showed that the optimum pH for MO removal is 7.0 for all three types of biofilm.

16.3.2.3 Effect of Different Contact Times and Initial Concentrations of Methyl Orange

To understand the phenomenon of adsorption kinetics, it is important to investigate the effects of initial concentration. It is a state of dynamic equilibrium between the amount of dye adsorbed on the surface of an adsorbent and amount of desorbed dye from the adsorbent (Kadam et al., 2011; Lin et al., 2011). Four different concentrations were selected and sampled every 24 h. The equilibrium uptake of MO was amplified with increases in dye concentration (Fig. 16.5). Initially, the value of q_t increased with time until an equilibrium stage was attained (within 24–48 h). The q_t values are higher for higher dye concentrations, while the rate of adsorption was slower at higher initial concentrations of MO. Higher initial concentrations of MO result in higher numbers of MO ions surrounding the active sites of biofilm. This increases the competition between ions for adsorption sites and affects the adsorption capacity of biofilms. The competition for active sites results in higher values of q_t at higher initial concentrations of MO. The q_t of metaphyton, epiphyton, and epilithon were similar at different initial concentrations indicating similar competitive phenomena leading to lower rates of dye adsorption. At higher MO concentrations, there are insufficient available active sites for the adsorption of MO, therefore the dye molecules try to adsorb on the surface of biofilm by intraparticle diffusion from the surface to micropores. This process of adopting other pathways of adsorption has already been reported in many studies (Ghaedi et al., 2014; Mahmoud et al., 2012).

16.3.2.4 Bioadsorption Kinetics

Kinetic models are useful in depicting the rates of adsorption in a process and help in designing optimum conditions for an adsorption process by understanding the possible mechanisms involved in the process. Pseudo-first-order kinetic, pseudo-second-order kinetic, intraparticle diffusion, and Elovich models were used to understand the adsorption mechanism of MO to

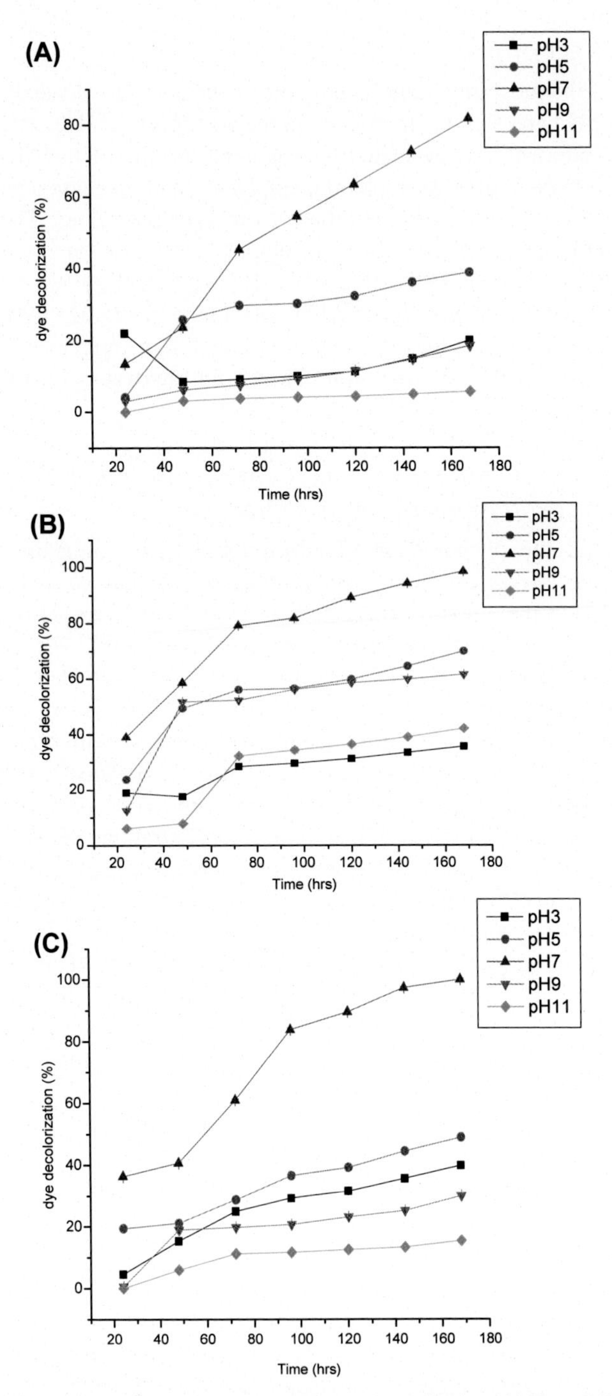

FIGURE 16.4 Effect of different pH values on dye decolorization of MO at different time intervals (experimental conditions: initial concentration = 25 mg L^{-1}, temperature = 30°C, biomass concentration = 0.4 mg L^{-1}).

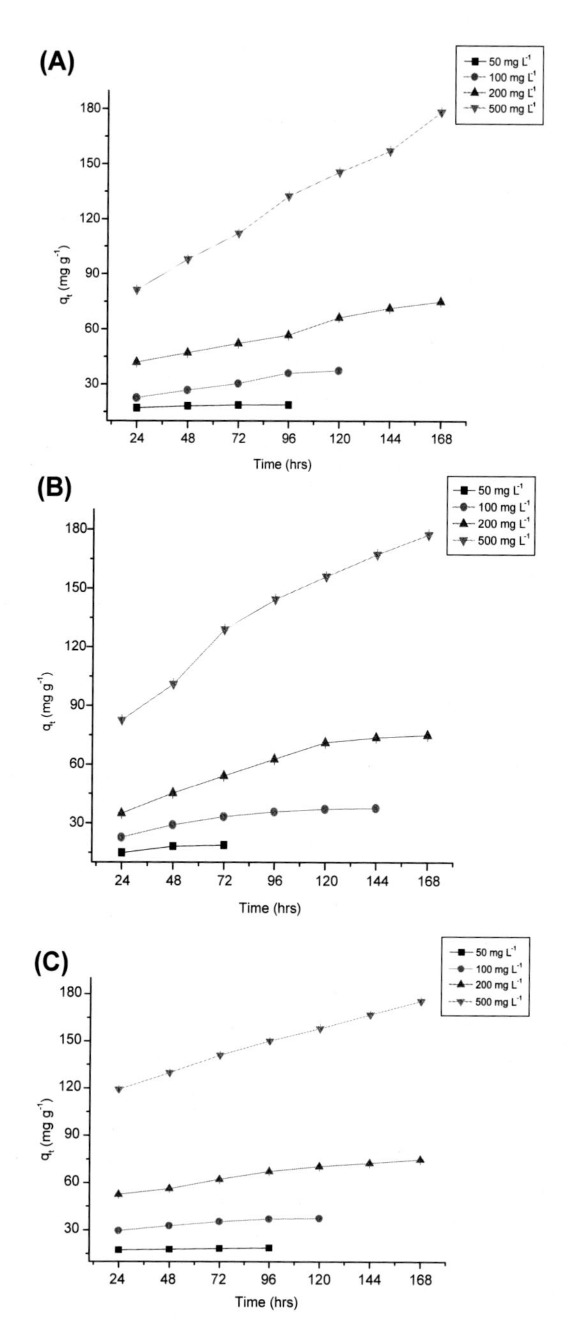

FIGURE 16.5 Effect of different initial concentration of MO at different contact times on specific decolorization by (A) metaphyton (B) epilithon, and (C) epiphtyon (experimental conditions: pH = 7, temperature = 30°C, $m = 0.4$ mg L^{-1}).

periphytic biofilms. Metaphyton, epiphyton, and epipelon experienced different concentrations of dyes depending on contact time.

16.3.2.5 Pseudo-First-Order and Pseudo-Second-Order Models

The pseudo-first-order model is applicable when adsorption is a function of the concentration gradient of an external coefficient of mass transfer. The prerequisite of pseudo-first order is a linear plot between time and log $(q_e - q_t)$ and the k_1 and q_e are calculated from the slope and intercept, respectively. In the pseudo-second-order model the values of k_2 and q_e are obtained from the slope and intercept of a linear graph between t/q_t and time (Fig. 16.5). The best model for adsorption is the one with the highest regression coefficient (R^2). For all three kinds of biofilm and different concentrations of MO, the R^2 was higher for the pseudo-second-order model (Table 16.2). The calculated values of q_e are different from the experimental uptake values. This can be explained by the fact that MO removal is accompanied by multiple pseudo-second-order processes. The first process is always faster than the others; therefore, MO was first physically adsorbed by the microbial community of biofilms and later degraded by the enzymes produced by biofilms in a relatively slower step. According to Gupta et al. (2006), structural properties, composition, topography, surface area, and charge density of surfaces can influence the rate of adsorption and ultimately result in differences in adsorption dynamics.

16.3.2.6 Elovich and Intraparticle Diffusion Models

It is appropriate to apply the Elovich model to an adsorption process that follows pseudo-second-order models and chemical adsorption on heterogeneous adsorbents. As the concentration of MO increases, the value of α also increases (Table 16.2). This suggests that increases in concentration increase the rate of adsorption of MO by biofilm. The regression coefficients are relatively lower for all three types of biofilm, suggesting that this model does not demonstrate the adsorption process of MO to periphytic biofilms.

Adsorption is a multistep process involving (1) the transfer of liquid phase to solid adsorbent followed by (2) a slow diffusion process into pores on the surface of adsorbent and (3) the attainment of equilibrium after adsorption of dye onto active sites. It is a rate determining process; therefore, the intraparticle diffusion model was used to study the adsorption of MO on different biofilms. If intraparticle diffusion is the rate-limiting step in the adsorption process, then a linear graph would result from a plot of q_t versus square root of contact time. The regression coefficients increased with the concentration of dye, indicating that intraparticle diffusion and other processes are involved in controlling adsorption rate. Furthermore, greater regression coefficients ($R^2 > 0.99$) indicate that intraparticle diffusion was influenced by the concentration of MO during the adsorption of dye by periphytic biofilms, and this influence was highest in epilithon.

TABLE 16.2 Kinetic Constants for Methyl Orange Onto Epiphyton, Metaphyton, and Epilithon

Biofilm	Initial Concentration	First-Order Kinetic Model			Second-Order Kinetic Model			Elovich Model			Intraparticle Diffusion		
	C_0 (mg L^{-1})	k_1	q_e	R^2	k_2	q_e	R^2	α	β	R^2	C (mg g^{-1})	k_{id} (mg g^{-1} $min^{-0.5}$)	R^2
Metaphyton	50	0.032	1.12	0.85	0.051	0.17	1.00	1.24	13.23	0.95	15.55	0.35	0.90
	100	0.013	1.60	0.79	0.021	0.72	0.99	9.63	9.30	0.93	9.11	2.61	0.97
	200	0.008	1.83	0.85	0.011	0.50	0.99	17.42	17.873	0.89	18.25	4.32	0.97
	500	0.006	2.17	0.97	0.004	0.30	0.97	48.21	82.69	0.91	17.43	11.85	0.98
Epilithon	50	0.019	1.07	0.95	0.046	0.48	0.99	3.71	3.24	0.90	9.63	1.12	0.96
	100	0.015	1.65	0.95	0.023	0.53	0.99	8.67	4.62	0.99	13.55	2.14	0.97
	200	0.012	2.08	0.90	0.010	0.53	0.99	22.31	38.54	0.97	9.12	5.35	0.99
	500	0.008	2.234	0.967	0.004	0.231	0.98	50.35	84.57	0.97	22.39	12.12	0.99
Epiphyton	50	0.011	0.412	0.865	0.052	0.166	0.99	0.938	0.938	0.91	16.07	0.27	0.96
	100	0.017	1.41	0.91	0.025	0.25	0.99	5.12	4.80	0.98	23.15	1.38	0.97
	200	0.008	1.63	0.97	0.012	0.23	0.99	11.86	12.07	0.96	2.92	37.79	0.99
	500	0.006	1.97	0.942	0.005	0.11	0.99	28.66	28.67	0.94	82.59	7.00	0.99

16.3.2.7 Adsorption Isotherms

Isotherms are widely used to describe the relationship between biosorbed and aqueous concentrations of dye (C_e). Isotherms are obtained by fixing the biomass of heterogeneous adsorbent and concentration of dye. The Langmuir and Freundlich isotherms were used to describe the biosorption of different concentrations of MO by heterogeneous periphyton surfaces, that is, metaphyton, epiphyton, and epilithon. Plots of C_e versus C_e/q_e were determined at different MO concentrations (50–500 mg L^{-1}). The values of q_{max} evaluate the approximate amount of MO adsorption onto different biofilms. These values showed high adsorption capacity of biofilms for MO (Table 16.3). In the Freundich equation, $1/n$ was helpful in determining the intensity of adsorption of MO on biofilms. If the value of $1/n$ lies between 0.1 and 0.5, it indicates excellent adsorption capability of the adsorbant. In our experiment all values of $1/n$ lay within the range, proving the biofilms excellent adsorbers for MO (Table 16.3). The adsorption equations in this study are better fitted to the Langmuir isotherm with higher adsorption coefficient values (R^2) than the Freundlich isotherm (Table 16.3). The adsorption capacity of periphyton was also confirmed by R_L values obtained by using Eq. (16.15). The values for

TABLE 16.3 Parameters of Adsorption Models for Four Different Initial Dye Concentrations

Biofilm	Concentration mg L^{-1}	Langmuir Equation			Freundlich Equation		
		K_L	q_m	R^2	logK_F	$1/n$	R^2
Epilithon	50	0.002	0.058	0.999	2.888	0.022	0.668
	100	0.077	0.043	0.972	3.879	0.186	0.812
	200	0.120	0.023	0.971	4.834	0.223	0.851
	500	0.498	0.013	0.940	6.259	0.297	0.763
Metaphyton	50	0.010	0.068	0.998	2.954	0.110	0.767
	100	0.073	0.044	0.976	4.640	0.188	0.710
	200	0.170	0.027	0.946	3.713	0.126	0.669
	500	0.449	0.012	0.951	6.336	0.311	0.802
Epiphyton	50	0.001	0.057	0.999	2.914	0.042	0.960
	100	0.015	0.034	0.994	3.652	0.073	0.765
	200	0.045	0.019	0.990	4.582	0.138	0.871
	500	−0.169	0.009	0.985	5.965	0.216	0.891

epiphyton were 0.012–0.576; metaphyton, 0.004–0.573; and epilithon, 0.004–0.167.

16.3.2.8 Biodegradation of MO

As previously mentioned, all three types of periphytic biofilm used in this study were composed of cyanobacteria, algae, protozoa, and detritus. Bacteria and algae are already known for their dye degradation capabilities; therefore, the microbial aggregates must be able to degrade the toxic materials by the production of certain intracellular and extracellular enzymes. The adsorption kinetics described the adsorption of MO as a second-order process, and MO was completely adsorbed by periphytic biofilms in the second step. The degradation of MO by metaphyton, epiphyton, and epilithon was confirmed by FTIR analysis. The comparison of the control dye with the extracted metabolites clearly indicated that MO was degraded by periphyton (Fig. 16.6). The 3457 cm^{-1} band in the MO control spectra (Fig. 16.6A) indicates the –NH stretching vibration.

Peaks at 2917 and 2848 cm^{-1} display asymmetrical stretching of methylene C–H and $-CH_3$, respectively. The peak at 1598 cm^{-1} is attributed to –N=N– of azo group stretching vibrations, the peak at 1444 cm^{-1} for –C=CH vibrations, the peak at 1359 cm^{-1} for S=O stretching confirming the presence of sulfonated groups in MO, peaks at 1189 and 1159 cm^{-1} for –C–N stretching, the peak at 1097 cm^{-1} for cyclohexane ring stretching vibrations, and peaks at 1004, 943, and 850 cm^{-1} for ring vibration. The peak at 819 cm^{-1} confirms the aromatic nature of MO and displays disubstituted benzene ring stretching. After degradation by epiphyton, metaphyton, and epiphyton for 72 h, extracted metabolites showed a number of new peaks, including a peak at 3598 cm^{-1} showing –OH stretching vibrations, 1517–1519 cm^{-1} associated with C=C–C stretching vibrations of aromatic rings, 2159–2161 cm^{-1} for aromatic combinations of bonds, and 1650 cm^{-1} for stretching vibrations for formation of benzaldehyde and benzoic acid. A peak at 1444 cm^{-1} for –C=CH stretching vibrations disappeared, thus predicting the deformation of MO. The formation of alcohols could be explained by the new peaks at 1315 cm^{-1} for –OH stretching of phenol and 1118 cm^{-1} for C–O stretching of secondary alcohol. Further, the peak showing the presence of azo bonds (1598 cm^{-1}) and the peaks at 1189 cm^{-1} and 1159–1160 cm^{-1} for –C–N stretching disappeared, and a new peak in the region 1604 cm^{-1} shows N–H bending of a secondary amine, confirming the breakage of the azo-bonds to secondary amines. The FTIR spectra of the three biofilms showed certain differences: (1) some of the peaks have slightly different values and are more or less sharper than others, such as peaks at 3461, 3465, and 3465 cm^{-1} in spectra of epiphyton, epilithon, and metaphyton for stretching vibrations of aromatic amines, and (2) some spectra have more peaks than others such as peaks at 1390 and 1186 cm^{-1} in metaphyton showing stretching vibrations of –OH bend of phenol and C–N stretching, respectively.

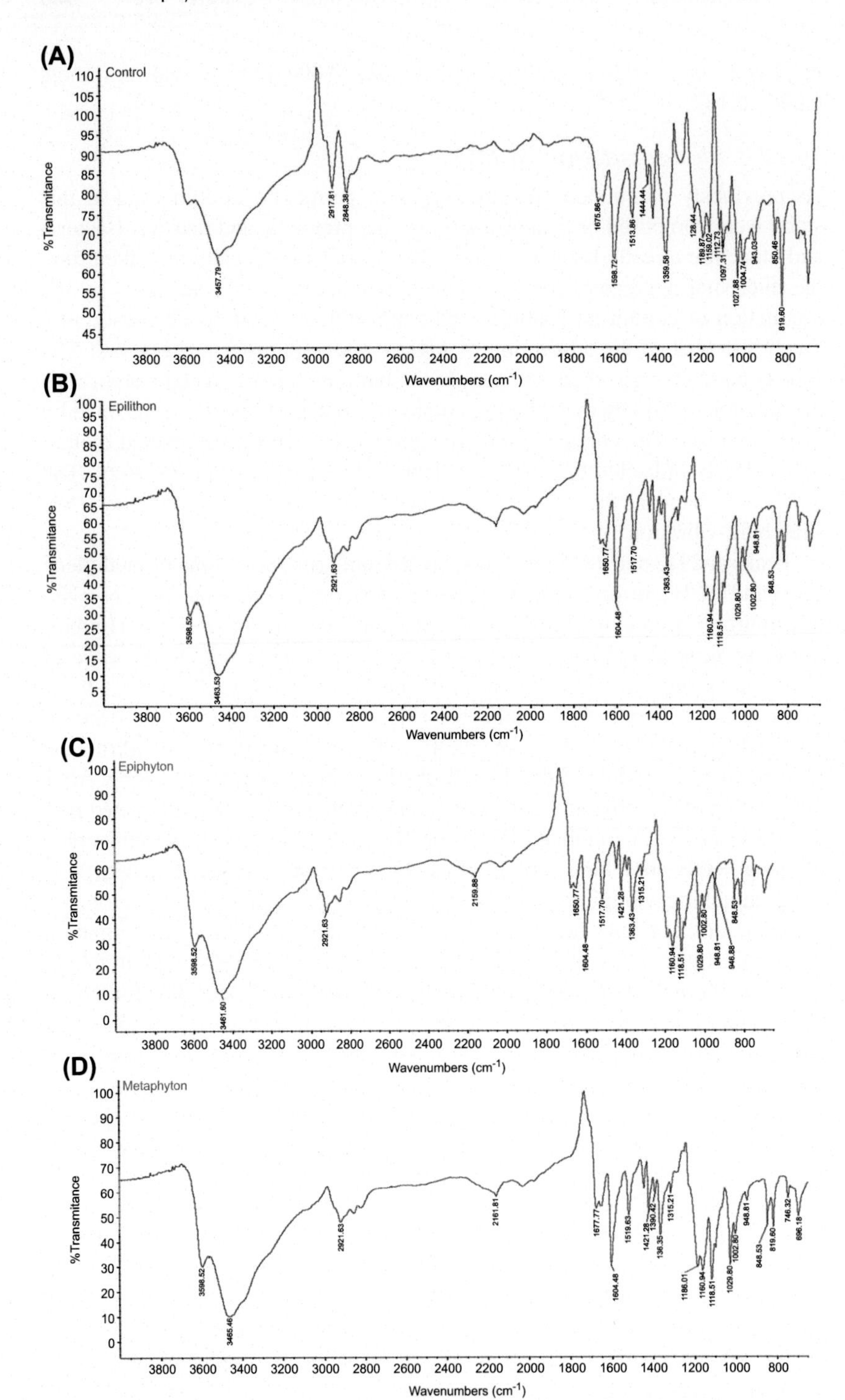

FIGURE 16.6 FTIR spectra of methyl orange and its degradation metabolites by (A) the control, (B) metaphyton, (C) epilithon, and (D) epiphyton.

16.4 CONCLUSIONS

This study has demonstrated a novel and promising eco-friendly, cost-effective, and functional technique for the dual treatment of an azo dye, MO, by naturally occurring periphytic biofilm, that is, epiphyton, metaphyton, and epilithon. The bioadsorption and biodegradation processes were studied simultaneously; the first step in the removal process was bioadsorption followed by degradation of MO. The removal efficiency and biodegradation decreased from epiphtyton > epilithon > metaphyton and resulted in complete removal of dye from solution. A significant effect of different concentrations of MO, pH, and biomass was observed with increasing dye concentration, biomass, and contact time resulting in increased dye adsorption. The optimum pH was 7. The adsorption kinetics of MO by periphyton followed pseudo-second-order reactions, and the data fit well into the Langmuir isotherm equation. The changes observed by FTIR spectral analysis confirmed the degradation of the azo-bond in MO and the formation of new products. These results indicate that periphytic biofilm is a viable biomaterial for the removal of MO and can be used for the removal of azo dyes from textile wastewater.

REFERENCES

Arunarani, A., Chandran, P., Ranganathan, B.V., Vasanthi, N.S., Sudheer Khan, S., 2013. Bioremoval of basic violet 3 and acid blue 93 by *Pseudomonas putida* and its adsorption isotherms and kinetics. Colloids and Surfaces B: Biointerfaces 102, 379–384.

Balser, T.C., Wixon, D.L., 2009. Investigating biological control over soil carbon temperature sensitivity. Global Change Biology 15 (12), 2935–2949.

Battin, T.J., Kaplan, L.A., Newbold, J.D., Hansen, C.M.E., 2003. Contributions of microbial biofilms to ecosystem processes in stream mesocosms. Nature 426, 439–442.

Dafale, N., Rao, N.N., Meshram, S.U., Wate, S.R., 2008a. Decolorization of azo dyes and simulated dye bath wastewater using acclimatized microbial consortium – biostimulation and halo tolerance. Bioresource Technology 99 (7), 2552–2558.

Dafale, N., Wate, S., Meshram, S., Nandy, T., 2008b. Kinetic study approach of remazol black-B use for the development of two-stage anoxic–oxic reactor for decolorization/biodegradation of azo dyes by activated bacterial consortium. Journal of Hazardous Materials 159 (2–3), 319–328.

Daneshvar, N., Ayazloo, M., Khataee, A.R., Pourhassan, M., 2007. Biological decolorization of dye solution containing Malachite Green by microalgae *Cosmarium* sp. Bioresource Technology 98 (6), 1176–1182.

Dong, D., Li, Y., Zhang, J., Hua, X., 2003. Comparison of the adsorption of lead, cadmium, copper, zinc and barium to freshwater surface coatings. Chemosphere 51 (5), 369–373.

Eichlerová, I., Homolka, L., Nerud, F., 2006. Synthetic dye decolorization capacity of white rot fungus *Dichomitus squalens*. Bioresource Technology 97 (16), 2153–2159.

El-Rahim, W.M.A., El-Ardy, O.A.M., Mohammad, F.H.A., 2009. The effect of pH on bioremediation potential for the removal of direct violet textile dye by *Aspergillus niger*. Desalination 249 (3), 1206–1211.

Feng, Y., Zhou, H., Liu, G., Qiao, J., Wang, J., Lu, H., Yang, L., Wu, Y., 2012. Methylene blue adsorption onto swede rape straw (*Brassica napus* L.) modified by tartaric acid: equilibrium, kinetic and adsorption mechanisms. Bioresource Technology 125, 138–144.

Fu, Y., Viraraghavan, T., 2001. Fungal decolorization of dye wastewaters: a review. Bioresource Technology 79 (3), 251–262.

Gaiser, E.E., Trexler, J.C., Richards, J.H., Childers, D.L., Lee, D., Edwards, A.L., Scinto, L.J., Jayachandran, K., Noe, G.B., Jones, R.D., 2005. Cascading ecological effects of low-level phosphorus enrichment in the Florida everglades. Journal of Environmental Quality 34 (2), 717–723.

García-Montaño, J., Domènech, X., García-Hortal, J.A., Torrades, F., Peral, J., 2008. The testing of several biological and chemical coupled treatments for Cibacron Red FN-R azo dye removal. Journal of Hazardous Materials 154 (1), 484–490.

Ghaedi, M., Ansari, A., Habibi, M.H., Asghari, A.R., 2014. Removal of malachite green from aqueous solution by zinc oxide nanoparticle loaded on activated carbon: kinetics and isotherm study. Journal of Industrial and Engineering Chemistry 20 (1), 17–28.

Gingell, R., Walker, R., 1971. Mechanisms of azo reduction by *Streptococcus faecalis* II. The role of soluble flavins. Xenobiotica 1 (3), 231–239.

Gong, R., Zhang, X., Liu, H., Sun, Y., Liu, B., 2007. Uptake of cationic dyes from aqueous solution by biosorption onto granular kohlrabi peel. Bioresource Technology 98 (6), 1319–1323.

Gupta, V.K., Mittal, A., Gajbe, V., Mittal, J., 2006. Removal and recovery of the hazardous azo dye acid orange 7 through adsorption over waste materials: bottom ash and de-oiled soya. Industrial & Engineering Chemistry Research 45 (4), 1446–1453.

Hameed, B.H., Mahmoud, D.K., Ahmad, A.L., 2008. Sorption of basic dye from aqueous solution by pomelo (*Citrus grandis*) peel in a batch system. Colloids and Surfaces A-Physicochemical and Engineering Aspects 316 (1), 78–84.

Jadhav, U.U., Dawkar, V.V., Jadhav, M.U., Govindwar, S.P., 2011. Decolorization of the textile dyes using purified banana pulp polyphenol oxidase. International Journal of Phytoremediation 13 (4), 357–372.

Kadam, A.A., Telke, A.A., Jagtap, S.S., Govindwar, S.P., 2011. Decolorization of adsorbed textile dyes by developed consortium of *Pseudomonas* sp. SUK1 and *Aspergillus ochraceus* NCIM-1146 under solid state fermentation. Journal of Hazardous Materials 189 (1–2), 486–494.

Khataee, A.R., Zarei, M., Dehghan, G., Ebadi, E., Pourhassan, M., 2011. Biotreatment of a triphenylmethane dye solution using a Xanthophyta alga: modeling of key factors by neural network. Journal of the Taiwan Institute of Chemical Engineers 42 (3), 380–386.

Lin, Y.-F., Chen, H.-W., Chien, P.-S., Chiou, C.-S., Liu, C.-C., 2011. Application of bifunctional magnetic adsorbent to adsorb metal cations and anionic dyes in aqueous solution. Journal of Hazardous Materials 185 (2–3), 1124–1130.

Lu, H., Yang, L., Shabbir, S., Wu, Y., 2014a. The adsorption process during inorganic phosphorus removal by cultured periphyton. Environmental Science and Pollution Research 21 (14), 8782–8791.

Lu, H., Yang, L., Zhang, S., Wu, Y., 2014b. The behavior of organic phosphorus under non-point source wastewater in the presence of phototrophic periphyton. PLoS One 9 (1), e85910.

Mahmoud, D.K., Salleh, M.A.M., Karim, W.A.W.A., Idris, A., Abidin, Z.Z., 2012. Batch adsorption of basic dye using acid treated kenaf fibre char: equilibrium, kinetic and thermodynamic studies. Chemical Engineering Journal 181–182, 449–457.

Patel, R., Suresh, S., 2008. Kinetic and equilibrium studies on the biosorption of reactive black 5 dye by *Aspergillus foetidus*. Bioresource Technology 99 (1), 51–58.

Pearce, C.I., Lloyd, J.R., Guthrie, J.T., 2003. The removal of colour from textile wastewater using whole bacterial cells: a review. Dyes and Pigments 58 (3), 179–196.

Phugare, S.S., Kalyani, D.C., Patil, A.V., Jadhav, J.P., 2011. Textile dye degradation by bacterial consortium and subsequent toxicological analysis of dye and dye metabolites using cytotoxicity, genotoxicity and oxidative stress studies. Journal of Hazardous Materials 186 (1), 713–723.

Pointing, S., 2001. Feasibility of bioremediation by white-rot fungi. Applied Microbiology and Biotechnology 57 (1–2), 20–33.

Ramalho, P.A., Scholze, H., Cardoso, M.H., Ramalho, M.T., Oliveira-Campos, A., 2002. Improved conditions for the aerobic reductive decolourisation of azo dyes by *Candida zeylanoides*. Enzyme and Microbial Technology 31 (6), 848–854.

Robinson, T., McMullan, G., Marchant, R., Nigam, P., 2001. Remediation of dyes in textile effluent: a critical review on current treatment technologies with a proposed alternative. Bioresource Technology 77 (3), 247–255.

Shah, M.P., 2013. Microbial degradation of textile dye (remazol black B) by *Bacillus spp.* ETL-2012. Journal of Applied & Environmental Microbiology 1 (1), 6–11.

Stolz, A., 2001. Basic and applied aspects in the microbial degradation of azo dyes. Applied Microbiology and Biotechnology 56 (1), 69–80.

Tuttolomondo, M.V., Alvarez, G.S., Desimone, M.F., Diaz, L.E., 2014. Removal of azo dyes from water by sol–gel immobilized *Pseudomonas* sp. Journal of Environmental Chemical Engineering 2 (1), 131–136.

Vandevivere, P.C., Bianchi, R., Verstraete, W., 1998. Review: treatment and reuse of wastewater from the textile wet-processing industry: review of emerging technologies. Journal of Chemical Technology & Biotechnology 72 (4), 289–302.

Wong, Y., Yu, J., 1999. Laccase-catalyzed decolorization of synthetic dyes. Water Research 33 (16), 3512–3520.

Writer, J.H., Ryan, J.N., Barber, L.B., 2011. Role of biofilms in sorptive removal of steroidal hormones and 4-nonylphenol compounds from streams. Environmental Science & Technology 45 (17), 7275–7283.

Wu, Y., He, J., Yang, L., 2010. Evaluating adsorption and biodegradation mechanisms during the removal of microcystin-RR by periphyton. Environmental Science & Technology 44 (16), 6319–6324.

Zimmermann, T., Kulla, H.G., Leisinger, T., 1982. Properties of purified orange II azoreductase, the enzyme initiating azo dye degradation by *Pseudomonas* KF 46. European Journal of Biochemistry 129 (1), 197–203.

Index

'*Note:* Page numbers followed by "f" indicate figures, "t" indicate tables.'

C

I

Q

R

S

T